Advances in Physical Geochemistry

Volume 6

Advances in Physical Geochemistry

Series Editor: Surendra K. Saxena

Volume 1 R.C. Newton/A. Navrotsky/B.J. Wood (editors)
Thermodynamics of Minerals and Melts
1981. xii, 304 pp. 66 illus.
ISBN 0-387-90530-8

Volume 2 S.K. Saxena (editor)
Advances in Physical Geochemistry, Volume 2
1982. x, 353 pp. 113 illus.
ISBN 0-387-90644-4

Volume 3 S.K. Saxena (editor)
Kinetics and Equilibrium in Mineral Reactions
1983. vi, 273 pp. 99 illus.
ISBN 0-387-90865-X

Volume 4 A.B. Thompson/D.C. Rubie (editors)
Metamorphic Reactions: Kinetics, Textures, and Deformation
1985. xii, 291 pp. 81 illus.
ISBN 0-387-96077-5

Volume 5 J.V. Walther/B.J. Wood (editors)
Fluid–Rock Interactions during Metamorphism
1986. x, 211 pp. 59 illus.
ISBN 0-387-96244-1

Volume 6 S.K. Saxena (editor)
Chemistry and Physics of Terrestrial Planets
1986. x, 405 pp. 94 illus.
ISBN 0-387-96287-5

Chemistry and Physics of Terrestrial Planets

Edited by
Surendra K. Saxena

With Contributions by
P.A. Candela M. Catti G. Eriksson
R.F. Galimzyanov R. Jeanloz I.L. Khodakovsky
E. Knittle O.L. Kuskov M.I. Petaev
V.S. Safronov S.K. Saxena Yu.I. Sidorov
A.V. Vitjazev V.P. Volkov
D.J. Weidner M.Yu. Zolotov

With 94 Illustrations

Springer-Verlag
New York Berlin Heidelberg Tokyo

Series Editor
Surendra K. Saxena
Department of Geology
Brooklyn College
City University of New York
Brooklyn, New York 11210
U.S.A.

Library of Congress Cataloging-in-Publication Data
Chemistry and physics of terrestrial planets.
(Advances in physical geochemistry ; v. 6)
Bibliography: p.
Includes index.
1. Earth. 2. Planets. 3. Cosmochemistry.
4. Geochemistry. 5. Earth—Mantle. I. Saxena,
Surendra K. (Surendra Kumar), 1936–
II. Candela, P. A. III. Series.
QB631.C49 1986 551 86-1804

Typeset by Asco Trade Typesetting Ltd., Hong Kong.
Printed and bound by R.R. Donnelley and Sons, Harrisonburg, Virginia.
Printed in the United States of America.

9 8 7 6 5 4 3 2 1

ISBN 0-387-96287-5 Springer-Verlag New York Berlin Heidelberg Tokyo
ISBN 3-540-96287-5 Springer-Verlag Berlin Heidelberg New York Tokyo

Preface

The purpose of this volume is to present the latest planetary studies of an international body of scientists concerned with the physical and chemical aspects of terrestrial planets.

In recent years planetary science has developed in leaps and bounds. This is a result of the application of a broad range of scientific disciplines, particularly physical and chemical, to an understanding of the information received from manned and unmanned space exploration. The first five chapters expound on many of the past and recent observations in an attempt to develop meaningful physical–chemical models of planetary formation and evolution.

For any discussion of the chemical processes in the solar nebula, it is important to understand the boundary conditions of the physical variables. In Chapter 1, Safranov and Vitjazev have laid down explicitly all the physical constraints and the problems of time-dependence of nebular evolutionary processes. Planetary scientists and students will find in this chapter a collection of astrophysical parameters on the transfer of angular momentum, formation of the disk and the gas envelope, nebular turbulence, physical mixing of particles of various origins and growth of planetesimals. The authors conclude their work with important information on evolution of terrestrial planets. Although symbols are defined in the text of the article, readers who are not familiar with the many symbols and abbreviations in astrophysical literature will find it useful to consult the Appendix for explanations.

The second chapter, by Saxena and Eriksson, provides data on phase equilibrium in a hot gaseous nebula of solar composition. While the role of equilibrium condensation in forming all or most of the planetary and meteoritic material is debatable, few would deny the importance of the data as reference for a comparative study. In the third chapter, Pataev and Khodakovsky discuss that high temperature equilibration of some phases in enstatite meteorites has taken place under reducing conditions corresponding to a C/O ratio close to unity in the solar nebula.

In Chapter 4, Volkov, Zolotov, and Khodakovsky have reviewed the recent studies on the lithospheric–atmospheric interaction on Venus. The authors use the latest information on the composition of the atmosphere and on the surface conditions to model the chemistry of the Venusian clouds and the recycling of the outer planetary shell.

In the fifth chapter, Sidorov and Zolotov have discussed the available data on Martian surface and atmosphere. Their conclusion is that Martian regolith is a mixture of mechanically derived minerals of magmatic rocks of basic and/or ultrabasic compositions, and products of chemical weathering. The latter process is not significant at present being largely replaced by mechanical weathering.

The planetary missions and observations have greatly enhanced our curiosity about our own planet Earth. The next five chapters are devoted to the study of the earth's interior. In Chapter 6, Catti argues in favor of theoretical models for calculating physical properties of crystalline phases. It appears that, at present, theory can be used with reasonable accuracy to obtain physical properties of some phases with simple structure. Further development in determining the form of interatomic potentials and the optimization procedures are awaited with great expectations.

In Chapter 7, Weidner has addressed a classical problem in the study of the Earth's interior. Do the seismic data reflect the compositional inhomogeneity of the mantle or the effect of physical changes in a homogeneous mantle? His conclusion is that from the available data, pyrolite produces an acoustic velocity model that agrees quantitatively with seismic data. However, such a study is critically dependent on the accuracy of the pressure- and temperature-derivatives of the elastic moduli of the mantle phases. Jeanloz and Knittle are also concerned with the interior of the Earth. Their approach, as outlined in Chapter 8, is to use an equation of state relating the seismologically determined wave velocities and densities, and then compare the recently obtained physical data on minerals with the adiabatically decompressed mantle properties to zero pressure. It appears that the Eulerian formulation of the equation of state is best among those that are currently used (e.g., Murnaghan equation).

In Chapter 9, Kuskov and Galimzyanov have adopted a thermochemical approach to study phase equilibrium at high pressures and understand the physical properties of the mantle. The authors propose an equation of state in the form of Mie–Gruneisen and find the equation useful to a pressure of 1 Mbar and to a temperature of 3000 K. The authors find that all chemical transformations are completed within a depth of 700–800 km and the mantle that follows has a simple mineralogy.

Finally, in Chapter 10 Candela has discussed the mathematical basis for understanding the formation and evolution of magmatic vapors. The emanation of vapors leads to many profound crustal changes which also affect the atmosphere.

The volume contains a large number of original articles from Soviet planetary scientists. While all chapters were reviewed for their scientific content, the translation of the original article into English has been the responsibility of the author.

The production of this volume was possible through an enthusiastic cooperation by the contributors, reviewers and the editorial staff of Springer-Verlag. I thank S. Ghose, R. Jeanloz, E. Olsen, R.F. Mueller, and many members of the editorial board of the A.P.G. for providing reviews, support, and advice. R. Palestino made language improvements in the chapters contributed by the Soviet scientists and Julia A. Sykes prepared the Subject Index.

S.K. SAXENA

Contents

Contributors

Candela, P.A.	Department of Geology, University of Maryland, College Park, Maryland 20742, U.S.A.
Catti, M.	Dipartimento di Chimica Fisica ed Elettrochimica, Universita di Milano, Via Golgi, 19, 20133 Milano, Italy
Eriksson, G.	Department of Inorganic Chemistry, University of Umeå, S-901 87 Umeå, Sweden
Galimzyanov, R.F.	V.I. Vernadsky Institute of Geochemistry and Analytical Chemistry, U.S.S.R. Academy of Sciences, Moscow, V. 334, U.S.S.R
Jeanloz, R.	Department of Geology and Geophysics, University of California, Berkeley, California 94720, U.S.A.
Khodakovsky, I.L.	V.I. Vernadsky Institute of Geochemistry and Analytical Chemistry, U.S.S.R. Academy of Sciences, Moscow, V. 334, U.S.S.R.
Knittle, E.	Department of Geology and Geophysics, University of California, Berkeley, California 94720, U.S.A.
Kuskov, O.L.	V.I. Vernadsky Institute of Geochemistry and Analytical Chemistry, U.S.S.R. Academy of Sciences, Moscow, V. 334, U.S.S.R.

PETAEV, M.I. — V.I. Vernadsky Institute of Geochemistry and Analytical Chemistry, U.S.S.R. Academy of Sciences, Moscow, V. 334, U.S.S.R.

SAFRONOV, V.S. — O. Yu. Schmidt Institute of Physics of the Earth, U.S.S.R. Academy of Sciences, Moscow, U.S.S.R.

SAXENA, S.K. — Department of Geology, Brooklyn College, Brooklyn, New York 11234, U.S.A.

SIDOROV, YU.I. — V.I. Vernadsky Institute of Geochemistry and Analytical Chemistry, U.S.S.R. Academy of Sciences, Moscow, V. 334, U.S.S.R.

VITJAZEV, A.V. — O. Yu. Schmidt Institute of Physics of the Earth, U.S.S.R. Academy of Sciences, Moscow, U.S.S.R.

VOLKOV, V.P. — V.I. Vernadsky Institute of Geochemistry and Analytical Chemistry, U.S.S.R. Academy of Sciences, Moscow, V. 334, U.S.S.R.

WEIDNER, D.J. — Department of Earth and Space Sciences, State University of New York, Stony Brook, New York 11794, U.S.A.

ZOLOTOV, M.YU. — V.I. Vernadsky Institute of Geochemistry and Analytical Chemistry, U.S.S.R. Academy of Sciences, Moscow, V. 334, U.S.S.R.

Chapter 1
The Origin and Early Evolution of the Terrestrial Planets

V. S. Safronov and A. V. Vitjazev

Introduction

The formation of the presolar nebula, of the sun, and of the preplanetary disk, the evolution of the disk, and the formation of the planets are the topics considered in this chapter. Main attention is paid to the processes of formation of the terrestrial planets. The initial state and early evolution and differentiation of the planets is discussed. Present-day models lead to the conclusion that differentiation began during the formation of the planets.

Main ideas in the Earth sciences on the general features of the evolution of the Earth have always been dominatingly cosmogonical. Laplace's and Jeans' hypotheses have for a long time supported the idea of an initially thoroughly melted and gradually cooled Earth. During the 1940s O. Y. Schmidt, based on the idea of the accumulation of the planets from solid particles and bodies, came to the opposite conclusion, of the relatively cold initial state of Earth and its consequent heating by the energy of radioactive elements. Similar ideas were developed by V. I. Vernadsky and H. C. Urey, founders of the cosmochemical approach. However, the cold Earth model caused difficulties for geologists, geochemists, and geophysicists when, in the 1960s and 1970s, evidence on active tectonic processes during the first hundred million years of the Earth's life was obtained. Several ad hoc ideas concerning the cosmogonic processes were suggested. Some authors assumed a very short time scale of accumulation—a thousand times shorter than the 10^8 years obtained from dynamical considerations. Others suggested various versions of "inhomogeneous" accumulation, in which at first the iron core of the planet was formed and only afterwards, was the silicate mantle accreted. It has since been shown that such ideas are groundless. At the same time, the development of the accumulation theory has revealed the important role of large bodies in the formation of the planets, comprising a considerable part of the mass of all accreted bodies. Calculations of the heating of Earth during its growth have shown that because of the impacts of large bodies, the upper mantle was heated, on the average, to the melting temperature of the material. On the sites of the impacts of the largest bodies, vast regions of partial melting appeared and a differentiation of substances according to their densities occurred.

The theory of accumulation has explained the main features of our planetary system—the masses and distances of the major planets, their densities and total number, peculiarities of their rotation, etc. Several problems remain unsolved, however, and one of the most important is probably the explanation of appreciable variations in the chemical composition of planets and satellites. It is difficult in one chapter to describe all the results and problems in the planetary cosmogony. We will try to choose some of these that make up the main content of the theory and are confirmed by present data in astrophysics, comparative planetology, and Earth sciences. We begin by summarizing the results of current investigations into the earliest stage of formation of the solar nebula—collapse of a fragment of a molecular cloud (presolar nebula), which led to the formation of the Sun and circumsolar gas–dust disk. In the following text we consider the main stages of evolution of the preplanetary disk—separation of the dust component from the gaseous one and the formation of the planets, the latter being preceded by establishment of equilibrium spectra of the masses and the velocities of the preplanetary bodies in the process of accumulation. Cosmochemical aspects of the evolution of the preplanetary disk are discussed. Finally, we discuss the initial temperatures and the early evolution of the terrestrial planets. See Appendix 1 at the end of the book for explanation of symbols.

Formation of the Sun and the Circumsolar Disk

The Sun is thought to be a second-generation star. It is at least half as young as the old stars of the first generation. It should have formed at conditions not much different from those of the present formation of solar-type stars. More than some 30 years ago V. A. Ambarzumjan called attention to the fact that stars are formed in groups (associations). Later it was shown that the formation of stars took place in giant gas–dust complexes and, more exactly, in cold and dense clouds. These were named molecular clouds because infrared observations revealed molecules of CO, OH, H_2CO, and many other compounds. Of course, molecules (H_2) of the most abundant element, hydrogen, should be dominant among them. If the content of atomic hydrogen is taken to be 10^5, then the elements next in abundance are approximately as follows: He, 10^4; 0, 10^2; C, 60; N and Ne, 20; Si, Mg, and Fe, 10; S, about 5; Ca, Ar, and Al, 1; all remaining elements are also about 1. Most of the nonvolatiles are in the solid state, in the form of small grains 10^{-5} cm in size, accounting for about 1% of the total mass of the cloud. A remarkable example of the region of the present-day star formation is the well-known Orion nebula. The cloud complexes are located mainly in the galactic spiral arms. This indicates that their initial contraction was triggered by spiral density waves.

Estimates show that interstellar clouds of moderate density can begin to contract under the action of self-gravitation only if their masses reach several thousand solar masses. Giant molecular clouds have masses on the same order of magnitude. However, they possess large angular momentum resulting from galactic rotation and cannot contract until they attain large densities because of the increase of centrifugal force. At some stage the contraction is stopped, the cloud disintegrates into smaller fragments, a part of the angular momentum transforms into an orbital one, and the galactic magnetic field slows down the rotation of the fragments so they can continue to contract further. Such fragmentation is repeated several times, producing finally fragments able to contract up to densities $\sim 1\ g \cdot cm^{-3}$ and transforms into stars. One of such fragments was the presolar nebula, precursor of our solar system.

Investigations of the decay products of radioactive elements with different half-lives has allowed us to estimate the time of separation of the presolar nebula. It was found that the elements with long half-lives, ^{238}U, ^{235}U, ^{232}Th, and ^{40}K, entered the solid substance of meteorites 4.6 billion years ago. This figure is considered to be the age of the solar system. Xenon isotopes—decay products of the extinct radioactive elements ^{129}I and ^{244}Pu (half-lives of 17 and 82 million years)—show that synthesis of these elements took place ~ 100 million years before the formation of the meteorites and that the differences in the ages among latter are about 10 million years. Discovery of an appreciable excess of ^{26}Mg, a decay product of short-lived, extinct radioactive ^{26}Al, in refractory inclusions of the Allende meteorite indicates that an even later event of nucleosynthesis occurred only a few million years before ^{26}Al entered the solids. It was suggested that the last synthesis took place during the explosion of a nearby supernova and that the same supernova was responsible for triggering the collapse of the presolar nebula (Cameron and Truran, 1977). One explosion event cannot account for all the varieties of isotopic anomalies in meteorites. However, this fact was not considered a difficulty. Reeves (1978) pointed out that during the time interval in which the planets formed (10^8 years) about a dozen supernova explosions could have taken place in the solar neighborhood. The idea was attractive but now it is doubted. Supernovae have very low $^{26}Al/^{27}Al$ production ratios (about 10^{-3}). Clayton (1983) found that they could not account for the ^{26}Al γ-ray line in galactic interstellar medium radiation. He suggested that Nova eruptions accounted for the presence of the necessary quantity of ^{26}Al. Cameron (1984) now proposes that red giant stars can do this as well. He assumes that the expanding envelope of a red giant star entering the molecular cloud will produce considerable disturbance in the cloud, leading to the formation of a dense core, which collapses to form a star.

Numerical computations of gravitational contraction (collapse) of cold clouds with the masses about that of the sun were performed by many specialists (R. Larson, P. Bodencheimer, V. Tscharnutter, J. Tohline, P. Cassen, etc.) and are still going on. They are very laborious and require a very large computing time, especially when the collapse of rotating clouds is studied. The initial size of the cloud is usually taken from the condition that the cloud is on the verge of gravitational (Jeans') instability. This takes place for the radius R_0 at about 10^{17} cm (one tenth of the distance to the nearest stars), and correspondingly for the cloud density $\approx 10^{-19}$ to 10^{-18} $g \cdot cm^{-3}$, i.e., 10^5 atoms per 1 cm^3. From calculations one can see that the final result of collapse substantially depends on the initial rate of rotation of the cloud. Formation of a sun with a planetary system is possible at an initial angular momentum $J \approx 10^{52}$ $g \cdot cm^2 \cdot s^{-1}$ that is 50 times higher than that of the present solar system (Safronov and Ruzmaikina, 1978; Ruzmaikina, 1981). At 10 times higher angular momentum a double star should form, whereas for 10 times smaller, a single star without a planetary system would form. Later we accept the model of a slowly rotating presolar nebula ($J \approx 10^{52}$ $g \cdot cm^2 \cdot s^{-1}$) and a preplanetary disk (solar nebula) of a small mass ($M_d \approx 0.03 M_\odot$, where $M_\odot$ is the mass of the sun). Such a nebula is initially transparent and its contraction proceeds isothermally with a temperature $T \approx 10$K. Its inner part contracts much faster, and in the center, rather early, a dense opaque core forms ($\rho \gtrsim 10^{-12}$ $g \cdot cm^{-3}$). Its heats up rapidly and approaches a state of hydrostatic quasi-equilibrium. The main mass of the gas is still in the envelope, which collapses with nearly a free-fall velocity. Its density increases to the center as $R^{-3/2}$. The gas falls onto the core, creating at its surface an accretion bow shock. Accretion continues for 10^5 to 10^6 years. In order for the preplanetary gas disk to form around the core, an effective transfer of the angular momentum from the inner part of the nebula outward is needed. This problem came to be one of the most difficult in the cosmogony of the solar system.

The total mass of the planets is ~ 750 times less than that of the sun, but they possess almost all the angular momentum of the solar system and only 2% of it is contained in the sun. A century ago the main reason for the rejection of Laplace's hypothesis was its apparent inability to solve this problem. In the 1960s a number of mechanisms for the angular momentum transfer were suggest (F. Hoyle, A. Cameron, E. Schatzman) but all the models proved to be insufficient. During the last decade in astrophysics, a theory of evolution of viscous turbulent disks rotating around stars has been actively developed in connection with the study of close binary systems, black holes, etc. (e.g., Shakura and Syunyaev, 1973; Lynden-Bell and Pringle, 1974). At high viscosities the angular momentum is transferred in the disk quite effectively. It is believed that such a viscosity is produced by turbulent motions of the gas in the disk, although the mechanisms of their excitation and the sources of energy maintaining the turbulence have not yet been investigated in detail.

In a similar way a study of the formation and evolution of a circumsolar preplanetary disk has proceeded. It is based to a considerable degree on the ideas of von Weizsäcker (1948) and S. Chandrasekhar and ter Haar (1950) concerning the general features of turbulence in systems with a differential rotation. However, the principal question of which mechanism maintained the turbulent motions remains unclear. Weizsäcker supposed that the turbulence was induced by thermal convection in a radial direction, which was established via the decreases of temperature with the distance from the sun. However, it was shown (Safronov and Ruskol, 1957; Safronov, 1958) that such a convection could not arise in the rotating circumsolar disk because the increase of the angular momentum with R in the disk ($\omega R^2 \propto R^{1/2}$) makes it stable against convection. Now the mechanism of turbulence by convection in a z direction, perpendicular to the central plane of the disk, has been suggested (Lin and Papaloizou, 1980). It is accepted that the thermal energy which escaped the disk by the convection was restored from the turbulent energy that dissipated in the smallest eddies. It is not yet clear how stable such a state of the disk may be, as well as when and how the conditions necessary for establishing such a convective regime may arise. The first of the two conditions is that the disk be very opaque and hot in the central plane (with a superadiabatic temperature gradient). For the disk to have a high opacity a large number of small particles is needed. If the particles stick together on colliding and grow to a size of ~ 1 cm, however, the opacity becomes low. At high temperatures the particles evaporate and the opacity is also decreased. Therefore, for the inner part of the disk ($R \lesssim 0.1$ a.u.), convection seems unlikely.

The problem of angular momentum transfer for the central core is solved more easily. Because of high temperatures a remnant magnetic field is conserved in the core. It increases during the contraction and effectively transports the angular momentum outward, including the formation of the embryo disk at the equatorial edge of the core (Ruzmaikina, 1981; Safronov and Ruzmaikina, 1985a, b). It seems reasonable to suppose that this disk can be maintained in a turbulent state by the energy added by the envelope material falling into it (Safronov, 1982; Ruzmaikina, 1982). For the viscosity $\nu \approx 10^{16}$ cm$^2 \cdot$ s^{-1} the disk can grow up to the size of the present planetary system during the time (10^5 to 10^6 years) of the collapse of the solar nebula. Such a viscosity is reached in a boundary layer between the disk and the accreted envelope because of the large differences in their rotational velocities. An upper part of the layer, where the matter of the disk and envelope is mixed (layer A), moves along the disk's surface to its center. A lower layer, also made turbulent but poorly mixed with the envelope material (layer B), moves away from the center (Safronov, 1982; Safronov and Ruzmaikina 1985a, b). Layer B is especially interesting because solid particles entering it in the inner hot part of the disk can be transported to considerable distances from the protosun. Such a direct transport of material in the sublayer seems to be more effective than that by turbulent diffusion which has been suggested in a number of studies (Morfill, 1983; Clayton and Wieneke, 1983).

Gas of the envelope accreted by the innermost part of the disk (inside the Mercury's orbit) is heated up to 1500–2000K and the dust (interstellar) particles are evaporated. During the outward motion of gas in layer B, condensation occurs and new particles form until they penetrate under the layer. The matter accreted far from the core undergoes only a moderate heating. According to estimates by Wood (1983) and by Vityazev and Pechernikova (1985) particles are heated behind the shock front to about 1000K at distances 1–2 a.u. from the core and to 200–300K at 5–7 a.u. In the region of the terrestrial planets, therefore, interstellar particles should have lost icy mantles and parts of volatiles. The main composition of interstellar particles could remain at these distances. Therefore in the region of the terrestrial planets the condensed matter could contain a mixture of interstellar particles (partially metamorphosed) with that condensed anew in the preplanetary disk.

In our early model (Safronov, 1969) the study of the accumulation of particles into large bodies began from the moment the gas–dust disk had formed. It was assumed that the turbulence decayed rapidly (for several revolutions around the sun) and particles settled to the disk's central plane. Their sticking at collisions accelerated the settling. During on the order of 10^3 to 10^4 years in the zone of the terrestrial planets a dust layer is formed near the equatorial plane. When its density exceeds the critical value ρ_c, where:

$$\rho_c \approx 3M_\odot/2\pi R^3 \tag{1}$$

the layer becomes gravitationally unstable and breaks up into numerous dust condensations which coalesce on collision, become more dense, and transform rather early into usual solid bodies of kilometer size. A similar result was obtained by Goldreich and Ward (1973). The following evolution of this swarm of bodies proceeded with their coalescence or fragmentation on collision, an increase of their relative velocities caused by mutual gravitational perturbations, and a final sweeping up of all bodies by the largest ones until the formation of small number of planets moving in orbits separated by distances that secured the stability of the system for billions of years. The whole process lasted 10^8 years.

To understand the main features of the formation process of the disk itself, we should also try to consider the possibility of growth of particles at this stage, lasting 10^5 to 10^6 years. One can easily estimate that in the collapsing envelope the growth was very slow because of a very low density of the presolar nebula—particles of 10^{-5} cm in diameter could not grow to more than 10^{-4} cm. Conditions were more favorable in the disk. In the absence of turbulence and with effective sticking, kilometer-sized bodies could grow for 10^4 years. Turbulence increases relative velocities of particles and accelerates their growth. However, the larger the particles, the higher become their relative velocities. At some size, instead of sticking, particles begin to disintegrate at collisions. For the strength 10^4 erg $\cdot$ cm^{-3} their maximum radii are 0.1–1 cm (Weidenschilling, 1984). Formation of a dense dust layer is also prevented by the turbulence. On the other hand, the growth of particles decreases the disk's opacity, weakens the convection (if it exists) generating the turbulence, and can lead to the decay of the latter. It should be added that the turbulence of the disk caused by accretion flow is most intensive near the surface and, if it does not reach the central plane, conditions for further particle growth near this plane become favorable. At the same time accretion brings new particles into the upper layer, making it more opaque. The problem of convection and turbulence in the disk therefore is not yet solved. To maintain convection a permanent heating of the central part of the disk (small z coordinate) is needed, and the source of energy is not yet known. A resumption of turbulence cannot stop the growth of planetesimals if they have formed during a stage when the turbulence has decayed. In this case several kilometer-sized bodies can grow to the end of the collapse.

Growth of Planetesimals and Formation of the Planets

It is well known that terrestrial planets and giant planets differ substantially in several respects. The former consist mainly of silicates and metals and have almost no volatiles. The latter, on the contrary, contain mainly volatiles, the chemical compositions of Jupiter and Saturn being close to cosmical abundances (mainly hydrogen and helium). They have higher rates of rotation and are surrounded by numerous satellites. The composition of the terrestrial planets and, especially, a strong deficiency of noble gases are evidence in favor of their accumulation from much smaller solid bodies that could not retain gases from the very beginning. The majority of specialists share this point of view.

At the same time, an old theory of planet formation from massive gaseous protoplanets with about the mass of Jupiter is being developed (Cameron, 1978; Cameron *et al.*, 1982). Calculations showed that if such a gaseous bubble could form in some way it could contract, heat, and differentiate in density, transforming into a giant planet. The formation of the terrestrial planets is similarly explained. It is supposed that after settling of the heavy elements to the center of the protoplanet, its outer envelope, consisting of light gases, is removed by tidal forces or by solar radiation. However, the process of formation of protoplanets themselves has not been studied and the reliability of this theory has not been proved. In the disk with Keplerian rotation (after the collapse of the nebula) formation of protoplanets is possible as a result of gravitational instability in the gas but only at a large mass of the disk—about that of the Sun (Ruskol, 1960; Cassen *et al.*, 1981). Such a disk, however, would be unstable with respect to nonaxisymmetric perturbations and should break up much earlier into two or three large objects with masses many times that of Jupiter. Attempts to get rid of these difficulties by placing the formation of protoplanets at an earlier stage (in the collapsing envelope) and to larger distances from the center give rise to other difficulties.

Because protoplanets cannot effectively be driven outward by the turbulent gas in the disk, the protosolar nebula should have rapid initial rotation ($J = 10^{53}$ to 10^{54} $g \cdot cm^2 \cdot s^{-1}$). In this case, however, the problem of the loss of angular momentum from an extended protosun necessary for its contraction into a star becomes very difficult. It remains unclear why Uranus and Neptune contain almost no gas. It is not evident that the contracting protoplanet, when entering the Keplerian orbit, becomes so dense that it can be stable against tidal disruption. This is of especial concern in the zone of the terrestrial planets, where tidal forces are much greater. The model of a rapidly rotating nebula also meets with difficulties from a cosmochemical point of view because it lead to lower temperatures of the preplanetary disk. The model of a slowly rotating nebula that we accept here seems to be preferable. In this model, formation of giant protoplanet in the region of the terrestrial planets seems to be impossible.

Stage of Dust Condensations

As was pointed out above, after collapse and decay of the turbulence in the solar nebula is complete a dust layer forms in its central plane which breaks up into numerous masses of condensates because of gravitational instability. In the zone of the terrestrial planets, the dust condensations were about 10^{16} g and their densities about 10^{-6} $g \cdot cm^{-3}$. Because of self-gravitation they did not disintegrate and because of centrifugal forces generated by rotation they did not contract. However, their coalescence was usually accompanied by contraction.

In the region of the outer planets Uranus and Neptune, initial densities of the condensates were much lower ($\sim 10^{-10}$ $g \cdot cm^{-3}$) and the duration of the condensation stage was long. Because of their much larger sizes compared to solid bodies, the condensates grew faster, accelerating the accumulation of preplanetary bodies at an early stage. However, in the region of the terrestrial planets the condensation stage was quite short and it did not influence the accumulation process substantially. In less than 10^4 years, their masses increased to about 10^2 times and the condensates transformed into usual bodies of several kilometers in size.

Dynamical Evolution of Preplanetary Swarm of Bodies: Distribution of Masses and Velocities

The most important dynamical characteristics of the swarm of preplanetary bodies (planetesimals)—the distribution of their masses and velocities—are closely interconnected. The velocities of bodies are determined by their mutual gravitational perturbations and collisions and depend directly on their masses. Their mass distribution in turn depends on their velocities. The velocities determine the frequency of collisions and the character of the process: at small velocities the bodies coalesce and at larger velocities smaller bodies disintegrate, changing the distribution of masses. In principle, the coupled evolution of these two distributions should be considered. Analytical solution of such a problem is not possible. Numerical computations also meet with considerable difficulty and are still performed for simplified models only. The results for an initial stage of accumulation (Greenberg *et al.*, 1978) differ considerably from that for the final stage (Wetherill, 1978, 1980).

From the general character of gravitational perturbations, one can expect random velocities of bodies to depend only slightly on their sizes. Hence an analytical study of the mass distribution of planetesimals was done assuming independent velocities of the masses. The velocities were then found using the distribution of masses so obtained (Safronov, 1969). We omit here the complicated integrodifferential equation of coagulation from which the distribution of masses was determined and present only a short discription of the results.

With the assumption that bodies coalesce at any collision, for the case when the cross section of the collision of two bodies is proportional to the sum of their masses and taking the initial mass distribution of the form $n(m, 0) = ae^{-bm}$ where n represents distribution, m mass and a and b are constants the exact solution of the equation was found to be:

$$n(m, t) = c(t)m^{-q}e^{-bm} \tag{2}$$

where t is time and q is a dynamical parameter. Later it was shown that the same type of the distribution also took place for other initial distributions. According to Eq. (2) in the region of small masses m the distribution is satisfactorily described by the inverse power function, $m^{-q}\,\delta m$, for the number of bodies with the masses m and $m + \delta m$. With the lapse of time the bodies grow, m increases, and the power function m^{-q} extends to the region of larger masses. However, in the region of the largest masses $m \sim b^{-1}$, the distribution function steeply (exponentially) decreases.

The coagulation efficiency taken above $[\propto (m + m')]$ is an intermediate one between its value for small bodies (geometrical cross section $\propto m^{2/3}$) and that for large bodies (gravitational cross section $\propto m^{4/3}$). A qualitative study of the coagulation equation was done for a more general form of the coagulation efficiency, valid for the whole range of mass variation in the swarm. The study led to an asymptotic solution (when $t \to \infty$). For long durations the mass distribution should tend to the power function:

$$n(m, t) \approx c(t)m^{-q} \tag{3}$$

which is valid over a wide range of masses except the region of largest bodies. In the case of pure coagulation without fragmentation of bodies ($q \approx 1.6$). Partial fragmentation at collision makes the process much more complicated. For a smooth dependence of a degree of fragmentation on masses $q \approx 1.8$ has been found (Zvyagina *et al.*, 1973). The results were confirmed by numerical computations.

On the other hand, when there is only fragmentation (as, for example, in the asteroid belt) a solution exists with $q = 11/6$ for the case with the fragmentation independent of the masses of bodies. These values of q agree satisfactorily with the values estimated for the distribution of masses of the asteroids estimated from observations as well as from the population of bodies responsible for craters on Moon, Mercury, and satellites of the planets.

For the power law of mass distribution the exponent q is an important dynamical parameter. When $q < 2$ the main mass is concentrated in larger bodies. These bodies effectively increase the random velocities of all other bodies by gravitational perturbations. In contrast, when $q > 2$, the main mass is comprised in small bodies and particles. Because of frequent collisions their random velocities are relatively small. In this case the system is in an unsteady state that rapidly changes and, because of a sweeping up of small bodies by large ones, turns into a state with $q < 2$ long before the end of accumulation.

In the rotating system random velocities of particles increase via transformation of the energy of regular (ordered) motion into the energy of chaotic (disordered) motion. In thc prcplanctarty swarm a rcgular motion of bodies (Keplerian rotation) is restored at the expense of their potential energy, i.e., of the radial contraction of an inner part of the swarm. Random velocities v, measured relative to the Keplerian circular velocity, increase via gravitational perturbations of bodies during their encounters. In contrast, inelastic collisions decrease the velocities. As a result some "equilibrium" values of velocities should be established, when the contribution of collisions balances that of perturbations.

There is an analogy between random motions of bodies in the swarm and the chaotic thermal motions of molecules in a gas or a liquid. Therefore, the increase of velocities can be estimated in the same way as used to evaluate heating of a viscous fluid moving with the velocity gradient perpendicular to the motion. It is known from hydrodynamics that for an axisymmetrical rotation with the angular velocity $\omega(R)$ the thermal energy liberated per gram per second equals:

$$\varepsilon_1 = \nu R^2 \left(\frac{d\omega}{dR}\right)^2 \tag{4}$$

where ν is the kinematic viscosity and R is the distance from the axis of rotation. It was found that for large mean free path λ, Eq. (4) needs a substancial modification (Safronov, 1969). First, the usual expression for the viscosity, $\nu = \lambda v/3$, should be changed to:

$$\nu = (1/6)\overline{\Delta R^2}\tau = (1/3)e^2 R^2/\tau \approx v^2/3\omega^2\tau \tag{5}$$

where e is the orbital eccentricity of the body and τ is its relaxation time, i.e., an average time interval during which the vector v substantially changes because of encounters with other bodies. The correction takes into account the fact that even during a long motion of the body along its orbit it cannot move away from its initial radial distance R farther than $\Delta R_m \approx eR$. Second, in Eq. (4) an additional factor $\beta < 1$ should be inserted that takes into account the fact that between collisions the angular momentum ωR^2 of the body is conserved instead of the angular velocity. As a result the body acquires the energy of random motion per gram per second equals:

$$\varepsilon_1 = \beta v^2/\tau \tag{6}$$

It was estimated that $\beta \approx 0.1$–0.2. The energy lost by the body per unit mass per second via collisions can be written in the form:

$$\varepsilon_2 = \zeta v^2/2\tau_s \tag{7}$$

where $\zeta < 1$ and 2β and τ_s an average time interval between collisions of the body. The maximum loss takes place at completely inelastic collisions, when bodies coalesce, which is characteristic for the effective accumulation. Then

$\zeta \approx 0.4$. It was revealed that the velocities were nearly independent of mass and could be expressed conveniently in terms of the mass m and radius r of the largest body in the zone:

$$v^2 = Gm/\theta r \tag{8}$$

where θ is a dimensionless parameter on the order of several units which depends on q and slowly changes in the course of accumulation.

The velocities of bodies are therefore proportional to (but several times less than) the velocity of escape from the surface of the largest body m. In this case we have:

$$\varepsilon = v\,dv/dt \approx v^2/3\tau_s = \varepsilon_1 - \varepsilon_2 \tag{9}$$

from which the parameter θ, characterizing the velocities of planetesimals according to Eq. (8), can be found. We shall not give complicated expressions for θ here. In the simplest case of equal bodies in the system $\theta \approx 1$. When bodies grow, their velocities increase proportionally to their radii. For the mass distribution Eq. (3) with the exponent $q < 2$ the expressions for τ and τ_s becomes more complicated but the conclusion remains mainly the same.

The random velocities of bodies are almost independent of their masses and can be found from Eq. (8). At $q \approx 1.8$ we obtain $\theta \approx 3$ to 5. For the radius of the largest body $r \approx 200$ km the velocities of other bodies in its feeding zone are about 0.1 km $\cdot$ s^{-1}. At the final stage of accumulation gravitational perturbations become more effective, whereas collisions become less frequent and θ decreases. For $r \sim 6000$ km and $\theta \approx 1$ in the region of the Earth and Venus, we obtain from Eq. (8) $v \approx 7$ km $\cdot$ s^{-1} and eccentricities of orbit $e \approx 0.25$ for all bodies except the largest ones.

In the region of the giant planets for the same sizes of the largest bodies, velocities are also the same. Because of smaller Keplerian velocities, however, the eccentricities of orbits are larger, 0.3–0.4, and the ejection of bodies from the region becomes appreciable. In the energy balance for random velocities ε_1 and ε_∞, the energies taken away by escaping bodies, now dominate. A further growth of bodies does not lead to an increase in mean velocities. They become almost constant after reaching about a half of the velocity of escape from the solar system.

At the early stage, the accumulation of planetesimals proceeded in the presence of gas which represented almost all the mass of the disk. The gas affected the motion of bodies in two ways. First, it decreased random velocities of bodies. Second, being partly supported by the radial pressure gradient, the gas rotated around the Sun more slowly than the bodies and resisted their motion, diminishing their angular momenta and thus their orbital radii R. A drag force experienced by the body is described differently depending on the values of Reynolds and Knudsen numbers. For $Re \geqslant 10^3$, i.e., for bodies larger than about 20 m, we have:

$$F = 0.22\pi r^2 \rho_g v^2 \tag{10}$$

where ρ_g is the gas density. One can compare the loss of energy ε_3 resulting from the gas drag with its loss ε_2 from collisions. We obtain:

$$\frac{\varepsilon_3}{\varepsilon_2} \approx \frac{0.2}{1+\theta} \frac{\sigma_g}{\sigma_p} \frac{v}{v_g} \sim \frac{1}{(1+\theta)\sqrt{\theta}} \frac{\sigma_g}{\sigma_{go}} \left(\frac{r_1}{20 \text{ km}}\right) \tag{11}$$

where σ_p is the surface density of the solid matter, v_g is the thermal velocity of molecules in the gas, σ_g/σ_{go} is the ratio of the surface density of the gas to its initial value (preceding the gas dissipation from the disk), and r_1 is the radius of the largest body in the power-law mass distribution (3) which determines the random velocities according to Eq. (8).

We see from Eq. (11) that for $r < 200$ km the decrease of random velocities of bodies by the gas is unimportant. For large r_1 it can be considerable if there is no dissipation of the gas. However, there are serious indications (see next section) that when the largest bodies have grown to $r_1 \sim 100$ to 1000 km the gas has already dissipated from the region of the terrestrial planets.

Diminishing of the orbital radius of the body with the mass m, radius r, and density δ is:

$$v_R = \frac{dR}{dt} = -2F/m\omega = -\frac{\rho_g \Delta V^2}{3\delta r\omega} = -\frac{2}{3\pi\delta} \frac{\sigma_g \Delta V^2}{v_g r} \tag{12}$$

where $\Delta V \sim 10^3$ to 10^4 cm $\cdot$ s^{-1} is the velocity of the body relative to the gas. The radial drift of bodies accelerates the accumulation of small bodies. The maximum radial velocity v_R takes place for about meter-sized bodies and almost reached ΔV. However, such rapidly drifting bodies comprised less that 1% of the total mass of the swarm and, besides, most of them were swept up rather early by large bodies.

The growth of bodies decelerates their drift. According to Eq. (12) $v_R \sim r^{-1}$. The rate of accumulation of large bodies is determined by the relation:

$$\frac{dm}{dt} = 4\pi r^2 \delta \frac{dr}{dt} \approx (1 + 2\theta r^2/r_1^2)\rho_p(v^2 + v_R^2)^{1/2}\pi r^2 \tag{13}$$

Eliminating the time t from Eqs. (12) and (13) and integrating the equation obtained, R is found to depend on r (Safronov and Ruzmaikina, 1978). For $\sigma_g/\sigma_p = 300$, $T = 600$, $r_0 = 100$ m, $r = 500$ km, and $R_0 = 1$ a.u., we obtain $R/R_0 \approx 2/3$. In fact the radial drift could be slower becuase of the early dissipation of the gas and the decrease of σ_g.

Formation and Growth of Planetary Embryos

There was a considerable difference in the accumulation of the largest bodies. The distribution of their masses differed from the power law (3) found for smaller bodies. They swept up all bodies that collided with them without losing all fragments ejected at impacts because of this gravitational fields.

When the embryos grew the velocities of bodies increased but the absolute value of their potential energy also increased and always exceeded the kinetic energy. Transformation of a considerable fraction of the latter into heat strengthened the relative dominance of gravitation. Consider, for example, a head-on collision of two equal bodies with the masses m and radii r in the zone of the embryo with radius r_e. The energy liberated on impact is:

$$E = \frac{Gm^2}{r}(0.7 + r_e^2/\theta r^2) \qquad 1 < \theta < 5 \tag{14}$$

A part of this energy $\varepsilon' E$ goes into crushing up and diminution of the material, a part $\varepsilon'' E$ goes into heating and melting, and the remainder $\varepsilon''' E$ goes into ejection of the fragments. Complete scattering of the whole mass $2m$ into separate pieces is possible if r is less than the critical value r_c:

$$r_c = r_e \theta^{-1/2} [\varepsilon'''/(1.9 - 0.7\varepsilon''')]^{1/2} \tag{15}$$

For $\varepsilon''' = 0.1$, $\theta = 2$ we obtain $r_c = 0.16 r_e$. Smaller bodies colliding with those of a comparable size lost a large part of their mass, whereas larger bodies did not lose at all.

The largest bodies grew relatively more rapidly than the other bodies in their feeding zones because of their larger gravitational cross sections of collisions, which increased in proportion to the fourth power of their radii. They "ran away" from an overall mass distribution in the zone and formed planet embryos, although only few of them became the planets. The eccentricities of their orbits were several times smaller than the average eccentricities $\bar{e}$ of the orbits of other bodies. The half-width of the ring-shaped feeding zone of the embryo was:

$$\Delta R_f = (\bar{e} + e)R = \frac{1 + e/\bar{e}}{\omega}\left(\frac{8\pi G\delta}{3\theta}\right)^{1/2} r \tag{16}$$

All bodies with the semimajor axes of orbits in the range $2\Delta R_f$ could have fallen onto the embryo. The half-width of the zone of gravitational influence of the embryo moving along a circular orbit is:

$$\Delta R_{go} = \kappa r_{L1} = \kappa\left(\frac{4\pi G\delta}{g\omega^2}\right)^{1/3} r \tag{17}$$

where $r_{L1} = (m/3M_\odot)^{1/3}$ is the distance from the embryo to its first Lagrangian point of liberation and $\kappa \approx 3$–4. Gravitational interaction of two neighboring embryos between their average distances from the Sun is less than:

$$\Delta R_g = \Delta \mathrm{R}_{go} + (e + e')R \tag{18}$$

which is about one half of ΔR_f. While the embryos grew, their zones widened and overlapped with neighboring zones. The smaller of the two embryos, which happened to be inside the feeding zone of the latter, began to grow more slowly. Occurring inside the gravitational influence zone ΔR_g of the neighbor, it acquired a more eccentric orbit and began to "walk" randomly between other embryos. Finally it either fell onto one of them or passed very near it inside its Roche limit and distintegrated into smaller fragments. Hence the distances ΔR_{fg} between neighboring embryos varied between ΔR_f and ΔR_g. So at least three embryos were placed in the same feeding zone with $2\Delta R_f$. The number of the embryos in the whole zone of the planet was:

$$N \approx 2\Delta R_p/\Delta R_{fg} \approx (m_p/m)^{1/3}(\theta/\theta_p)^{1/2} \qquad (19)$$

where $2\Delta R_p$ is the present width of the zone of the planet m_p and θ_p is the value of θ at the end of the accumulation.

At earlier stages there were many embryos (for $m \approx 10^{-3} m_p$ $N \gtrsim 10$) but they contained only a small fraction μ_e of the whole mass in the zone:

$$\mu_e = N_m/m_p \approx (m/m_p)^{2/3}(\theta/\theta_p)^{1/2} \qquad (20)$$

Their contribution to increasing the random velocities of bodies was small, and effective values of m and r in Eq. (8) almost coincide with m_1 and r_1. As long as m did not differ much from m_1 the substitution m,r for m_1,r_1 did not change θ appreciably. This led to "runaway" models assuming that later on θ could increase to high values $\sim 10^2$ to 10^3. However, estimates have shown (Pechernikova and Vityazev, 1979; Safronov, 1982) that the runaway growth of the embryos was rather slow. When m/m_p amounted to $\sim 10^2$, m/m_1 increased to between 10 and 20 and the parameter θ reached its maximum value $\theta_m \approx 10$–15. Then θ diminished to about 1. The conclusion that at the end of accumulation θ approaches 1–2 was derived also by Wetherill (1978, 1980) from numerical modeling of the process in the region of the terrestrial planets.

Dissipation of Gas

It is known that the terrestrial planets consist primarily of iron–magnesium silicates and Fe–Ni and that the iron content of Mercury could be $\approx$65%. If we start from the cosmic elemental abundances, then it is necessary to explain a loss of more than 99% of the matter in this zone. As judged by the densities of Uranus and Neptune, almost all the hydrogen and helium were lost from this zone. Only Jupiter and Saturn were able to accrete a considerable amount of the gas located in their zones. The determination of the time of the gas removal is important for understanding the process of the planets formation. According to the Kyoto model (Hayashi *et al.*, 1977, 1979, 1985), the entire process of accumulation of the terrestrial planets occured in the presence of the gas (the time of the gas dissipation from the zone of the Earth is supposed to be greater than the accumulation time). In this model the Earth had a very thick, opaque atmosphere, which prevented the energy brought to the planet by falling bodies from radiating away.

According to the estimates made by Hayashi *et al.* (1985), at the end of the process, the mass of the Earth's atmosphere was more than 10^{26} g, while the temperature of the Earth's surface exceeded 4000K. It is not clear how such a high temperature for primordial Earth can be matched with data from the Earth sciences. Moreover the Kyoto model meets serious difficulties in the dynamical and cosmochemical aspects. In the presence of gas the values of parameter θ in the zone of terrestrial planets are several times greater than estimated above. In this case the computed number of planets forming in the terrestrial zone is about 10. According to Herndon and Wilkening (1978) and Wood and McSween (1977), during the formation of the meteorite parent bodies the ratio of dust to gas (ρ_p/ρ_g) was orders of magnitude greater than its initial (interstellar) ratio ($\sim 10^{-2}$). Their conclusion about the considerable deficit of the gas in the zone of asteroids, where the parent bodies of meteorites formed, is inferred form mineralogical features of the chondrites. Vityazev and Pechernikova (1985) pointed out that the existence of preaccumulation tracks of solar cosmic rays in some meteoritic grains also shows the absence of gas during the formation of asteroids. One thousandth of the initial mass of the gas between the Sun and the bodies in the asteroidal zone was enough to shield the surfaces of the bodies from the solar wind particles. However, irradiated grains were found in deep parts of meteorites. One can estimate the sizes of bodies during the irradiation process. Vityazev and Pechernikova concluded that in the zone of the terrestrial planets gas was practically absent when the 100-km-size bodies had formed.

There were many attempts to estimate the time scale of gas removal through some physical mechanisms of dissipation. Three main mechanisms were examined: removal of gas by intense solar wind in the T-Tauri stage, dissipation of gas from the disk exosphere heated by ultraviolet radiation from the Sun, and removal of gas by turbulent scattering in the preplanetary disk.

Elmegreen (1978) showed that the solar wind flowing along the disk created turbulence in its surface layer. Part of the gas in the preplanetary disk therefore could dissipate; the larger part of this slowly drifted toward the Sun and accreted on it. According to Elmegreen the characteristic time of scattering of the mass M is about:

$$t = \left(\frac{M}{10^{-2}M_\odot}\right)^{4/3}\left(\frac{\dot{M}_\odot v_w}{10^{26}\ \mathrm{g\cdot cm\cdot s^{-2}}}\right)^{-4/3}\left(\frac{\xi}{0.3}\right)^{-1} 10^5 \text{ years}$$

where $\dot{M}_\odot$ is the solar mass loss rate caused by intense wind ($\sim 10^7$ of the current value), $v_w \sim 400\ \mathrm{km\cdot s^{-1}}$ is the solar wind velocity, ξ is the efficiency parameter of the transformation of the wind energy into the energy of turbulent motion. Assuming a rather high value of $\xi \approx 1/3$, we obtain $t \approx 10^5$ years. In this case it is difficult to understand how Jupiter and Saturn have amassed the gas. According to Safronov (1969) the efficient accretion of the gas begins when the solid core is three to five Earth masses. The growth time of such cores in the region of Jupiter is almost 10^8 years.

Assuming lesser values of ξ we can formally obtain $t \approx 10^8$ years. However, observations of the young stars in the T-Tauri stage show that their ages are only $\approx 10^6$ years. The time of turbulent scattering of the gaseous disk is about $T = R^2/v_t$, where v_t is the turbulent viscosity. Assuming for Jupiter's zone that $v_t = 10^{12}\ \mathrm{cm^2\cdot s^{-1}}$, we obtain $t \approx 10^8$ years. However, it is necessary to show that this value is physically acceptable and is neither more (fast gas removal) nor less (small efficiency).

The gas heating in the exosphere of the disk by solar UV radiation was discussed by Pechernikova and Vitjazev (1981). Assuming the solar UV flux to be the same as at present, they showed that the temperature in the exosphere was sufficient for gas dissipation from Jupiter's and Saturn's zone with characteristic time $\approx 10^8$ years and from Uranus, and Neptune's zones $\approx 10^7$ years. For zone of the terrestrial planets thermal dissipation is effective only with UV luminosity that is an order of magnitude greater than its current value. In the latter case gas removal from the preplanetary disk occurs not because of Jean's escape mechanism, but because of gas dynamic efflux into interstellar space.

Some combinations of these mechanisms and their efficiencies can explain early scattering of the gas from the inner and outer regions of preplanetary disk and less intensive removal of the gas from the zone of Jupiter and Saturn. It is desirable to investigate the common action of these mechanisms, but to date we have too few data on parameters of the young active Sun. Nevertheless meteoritic and Earth sciences data coupled with the results of dynamic theory give strong evidence for absence of the gas in the zone of terrestrial planets by the end of their accumulation.

Masses, Accumulation Times, and Distances of the Planets

The rate of growth of the embryo, sweeping up all bodies that collided with it, is determined by the expression:

$$\frac{dr}{dt} = \frac{(1 + 2\theta)\sigma_o\omega}{2\pi\delta}(1 - \mu_e) = \frac{(1 + 2\theta)\sigma_0\omega}{2\pi\delta}(1 - r^2/r_M^2) \tag{21}$$

For simplicity we assume here that the initial surface (column) density σ_o of the solid material is independent of R and that the feeding zone of the planet embryo is not limited by the presence of neighboring embryos. Overlapping of the zones narrowed the feeding zone, but the motion along the eccentric orbits widened it. Equation (21) shows that the radius of the planet can not exceed r_M because at $r \to r_M$, $dr/dt \to 0$. Inserting $u_e = m/Q = m/4\pi\sigma_o R\Delta R_f$, where Q is the mass contained in the feeding zone, we find (Vityazev *et al.*, 1978):

$$r_M \approx (2\sigma_o)^{1/2} R^{5/4} (6\pi/\theta\delta M_\odot)^{1/4} \tag{22}$$

and accordingly the maximum mass of the planet:

$$m_M = \tfrac{4}{3}\pi\delta r_M^3 \approx \tfrac{4}{3}\pi\delta(2\sigma_o)^{3/2} R^{15/4} (6\pi/\theta\delta M_\odot)^{3/4} \tag{23}$$

The upper limit to the mass of the planet has a simple explanation. The mass in the feeding zone is in proportion to the width of the zone, $2\Delta R_f \propto r \propto m^{1/3}$, and increases much more slowly than the mass m of the planet.

For $\sigma_o = 10\ \mathrm{g\cdot cm^{-2}}$ and constant $\theta = 3$ the Earth grow to 98% of its present mass in 10^8 years (Safronov, 1969). An increase of θ at an intermediate stage shortened its duration, but the decrease of θ at the final stage made it longer.

On the whole, the time scale of accumulation remains practically the same. It agrees with the estimates of Weidenschilling (1976) and Wetherill (1978).

The maximum distance between the neighboring planets (when their feeding zones are not overlapping) is:

$$R_{n+1} - R_n = \Delta R_{fn} + \Delta R_{f,n+1} \tag{24}$$

or in another form:

$$R_{n+1}/R_n = b \approx (1 + e)/(1 - e) \tag{24a}$$

For constant e this gives the Titius-Bode relation, but it does not take into account differences in the masses of the planets. For the terrestrial planets at $\theta = 2$ it gives $R_{n+1}/R_n \approx 1.5$ and for the outer planets, at $e \approx 0.4$ (see above), $R_{n+1}/R_n \approx 1.8$ to 1.9 in a satisfactory agreement with observations. This result is obtained without an assumption (made by some authors) that eccentricities of bodies in the region of the terrestrial planets were increased by perturbations of Jupiter.

Origin of the Asteroids

The very specific structure of the asteroid belt, consisting of a great number of small planets (more than 3000 already are known), can be well understood in the framework of the theory of accumulation of the planets from solid bodies without additional exotic hypotheses. The Olbers hypothesis, that a normal planet existed originally in this zone (Phaeton), which then disintegrated in some way, could not explain the dynamical and physical properties of the system. In the modified hypothesis it was asserted that an explosion of a massive planet (90 times the mass of the Earth) had occured 16 million years ago (Ovenden, 1972) or 6 million years ago (Van Flandern, 1977). The hypothesis was sharply criticized by Öpik (1977), who has shown how dramatic would be the consequences of such an explosion for life on the Earth.

The idea of Schmidt (1958) that the asteroids are the result of an interruption of the accumulation of a planet at an intermediate stage because of the presence of massive Jupiter is accepted now by many specialists. The temperature of the preplanetary disk decreased with the distance from the Sun. This change of the temperature has led to important differences between the asteroids' and Jupiter's zones because in the former the abundant volatiles H_2O, NH_3, and CH_4 were in the gaseous state, whereas in the latter they have condensed into icy particles.

Space density of solids in the Jupiter zone was higher, so bodies grew faster, and accordingly their random velocities and the eccentricities of their obits also increased faster. For a mass of the Jupiter embryo $\approx 10^{27}$ g, the bodies of its zone began to penetrate the zone of asteroids. Being much more massive than the asteroids, they swept up the latter on collision and returned to the zone of

Jupiter. This way the asteroid zone lost more than 99% of the mass of its bodies. Larger bodies of the Jupiter zone ($\approx 10^{26}$ g) at close encounters with the remaining asteroids increased their velocities because of their gravitational perturbations to the present values of about 5 km · s^{-1} (Safronov, 1979). At such velocities the asteroid bodies could not coalesce on collision. Their kinetic energy became higher than the potential energy and the process of accumulation changed to erosion and disintegration. This is evidenced by the presence of many "families" of asteroids, containing about a half of all asteroids. Perturbations by Jupiter expelled all asteroids of the outer zone with semimajor axes >3.4 a.u. and produced the gaps (Kirkwood gaps) for the periods of revolution commensurable with the period of Jupiter (1 : 2, 1 : 3, 2 : 5, etc.). Continuing collisional evolution, together with an increase in the eccentricities of resonant asteroids, replenished the bodies with orbits crossing the orbits of the Earth and Mars (Apollo and Amor asteroids) which are a main source of meteorites falling onto the Earth. A study of the complicated structures of various types of meteorites provides a favorable opportunity to reconstruct the physical conditions in the preplanetary disk and to reveal the main features of the accumulation process at an early stage, which has been recorded in the unique formation—the belt of the asteroids.

Accumulation Process in the Outer Part of the Disk

As mentioned before, the gas played an important role in the zones of Jupiter and Saturn. It diminished the random velocities of bodies. At a later stage, when the cores (the embryos with masses about three to five Earth masses, $m_{\oplus}$) had accreted, the accretion of gas onto the cores began. Such an accretion of gas provided the large masses of Jupiter ($\approx 300 m_{\oplus}$) and Saturn ($\approx 100 m_{\oplus}$). In the formation of outer planets, a long-lasting earlier stage of growth of condensations of dust and ice particles was very important. When more massive condensations reached 10^{-3} to 10^{-2} of the present masses of Uranus and Neptune, they became rather compact and began to scatter other bodies at large distances. Later on, when the giant, planets achieved masses comparable to their present values, they effectively ejected bodies from the solar system and to its periphery by their gravitational perturbations. At this stage, the so-called Oort cometary cloud formed. The estimated total mass of the comets ejected into the cloud could have been a few Earth masses. After encounters with stars or molecular clouds, some perturbed comets returned to the inner part of the solar system. The study of comets and related phenomena is important because comets are relicts of the preplanetary matter.

During the accumulation of the planets circumplanetary swarms formed. Via inelastic collisions, particles and bodies were captured by the planet and accumulated mainly at distances of several planetary radii. A process of formation of satellites in these swarms followed, which was similar in many respects to that of the formation of the planets in the preplanetary disk (Ruskol, 1975;

Safronov *et al.*, 1985). Investigation of the satellites as well as planetary rings is important for understanding the processes in an early preplanetary disk. It allows us to obtain additional restrictions on a general character of accumulation and accretion.

Cosmochemical Aspects of Evolution

Until recently dynamical and cosmochemical researches were carried out independently to a certain degree. Important cosmochemical evidence is not always taken into account in dynamic models. Cosmochemists often used exotic schemes of solar nebula formation or oversimplified dynamic processes. An example is a scenario of solar system formation directly from the debris of a single supernova (Manuel and Sabu, 1977; Bradley *et al.*, 1978; Lavruchina, 1980). In this scenario the Sun accumulated from the supernova core, the iron meteorites and cores of the terrestrial planets were formed primarily from elements synthesized in the hot stellar envelope, and the outer planets and the carbon phases of the chondrites were condensed from elements of the cooler, outer zone. However from the dynamical point of view such a scenario is inconceivable.

In spite of the continuously increasing amounts of data on the composition of interstellar and interplanetary dust and on the planets and their atmospheres and new research on asteroids and comets, modern models and their verification are still based mainly on the meteoritic data. A good introduction in this complex area is given in the books of Wasson (1974) and Dodd (1981).

In the 1970s the Cameron model of a massive solar nebula cooling from high ($\approx$2000K) temperatures was popular. It was used as a basis for the theory of successive condensation of matter of undifferentiated meteorites from a gas with cosmic composition (Larimer, 1967; Anders, 1972; Lewis, 1972). Later it became clear that the theory meets with difficulties in explaining many properties of chondritic meteorites. In particular, the composition and structure of chondrules and the process by which the volatile and moderately volatile elements fractionate remain unexplained. The difficulties in explaining unusual Ca–Al inclusions observed in some carbonaceous chondrites (Allende, Murchison), which conserved a number of uncorrelated isotopic anomalies led Clayton (1977, 1983) to a hypothesis of the presence of primitive interstellar particles in chondrites. However, this hypothesis gives no explanation of the large differences in the properties of different types of chondrites and even of the fact of their existence. Moreover, according to recent data (see Nagahara, 1983), the chondrules consist of remelted precursor material. Many authors believe that this precursor material is similar in chemical composition to the chondritic matrix in which the chondrules are now embedded.

Clayton does not point out processes that could rework a part of interstellar material into spherules of millimeter size. According to Wood (1983), the heating and melting of interstellar 1-mm grains could occur during accretion of the gas of the presolar nebula onto the preplanetary disk. However, unsolved questions remain:

1. How could particles of about 1 mm form? According to well-known astrophysical data the bulk mass of interstellar dust is contained in grains with sizes $\approx 10^{-5}$ cm.
2. Wood overestimates the heating at the shock front. Really the particles could be melted only in the central part of the disk with $R < 1$ a.u.

Morfill (1983) and Clayton and Wieneke (1983) have studied mass transfer in a turbulent preplanetary disk. According to their models some part of material in the disk moved from the cold region to the hot one, and vice versa. Particles underwent heating, melting, cooling, etc. However, as mentioned before, the existence of effective turbulent motion at the stage of planetesimal formation remains an open question.

Besides the reworking of interstellar dust at the stage of formation of the preplanetary disk a significant (and maybe more important) reworking of material should occur during collisions of bodies at the stage of planet formation. It follows from dynamical estimates that the relative velocities at the stage of dust condensations and small bodies were not high enough to produce strong heating of the bulk mass of these bodies. Some heating in the first 10^6 years could result from ^{26}Al if its ratio to ^{27}Al to this moment was $\sim 10^{-5}$ (which is in accordance with modern astrophysical and meteoritic data). The velocities increased on the average proportionally to the radii of the largest bodies. According to Eq. (8) $v \approx 5 \times 10^{-4} r$ km $\cdot$ s^{-1}. The velocities reached 0.5 km $\cdot$ s^{-1} only after the formation of large ($\approx$1000 km) bodies. Under these conditions a fragmentation and slight heating of bodies occurred. In the asteroid belt, big bodies penetrating from the Jupiter's zone could increase the velocities to $\sim$5 km $\cdot$ s^{-1}. This stage was characterized by a transition to intense impact reprocessing, strong metamorphism, local melting, and outgassing (Vityazev, 1982; Safronov and Vityazev, 1985). Such an approach allows to understand some important features of the meteoritic matter. It seems to us that primitive carbonaceous chondrites simply represent fragments of bodies that were not subjected to high-velocity impact heating. The C2–C3 types are fragments of bodies that formed at the stage of a moderate hydrothermal metamorphism and outgassing.

The formation of chondrules formation and the ordinary chondrites occurred at the stage of larger velocities $\sim$3 to 5 km $\cdot$ s^{-1}. Simultaneously, impact melting and differentiation in some bodies can lead to the formation of achondrites and iron meteorites. Such a model can explain the well-known cases of complementarity in composition between some chondritic and achondritic meteorites (Safronov and Vityazev, 1985). Recently Wood (1985) summarized limitations on the formation of chondrites. He concluded that the formation of chondrules proceeded at the same place and time that the parent bodies of meteorites were formed. It seems to us that this conclusion is convincing confirmation of the collisional model, although in future the synthesis of different approaches will be necessary.

Initial State and Early Evolution of the Planets

In the 1950s the dominating concept of a molten primordial and gradually cooling Earth was rejected. A new scenario of a primary cold Earth gradually heated by radioactive sources and the gravitational energy liberated by differentiation was proposed. Later, strong evidence of an early active evolution of the terrestrial planets appeared: the absence of the ancient Earth rocks, signs of a considerable early heating of the Moon and the parent bodies of meteorites, etc. At the same time important results in the theory of accumulation were obtained concerning the important role of large bodies in the formation and early evolution of terrestrial planets. In the 1960s the first models of impact heating of the growing Earth were calculated and an initial temperature distribution of the Earth was obtained.

Initial Temperature of the Earth

Let us consider the theory that the Earth formed from particles and small bodies ($r < 1$ km). In this case, the Earth's temperature would be determined by their rate of accumulation, and all energy of impacts would be liberated on the surface of the growing Earth. With an accumulation time of 10^8 years the Earth's interior would be heated to less than 400K. The situation becomes quite different if the planet formed from large bodies. In this case the energy would be liberated at a depth on the order of the diameter of the falling body and the fraction of the energy stored beneath the bottom of the crater would increase.

The heating of the Earth during its accumulation was estimated using the equation of heat conduction with relatively simple relations for the impact energy sources and for the intensity of impact mixing (Safronov, 1969, 1978; Safronov and Kozlovskaya, 1977). The main factor determining the heating is the sizes of the largest bodies falling onto the planet. For a power-law mass distribution with $q = 1.8$ and a radius of the largest body $r_1 = 100$ km, the temperature at the center of the Earth would be about ~1000K. The upper layer (at depths of approximately 100 to 1000 km) was heated just to the melting point of the material. For the same value of q, Kaula (1979) used higher upper limit for r_1 (500 km) and found that the thickness of the layer in which matter would start to melt would be about 3000 km. Actually the energy of the impacts was high enough to produce much stronger heating, but the heating was terminated by the onset of convection. Of course, in the direct vicinity of the impacts of large bodies, matter was melted and even partially vaporized. Considerable outgassing also occurred at this time. However, the average surface temperature of the growing Earth was less than the boiling temperature of water. At earlier stages the Earth's embryo could collide with bodies of a comparable size. Collision of the proto-Earth, at a radius 2000–2500 km, with two or three bodies of lunar size is enough to heat it to the average temperature of its interior, ~2000K. Because of the adiabatic compression of the growing planet, this value increased 1.5–2 times. In such a case, differentiation in terrestrial planets could begin and effectively proceed along a course of further accumulation.

The Initial Temperature of Mercury, Mars and the Moon

An estimate of the heating of Mercury and Mars using the same technique shows that up to the end of their accumulation periods the matter in their interior also underwent a high-temperature phase. Mercury is much smaller than the Earth. Nevertheless the relative velocities in its zone resulting from its eccentric orbit and a greater Keplerian velocity at this distance from the Sun are rather high. The surface temperature of Mercury would be always much higher than the boiling temperature of water even without its heating by the Sun. The surface of Mercury retains the traces of ancient bombardment. The idea of its early differentiation will be confirmed if the existence of a Mercurian core is proved.

Mars has two times the mass of Mercury. Its formation and heating substantially depended on collision with large bodies that flew from the zones of the asteroids and Earth. Data on abundances and isotopic compositions of rare gases in its atmosphere are evidence in favor of the differentiation of Mars' interior and of formation of its atmosphere at a rather early stage (Vityazev, 1985).

Primary heating of the Moon was determined by the means of its formation. It seems that a model in which the Moon has grown in a presatelite swarm of bodies orbiting around the Earth can satisfy both dynamical and cosmochemical restrictions.

According to a multimoon model, collisions of a few satellites, moving outwards under the action of tidal forces and their coalescence into a single body, the Moon, could lead to a rather high temperature at its interior ($\approx$1500K according to Ruskol, 1979). The "late heavy bombardment" also created substantial heating and melting of the upper layer of the Moon.

Early Evolution of the Terrestrial Planets

The whole complex of data about structures of the Earth, Mars, and Moon; as well as isotopic data, are strong arguments favoring an early differentiation of the interiors of the terrestrial planets. During the accumulation resulting from intense outgassing of material, the primitive atmospheres of the Earth, Venus, and Mars were formed. Mercury and the Moon formed mainly from bodies that lost their volatiles before their collisions with growing planets. Bodies in the Mercury zone had lost their volatiles because of the high temperatures at their small distances from Sun; the Moon lost its volatiles because of a supplementary reprocessing of the matter in the circumterrestrial swarm. In addition, dissipation of gas from these smaller planets was very efficient. [The problem of differences in composition and structure of the upper shells of Earth and Venus is still far from solution.] The problem of formation and evolution of the atmosphere and hydrosphere of the Earth was reviewed by Henderson-Sellers (1983). Below we discuss briefly the model of the formation of the shell structure of the interior of the Earth. The same approach can be used for the other terrestrial planets.

The internal constitution of a planet with assumed mass and distributions of composition and temperature may be modelled with the aid of the equation of hydrostatic equilibrium and the equations of state. The latter relate densities of substances with pressure and temperature. The thermal history of a planet can be investigated by adding the equations of heat transfer. Models of such types with an assumed radial distribution of radioactive elements (^{235}U, ^{238}U, ^{232}Th, and ^{40}K) were calculated up to the 1970s. Then it became clear that heat transfer in the moderately viscous interior occurred mainly by thermal convection. Calculations of three-dimensional convection require a lot of computer time. Therefore in evolution models, usually some parametrization is used.

In order to describe the influence of thermal convection one can introduce an effective coefficient of thermal diffusivity $\kappa_c = \text{Nu}\,\kappa$, where κ is the sum of molecular, radiative, and excitation thermal diffusivities (or, in the absence of the others, mass-transfer currents), the Nusselt number Nu is related to the Rayleigh number Ra, its critical value Ra_{cr}, and other physical parameters by the relations:

$$\text{Nu} = \left(\frac{\text{Ra} - \text{Ra}_{cr}}{\text{Ra}_{cr}},\right)^{1/n}, \quad \text{Ra} = \left\langle\frac{\alpha g \nabla T}{\kappa \nu}\right\rangle H^4, \quad \text{Ra}_{cr} \approx 10^3, \quad 2 \lesssim n \lesssim 4 \tag{25}$$

where α is the coefficient of thermal expansion (typically $3 \times 10^{-5}\ \text{K}^{-1}$) $g(\approx 10^3\ \text{cm} \cdot \text{s}^{-2})$ is the acceleration of gravity, ν is the kinematic viscosity, ∇T is the superadiabatic temperature gradient, and H is the layer thickness. Impact stirring of the near-surface layer with a thickness H_1 in a growing planet gives the effective coefficient of "impact" diffusivity as used by Safronov (1969, 1978):

$$\kappa_i = \kappa_o \Gamma r' \dot{r}_p / [1 + (r'/r_1)^2]^{1/2} \tag{26}$$

where $\kappa_o \approx 1/3$, Γ ($\sim 10^2$ to 10^3) is the ratio of the mass ejected from the crater to the mass m' of the fallen body, and $m' = \frac{4}{3}\pi\bar{\rho}r'^3$. According to Safronov, $H_1 \sim 4r'$, $r_1 \approx 1$ km. For the radius of the planet $r_p \sim 0.8 r_\oplus$ $\dot{r}_p \sim 3 \times 10^{-7}\ \text{cm} \cdot \text{s}^{-1}$ and $\kappa_i \approx 1$ to $10\ \text{cm}^2 \cdot \text{s}^{-1}$.

The planets formed from bodies that accumulated at different distances and had slightly different compositions. On the average the variations in their composition and density were the same as in neighboring planets:

$$|\delta c_0|/\bar{c} \approx |\delta \rho_0|/\bar{\rho} \lesssim 0.1$$

Impact stirring decreased these fluctuations of composition and density

$$|\delta \rho| \approx |\delta c| \approx |\delta c_0|/\Gamma \approx 10^{-3} \text{ to } 10^{-4}$$

The mass distribution of such inhomogeneities with sizes $\lambda' \approx \Gamma^{1/3} r'$ was determined by the mass distribution of falling bodies. Using the Stokes-Batchelor approximation one can obtain an estimate of velocities:

$$|v_d| \approx \frac{|\delta\rho| \cdot g\lambda'^2(1 - \frac{5}{2}c')}{4\eta}, \qquad \eta = \bar{\rho}v \tag{27}$$

where c' is the fraction of volume occupied by inhomogeneities with $\lambda' - \lambda'/2 < \lambda < \lambda' + \lambda'/2$. The heat transfer resulting from these motions can be estimated by introducing the new effective coefficient of thermal diffusivity (Vityazev and Majeva, 1980):

$$\kappa_d \approx \frac{c'}{3}|v_d|\lambda' \approx \frac{c'(1 - \frac{5}{2}c')|\delta\rho_o|gr'^3}{12\eta} \tag{28}$$

Assuming $c' = 1/5$, $\delta\rho_o = 0.1\ \mathrm{g \cdot cm^{-3}}$, and $r' = 10^6$ cm, we find for $\eta \lesssim 10^{18}$ poise, $\kappa_d \gtrsim \kappa_i$.

One can write the spherically symmetric equation of heat transfer for the growing planet:

$$\frac{\partial T}{\partial t} = \frac{1}{r^2}\frac{\partial}{\partial_r}r^2(\Sigma\kappa_j)\frac{\partial T}{\partial r} + \frac{\gamma T}{\bar{\rho}}\frac{\partial \rho}{\partial t} + \Sigma\varepsilon_j/\bar{\rho}C_p \tag{29}$$

where the sum $\Sigma\kappa_j$ includes all types of thermall diffusivity, $\Sigma\varepsilon_j$ includes sources of energy, and c_p is the heat capacity at constant pressure. The term with the Grüneisen coefficient $\gamma \approx 1.5$ describes the heating from contraction. Three main sources are: the energy of impacts, $\sim 10^{39}$ erg; the energy of radioactive elements; and energy liberated by gravitational differentiation. The latter are both about 10^{38} erg. The radioactive energy liberated during billions of years in primitive chondrite-like material is $\varepsilon_r \approx 10^{-6}\ \mathrm{erg \cdot cm^{-3} s^{-1}}$. According to Safronov (1969, 1978) for the energy liberated resulting from impacts in a near-surface layer with thickness $H_1 \approx 100$ km one can write:

$$\varepsilon_i \approx \frac{v_i^2 \bar{\rho} \dot{r}_p}{8r'_M} \tag{30}$$

where v_i is the impact velocity and r'_M is the size of the largest body in the mass distribution of falling bodies. If $\dot{r}_p = 3 \times 10^{-7}\ \mathrm{cm \cdot s^{-1}}$ ($\dot{r}_p = 0.8 r_\oplus$), $v_i = 8 \times 10^5\ \mathrm{cm \cdot s^{-1}}$, and $\bar{\rho} = 4\ \mathrm{g \cdot cm^{-3}}$, then:

$$\varepsilon_i = 2 \times 10^{-2}\left(\frac{5 \times 10^6\ \mathrm{cm}}{r'_M}\right)\mathrm{erg \cdot cm^{-3} s^{-1}}$$

Consider first the situation in the upper layer $H_1 \approx 4r'_M$. From (29) in the order of magnitude one can write a simple energy balance relation:

$$\varepsilon_i \mathrm{H}_1 \approx \bar{\rho}C_p \nabla T(\kappa_i + \kappa_d)(1 + \mathrm{Nu}) \tag{31}$$

where Nu now is the function of κ_i, κ_d also, and we assume $n = 3$. At small depths viscosity is high, $\kappa_i \gg \kappa_d$, and $\mathrm{Ra} < \mathrm{Ra}_{cr}$. From Eqs. (26) and (30) one can find that at the depths ≈ 100 km the melting temperature $T_m \sim 1500$K must be achieved. At $\kappa_i \approx 1$ to $10\ \mathrm{cm^2 \cdot s^{-1}}$, $\mathrm{H}_1 = 4r'_M = 200$ km, $C_p = 10^7\ \mathrm{erg \cdot g^{-1} grad^{-1}}$, $\bar{\rho} = 4\ \mathrm{g \cdot cm^{-3}}$, $\bar{m}' = (2 - q)m'_M/(3 - q) \approx 0.1\ m'_M$ and

$\nabla T \approx \nabla T_m \approx 10^{-5}$ grad·cm^{-1} from Eqs. (25), (28), and (31) we find:

$$\kappa_d \approx 30 \text{ cm}^2\cdot\text{s}^{-1}, \eta \approx 3 \times 10^{16} \text{ poise, Ra} \approx 5 \times 10^4, \text{ Nu} \approx 3$$

i.e., relaxation of primary inhomogeneities and weak thermal convection occurred. Because of the relaxation of inhomogeneities, the energy of gravitational differentiation could begin liberating:

$$\varepsilon_d \approx c'\delta\rho g v_d \approx 5 \times 10^{-5} \text{ erg}\cdot\text{cm}^{-3}\text{s}^{-1} \tag{32}$$

where we have used the same values of parameters as in the above estimates, i.e., for the largest inhomogeneities. The smaller inhomogeneities began relaxing after the largest ones disappeared. The small gradient of density appeared, which prohibited the thermal convection on the scale $H < \delta\rho/\bar{\rho}\alpha\nabla T$.

Some problems arise when the separation of a heavy (core) component from a light one is considered. The first problem is their composition. According to experiments at pressures $\lesssim$30 kbar, the heavy fraction formed during melting of chondrite material had contained Fe + Ni + S alloy. The composition of remnant light ferrobasaltic magma is similar to some achondritic and lunar rocks. How large a pressure is required to change the results is not yet known, however. Many authors believe that some fraction of FeO can enter into the heavy component at large pressures. The fugacities of volatile components (O_2, H_2O, etc.) influencing the compositions of main phases are also poorly known.

The second problem is the hydrodynamics of the separation process. It is not clear how effective is the segregation of two immiscible fractions in rather large volumes. The simple model of spherical blob sedimentation is valid in the narrow range of drops sizes a:

$$\sqrt[3]{D\eta/\Delta\rho g} < a < \sqrt[2]{\sigma'/\Delta\rho g}$$

where D is the diffusion coefficient, $\Delta\rho$ is the difference of density between heavy and light components (~ 4.5 g·cm^{-3}) and σ' is the surface tension. Usually $10^{-2} < a < 1$ cm. For smaller sizes, diffusion must be taken into account. For larger ones, the alteration of the forms of drops is essential, and their average size $[a \sim (a_1 \cdot a_2 \cdot a_3)^{1/3}]$ in some statistical sense should be considered in the same manner as the eddies in a turbulence. Therefore all estimates for much larger sizes of sinking heavy masses with Stokes formula have illustrative character. Many questions remain about other aspects of the problem. In a melted region produced by impacts with viscosity of $\sim 10^2$ to 10^6 poise the separation occurs rapidly and large masses of a heavy component form. However, because of repeated impacts and mixing, the sizes of heavy "drops" are difficult to estimate. From the energy balance relation:

$$(\varepsilon_r + \varepsilon_d)H \approx \rho C_p \nabla T(\kappa + \kappa_d) \tag{33}$$

and Eq. (27), (28), and (32), where the sizes a and volume fraction of heavy component $c \approx 0.16$ are used, it can be determined that an appreciable differentiation began when $a \gtrsim 10^2(\eta/10^{16} \text{ poise})^{1/2}$ cm. The essential heating caused by liberation of the energy of gravitational differentiation accelerated the process of separation (Vityazev, 1980). The rate of differentiation increased exponentially because of the temperature dependence of viscosity ($\eta \approx \eta_o \exp(bT_m/T)$), $b \approx 20$–30). Such a phenomenon is similar to a thermal explosion in chemical physics. From Eq. (33) one can obtain the lower estimate of the rate of differentiation $v_d \gtrsim 10^{-7}$ to 10^{-8} cm·s^{-1}. The time of segregation in the layer with thickness $H \approx 1000$ km is about $H/v_d \approx 3 \times 10^7$ to 3×10^8 years. Of course, this is not a rigorous theory. Further work in this area is necessary. The results of some calculations of the thermal history of the Earth in such a model are given by Vityazev and Mayeva (1977).

An important problem is the formation of the primitive crust. It is interesting to investigate its reprocessing by late bombardment, its peneplaining by lunar tides, and its evolution during differentiation.

The origin and evolution of the core, the mantle, and the crust and the formation of the oceans and the atmosphere present a range of problems wherein joint efforts of specialists in the fields of planetary cosmogony, comparative planetology, and Earth sciences are necessary.

Acknowledgments

We extend appreciation to Prof. S. K. Saxena for his suggestion that we write this paper and for help in improving its English.

References

Anders E. (1972) Physico-Chemical processes in the solar nebula as infered from meteorites, in *The Origin of the Solar System*, edited by H. Reeves, pp. 179–201, CNRS, Paris.

Bradley, J. G., Huneke, J. C., and Wasserburg, G. J. (1978) Ion microprobe evidence for the presence of excess ^{26}Mg in a Allende anortite, J. Geophys. Res. **83**(B1), 244–254.

Cameron, A. G. W. (1978) Physics of the primitive solar nebula and of giant gaseous protoplanets, in *Protostars and Planets*, edited by T. Gehrels, pp. 453–487, Univ. Arizona Press, Tucson.

Cameron, A. G. W. (1984) Star formation and extinct radioactivities, *Icarus* **60** (2), 416–427.

Cameron, A. G. W., and Truran, J. W. (1977) The supernova trigger for formation of the solar system, *Icarus* **30**(3), 447–461.

Cameron, A. G. W., De Campli, W. M., and Bodencheimer, P. (1982) Evolution of giant gaseous protoplanents embedded in the primitive solar nebula, *Icarus* **49**(3), 298–312.

Cassen, P., Smith, B. F., Miller, R. H., and Reinolds, R. T. (1981) Numerical experiments on the stability of preplanetary disks, *Icarus* **48**, 377–392.

Chandrasekhar, S., and ter Haar, D. (1950) The scale of turbulence in a differentially rotating gaseous medium, *Astrophys. J.* **111**, 187.

Clayton, D. D. (1983) Chemical state of presolar matter, in *Chondrules and Their Crigins*, edited by E. A. King, pp. 26–36, Lunar and Planetary Institute, Houston.

Clayton, D. D., and Wieneke, B (1983) Aggregation of grains in a turbulent presolar disk, in *Chondrules and Their Origins* edited by E. A. King, pp. 377–387, Lunar and Planetary Institute, Houston.

Dodd, R. T. (1981) *Meteorites: A Petrological-Chemical Synethesis*, Cambridge Univ. Press, Cambridge, London, 368 pp.

Elmegreen, B. J. (1978) On the interaction between a strong stellar wind and a sorrounding disk nebular. *Moon Planets*, **19**, 261–277.

Goldreich, P., and Ward, W. R. (1973) The formation of the planetesimals *Astrophys. J.* **183**, 1051–1061.

Greenberg, R., Wacker, J., Hartmann, W., and Chapman, C. (1978) The accretion of planets from planetesimals, in *Protostars and Planets*, edited by T. Gehrels, pp. 599–610, Univ. Arizona Press, Tucson.

Hayashi, C., Nakazawa, K., and Adachi, I. (1977) Long-term behavior of planetesimals and the formation of the planets, *Publ. Astron. Soc. Japan* **29**, 163–196.

Hayashi, C., Nakazawa, K., and Mizuno, H. (1979) Earth's melting due to blanketing effect of the primordial dense atmosphere, *Earth Planetary Sci. Lett.* **43**, 22–28.

Hayashi, C., Nakazawa K., and Nakagawa Y. (1985) in Formation of the Solar System. *Protostars and* Planets II, edited by D. Black, pp. 1100–1154, Univ. Arizona Press, Tucson.

Henderson-Sellers, A. (1983) The chemical composition and climatology of the Earth's early atmosphere, in *Cosmochemistry and the Origin of Life* edited by C. Ponnamperuma, pp. 175–212, NATO Advanced Study Institute, D. Reidel PC, Dodrecht, Holland.

Herndon, J. M., and Wilkening, L. L. (1978) Conclusion derived from the evidence on accretion in meteorities in *Protostars and Planets*, edited by T. Gehrels, pp. 502–515, Univ. Arizona Press, Tucson.

Kaula, W. M. (1979) Thermal evolution of earth and moon growing by planetesimals impacts, *J. Geophys. Res.* **84**, 999–1008.

Lavruchina, A. K. (1980). On the nature of the isotopic anomalies in meteorites, *Nukleonika*, **25**(11–12), Warszawa, 1495–1515.

Larimer, J. W. (1967). Chemical fractionation in meteorites. I. Condensation of the elements, *Geochim. Cosmochim. Acta* **31**, 1215–1238.

Lewis, J. S. (1972) Origin and composition of terrestrial planets and satellites of the outer planets, in *The Origin of the Solar System*, edited by H. Reeves, pp. 202–205, CNRS, Paris.

Lin, D. N. C., and Papaloizou, J. (1980) On the structure and evolution of the primordial solar nebula, *Monogr. Not. Roy. Astron. Soc.* **191**, 37–38.

Lynden-Bell, D., and Pringle, J. (1974) The evolution of viscous disks and the origin of the nebula variables, *Monogr. Not. Roy. Astron. Soc.* **168**, 603–609.

Morfill, G. E. (1983) Some cosmochemical consequences of a turbulent protoplanetary cloud, *Icarus* **53**(1), 41–54.

Nagahara, H. (1983) Chondrules formed through incomplete melting of the preexisting mineral clusters, in *Chondrules and Their Origins*, edited by E. A. King, pp. 211–222. Lunar and Planetary Institute, Houston.

Öpik, E. J. (1977) Origin of asteroids and the missing planet, *Irish Astron. J.* **13**, 22–39.

Ovenden, M. W. (1972) Bode's law and the missing planet, *Nature* (*London*), **239**, 508–509.

Pechernikova, G. V., and Vityazev, A. V. (1979) The masses of the largest bodies and relative velocities during planets formation, *Sov. Astron. J. Lett.* **5**(1), 703–710 (in Russian).

Pechernikova, G. V., and Vityazev, A. V. (1981) Thermal dissipation of gas from the protoplanetary cloud, *Adv. Space Res.* **1**, 55–58.

Reeves H. (1978) The Bing-bang theory of the origin of the solar system, in *Protostars and Planets*, edited by T. Gehrels, pp. 399–426, Univ. Arizona Press, Tucson.

Ruskol, E. L. (1960) On the problem of protoplanet formation, *Voprosy Kosmogonii*, **7**, 8–14 (in Russian).

Ruskol, E. L. (1975). *The Origin of the Moon*, Nauka, Moscow, 186p.; NASA TTF 16623, Washington, D.C.

Ruskol, E. L. (1979) On the collision of two massive bodies and the initial temperature in the Moon, Lunar and Planetary Science Conference X, March Houston, pp. 1042–1044.

Ruzmaikina, T. V. (1981) On the role of the magnetic field and turbulence in the evolution of the presolar nebula, *Adv. Space Res.* **1**, 49–53.

Ruzmaikina, T. V. (1982). Statement in Diskussionforum, in *Ursprung des Sonnen-systems*, edited by H. Völk, *Mitt. Astron. Ges.* **B57**, 49–54.

Safronov, V. S. (1958) On the growth of the terrestrial planets, *Voprosy Kosmogonii* **6**, 80–94 (in Russian).

Safronov, V. S. (1969) *Evolution of the Protoplanetary Cloud and Formation of the Earth and Planets*. Nauka, Moscow, 244 pp. NASA, F-677, 1972, transl., Washington, D.C.

Safronov, V. S. (1978) The heating of the Earth during its formation *Icarus* **33**(1), 3–12.

Safronov, V. S. (1979) On the origin of asteroids, in *Asteroids*, edited by T. Gehrels, pp. 975–991, Univ. Arizona Press., Tucson.

Safronov, V. S. (1982) The present state of the theory of the origin of the Earth. *Izv. Acad. Sci. USSR, Physics of the Solid Earth*, **18**(6), 399–415.

Safronov, V. S., and Kozlovskaya S. V. (1977) The heating of the Earth by impacts of the bodies which had formed it, *Izv. Acad. Sci. USSR, Physics of the Solid Earth*, **13**,(10), 677.

Safronov, V. S., and Ruskol, E. L. (1957) On the hypothesis of tubulence in a protoplanetary cloud, *Voprosy Kosmogonii*, 22–46, (in Russian).

Safronov, V. S., and Ruzmaikina, T. V. (1978) On angular momentum transfer and accumulation of solid bodies in the solar nebula, in *Protostars and Planets*, edited by T. Gehrels, pp. 546–564, Univ. Arizona Press, Tucson.

Safronov, V. S., and Ruzmaikina, T. V. (1985) Primitive particles in the solar nebula, Lunar and Planetary Science Conference XVI, March, Houston.

Safronov, V. S., and Ruzmaikina, T. V. (1985) Formation of the solar nebula and the planets in *Protostars and Planets II*, edited by D. Black, pp. 959–980, Univ. Arizona Press, Tucson.

Safronov, V. S., Pechernikova, G. V., Ruskol, E. L., and Vityazev, A. V. (1985) Protosatellites swarm, in *Satellites of the Solar System*, edited by J. Burns, Univ. Arizona Press, Tucson.

Safronov, V. S., and Vityazev, A. V. (1985) Origin of the solar system, in *Sov. Sci. Rev. E Astrophys. Space Phys.*, Vol. 4, edited by R. A. Syunyaev, pp. 1–98,

Schmidt, O. Yu. (1958) *A Theory of the Earth's Origins: Four Lectures*, Foreign Languages Publishing House, Moscow.

Shakura, N., and Syunyaev, R. (1973) Black holes in binary systems. Observational appearance, *Astron. Astrophys.* **24**, 337–355.

Van Flandern, T. C. A. (1977) A former major planet of the solar system, in *Comets, Asteroids, Meteorites*, edited by A. H. Delsemme, pp. 475–481, Univ Toledo Press, Ohio.

Vityazev, A. V. (1980) Heat generation and heat-mass transfer in the early evolution of the Earth, *Phys. Earth Planet. Inter.* **22**, 289–295.

Vityazev, A. V. (1982). Fractionation of the matter during formation and early evolution of the Earth, *Izv. Acad. Sci. USSR, Physics of the Solid Earth*, **18**(6), 425–433.

Vityazev, A. V., and Majeva, S. V. (1977) A model of the early evolution of the Earth, *Tectonophysics* **41**, 217–225.

Vityazev, A. V., and Majeva, S. V. (1980) Simulation of the Earth's core and mantle formation, *Phys. Earth Planet. Inter.* **22**, 296–301.

Vityazev, A. V., Pechernikova, G. V. and Safronov, V. S. (1978) The limited masses, distances and times of formation of the terrestrial planets, *Sov. Astron.* **25** 494–499.

Vityazev, A. V. (1985) Solar wind and rare gases in the atmospheres of Mars and Earth, Lunar and planetary Science Conference XVI, March, Houston.

Vityazev, A. V., and Pechernikova, G. V. (1985) On the evaporation interstellar dust during the preplanetary disk formation, Lunar and Planetary Science Conference XVI, March, Houston.

von Weizsäcker, C. F. (1948) Rotation kosmischer Gasmassen, *Z. Naturforsch.* **3a**, 524.

Wasson J. T. (1974) *Meteorites*, Springer-Verlag, Berlin.

Weidenschilling, S. J. (1976) Accretion of the terrestrial planets. II, *Icarus*, **27**(1), 161–170.

Weidenschilling, S. J. (1984) Evolution of grains in a turbulent solar nebula, *Icarus* **60**(3), 553–564.

Wetherill, G. W. (1978) Accumulation of the terrestrial planets, in *Protostars and Planets*, edited by T. Gehrels, pp. 565–598, Univ. Ariz. Press, Tucson.

Wetherill, G. W. (1980) Formation of the terrestrial planets, *Ann. Rev. Astron. Astrophys.* **18**, 77–113.

Wood, J. A. (1983) Formation of chondrules and CAI's from interstellar grains accreting to the solar nebula. *Proceedings of the Eighth Symp. on Antarctic Meteorites*, No. 30, pp. 84–92, National Inst of Polar Research, Tokyo.

Wood, J. A. (1985) Meteoritic constraints on processes in the solar nebula, in *Protostars and Planets II*, edited by D. Black, pp. 687–702, Univ. Arizona Press, Tucson.

Wood, J. A., and McSween, H. J. (1977) in *Comets, Asteroids, and Meteorites*, edited by A. Delsemme, pp. 365–373, Univ. of Toledo Press, Ohio.

Zvyagina, E. V., Pechernikova, G. V., and Safronov, V. S. (1973) Qualitative solution the equation of coagulation with fragmentation, *Astron. Zh.* **50**, 1261 (in Russian, *Sov. Astron.* **17**, 793–799).

Chapter 2
Chemistry of the Formation of the Terrestrial Planets

S. K. Saxena and G. Eriksson

Introduction

Many planetary scientists have successfully argued that planets and meteorites are genetically related and have formed, directly or indirectly, from the solar nebula. The physical and chemical evolution of the solar nebula is, therefore, of great importance and many astrophysicists and astrochemists have devoted considerable effort to elucidating the process. A chemical study of the solar nebula requires information on the nebular distribution of pressure and temperature with time. Such information can be deduced partly from present-day astronomical observations, assuming the principle of uniformitarianism, and partly from theoretical models. The uncertainties in these models are large, resulting in a variety of possible physical conditions as is evident in many works of Hoyle (e.g., Hoyle, 1960; Hoyle and Wickramasinghe, 1968) and Cameron (e.g., Cameron, 1971, 1978; and Cameron and Pine, 1973). Depending on the physical state of the gas, totally different models of the nebula, such as the band model of the plasmatic nebula (Alfvén, 1978), may result. The flexibility of the physical models results from lack of discriminatory criteria that can finitely rule out some models and permit a preference of one or more over others. In view of the difficulties associated with the construction of a model of a physically evolving solar nebula, can one proceed to understand the chemistry of formation of the meteorites and terrestrial planets in a general way without invoking any one particular physical model? To answer this question, we must briefly look into the various classes of physical models as reviewed by Reeves (1978).

The available physical models distinguish between two possibilities. The first is that the sun and planets formed the interstellar or "cosmic" material. Models that entertain this possibility include the classic Kant-Laplace hypothesis. The second possibility, that the sun and planets formed from stellar material (processed by passage through a star), is entertained by the models of binary star (e.g., Lyttleton, 1982) and stellar collision (Woolfson, 1978). Reeves (1978) quotes the evidence that the deuterium to hydrogen ratio in the atmosphere of Jupiter is similar to that in interstellar space and is different from that in the solar photosphere. This evidence favors the first group of models, requiring the formation of the sun and planets from the "cosmic" gas with chemical composition as discussed by Cameron (1973, 1982) and recently by Anders and Ebihara (1982). Within this group of models, physical conditions may vary widely. We may consider the plasma theory of Alfvén and Arrhenius (1976), which requires the formation of planets in a chemically differentiated banded hot nebula. A homogeneous, well-stirred nebula (but isotopically variable) figures in many models of Cameron. The Laplacian protosolar nebula may be hot (Cameron and Pine, 1973) or initially cold and subsequently heated in part (Cameron, 1978). Chemically the consequences of an initial hot or an initially cold nebula will be very different, because the thermal history of a gaseous nebula determines the composition of the phases forming planetary bodies.

Cameron's model (e.g., Cameron and Pine, 1973) of a massive hot nebula exercised a great influence on geochemical thinking (Lewis, 1974; Turekian and Clark, 1975). The massive character helps in the dissipation of angular momentum outwards in an initially rotating nebula that has collapsed to a disk form. Temperatures as high as 2000K developed in the inner region within the heliocentric distance of Mars. This temperature is sufficiently high to vaporize all substances. As the nebula cooled, selective condensation and accretion dominated the formation of layered planets.

Cameron (1978), finding that there could be no energy source available for an initially hot nebula, argued against his previous model. In Cameron's (1978) model of a possible evolutionary history of the primitive solar nebula, the temperature, which is initially very low in each of the regions of planetary formation, gradually rises with time to a maximum, where upon the transition takes place from mass accretion onto the disk to mass loss from the disk. According to this model the temperatures in the nebula were then high enough to vaporize the interstellar grains only within the heliocentric distance of Mercury. Elsewhere temperatures varied from perhaps a high of 900K to a few tens of degrees Kelvin.

As discussed above there are different physical models possible and, except for the plasma model, the range of temperatures from 0 to 2000K is useful for chemical calculations concerning silicates and oxides that might form in a solar nebula, either heating or cooling. The essential difference between the two physical models discussed above is in the nature of the starting material. In a model in which temperature has reached 2000K over a large part of the nebula, the starting material is a gas mixture. In the model in which a temperature gradient of some sort is assumed, there is assumed a starting material of mixtures of interstellar grains and gases. In applying the results of thermochemical calculations to an understanding of the origin of planets, a major question would then be, How closely was equilibrium achieved in the primitive solar nebula? The main difficulty would be the equilibrium of the early formed condensate grains with a cooling gas. Tozer (1978) argues that any condensate that has grown few microns in size would become "such a good radiator of its thermal energy to the 'cold' universe outside the nebula that its temperature would rapidly fall to a temperature perhaps only a tenth of the surrounding gas." Similar difficulties would arise during heating of the interstellar grains to maximum temperatures as envisaged in Cameron's model. Barshay and Lewis (1976) have discussed the important differences that would be observed in the condensation sequence if gas–solid equilibrium was not attained. We shall be very much concerned with such differences in this chapter. We also note that, irrespective of the physical model, computation of stable-phase assemblages, starting from a total gas or from a mixture of gas and interstellar grains of the "cosmic" composition, forms the basis of all physicochemical models of planetary formation. The computational results alone do not tell us how the planets have formed, but they represent the basic data on which all models of planetary formation must be based. Nonequilibrium processes can then be studied by referring to the extent of departure required from the state of equilibrium.

Calculations of equilibrium condensation products in a gas of solar composition have been done by several chemists in the past (e.g., Lord, 1965; Larimer, 1967, 1973; Grossman, 1972; Lewis, 1972a, b, 1974; Sears, 1978). In recent years these calculations have improved steadily, both in computational technique and in their thermochemical data base. Latterly, the thermochemical data base on silicate minerals and oxides has been considerably enlarged and improved through several studies, which have aimed particularly at obtaining data on solid solution behavior (activity–composition relationships). Both the newly added data and the computer technique of introducing activity–composition data into the calculation of the stable assemblage were not available to previous workers. This chapter aims at providing the readers with computations of equilibrium assemblages forming in gases or in mixtures of gases and solids of solar and nonsolar composition over a wide range of pressure (1 to 10^{-6} bar) and temperature (423 to 1973K). For a given pressure and temperature the computed results consist of (a) the description, composition, and amounts (in moles) of each phase (gas mixture, solid solutions, pure solids); (b) the average density of the solid assemblage; and (c) the total mass of the condensates. The results have been used to construct petrologically and geophysically feasible models of Earth.

The system used in this work consists of 14 elements: Si, Al, Mg, Fe, Ca, Na, K, Ti, Ni, C, H, O, S, and N. Our main concern was the mineralogy of the silicates. This work should complement the recent studies of Fegley and Lewis (1980) on the system with Na, K, F, Cl, Br, and P and of Lewis *et al.* (1979) on the system with Fe–Ni–C–S–P. Many geochemical arguments are based on the compositional data on trace elements in meteorites and peridotites, the latter thought to represent Earth's mantle. Grossman and Larimer (1974) reviewed the data on condensation of several trace elements and Boynton (1975) studied the condensation of yttrium and rare earth elements. These studies require thermochemical data on end members of solid solutions and the properties of solid solutions. Such data are meager at present. Note that the absence of such data in our computation does not affect the results presented in this book. Their inclusion, however, would have provided us with important geochemical constraints on the models of planetary origin.

A Review of Calculations of Chemical Equilibrium in Solar Gas

It is widely accepted that planetary objects formed directly or indirectly in a well-stirred nebula of solar composition. However, the turbulence did not homogenize the solar nebula with respect to isotopes (Black, 1978; Clayton *et al.*, 1973 Drozd and Podosek, 1976). There is also some uncertainty in the determination of the major, minor, and trace element composition of the primitive nebula. If the starting nebular composition could be fixed, as discussed by Cameron (1973), the consideration of chemical reactions in space becomes a more or less direct application of thermochemistry, which has been attempted by many chemists and geochemists following Urey (1952). Whereas there have been some notable changes in the estimation of solar abundances, major developments have taken place in thermochemical data on silicates and solid solutions and in computer techniques. These developments are described and discussed in the next section. In this section, we review some of the important conclusions drawn by previous workers on the basis of gaseous and solid–gas reactions.

The C–H–O System

Hydrogen, oxygen, and carbon are three of the most abundant elements in the solar gas. Lewis *et al.* (1979) have calculated the C–H–O system with results, shown in Fig. 1, indicating that graphite would not be a condensate under the nebular adiabat. However, if the initial carbon content of the nebula is higher, the situation may be different. Graphite, along with organic substances, is an important constituent of carbonaceous chondrites.

Larimer and Bartholomay (1979) studied the carbon chemistry and came to some interesting conclusions. In view of the possible uncertainty in determining the C/O ratio in solar gas (Cameron, 1973), they investigated several systems with C/O varying from the solar value of 0.57 to > 1. As shown for a pressure of 10^{-4} atm in Fig. 2, at high temperatures the ratio CO/CH_4 decreases as C/O increases. It is interesting to note that when C/O is greater than 1, the excess carbon that is not included in CO or CH_4 must precipitate at high temperature as graphite. This happens because CH_4 is unstable at high temperature and all oxygen has been used up by CO. As the temperature increases, the reaction

$$nC + (n + 1)H_2 \rightarrow C_nH_{2n+2} \quad \text{(a)}$$

proceeds, leading to vaporization of graphite. Thermochemical calculations

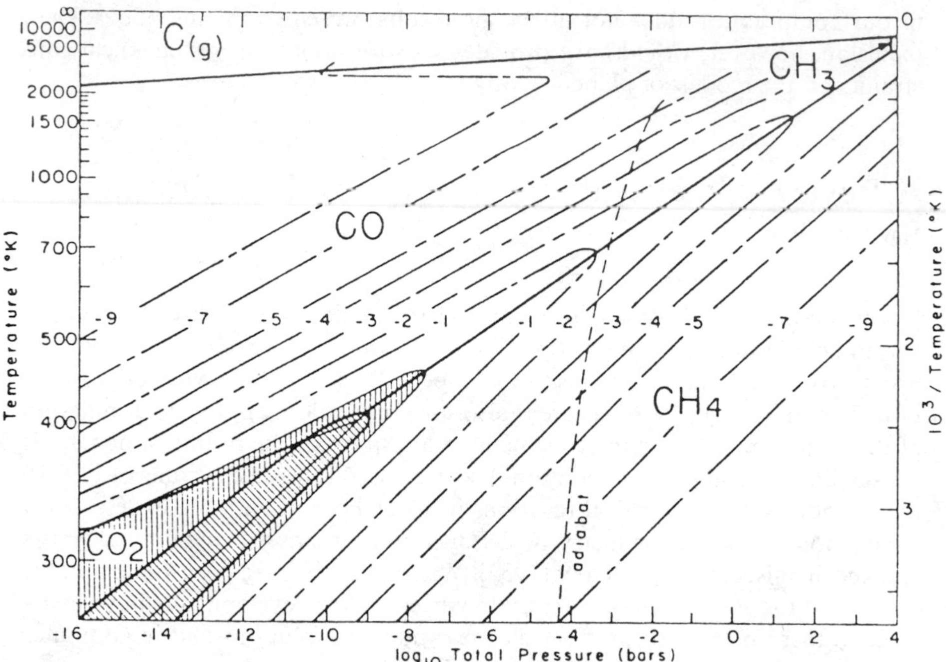

Fig. 1. The dominant carbon-bearing species, and the activity of graphite, in a solar composition system under equilibrium conditions. Heavy lines separate the regions in which gaseous C, CO, CH_4, CO_2, or solid graphite is the most abundant carbon-containing species. The broken lines are contours of constant log $a(C, s)$ values. Graphite is stable within the crosshatched regions and is the most abundant carbon-containing species within the diagonally crosshatched region. Through this region of graphite dominance, the line where $p(CO_2) = p(CH_4)$, and parts of the lines where $p(CO) = p(CO_2)$ and $p(CO) = p(CH_4)$, are lightly drawn. Note the three separate "triple points" where $p(CO) = p(CO_2) = p(CH_4) \sim \frac{1}{3}a(C, s)$, $p(CO) = p(CH_4) = a(C, s)$ and $p(CO) = p(CO_2) = a(C, s)$. A representative adiabat for the primitive solar nebula is drawn as a dashed line. Note that a maximum graphite activity near 0.1 is expected for almost any plausible nebular model. We have made no attempt to calculate or describe the chemistry of the gas at very high temperatures, where ions are dominant. The amount of oxygen condensed in silicates and other minerals (up to ~15%) was taken into account. After Lewis *et. al.* (1979).

show that the equilibrium constant for reaction (a) increases greatly with falling temperature so that in any equilibrium mixture the higher polymers increase in proportion. However, for most temperatures at which equilibrium is readily attainable light species dominate.

The Mg–Fe– Si–O System

Meteorites are on the average more reduced than terrestrial rocks are, but the range of oxidation is quite large. It is, therefore, important to understand the

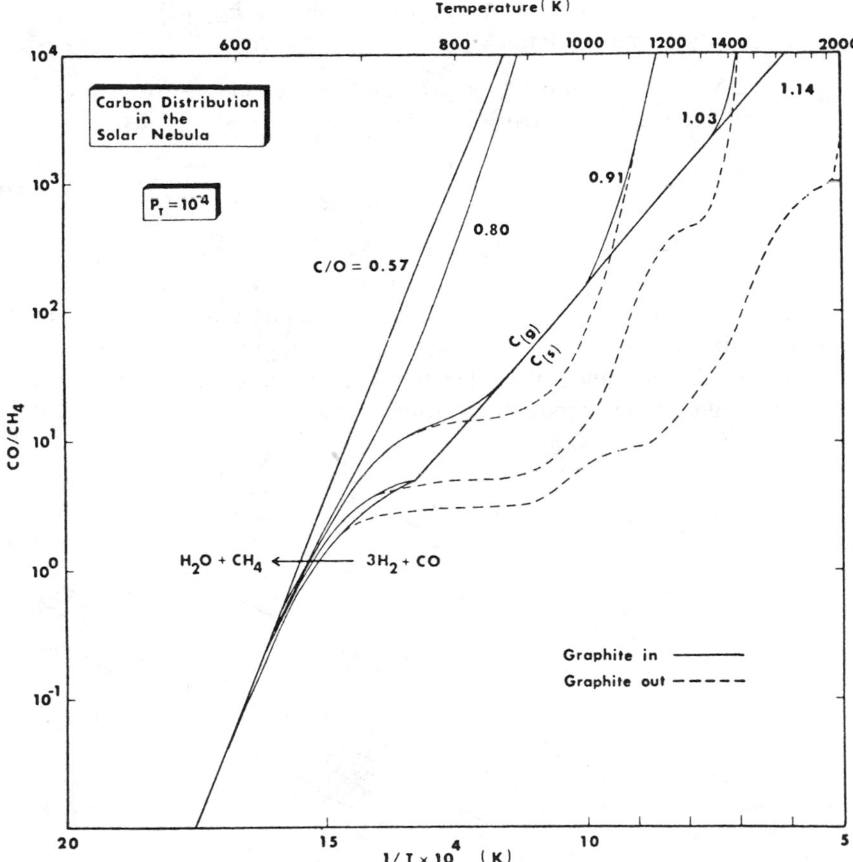

Fig. 2. The C–H–O system from Larimer and Bartholomay (1979). At high temperature at constant pressure, the amount of CH_4 in the gas increases as C/O increases. However if equilibrium is assumed, at low temperature all carbon is present as CH_4. To the right of the C(g)–C(s) line graphite is a potentially stable condensate. If it does not condense, then the equivalent amount of carbon remains in gas as CH_4 (dashed curves). Permission Pergamon Press Limited, U.K.

oxidation–reduction equilibrium in space, particularly its role in determining the compositions of the ferromagnesian silicates. Later we shall compute compositions of minerals in the many-element system, but for now we follow Mueller's (1963) results on the four-element system. In view of the low concentrations of elements other than Mg and Fe in orthopyroxenes and olivines, the results discussed here are not greatly different from those of the multicomponent solutions. Several reactions are possible among the olivine, pyroxene, metal, magnetite, and gas phases. Mueller (1963) considered the following:

$$\underset{\text{Olivine}}{\tfrac{1}{2}Mg_2SiO_4} + \underset{\text{Pyroxene}}{FeSiO_3} \rightleftharpoons \underset{\text{Pyroxene}}{MgSiO_3} + \underset{\text{Olivine}}{\tfrac{1}{2}Fe_2SiO_4} \qquad \text{(b)}$$

$$\underset{\text{Pyroxene}}{2\,MgSiO_3} + \underset{\text{Metal}}{2\,Fe} + \underset{\text{Gas}}{O_2} \rightleftharpoons \underset{\text{Olivine}}{Mg_2SiO_4 + Fe_2SiO_4} \tag{c}$$

$$\underset{\text{Pyroxene}}{2\,MgSiO_3} + \underset{\text{Magnetite}}{\tfrac{2}{3}Fe_3O_4} \rightleftharpoons \underset{\text{Olivine}}{Mg_2SiO_4 + Fe_2SiO_4} + \underset{\text{Gas}}{\tfrac{1}{3}O_2} \tag{d}$$

$$\underset{\text{Metal}}{3\,Fe} + \underset{\text{Gas}}{2\,O_2} \rightleftharpoons \underset{\text{Magnetite}}{Fe_3O_4} \tag{e}$$

Although Mueller's (1963) set of thermochemical data was incomplete and some data were estimated from observations of naturally coexisting minerals, his calculated phase diagram, as shown in Fig. 3, needs no major changes. In this figure the stability fields of the dominant classes of chondritic meteorites are shown. The oxidation state of the solar nebula is shown by the dotted curve. As we shall see later in our new computations, hydrous minerals become stable

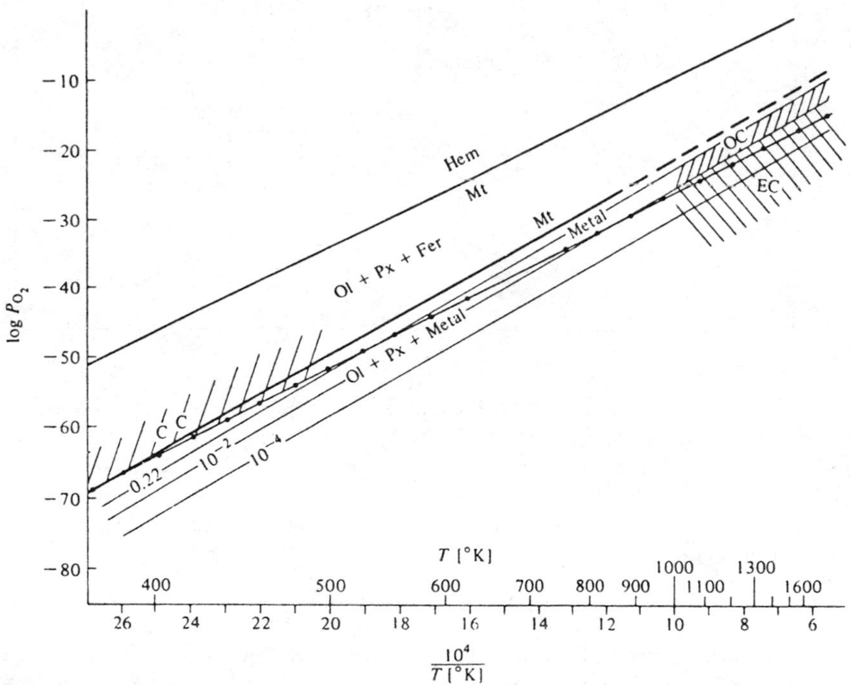

Fig. 3. Stability fields of important minerals in the system Mg–Fe–Si–O. Approximate fields of the three major classes of chondritic meteorites are shown as also the fields of the ordinary chondrites (OC), enstatite chondrites (EC), and carbonaceous chondrites (CC). The light, numbered curves indicate atomic fractions $Fe^{2+}/(Fe^{2+} + Mg)$ for the hypothetical assemblage olivine–pyroxene–metal–gas. However, olivine is usually replaced by free silica in the EC field. The dotted curve indicates a gas of solar composition with hydrogen pressures high enough to maintain most carbon as methane. After Mueller and Saxena (1977).

below 500K and are not shown here. The position of the ordinary chondrites (OC) well to the left of the solar curve at high temperature indicates their formation under conditions more oxidizing than those prevailing in the initial nebula. The enstatite chondrites (EC) lie on the reduced side. The solar curve approaches the field of carbonaceous chondrites as the temperature decreases, indicating that their more oxidizing nature may actually correspond to that of the low-temperature solar gas.

Computations in a Multielement System

Chemical equilibrium calculations involving several phases of multielemental composition possible in the solar nebula were initiated by Lord (1965) and Larimer (1967) and continued by Lewis (1972b, 1974) and Grossman (1972). The latter author did extensive calculations of equilibria at a pressure of 10^{-3} atm over a range of temperature extending from ~1100 to 1800K. Figure 4 shows the equilibrium condensation temperatures of elements and phases in a gas of solar composition at 10^{-4} atm as presented by Grossman and Larimer (1974). These results, which are from different sources (Larimer, 1973; Grossman, 1972; Grossman and Clark, 1973; Grossman and Olsen, 1974), are important in understanding the cosmochemical behavior of elements. The figure shows that the condensation of the major elements Al, Ti, Ca, Ni, Mg, Fe, and Si, and many

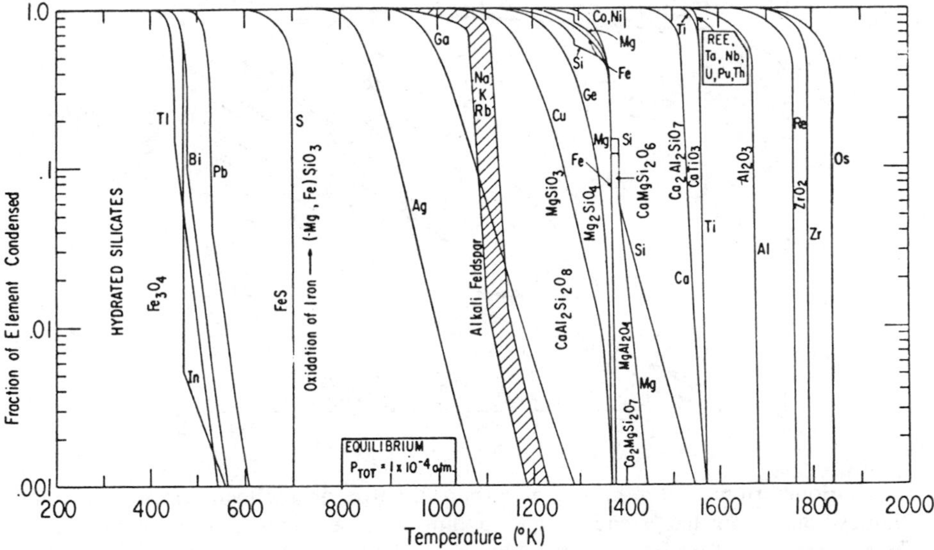

Fig. 4. Condensation of elements from a gas of solar composition at 10^{-4} atm. Chemical formulas of stable condensate phases are written below the temperatures at which they become stable. After Grossman and Larimer (1974). Permission American Geophysical Union.

trace elements, e.g., Os, Zr, Re, rare earths, Co, Ge, and Cu, may be mostly complete above 1200K. Of the major elements only Na and K remain in the vapor phase, but these too are removed from the gas by 900K. Gallium, which straddles the alkali feldspars curve, and Ag condense at medium temperature. Iron may react with Mg silicates and with S at lower temperatures. The trace elements Pb, Bi, In, and Tl remain in the gas until ~600K.

Lewis and co-workers (Lewis *et al.*, 1979; Fegley and Lewis, 1980; Lewis and Prinn, 1983) have performed extensive thermochemical equilibrium calculations aimed at understanding the formation of planetary atmospheres and many meteorite minerals. Fegley and Lewis' (1980) study of the Na–K–F–Cl–Br–P system yielded the interesting result that some of the minerals found in meteorites, e.g., schreibersite (Fe_3P), sodalite [$(Na_4AlSiO_4)_3Cl$], fluorapatite [$Ca_5(PO_4)_3F$], and whitlockite [$Ca_3(PO_4)_2$], formed only when equilibrium between gases and solids was maintained along an adiabatic path. If equilibrium is not maintained and elements are removed immediately following condensation, the condensation consequences are very different. The data of Fegley and Lewis (1980) show that in the latter situation, which may be identified as an extreme case of the heterogeneous condensation model of Turekian and Clark (1969), alkali sulfides, ammonium halides, and ammonium phosphates form, none of which has been reported from meteorites.

The studies by Lewis *et al.* (1979) and Larimer and Bartholomay (1979) have

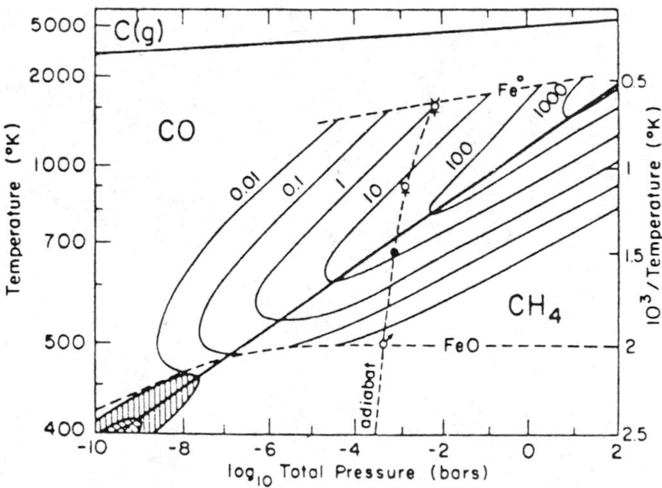

Fig. 5. Equilibrium concentrations of carbon dissolved in the metal phase in a solar composition system under equilibrium conditions, superimposed on parts of Fig. 1. Contours given are for parts per million (by weight) of dissolved carbon in the metal phase from 0.01 to 1000 ppm. These contours are truncated at high temperatures by the condensation curve of γ-iron (taenite), and at low temperatures by the endpoint for the oxidation of metal to FeO. Symbols on the adiabat mark the appropriate formation conditions of Mercury, Venus, Earth, and Mars. After Lewis *et al.* (1979).

important bearing in understanding the carbon content of the planets. Lewis *et al.* (1979) calculated isopleths (lines of equal concentration) of carbon dissolved in the metal phase (see Fig. 5). They also studied the stability of cohenite (Fe_3C), other carbides, and carbonates. Cohenite is thermochemically unstable below ~1100K and yet is a common mineral in meteorites. In solar gas Fe_3C is not a condensate at any temperature. As shown by Larimer and Bartholomay (1979), Fe_3C becomes stable in the nebula at high temperatures if C/O is greater than unity. Carbonates are unlikely to be important condensates in the nebula. The pressure–temperature regions of calcite stability and graphite stability are very similar.

All these studies confirm that a direct equilibrium precipitation of carbon-bearing species from the solar gas could not be an important source of carbon in the planets. The amount of carbon dissolved in metal is also insufficient to explain the carbon content of planetary bodies.

Nonequilibrium Calculations

Although we have considered mainly equilibrium condensation, there can be little doubt that nonequilibrium reactions in space are common. However, without equilibrium results the recognition of nonequilibrium processes would be impossible. Nonequilibrium reactions have been studied by Arrhenius (1978) and Blander and Katz (1967). The latter authors considered the kinetic barriers to homogeneous nucleation of condensed phases and suggested that these would lead to a supersaturation of the gas phase and consequently to lowering of condensation temperatures. The supersaturation also leads to much higher concentration of FeO in the condensing silicates than found by equilibrium calculations. If some minerals in meteorites can be considered as unaltered condensates, or if the alteration does not change the original composition, Blander and Katz (1967) may have a simple explanation of the variety of oxidation states observed in meteorite minerals. Whereas homogeneous nucleation may be inhibited, the possibility that incompletely vaporized interstellar grains have provided nuclei for heterogeneous nucleation is very high in the primitive nebula. Disequilibrium is more likely to have resulted from incomplete solid–gas reactions than from nucleation kinetics.

Method of Calculation

Calculations of chemical equilibria in solar gas have been done for selected reactions by many chemists, some of which were discussed in the preceding section. The equilibrium in a system is determined by solving simultaneous equations of mass balances and equilibrium constants. For example, for hydrogen we have:

$$n(\mathrm{H}) + 2n(\mathrm{H_2}) + 2n(\mathrm{H_2O}) + 3n(\mathrm{NH_3}) + \cdots = n^0(\mathrm{H}) \qquad (1)$$

where n and n^0 denote equilibrium and total amount, respectively. Now, each hydrogen-bearing species, such as NH_3, is related to hydrogen through the reaction:

$$N(g) + 3\,H(g) \rightleftharpoons NH_3(g) \tag{f}$$

for which we have

$$K_p = p(NH_3)/p(N)\cdot p(H)^3 \tag{2}$$

where K_p is the equilibrium constant and p represents partial pressure. For an ideal gas we may write:

$$n(NH_3) = K_p \cdot n(N) \cdot n(H)^3 \cdot (P/N)^3 \tag{3}$$

where N is the total amount in the gas phase and P is total pressure. Equations such as (1) are written for each element and ones such as (3) for each independently formed species. Combinations of Eq. (1) and (3) result then in mass balance equations containing the unknown values of n for each element. Solids and liquids may be added in a similar manner to the matrix of equations. This method, which was used by Grossman (1972), Lord (1965), and Larimer (1967), in principle minimizes the total Gibbs free energy empolying equilibrium constants. Alternate formulations are possible that do not require writing specific reactions.

The method of minimizing the total Gibbs free energy employing linear algebraic techniques as discussed by Eriksson (1975) and by Smith and Missen (1982) is particularly suitable for computers. It forms the basis of the computer program SOLGASMIX (Eriksson, 1975), which is used for all chemical equilibrium calculations in this chapter.

The total Gibbs free energy of a chemical system can be expressed as:

$$G = \Sigma \mu_i n_i \tag{4}$$

where μ is chemical potential. Chemical potential and activity are interrelated by the expression:

$$\mu = \mu^0 + RT \ln a \tag{5}$$

where μ^0 denotes the standard state potential, R is the gas constant, T absolute temperature, and a the activity. For a gas-phase species, a species in a nonideal solution phase and a stoichiometric phase, respectively, we find:

$$\mu_i = \mu_i^0 + RT \ln P + RT \ln X_i \tag{6}$$

$$\mu_i = \mu_i^0 + RT \ln \gamma_i + RT \ln X_i \tag{7}$$

$$\mu_i = \mu_i^0 \tag{8}$$

In these equations X is the mole fraction and γ is the activity coefficient. The minimization of G in Eq. (4) at constant pressure and temperature is achieved with the constraints imposed by the mass balance relationships represented as:

$$\Sigma a_{ij} n_i = b_j \qquad (j = 1, 2, \ldots, l) \tag{9}$$

where a_{ij} is the number of atoms of the jth element in a molecule of the ith species, l is the total number of elements, and b_j is the total amount of the jth element. This is a simple form of a constrained optimization problem that may be solved by the Lagrange method of undetermined multipliers. For this, we define a functions,

$$F = G + \sum_{j=1}^{l} \lambda_j (b_j - \Sigma a_{ij} n_i) \tag{10}$$

where λ_j denotes the Lagrangian multipliers, for which the necessary conditions are:

$$\left(\frac{\partial F}{\partial n_i}\right)_{n_{k \neq i}, \lambda} = \mu_i - \Sigma a_{ij} \lambda_j = 0 \tag{11}$$

$$\left(\frac{\partial F}{\partial \lambda_j}\right)_{\lambda_{k \neq j}, n} = b_j - \Sigma a_{ij} n_i = 0 \tag{12}$$

$$n_j \geq 0 \tag{13}$$

Combining Eq. (6)–(8) with Eq. (11) gives:

$$\mu_i^0 + RT \ln P + RT \ln X_i - \Sigma a_{ij} \lambda_j = 0 \tag{14}$$

$$\mu_i^0 + RT \ln \gamma_i + RT \ln X_i - \Sigma a_{ij} \lambda_j = 0 \tag{15}$$

$$\mu_i^0 - \Sigma a_{ij} \lambda_j = 0 \tag{16}$$

Equation (14) is valid for gas-phase species, whereas Eq. (15) and (16) hold for components of solution phases and stoichiometric phases, respectively.

The system consisting of Eq. (9) and (14)–(16) with the unkowns n_i and λ_j is nonlinear because of the logarithmic term in Eq. (14) and (15). The next step is therefore a linearization of these equations by expansion in a Taylor series around an estimated equilibrium composition up to and including the term of the first order. This is equivalent to making a quadratic approximation to the free-energy surface and we can obtain an expression that relates n_i linearly to λ_j and the estimated equilibrium amounts. Incorporation of this expression into Eq. (9) then gives the final linear system of equations. The number of unknowns is reduced to the sum of elements and phases assumed to be present at equilibrium.

Parametrized activity-coefficient expressions to be inserted into Eq. (15) are supplied by the user. In order to avoid the need of also specifying derivatives, $\ln \gamma$ is treated as constant when calculating the partial derivative with respect to n_i in the Taylor expansion. The partial derivative is therefore approximate as long as the estimated equilibrium composition does not correspond to a free-energy minimum.

The approximation to the free-energy surface implies an iterative algorithm and, if positive, the calculated n_i values are used as improved estimates in the subsequent iteration cycle. If some n_i values are negative, these are transformed before being used as the starting point for a new Taylor expansion. The iterative procedure ends when the calculated values coincide with the starting estimates.

The condensed phases included in the initial estimate are constrained by the Gibbs phase rule but need not necessarily be the correct ones of the final equilibrium state. Another phase combination may yield a lower free energy, and condensed phases need to be withdrawn from or added to the previous combination until the set of equilibrium phases is found. This set has the characteristic feature that the virtual activity for an omitted stoichiometric phase must be less than one, as must the sum of virtual mole fractions for an omitted solution phase.

For a more detailed description of the equations used in SOLGASMIX, see Eriksson and Rosén (1973).

Thermochemical Data

Selection of data

While *JANAF Thermochemical Tables* (1971–1982) and the tables complied by Barin and Knacke (1973) and Barin *et al.* (1977) remain the standards for gases, thermochemical constants for solids may be selected in more than one way. The two compilations generally used are by Helgeson *et al.* (1978) and Robie *et al.* (1978). The data compiled by Robie and co-workers are the thermochemical values as determined by direct calorimetric measurements and other experimental methods. These values, being independent of any experimental phase equilibrium data, may or may not reproduce the results obtained by experimental petrologists. If the errors in the thermochemical measurements for each phase participating in a chemical reaction are small, a discrepancy between the calculated and experimental phase equilibrium results may result from faulty petrological experiments. On the other hand, if the errors are large, it may be necessary to accept well-reversed experimental phase equilibrium data and use these for calculating missing thermochemical constants. This approach, which was used by Zen (1972), becomes particularly important if a large amount of internally consistent data can be generated from many well-reversed experiments, as done by Helgeson *et al.* (1978). It is important to use an internally consistent data set, even if some of the data do not seem to be close to the thermochemical measurements. Because calculated results have to be finally matched against phase equilibrium relationships, we decided to construct our set of thermochemical data largely from Helgeson *et al.* (1978). We have done some modifications within permissible errors in order to include experimental data published after 1978. Some of the data that were missing have been calculated from the following experimental phase equilibria:

1. Enstatite–diopside (Lindsley *et al.*, 1981; Saxena, 1981a)
2. $MgO–Al_2O_3–SiO_2$ (Danckwerth and Newton, 1978; Lane and Ganguly, 1980; Perkins *et al.*, 1981)
3. $CaO–MgO–Al_2O_3–SiO_2$ (Herzberg, 1978; Jenkins and Newton, 1979;

Kushiro and Yoder, 1966; Perkins and Newton, 1980; Yamada and Takahashi, 1982)
4. $CaO-MgO-FeO-SiO_2$ (Lindsley, 1980, 1983; Turnock and Lindsley, 1981)
5. $MgAl_2O_4-FeAl_2O_4$ (Fujii, 1977; Fujii and Scarfe, 1982)
6. Fe^{2+}–Mg distribution between garnet and other minerals (Ferry and Spear, 1978; Ganguly, 1979; Ganguly and Saxena, 1984)
7. Olivine–orthopyroxene (Bohlen and Boettcher, 1982)

The data on entropy and heat capacity for iron–biotite are from Helgeson *et al.* (1978), but the enthalpy has been adjusted so that the calculated Fe–Mg distribution between biotite and a coexisting mineral, e.g., orthopyroxene (Saxena, 1968), garnet (Saxena, 1968; Ferry and Spear, 1978), matches the observed distribution in natural or experimental systems. For iron–pargasite, iron–tremolite, and iron–anthophyllite we calculated the heat capacity by subtracting the heat capacity of MgO from the corresponding magnesian compound and adding that of FeO. The enthalpy and entropy data are then calculated by trying to reproduce the Fe–Mg distribution in coexisting minerals (e.g., Saxena, 1968, 1971).

Because all new data are calculated (or adjusted) by referring to experimental phase relationships, additional corrections for order–disorder in oxides (e.g., $MgAl_2O_4$) or in silicates (e.g., $MgSiO_3-Al_2O_3$) are not necessary.

After this work was completed, a new thermochemical data base on the $CaO-MgO-Al_2O_3-SiO_2$ system was created (Saxena and Chatterjee, 1986). This data base is largely calorimetric. However, because both data sets are constrained by the same experimental phase equilibrium data, the results of computation are virtually identical.

Table 1 shows the chemical species with name abbreviations and sources of data. Tables 2 and 3 contain the new data that are consistent with Helgeson *et al.* (1978) and may be regarded as a subset to their data base.

Calculation of Errors

Most reactions involving solid solutions are very sensitive to errors as small as 500 $J \cdot mol^{-1}$ in the thermochemical data. Because all the data used here are calculated or adjusted to reproduce experimental phase equilibria, the relative errors in ΔG° of the many reactions [e.g., Mg–Sp (spinel) + 4 En (enstatite) $\rightleftharpoons$ Py (pyrope) + Fo (forsterite)] involved cannot be more than a few hundred joules. An error of 500 J in ΔG° for the reaction 3 En + Al–Opx $\rightleftharpoons$ Py changes the mole percentage of Al_2O_3 in orthopyroxene to make it inconsistent with the experimentally determined value. Similarly all Fe^{2+}–Mg ion-exchange reactions are sensitive to small changes in the thermochemical data. Whereas no absolute errors can be calculated for individual compounds, therefore, the errors for the entire data set, as judged by comparing the calculated equilibrium relationships with the experimental and observed phase equilibria, are insignificant.

Table 1. Source[a] of thermochemical data and name abbreviations.

Gases

Al	AlS	Ca	H_2	KH	$Mg(OH)_2$[1]	NaH	S_2	SiH_4
AlH[1]	C	Ca_2[1]	HCN	KO	MgS[1]	NaO	S_8	SiO
AlO	CH_4	CaH	HCO	KOH[1]	N_2	NaOH[1]	SO	SiO_2[1]
AlO_2	CO	CaO[1]	H_2O	Mg	NH_2[1]	Ni	SO_2	SiS
Al_2O	CO_2	CaOH[1]	HS[1]	Mg_2[1]	NH_3	O	Si	Ti[1]
Al_2O_2	COS	$Ca(OH)_2$[1]	H_2S	MgH[1]	NS	O_2	Si_2	TiO
AlOH[1]	CS	Fe	K	MgO[1]	Na	OH[1]	Si_3	TiO_2[1]
AlO_2H[1]	CS_2	H	K_2	MgOH[1]	Na_2	S	SiH	TiS

		Temp. range
Liquid solution		
Fe-Ni-Si alloy	Fe[2], Ni[2], Si[2]	H
Solid solutions		
Spinel (Sp)	$MgAl_2O_4$, $FeAl_2O_4$	H
Olivine (Ol)	$MgSi_{0.5}O_2$, $FeSi_{0.5}O_2$, $NiSi_{0.5}O_2$	HL
Orthopyroxene (Opx)	$MgSiO_3$, $FeSiO_3$, $Ca_{0.5}Mg_{0.5}SiO_3$, Al_2O_3	HL
Clinopyroxene (Cpx)	$MgSiO_3$, $FeSiO_3$, $Ca_{0.5}Mg_{0.5}SiO_3$, $Na_{0.5}Al_{0.5}SiO_3$, Al_2O_3	HL
Garnet (Gar)	$MgAl_{2/3}SiO_4$, $FeAl_{2/3}SiO_4$, $CaAl_{2/3}SiO_4$	HL
Melilite (Mel)	$Ca_2Al_2SiO_7$, $Ca_2MgSi_2O_7$	H
Magnetite (Mt)	Fe_3O_4, $FeAl_2O_4$, Fe_2TiO_4	HL
Ilmenite (Ilm)	$Fe_{0.5}Ti_{0.5}O_{1.5}$, $Mg_{0.5}Ti_{0.5}O_{1.5}$, $FeO_{1.5}$	HL
Fe–Ni–Si alloy	Fe_2[2], Ni_2[2], Si[2]	H
γ-FeNi alloy	Fe[2], Ni[2]	L
α-FeNi alloy	Fe, Ni	L
Plagioclase (Plag)	$CaAl_2Si_2O_8$, $NaAlSi_3O_8$, $KAlSi_3O_8$	HL
Orthoclase (Or)	$NaAlSi_3O_8$, $KAlSi_3O_8$	L
Cordierite (Cord)	$MgAl_2Si_{2.5}O_9$, $FeAl_2Si_{2.5}O_9$	HL
Carbide (Car)	Fe_3C[2], Ni_3C[2]	H
Biotite (Bi)	$1/3(KMg_3AlSi_3O_{12}H_2)$, $1/3(KFe_3AlSi_3O_{12}H_2)$	L
Anthophyllite (Ant)	$1/7(Mg_7Si_8O_{24}H_2)$, $1/7(Fe_7Si_8O_{24}H_2)$	L
Tremolite (Tr)	$1/5(Ca_2Mg_5Si_8O_{24}H_2)$, $1/5(Ca_2Fe_5Si_8O_{24}H_2)$	L
Pargasite (Pg)	$1/4(NaCa_2Mg_4Al_3Si_6O_{24}H_2)$, $1/4(NaCa_2Fe_4Al_3Si_6O_{24}H_2)$	L
Muscovite (Ms)	$1/3(KAl_3Si_3O_{12}H_2)$, $1/3(NaAl_3Si_3O_{12}H_2)$	L
Epidote (Ep)	$Ca_2FeAl_2Si_3O_{12}(OH)$, $Ca_2Al_3Si_3O_{12}(OH)$	L
Pure phases		
AlN[2] (H)	MgS[2] (H)	
Al_2O_3 (H)	$MgSO_4$[2] (HL)	
Al_2SiO_5 (andalusite) (HL)	$Mg_3Si_2O_5(OH)_4$ (L)	

Table 1. (Continued)

$Al_2Si_2O_5(OH)_4$ (L)	$Mg_3Si_4O_{10}(OH)_4$ (L)
$Al_2Si_4O_{10}(OH)$ (L)	$Mg_4Si_6O_{15}(OH)_2(OH_2)_2 \cdot (OH_2)_4$ (L)
C[2] (HL)	$Mg_5Al_2Si_3O_{10}(OH)_8$ (7-Å) (L)
$CaAl_2Si_4O_{12} \cdot 2H_2O$ (L)	$Mg_5Al_2Si_3O_{10}(OH)_8$ (14-Å) (L)
$CaAl_2Si_4O_{12} \cdot 4H_2O$ (L)	$NiTiO_3$[2] (H)
$CaAl_4Si_2O_{10}(OII)_2$ (L)	SiC[2] (H)
$CaCO_3$ (calcite) (HL)	SiO_2 (quartz) (HL)
$CaMgC_2O_6$ (HL)	SiS_2[2] (H)
CaS[2] (HL)	Si_3N_4[2] (H)
$CaSO_4$[2] (HL)	TiC[2] (H)
$CaSiO_3$[2] (HL)	TiH_2[2] (H)
$CaTiO_3$ (H)	TiN[2] (H)
$CaTiSiO_5$[2] (HL)	TiO[2] (HL)
$Ca_2Al_2Si_3O_{10}(OH)_2$ (L)	TiO_2 (rutile)[2] (HL)
FeS (troilite) (HL)	Ti_2O_3[2] (HL)
FeS_2 (L)	Ti_3O_5 (HL)
$Mg(OH)_2$ (HL)	TiS[2] (H)

[a] All gases from Barin *et al.* (1973, 1977) except those with superscript (1) which are from JANAF Tables (1971–1982). All condensed species with superscript (2) are from Barin *et al.* (1977). Ni-olivine from Mah and Pankratz (1976). Ilmenite, geikelite, magnetite, hematite, troilite, pyrite, and perovskite from Robie *et al.* (1978). The C_p° for Fe_2TiO_4 is from Kubaschewski and Alcock (1979) and the H° and S° at 298 K are determined from the phase equilibrium data of Spencer and Lindsley (1981). Iron cordierite from Perchuk *et al.* (1981). All pure phases and components of solid solutions from Helgeson *et al.* (1978) except those that are discussed above and listed in Tables 2 and 3. Phases denoted by H are considered at temperatures above 1100K, whereas those denoted by L are considered below 1100K only.

Table 2. Calculated and adjusted thermochemical data on some species.

Species	$H^\circ(298)$ $(kJ \cdot mol^{-1})$	$S^\circ(298)$ $(J \cdot K^{-1} mol^{-1})$	$C_p^\circ(J \cdot K^{-1} mol^{-1})$		
			a	$b \times 10^3$	$c \times 10^{-5}$
$MgAl_2O_4$	−2293.800	80.40	154.00	26.80	40.60
$FeAl_2O_4$	−1943.201	106.27	138.66	43.40	0.00
$MgSiO_3$(Opx)	−1547.921	66.27	116.00	14.56	40.99
$Ca_{0.5}Mg_{0.5}SiO_3$(Opx)	−1580.565	71.96	0.00	173.00	0.00
$FeSiO_3$(Opx)	−1194.511	94.56	133.00	0.00	48.47
Al_2O_3(Opx)	−1634.500	57.70	80.36	37.00	0.00
$FeSiO_3$(Cpx)	−1175.000	102.70	133.00	0.00	48.47
Al_2O_3(Cpx)	−1652.500	51.80	80.36	37.00	0.00
$MgAl_{2/3}SiO_4$	−2094.800	88.00	148.39	16.83	40.55
$FeAl_{2/3}SiO_4$ ($T < 848$K)	−1769.868	95.48	136.05	47.00	26.14
$FeAl_{2/3}SiO_4$ ($T > 848$K)			149.36	31.00	14.83
$CaAl_{2/3}SiO_4$	−2208.811	84.89	145.07	23.73	38.10

Note: See text for the derivation. The heat capacity expression is $C_p^\circ = a + bT - cT^{-2}$.

Table 3. Calculated thermochemical data on iron minerals.

Species	$H^\circ(298)$ (kJ · mol^{-1})	$S^\circ(298)$ (J · K^{-1}mol^{-1})	C_p°(J · K^{-1}mol^{-1})				
			a	$b \times 10^3$	$c \times 10^{-5}$	$d \times 10^6$	T_{tr}(K)
Fe_2TiO_4	−1509.494	168.335	139.495	63.10	14.23	—	—
Fe-pargasite	−11172.804	805.440	901.068	170.32	330.08	—	—
Fe-tremolite	−10520.000	775.500	996.250	95.15	305.00	—	—
Fe-anthophyllite	−9547.986	774.690	831.600	246.4	169.68	—	—
Fe-biotite	−5164.504	398.317	445.303	124.56	80.79	—	—
α-Fe	−8.400	19.366	28.175	−7.318	2.895	25.04	800
			−263.454	255.8	−619.2	—	1000
			−641.905	696.3	—	—	1042
α-Ni	22.000	40.716	19.083	23.50	—	—	500
			−251.166	356.4	−259.5	—	631
			467.194	−678.7	—	—	640
			−385.698	404.2	−654.5	—	700
			−10.874	54.67	−56.48	−16.49	1400

Note: The heat capacity expression is $C_p^\circ = a + bT - cT^{-2} + dT^2$.

Solid Solutions

One of the major differences between the condensation calculations presented here and by others is the inclusion of the appropriate activity–composition relations in the computation. The solid solution properties may be described using the simple mixture parameters W_{ij} of Guggenheim (1967). If the number of components is three or less and if the binary joins are asymmetric, we use Wohl's (1953) formulation, according to which the activity coefficient of a component in a ternary solution is given by:

$$\begin{aligned} RT \ln \gamma_1 = {} & X_2^2[W_{12} + 2X_1(W_{21} - W_{12})] + x_3^2[W_{13} + 2X_1(W_{31} - W_{13})] \\ & + X_2X_3[0.5(W_{21} + W_{12} + W_{31} + W_{13} - W_{23} - W_{32}) \\ & + X_1(W_{21} - W_{12} + W_{31} - W_{13}) + (X_2 - X_3)(W_{23} - W_{32}) \\ & - (1 - 2X_1)C] \end{aligned} \tag{17}$$

where W_{ij} is the interaction parameter, X mole fraction, and C a ternary constant, not used in this work, because of lack of sufficient information (see Saxena, 1973). For solutions with four or more components, it is cumbersome to use the asymmetric expression. We use the following symmetric expression by suitably defining the composition limits of applicability arising from the neglect of the asymmetric property:

$$RT \ln \gamma_i = \Sigma X_j^2 W_{ij} + \Sigma X_j X_k (W_{ij} - W_{jk} + W_{ik}) \tag{18}$$

The solid solutions included in the calculations are listed in Table 1. Where available, W_{ij} parameters are used from the literature. Missing data on W_{ij} were initially estimated by comparison with data on similar compounds and then finally adjusted in the effort to reproduce the experimental phase relationships in the CaO–MgO–FeO–SiO_2 (Lindsley, 1980, 1983), diopside–jadeite–quartz (Gasparik, 1981), and the previously referred to CaO–MgO–Al_2O_3–SiO_2 systems. Bulk chemical compositions that are suitable for the mantle, in which the amount of FeO does not exceed more than 20% of the total, were used for estimating W_{ij} for binary solutions involving Fe. All W_{ij} data presented below are for mixing of the components on a one-cation basis and are given in kilojoules.

Some of the solutions described below do not appear in the calculated results. However, because the appearance of a phase could be critically affected by the solid solution data used, all data are presented here.

Spinel

The W for the binary $MgAl_2O_4$–$FeAl_2O_4$ solution in spinel is given by:

$$W_{12} = -21.460 + 0.00167(T \text{ in } K) \tag{19}$$

The W is calculated simultaneously with the thermochemical data of $FeAl_2O_4$ resulting in the best fit to the equilibrium relations observed by Fujii (1977) and

Fujii and Scarfe (1982). Note that the positive coefficient of T in the W expression signifies a negative excess entropy of mixing. This is unrealistic and W does not have real physical significance except as a fitting constant.

Olivine

The following data for the binary $MgSi_{0.5}O_2$–$FeSi_{0.5}O_2$ are from B. J. Wood and Kleppa (1981):

$$H^{Ex}/kJ \cdot mol^{-1} = 8.368 X_{Mg} X_{Fe}^2 + 4.184 X_{Mg}^2 X_{Fe} \tag{20}$$

Following these authors, excess entropy of mixing, (H^{Ex}) is considered negligible. The $NiSi_{0.5}O_2$ is included as a third component, mixing ideally with fayalite and nonideally with forsterite, the W_{ij} for the latter being the same as for forsterite–fayalite. Although olivine is a ternary solution with $CaMgSiO_4$, the calcium component is not included in the calculations. The small Ca content, although a potential indicator of pressure, does not affect the equilibrium compositions significantly.

Orthopyroxene

Alternative interpretations of the Fe–Mg order–disorder data in orthopyroxene yield a small positive deviation (Saxena and Ghose, 1971) or a small negative deviation (Sack, 1980) from ideal solution at 800°C. At temperatures above, orthopyroxene may be accepted as a closely ideal solution of Fe^{2+} and Mg. A fictive calcium component Odi (orthodiopside) is introduced to mix ideally with enstatite (Saxena, 1981b) and similarly Al_2O_3 with enstatite (Saxena, 1981a). The W for Odi–Al_2O_3 calculated from the experimental data on the CaO–MgO–Al_2O_3–SiO_2 system is 44.066. The mixing of ferrosilite with orthodiopside and Al_2O_3 is considered ideal. Note that in view of the low concentrations of orthodiopside and ferrosilite the ideal solution assumption is satisfactory.

Clinopyroxene

A general solution model for clinopyroxene must include at least five components representing each of Mg, Fe^{2+}, Al, Ca, and Na. There are many alternative sets of components possible. We find the following the most satisfactory for computation: clinoenstatite (Cen), diopside (Di), clinoferrosilite (Cfs), Al_2O_3–clinopyroxene (Al–Cpx), and jadeite (Jd), which are denoted below by the numbers 1 to 5, respectively.

For the binary Cen–Di solution, Lindsley *et al.* (1981) found W_{12} and W_{21} to be [25.484 + 0.0812 (P, in kbar)] and [31.216 − 0.0061 (P, in kbar)], respectively. Because we are restricted to a five-component symmetric solution model, and the calculated compositions are to be richer in diopside than in clinoenstatite, we use $W_{21} = W_{12}$. As for orthopyroxene, we assume the binary solutions Cen–Cfs, Di–Cfs and Cfs–Al–Cpx to be ideal. For the limited Mg : Fe composition range of interest, we found no difficulty with this assumption in reproducing available experimental data. For the binary solution Di–Jd, we found a satisfactory fit to the phase equilibrium data of Gasparik (1981) by using a W_{25} of [18.000 – 0.6(P, in kbar)]. This W does not yield satisfactory results beyond about 50% Jd. The W_{25} and W_{52} for complete calculations are available in Gasparik (1981). The W_{14} (Cen–Al–Cpx) and W_{24} (Di–Al–Cpx) obtained from the data on CaO–MgO–Al_2O_3–SiO_2 phase equilibria were found to be equal to 9.977 and [25.546 – 0.002(T, in K)], respectively. A value of 41.570 for W_{45} (Al–Cpx–Jd) was found to be consistent with Gasparik's (1981) data. The remaining data on W, by necessity in absence of any experimental data, are approximated as $W_{15} = W_{35} = W_{12}$.

Garnet

The following data on W are used:

$$W_{\text{Mg-Ca}} = 4.182 - 0.00626(T, \text{in K}) \tag{21}$$

$$W_{\text{Ca-Mg}} = 16.927 - 0.00626(T, \text{in K}) + 0.058(P, \text{in kbar}) \tag{22}$$

$$W_{\text{Fe-Ca}} = 19.330 - 0.00626(T, \text{in K}) + 0.058(P, \text{in kbar}) \tag{23}$$

$$W_{\text{Ca-Fe}} = -2.636 - 0.00626(T, \text{in K}) \tag{24}$$

The data for the Mg–Ca binary join are from Haselton and Newton (1980), who also considered the volume of mixing, which is a complex function of composition. The ternary analytical expression for activity coefficient requires a $W(P, T)$, and the Ws have been modified accordingly. However, this modification does not yield as high activities for grossular at high pressures (>40 kbar) as required (Haselton and Newton, 1980). An ideal Fe^{2+}–Mg solution has been inferred by B. J. Wood and Kleppa (1981) and Ganguly and Saxena (1984). Ganguly and Saxena (1984) also calculated W for the Ca–Fe mixing from the data of Cressey *et al.* (1978). Pressure modification on this is considered to be the same as for the Mg–Ca mixing.

Melilite

Charlu *et al.* (1981) measured the enthalpy of the gehlenite–åkermanite solid solution and found some excess enthalpy of mixing. When entropy of mixing is included, the result is best represented by the ideal solution model at 1000K.

Magnetite

The magnetite solid solution is considered as ternary, with Fe_3O_4, Fe_2TiO_4, and $FeAl_2O_4$ as components. The W_{ij} values are:

$$W_{12} = 20.798 - 0.019652(T, \text{in K}) \quad \text{(Spencer and Lindsley, 1981)} \tag{25}$$

$$W_{21} = 64.835 - 0.060296(T, \text{in K}) \quad \text{(Spencer and Lindsley, 1981)} \tag{26}$$

$$W_{13} = W_{31} = 9.816 \quad \text{(estimated from miscibility gap)} \tag{27}$$

The values for W_{23} and W_{32} have been set as approximately equal to W_{13}. Beyond 1100K, W_{12} and W_{21} become negative. Instead of the negative values, we have set both W as zero.

Ilmenite

The ilmenite solid solution is ternary with $Fe_{0.5}Ti_{0.5}O_{1.5}$, $Mg_{0.5}Ti_{0.5}O_{1.5}$, and $FeO_{1.5}$ as components. Because $Fe_{0.5}Ti_{0.5}O_{1.5}$ and $Mg_{0.5}Ti_{0.5}O_{1.5}$ are completely miscible, the W_{12} value is set to zero. The other values from Spencer and Lindsley (1981) are:

$$W_{13} = 51.187 - 0.035548(T, \text{in K}) \tag{28}$$

$$W_{31} = 18.409 - 0.0038857(T, \text{in K}) \tag{29}$$

Fe–Ni–Si

For both solid and liquid solutions, the data on W_{ij} are from Kubaschewski and Alcock (1979). For the liquid we have $W_{12} = -14.908$, $W_{21} = -5.338$, $W_{13} = -60.445$, $W_{31} = -113.283$, $W_{23} = -69.998$ and $W_{32} = -199.170$, and for the solid: $W_{12} = -10.135$, $W_{21} = -3.417$, $W_{13} = -38.720$, $W_{31} = -72.576$, $W_{23} = 44.847$ and $W_{32} = -127.608$. Below 1173K the alloy phase is considered binary with Fe and Ni. γ-FeNi is treated as ideal and the thermochemical data for the coexisting ideal α-FeNi phase have been calculated on this basis from the phase diagram (Kubaschewski, 1982); see Table 3.

Plagioclase and Orthoclase

There is a small positive deviation from ideal mixing for the binary high albite–anorthite solution (Blencoe *et al.*, 1982; Kerrick and Darken, 1975; Newton *et al.*, 1980; Orville, 1972; Saxena and Ribbe, 1972). The $W_{\text{Ab-An}}$ and $W_{\text{An-Ab}}$ based on Orville's (1972) experiments are 7.037 and 1.054, respectively. There is no significant difference in the calculations if the calorimetric data of Newton *et al.* (1980) are preferred. A ternary solution of plagioclase is used with $W_{\text{Ab-Or}}$ and $W_{\text{Or-Ab}}$ from Thompson and Waldbaum (see Saxena, 1973) and an estimated $W_{\text{Or-An}}$ and $W_{\text{An-Or}}$ from Saxena (1973). Orthoclase is considered a binary solution with the same $W_{\text{Or-Ab}}$ and $W_{\text{Ab-Or}}$ as in plagioclase.

We assume that amphiboles, cordierite, epidote, and carbide are ideal solutions. There are sufficient observational and/or experimental data on cordierite (Mueller, 1972; Perchuk and Lavrent'eva, 1983) to justify this assumption. Cordierite is considered anhydrous. Pargasite, anthophyllite, cordierite, and epidote do not appear in our results. If they are nonideal (positive deviation) then they will remain unstable. Tremolite may be close to ideal, but if not, the method used in calculating the thermochemical data on ferrotremolite makes it a fictive ideal component as defined by Saxena (1981b). Biotite is closely ideal (Mueller, 1972). A nonideal but symmetric model has been used for muscovite (Eugster *et al.*, 1972).

Table 4. Solar gas composition: atomic abundances per one Si atom.

Element	Sources[a] 1	2	3
Si	1.00	1.00	1.00
Al	0.0849	0.085	0.085
Mg	1.075	1.061	1.06
Fe	0.900	0.830	0.90
Ca	0.0611	0.0721	0.0625
Na	0.0570	0.060	0.060
K	0.003770	0.0042	0.0035
Ti	0.002400	0.002775	0.0024
Ni	0.0493	0.048	0.0478
C	12.1	11.8	11.1
H	2.72×10^4	3.18×10^4	2.66×10^4
O	20.1	21.5	18.4
S	0.515	0.50	0.50
N	2.48	3.74	2.31

[a] Sources: 1, Anders and Ebihara (1982); 2, Cameron (1973); 3, Cameron (1982).

Example of a Computed Result

Table 4 shows recent estimates of solar gas compositions from Cameron (1973, 1982) and Anders and Ebihara (1982). The abundances are presented on the basis of one Si atom. Table 5 shows results calculated in the 14-element (Si–Al–Mg–Fe–Ca–Na–K–Ti–Ni–C–H–O–S–N) system using the data from Anders and Ebihara (1982). In this particular example, the gas phase consisted of 72 species. The gases found with a vapor pressure below 10^{-10} bar are not listed in the table (except O_2). There were 17 solid solutions, out of which eight formed and 28 phases of fixed composition out of which only one formed. The temperature and pressure were fixed at 600K and 10^{-3} bar, respectively.

As shown later the modal distribution of the phases and the average density of the condensates are important in discussing the role of condensation in planet formation. Therefore, the output in Table 5 also contains such information.

Phase Equilibrium in a Gas of Solar Composition

Using the free-energy minimization method and the data as discussed in the preceding section, we have computed phase equilibria in a gas of solar composition over a wide range of pressures and temperatures. The data on the composition of the gas from Cameron (1973) were used in generating condensation curves

Table 5. Example of results from a phase equilibrium computation at 600K and 10^{-3} bar.

Gas phase[a] species	Amount (mol)	Pressure (bar)	Fugacity
CH_4	1.209E + 01	8.898E − 07	8.898E − 07
CO	7.449E − 03	5.482E − 10	5.482E − 10
H_2	1.356E + 04	9.978E − 04	9.978E − 04
II_2O	1.677E + 01	2.234E − 06	1.234E − 06
H_2S	2.713E − 01	1.997E − 08	1.997E − 08
NH_3	5.405E − 03	3.978E − 10	3.978E − 10
N_2	1.237E + 00	9.105E − 08	9.105E − 08
O_2	1.135E − 36	8.352E − 44	8.352E − 44

Solid solution	Component	Amount (mol)	Mole fraction	Activity
γ-FeNi	Fe	3.071E − 02	4.695E − 01	4.695E − 01
	Ni	3.469E − 02	5.305E − 01	5.305E − 01
α-FeNi	Fe	6.003E − 01	9.762E − 01	9.762E − 01
	Ni	1.460E − 02	2.375E − 02	2.375E − 02
Olivine	Fo	6.850E − 01	9.752E − 01	9.752E − 01
	Fa	1.741E − 02	2.478E − 02	5.722E − 02
	Ni-Ol	6.285E − 07	8.948E − 07	2.025E − 06
Orthopyroxene	En	3.372E − 01	9.719E − 01	9.718E − 01
	Odi	1.300E − 03	3.748E − 03	3.920E − 03
	Fs	6.667E − 03	1.921E − 02	1.921E − 02
	Al_2O_3	1.779E − 03	5.126E − 03	5.297E − 03
Clinopyroxene	Cen	6.206E − 04	6.084E − 03	9.118E − 01
	Di	1.005E − 01	9.850E − 01	9.856E − 01
	Cfs	1.069E − 04	1.048E − 03	1.024E − 03
	Al_2O_3	7.923E − 05	7.768E − 04	9.614E − 02
	Jd	7.218E − 04	7.075E − 03	2.432E − 01
Plagioclase	An	1.021E − 02	1.465E − 01	2.752E − 01
	Ab	5.661E − 02	8.123E − 01	7.742E − 01
	Or	2.873E − 03	4.123E − 02	9.736E − 01
Orthoclase	Ab	2.826E − 05	3.056E − 02	7.742E − 01
	Or	8.967E − 04	9.694E − 01	9.736E − 01
Ilmenite	$FeTiO_3$	2.330E − 03	4.853E − 01	4.853E − 01
	$MgTiO_3$	2.470E − 03	5.147E − 01	5.147E − 01
	Fe_2O_3	1.005E − 07	2.198E − 05	1.483E − 04

Pure phases	Amount (mol)
FeS	2.437E − 01

Modal distribution of condensed phase (mol %)

γ-FeNi	3.04	Orthopyroxene	16.13	Orthoclase	0.04
α-FeNi	28.59	Clinopyroxene	4.74	Ilmenite	0.22
Olivine	32.66	Plagioclase	3.24	FeS	11.33

Average density = 4.89 $g \cdot cm^{-3}$

[a]Partial listing out of 72 gas components. Except O_2, all species with vapor pressure less than 10^{-10} bar are not listed.

shown in the figures in this section. As shown later, the calculated results differ only slightly if the compositional data of Anders and Ebihara (1982) are used instead. The system is limited to 14 elements (Si, Al, Mg, Fe, Ca, Na, K, Ti, Ni, C, H, O, S, N).

Limitations of condensation calculations have been discussed by Grossman and Larimer (1974) and Goettel and Barshay (1978. The condensation temperature of a phase or element is the temperature at which it first appears as part of the equilibrium assemblage. Such temperatures are affected by errors in thermochemical data. When we use a data set that is internally consistent in reproducing experimental phase diagrams, then the errors in the calculation of equilibrium temperature with respect to the experimental phase relations are negligible. Therefore the errors in the experimentally determined curves are also the errors in the calculated temperatures and pressures. Such errors vary depending on the quality of reversal experiments and may be generally less than $\sim$30K. Besides the errors and other limitations pertaining to the applicability of equilibrium data to the nebular processes, there exists the possibility that important gaseous molecules and solids have not been included in the data. We are aware of the absence of a titanium end-member component in clinopyroxene. Clinopyroxenes rich in titanium have been described in the Allende meteorite by Grossman and Olsen (1974). Data on some pure phases found in meteorites are not available. An example is hibonite ($CaO \cdot 6Al_2O_3 \pm MgO, TiO_2$). Our system consists of only 14 elements, and certain important elements, such as phosphorus and manganese, were not included to keep the computation time reasonable. Although we miss some interesting information by not including these elements, the equilibrium assemblages are not significantly affected by such deficiencies.

Finally, we note that the thermochemical data on many Fe–Mg silicates are particularly adjusted to experiments conducted above 700K. The data at lower temperatures are extrapolations. The equilibrium calculations at low temperatures, therefore, are subject to considerable uncertainty.

Condensation Pressure and Temperature of Solids

Figures 6–10 show condensation and dissolution pressure and temperature curves for minerals in the solar nebula. For convenience the results are displayed separately in the first four figures before being combined together in the last figure. Figure 6 shows the aluminous phases. The first solid to condense from the vapor at all pressures under consideration (1 to 10^{-6} bar) is corundum. It is stable over a temperature range of $\sim$250K and yields spinel by reaction with magnesium in the gas. Melilite (and perovskite; see Fig. 8) condenses after corundum and remains stable over a temperature range of 150K by changing in composition because of the solid solution of åkermanite (Mg and Si replacing Al). Plagioclase condenses next, causing the disappearance of both spinel and melilite. The first plagioclase is nearly pure anorthite. It becomes increasingly albitic with decreasing temperature.

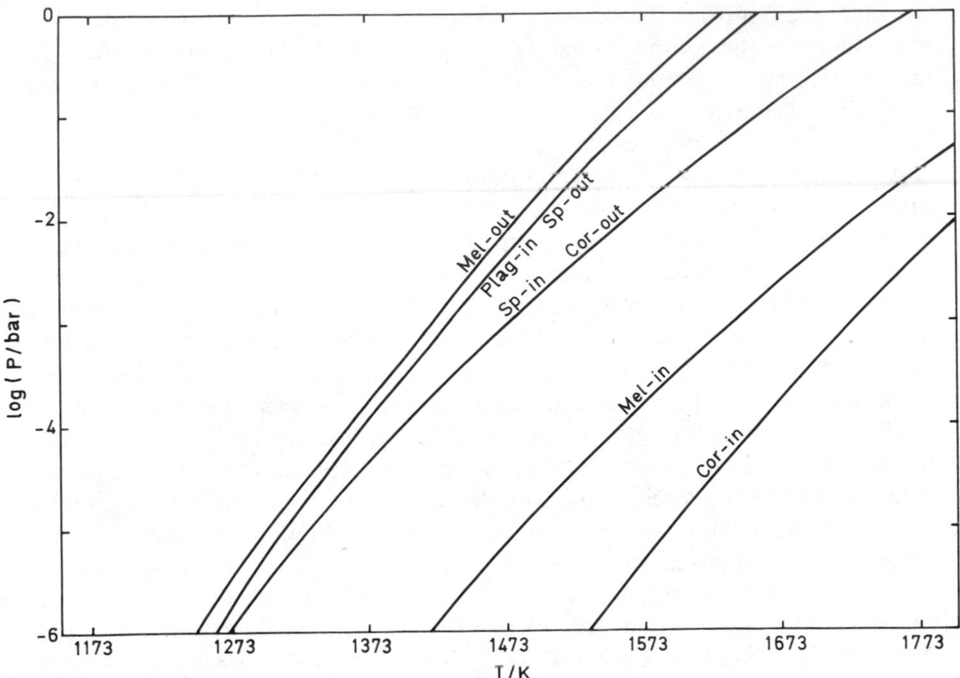

Fig. 6. Condensation and disappearance of aluminum-bearing phases in the high-temperature region. Although the two curves for the incoming spinel (Sp-in) and the incoming plagioclase (Plag-in) are also labeled corundum out (Cor-out) and spinel out (Sp-out), respectively, the latter are always lower in temperature, the temperature differing by less than 3K everywhere. After Saxena and Eriksson (1983a). Permission Pergamon Press.

Figure 7 shows the condensation behavior of Fe–Ni alloy and the Fe–Mg silicates. The Fe–Ni condensation curve crosses the olivine-in and the clinopyroxene-in curves at a pressure of about 10^{-4} bar. The early crystallization of Fe–Ni has been considered important in heterogeneous accumlation models of terrestrial planets (Grossman, 1972; Turekian and Clark, 1975). The sequence of crystallization Fe–Ni alloy–olivine–clinopyroxene at pressures higher than 10^{-4} bar becomes reversed below.

The condensation behavior of titanium compounds is shown in Fig. 8. The first titanium (and calcium) compound to condense is perovskite, which yields to Ti_2O_3, Ti_3O_5, or ilmenite depending on the pressure. Ti_2O_3 shows a complex pattern of condensation. It is restricted to the high-pressure ($>10^{-2}$ bar) region of the field. The condensation curve for clinopyroxene is also plotted in Fig. 8. Because a titanium component in clinopyroxene has not been included in the computations, the oxides Ti_2O_3 and Ti_3O_5 are not likely to be stable as independent pure phases but likely are part of the clinopyroxene solution. The clinopy-

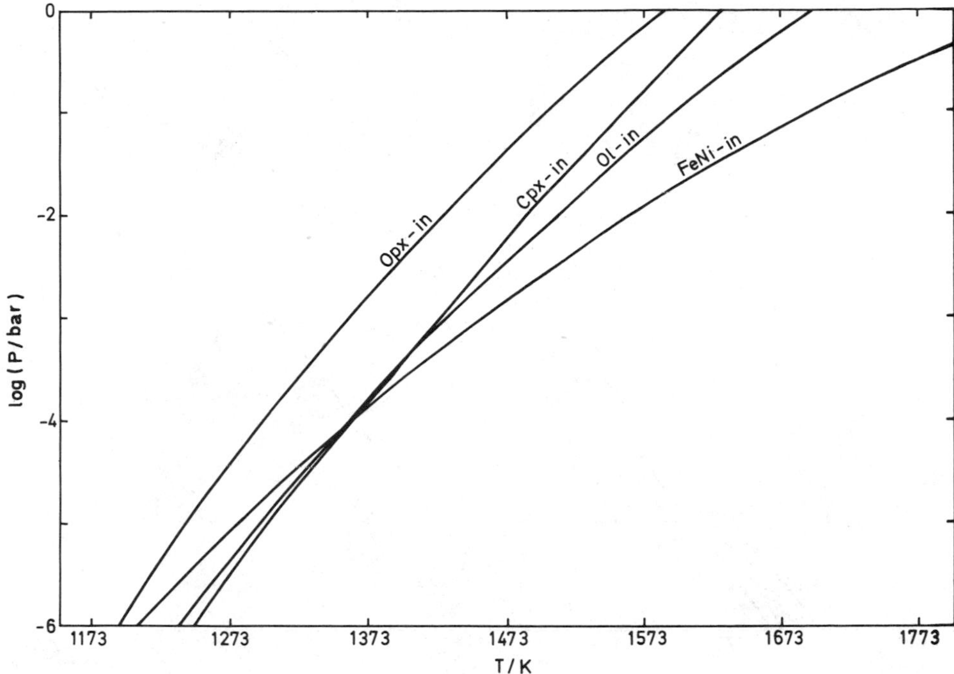

Fig. 7. Condensation curves for the metal alloy, olivine, and the two pyroxenes in the high-temperature region. At pressures below 10^{-4} bar, Fe–Ni metal condenses after both olivine and clinopyroxene and clinopyroxene condenses before olivine. After Saxena and Eriksson (1983a). Permission Pergamon Press.

roxene solution model also does not include $CaAl_2SiO_6$, which is not expected in any significant amount at the nebular pressures. Grossman (1972) and Larimer and Bartholomay (1979) included $CaAl_2SiO_6$ as a pure phase in their computations and did not find it as a condensate. If a titanium component and $CaAl_2SiO_6$ were both included in the clinopyroxene solution (a seven-component solution!), its condensation temperature would increase only slightly everywhere.

Figure 9 shows the stable mineral assemblages in the low-temperature region of study. Between 628 and 1017K, the assemblage γ-FeNi, α-FeNi, olivine, plagioclase, clinopyroxene, orthopyroxene, and ilmenite is stable, with rutile and orthoclase dropping in and out in parts of the pressure–temperature field. Troilite first appears at 621K, preceded by orthoclase and followed by almandine-rich garent at 628 and 486–533K, respectively. Almandine-rich garnet has not been found in meteorites but andradite and grossular do occur in Allende and other C3 meteorites (Dodd, 1981). The amounts of garnet are small (less than 1% by volume) and either this result is spurious or such a small amount may have been totally missed in petrographic studies of meteorites. The hydrous minerals biotite, tremolite, and talc also form in the low-temperature region,

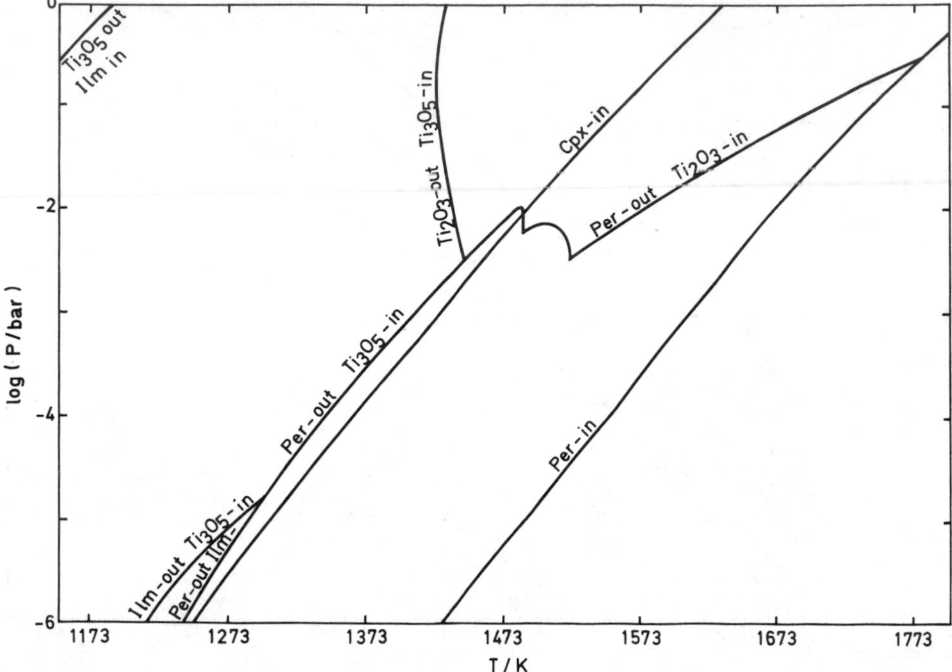

Fig. 8. Condensation equilibrium of titanium-bearing phases in the high-temperature region. Whereas the condensation temperature for perovskite changes regularly with pressure, the temperature for Ti_2O_3 varies irregularly. A narrow region for ilmenite condensation appears at low pressures. The clinopyroxene condensation curve is included to show the possibility that titanium may prefer this solution. After Saxena and Eriksson (1983a). Permission Pergamon Press.

where the results should be considered tentative because of possible errors in extrapolating the high-temperature thermochemical data to such low temperatures as 423K. At 10^{-8} bar graphite condenses in large amounts between 431 and 456K. Observed hydrous silicates in meteorites are kaolinite, montmorillonite, serpentine, and chamosite. If reliable thermochemical data were available, these phases might have appeared in the computed results at low temperatures.

Figure 10 shows the complete results of our calculations. As noted before the sequence of condensation of solids is important in considering fractional separation of compositionally different planetary material. In this respect, we note that the Fe–Ni curve (3 in Fig. 10) has a markedly different pressure–temperature slope and it straddles the other more or less parallel curves, namely perovskite in (5), corundum-out–spinel-in (7), olivine-in (9), spinel-out–plagioclase-in (10) and melilite-out–clinopyroxene-in (11). We note specifically

a. Fe–Ni alloy (3) condenses before all silicates except melilite (2) from 1 to $2 \cdot 10^{-4}$ bar.

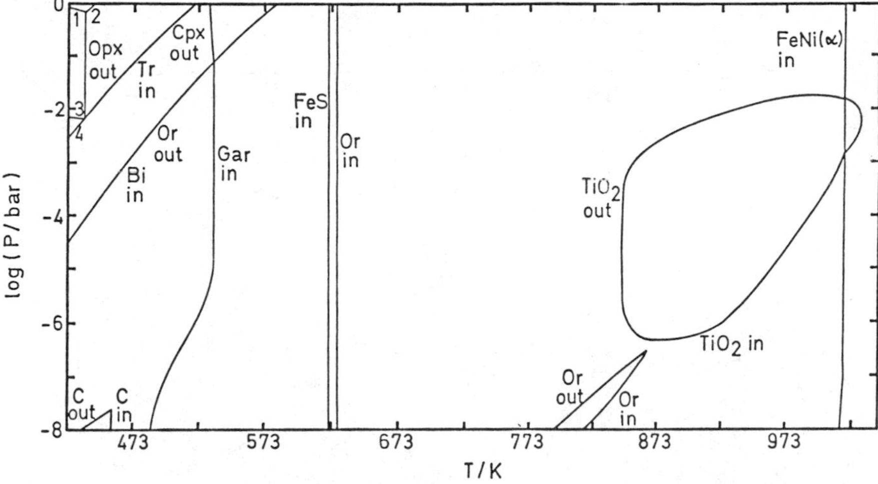

Fig. 9. Equilibrium condensation products in the solar gas in the low-temperature region. The assemblage of γ-FeNi, olivine, plagioclase, clinopyroxene, and orthopyroxene is stable throughout down to 520K. Rutile has an enclosed pressure–temperature field of stability where it replaces ilmenite, which remains stable in the rest of the P–T field. The α-FeNi starts as a coexisting Fe–Ni phase at 1017K. Orthoclase forms over a small high T–low P region. The latter appears again at 628K to yield to the hydrous mineral biotite. Small amounts of garnet ($\sim$1.5 mol %) form below 486–533K. At the lowest temperatures tremolite, talc and graphite become stable. Sulfur starts condensing first at 621K as FeS. 1, talc-out; 2, orthopyroxene-out, talc-in; 3, clinopyroxene-out; 4, orthopyroxene-out, tremolite-in. After Saxena and Eriksson (1983b). Permission Elsevier Science Publications.

b. Below 10^{-4} bar all the more abundant, and more important, silicates except orthopyroxene (12) condense before the Fe–Ni alloy.
c. Plagioclase (10) condenses before clinopyroxene (11) and clinopyroxene before orthopyroxene (12) everywhere in this pressure–temperature region.
d. Olivine (9) is the first silicate other than melilite (2) to condense after the Fe–Ni alloy (3) above 10^{-3} bar. Below this pressure it falls behind both plagioclase (10) and clinopyroxene (11).
e. The condensation temperature for orthopyroxene (12) is greater than that of the Fe–Ni alloy below about 10^{-7} bar (data not presented here).
f. Perovskite (5) condenses after melilite (2) at all pressures above 10^{-4} bar. The condensation temperature below this pressure is closely similar to that of melilite.
g. Finally the assemblage olivine–plagioclase–clinopyroxene–orthopyroxene is stable over very wide pressure and temperature ranges.

Figure 10 shows a close band of curves 6 to 16 in the middle of the pressure–temperature field. This band marks a major change in the character of the

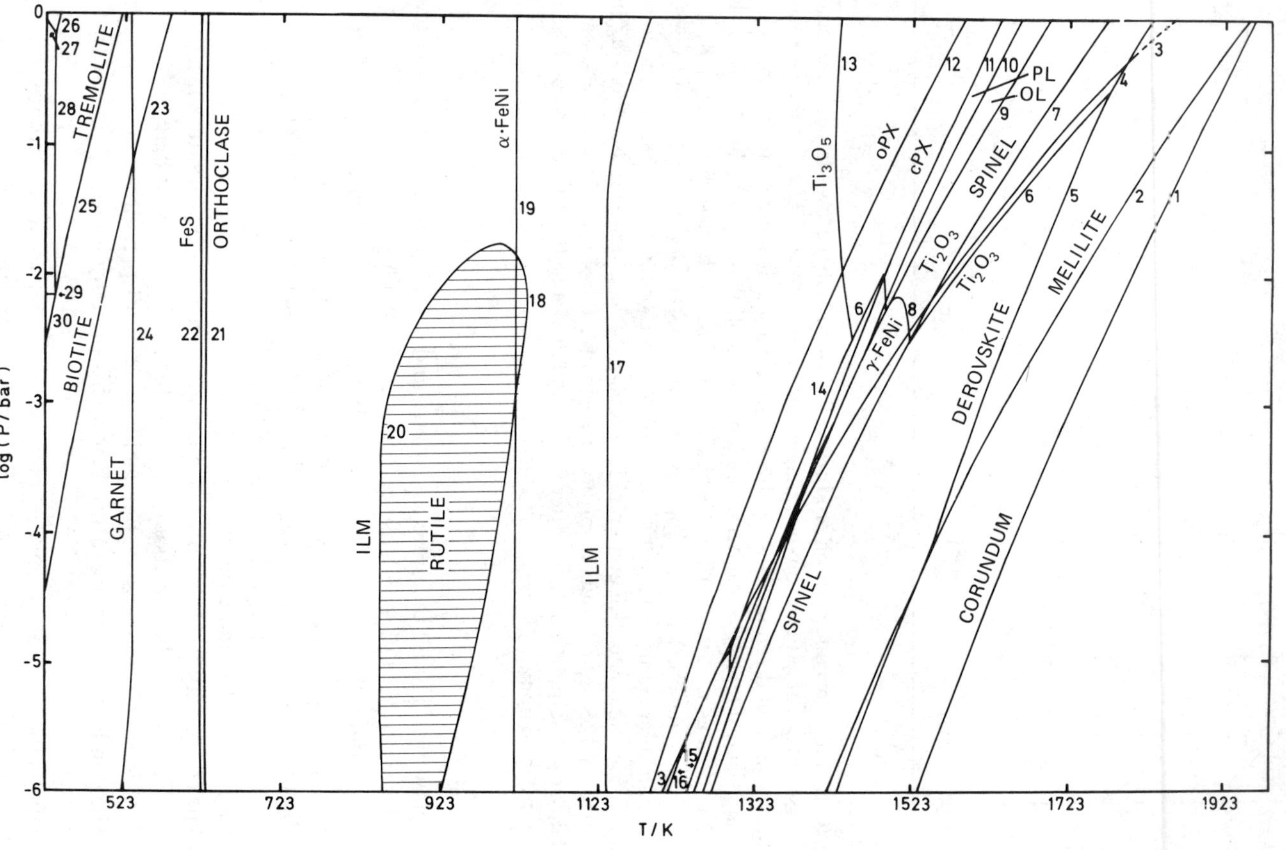
TREMOLITE
BIOTITE
GARNET
FeS
ORTHOCLASE
ILM
RUTILE
α-FeNi
ILM
Ti_3O_5
oPX
cPX
SPINEL
PL
OL
Ti_2O_3
Ti_2O_3
γ-FeNi
DEROVSKITE
MELILITE
SPINEL
CORUNDUM
log (P / bar)
T / K
0
-1
-2
-3
-4
-5
-6
523
723
923
1123
1323
1523
1723
1923

mineral assemblage. On the high-temperature side, the mineral assemblage consisting of one or more of corundum, melilite, perovskite, Fe–Ni alloy, and spinel is similar to those found as inclusions in meteorites (e.g., Allende). On the low-temperature side of this band, the mineral assemblage olivine–plagioclase–pyroxenes is of the "rock-forming" type.

Figure 4 showed Grossman's (1972) calculated temperatures for condensation of important solids. Table 6 shows a comparison of the temperatures in the two calculations. The two sets of results are generally similar. However, important differences in calculated condensation temperatures do occur in some cases, e.g., for plagioclase, rutile, troilite, and orthoclase. The different temperatures for the feldspars may result from to differences in the thermochemical data or from solution effects considered here. Another significant change is the presence of ilmenite and the total absence of magnetite in the present results. Note that complete calculations below 1125K comparable to those of the present study were not done before. Compared to the data of Barshay and Lewis (1976), the enstatite condensation between 10^{-2} and 10^{-3} bar is lower by about 50K, which may also be a result of different thermochemical data and solution of Al_2O_3 and diopside.

Modal Mineralogy of the Nebula

As shown in Fig. 10 many reaction boundaries crowd in the mid-diagonal region of the pressure–temperature field, where many silicates appear and all oxides (except those of titanium) disappear. Many equilibrium assemblages are possible in this region and we counted over 20 combinations of melilite, perovskite,

◁ *Fig. 10.* The condensation and dissolution pressure and temperature curves for minerals in the solar nebula. At high temperatures the curves are very close to each other in the pressure range of 10^{-2} to 10^{-6} bar. On the high-temperature side of this bundle of curves, the refractory phases, such as corundum (Al_2O_3), melilite (Ca_2Al_2SiO–$Ca_2MgSi_2O_7$) and perovskite ($CaTiO_3$), condense. On the low-temperature side of the bundle condensation of the major rock-forming minerals olivine, plagioclase, and pyroxenes occur. The Fe–Ni condensation curve straddles the bundle. The Fe–Ni condenses ahead of the silicates (other than melilite) above 2×10^{-4} bar and then falls behind them with decreasing pressure. 1, corundum-in; 2, melilite-in; 3, γ-FeNi-in (dashed part is liquid); 4, Ti_2O_3 in; 5, perovskite-in; 6, perovskite-out, Ti_2O_3-in; 7, corundum-out, spinel-in; 8, Ti_2O_3-out, perovskite-in; 9, olivine-in; 10, spinel-out, plagioclase-in; 11, melilite-out, clinopyroxene-in; 12, orthopyroxene-in; 13, Ti_2O_3-out, Ti_3O_5-in; 14, perovskite-out, Ti_3O_5-in; 15, perovskite-out, ilmenite-in; 16, ilmenite-out, Ti_3O_5-in; 17, Ti_3O_5-out, ilmenite-in; 18, ilmenite-out, rutile-in; 19, α-FeNi-in; 20, rutile-out, ilmenite-in; 21, orthoclase-in; 22, FeS-in; 23, orthoclase-out, biotite-in; 24, garnet-in; 25, clinopyroxene-out, tremolite-in; 26, orthopyroxene-out, talc-in; 27, talc-out; 28, orthopyroxene-out; 29, clinopyroxene-out; 30, orthopyroxene-out, tremolite-in.

Table 6. Condensation temperatures (K) of solids in the solar gas at 10^{-3} bar (this study) and 10^{-3} atm (Grossman, 1972).

	This study		Grossman (1972)	
Phase	In	Out	In	Out
Corundum	1726	1476	1758	1513
Melilite	1640	1421	1625	1450
Perovskite	1611	1409	1647	1393
Spinel	1476	1431	1513	1362
γ-FeNi	1457		1473	
Plagioclase	1433		1362	
Olivine	1431		1444	
Clinopyroxene	1425		1450	
Ti_3O_5	1409	1134	1393	1125
Orthopyroxene	1359		1349	
Ilmenite	1134	1015		
α-FeNi	1017			
Rutile	1015	854	1125	
Ilmenite	854			
Orthoclase	628	468	~1000	
FeS(troilite)	621		700	
Garnet	533			
Biotite	468			

Ti_2O_3, Fe–Ni alloy, spinel, olivine, plagioclase, and clinopyroxene. This pressure–temperature band is important because small fluctuations of pressure and/or temperature across it, result in drastic changes in the types and amounts of condensates. As examples, consider the changes (a) as pressure varies from $10^{-4.5}$ to 10^{-6} bar at 1273K (Fig. 11) and (b) as temperature varies from 1323 to 1383K at 10^{-4} bar (Fig. 12). The calculated modes of the condensed phases show how a relatively small pressure or temperature fluctuation may change the assemblage from one of high-temperature refractory minerals occurring as inclusions in meteorites to a "planet-forming" assemblage of common rock-forming silicates. As the gas cools through this transition band, it drops a high proportion of the major elements Si, Fe, Mg, and Ni as condensates (Fig. 13), mainly as Fe–Ni alloy and the silicates olivine and pyroxenes, which remain stable down to very low pressures and temperatures. Calcium and aluminum get locked in plagioclase in this band and remain there over a very wide pressure–temperature range. We note that the occurrence of this band may explain the variety of minerals found seemingly in disequilibrium in many meteorites.

Figure 14 shows the modal composition of the equilibrium assemblages in the low-temperature region at two different pressures 10^{-8} and 10^{-2} bar. At 10^{-8} bar, the amount of graphite (not shown) reaches a maximum of 41% at 442K. With decreasing temperature, there is a strong change in the amount of α-FeNi as it converts to troilite. Similary, olivine increases at the expense of orthopyroxene. Clinopyroxene increases in amount slightly from 6 to 7% and there

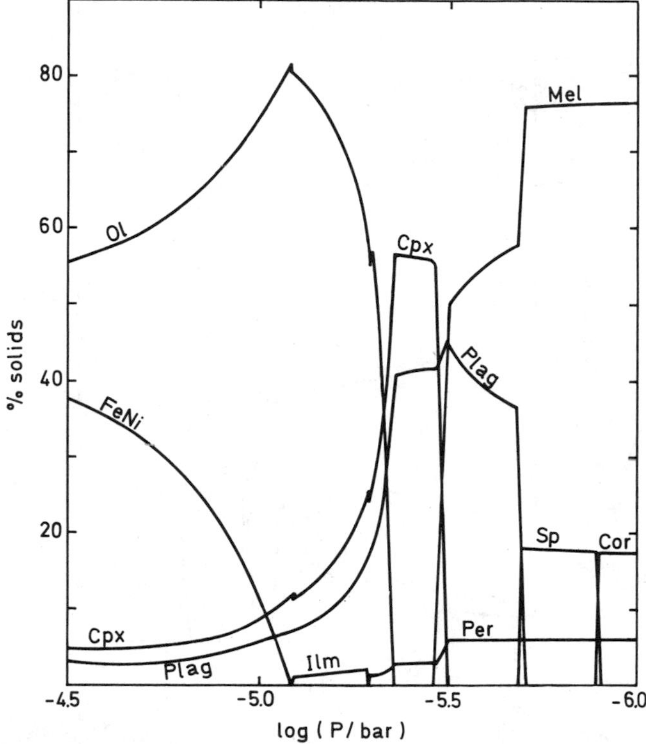

Fig. 11. Mineralogical changes across the band at 1273K. Many important changes take place within the pressure range given by log (*P*, in bar) of −5 to −5.5. The calculations were performed for every bar of pressure change in the vicinity of kinks and changes of slopes in the curves. After Saxena and Eriksson (1983a). Permission Pergamon Press.

are small variations in the amounts of γ-FeNi and plagioclase, both of which remain at about 3%. Small amounts of almandine-rich garnet (1.5%), ilmenite (0.3%), and orthoclase (<0.1%; not shown) complete the assemblages. At temperatures above 623K, the modes vary little.

At 10^{-4} bar and 663K, for γ-FeNi, α-FeNi, olivine, plagioclase, clinopyroxene, orthopyroxene, and ilmenite the amount (%) are 2.46, 40.26, 28.94, 3.56, 6.26, 18.25, and 0.27, respectively, which may be considered as a representive sample of the remaining pressure–temperature field not covered in Fig. 14.

Chemical Composition of Minerals Condensing in the Solar Gas

As mentioned earlier, one of the important aspects of our calculations is the documentation of compositions of minerals condensing in the solar gas. The practice of using compositions of minerals as indicators of physical and chem-

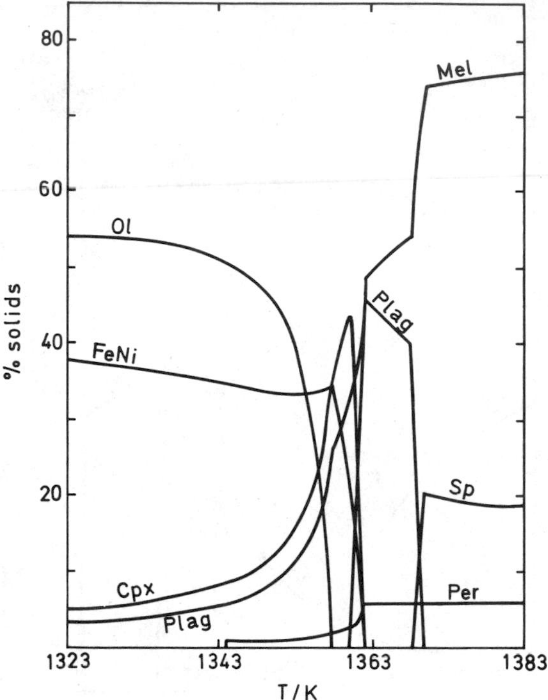

Fig. 12. Mineralogical changes across the band at 10^{-4} bar. Note the variation in percentage solids condensed between 1353 and 1363K. The calculations were performed to fraction of a degree of temperature change in the vicinity of the kinks and changes of slopes in the curves. After Saxena and Eriksson (1983a). Permission Pergamon Press.

ical conditions is well established in geochemistry. The basic premise is that in a system of constant composition, a solid solution may change in chemical composition as a function of temperature either by reactions that produce new assemblages or simply by exchanging certain cations with surrounding phases. In systems of constant total chemical composition, the concentration of a variable component in a solid solution indicates changes in the intensive thermodynamic variables. If the system has a varying composition, a concentration change may not necessarily be related to a change in pressure and/or temperature. In such systems we need activities of components and equilibrium constants to understand the variation of composition with pressure and temperature.

The equilibrium relations determined in a gas of solar composition are ideal for correlating the mineral compositions with pressure and temperature. Because the total composition is fixed, the mole fraction or atomic fraction of a component in a mineral is directly related to pressure and/or temperature. It is important to note, however, that such composition–pressure–temperature cor-

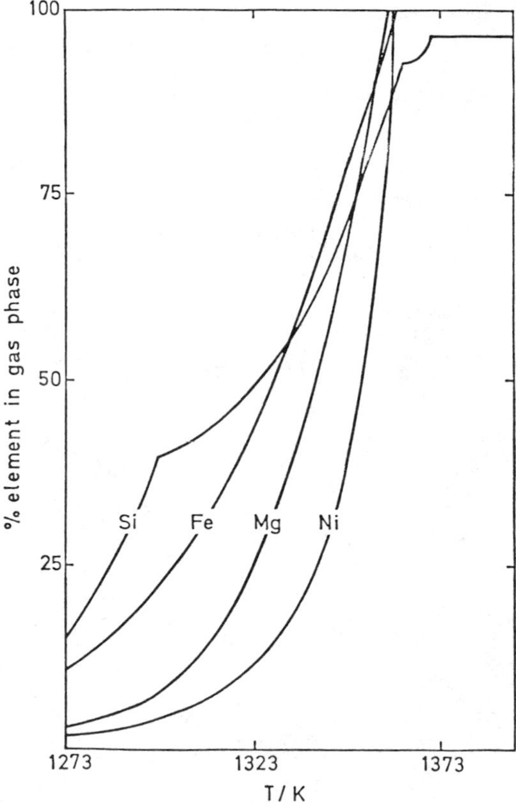

Fig. 13. Percentage element left in the gas phase as temperature changes at 10^{-4} bar across the band. Major condensation of the elements Si, Fe, Mg, and Ni takes place in this zone.

relations apply strictly only to *unaltered condensates* that are in chemical equilibrium with a gas of solar composition.

As will be shown subsequently, a change in solar gas composition affects the concentrations of some solid solution components more than those of others. Changes in the H/O and C/O ratios in solar gas profoundly affect the $Fe^{2+}/(Fe^{2+} + Mg)$ ratio in silicates but may not significantly affect the concentrations of some other components, e.g., Al_2O_3 or jadeite in pyroxenes or gehlenite in melilite. Therefore the compositional plots that follow may also be useful for nonsolar gas environments, e.g., the surface of an accreting body.

Melilite

Melilite is a solid solution of gehlenite ($Ca_2Al_2SiO_7$) and åkermanite ($Ca_2MgSi_2O_7$). Figure 15 shows the composition of melilite in the range of

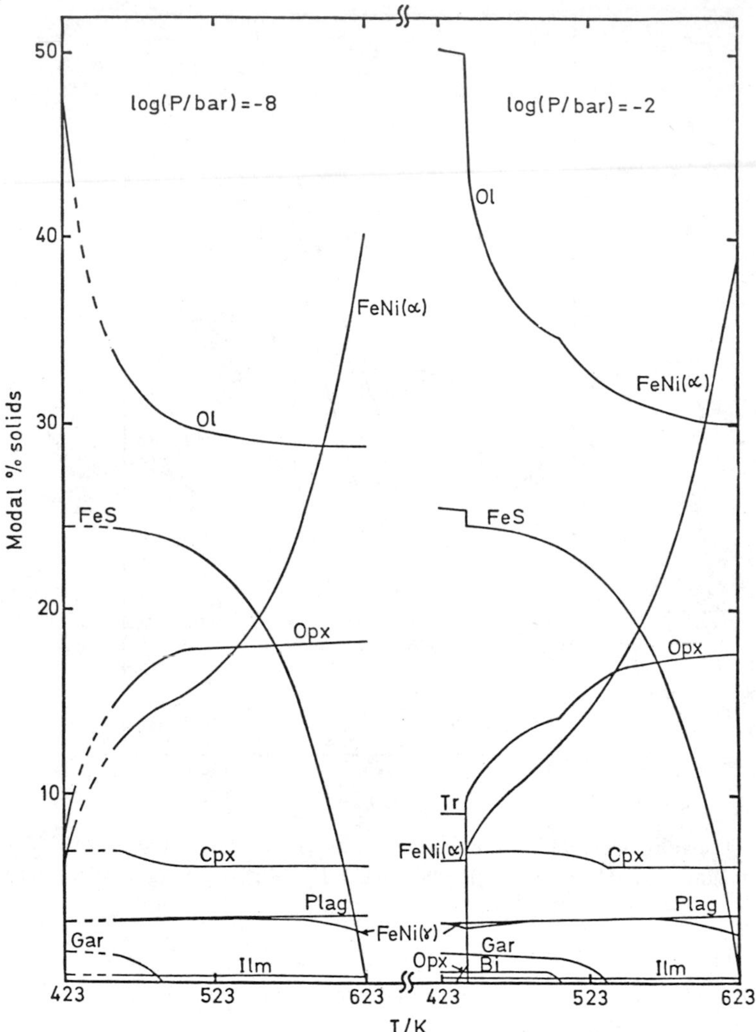

Fig. 14. Modal percentage variation of minerals between 423 and 623K. Although the minerals γ-FeNi, plagioclase, and clinopyroxene vary little with temperature, there is a significant change in the modal amounts of troilite and olivine, both increasing with decreasing temperature at the expense of α-FeNi and orthopyroxene, respectively. The effect of variation in pressure on the modal composition is not important except for graphite, which reaches as much as 41% at 442K (not shown) within the temperature range shown by the dashed curve. After Saxena and Eriksson (1983b). Permission Elsevier Science Publications.

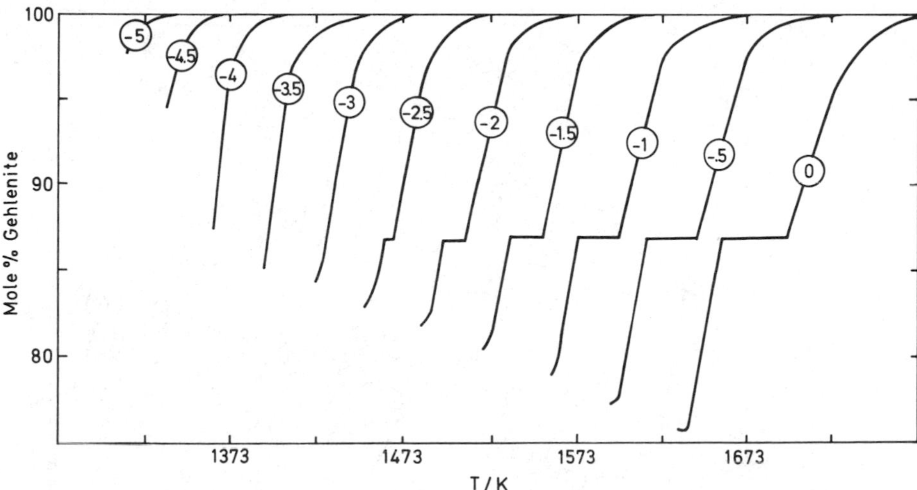

Fig. 15. Chemical composition of melilite as a function of pressure and temperature. The curves are isobars with numbers representing log (*P*, in bar). The inflections result from coexisting melilite, spinel, and olivine and the terminations represent melilite disappearance. After Saxena and Eriksson (1983a). Permission Pergamon Press.

pressure and temperature of its stability. The maximum åkermanite content of 24 mol % occurs at 1640K and 1bar. The isobars in the composition–temperature field have rather steep slopes, signifying that if melilite composition and temperature of condensation are known, the pressure can be estimated within narrow limits, or, if composition and pressure are known, the condensation temperature can be determined precisely. The inflections in the isobars are caused by the coexistence of melilite with spinel and olivine which makes the assemblage univariant. We note that the melilite compositions calculated by the present technique of introducing activities in the equilibrium calculations are at variance with Grossman's (1972) hand-calculated activities in an ideal melilite solid solution.

Clinopyroxene

The components included in our solution model for clinopyroxene are clinoenstatite ($MgSiO_3$), diopside ($Ca_{0.5}Mg_{0.5}SiO_3$), clinoferrosilite ($FeSiO_3$), Al_2O_3, and jadeite ($Na_{0.5}Al_{0.5}SiO_3$). The clinoenstatite content is plotted as isobars in a composition–temperature diagram in Fig. 16. The isobars bend sharply at the pressure–temperature at which orthopyroxene joins the equilibrium. Thus a mole percentage of 10 in Fig. 16 would indicate different temperatures depending on the presence or absence of orthopyroxene. At 1 bar the condensation temperatures are 1156 (orthopyroxene coexisting) and 1626K (orthopyroxene absent). Again, as for melilite, the isobars have steep slopes. If composition and pressure

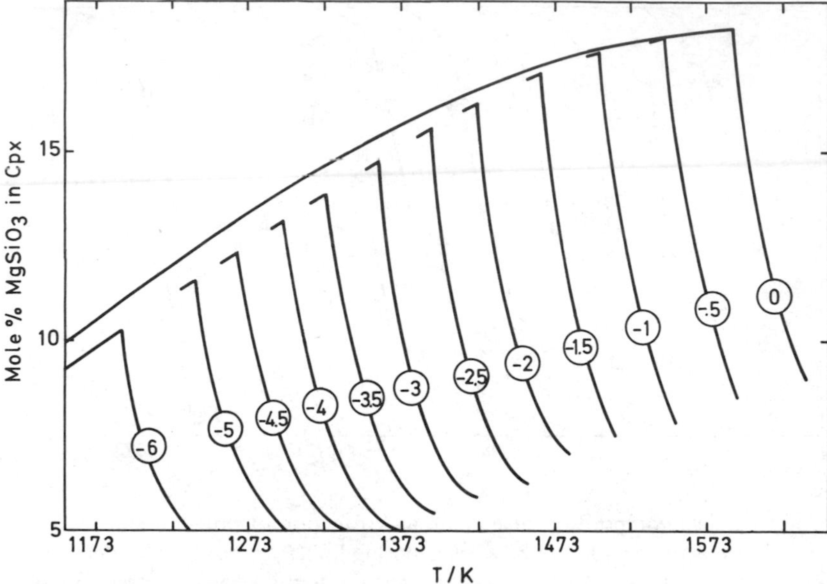

Fig. 16. Isobars of clinoenstatite mole percentage in clinopyroxene in the high-temperature region. The curves are labeled with log (P, in bar). Each isobar sharply bends at the temperature at which orthopyroxene joins the condensing clinopyroxene (details omitted after the bend). In the absence of orthopyroxene, excess magnesium is in the gas phase. See text for further explanation.

are known for an orthopyroxene-absent assemblage, the temperature can be estimated. Out of the five components, the solid solution composition is dominated by clinoenstatite and diopside. Sequentially if we follow the equilibrium change of composition at a pressure of 10^{-3} bar, we note that a diopside-rich clinopyroxene increases rapidly in clinoenstatite content, reaching close to 15 mol % within a temperature decrease of 50K, at which point orthopyroxene appears. As the temperature decreases, the clinopyroxene again starts toward diopside enrichment. Note that pressure has little effect on clinoenstatite composition of clinopyroxene coexisting with orthopyroxene. At temperatures lower than 1173K, the clinopyroxene remains mostly diopsidic. Iron content of clinopyroxene remains within 5 mol %, even at the lowest temperatures.

Figure 17 shows the jadeite content of clinopyroxene as a function of pressure and temperature. It appears that at low pressures (e.g., 10^{-3} bar) the jadeite content is small. A high content should be regarded as indicating high pressure. The maximum content of jadeite is reached in clinopyroxenes at 1 bar (maybe higher at higher pressures not studied here) and 1215K.

Orthopyroxene

The four components in orthopyroxene solid solution are enstatite ($MgSiO_3$), orthodiopside ($Ca_{0.5}Mg_{0.5}SiO_3$), ferrosilite ($FeSiO_3$) and Al_2O_3. At high tem-

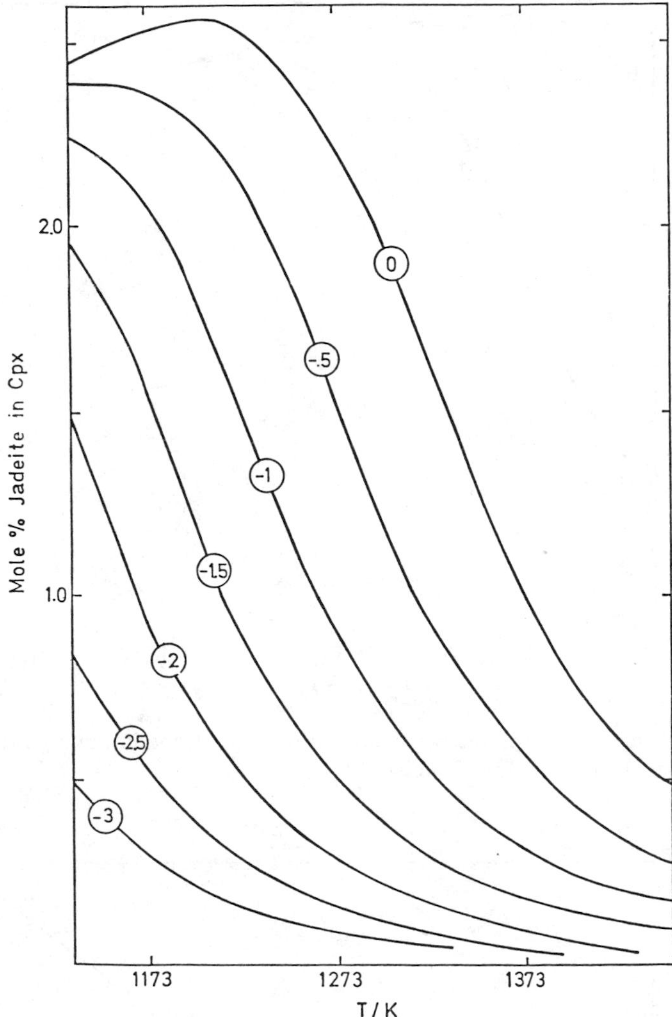

Fig. 17. Mole percentage of jadeite in clinopyroxene at different pressures in the high-temperature region. The curves are labeled with log (*P*, in bar).

peratures ferrosilite is negligible and the solution mainly consists of enstatite, orthodiopside, and Al_2O_3 in that order of concentration. Figures 18 and 19 show isobaric compositions plotted in composition–temperature diagrams. Figure 18 shows that the isobaric Al_2O_3 content reaches a maximum, which increases with decreasing pressure. At high temperatures (>1423K), the effect of pressure becomes negligibly small. Figure 19 shows the orthodiopside content. There is no effect of pressure on this relationship, which is similar to those discussed in the petrological literature (e.g., Lindsley *et al.*, 1981). For compositions with low iron content, this relationship should be considered as an excellent thermometer.

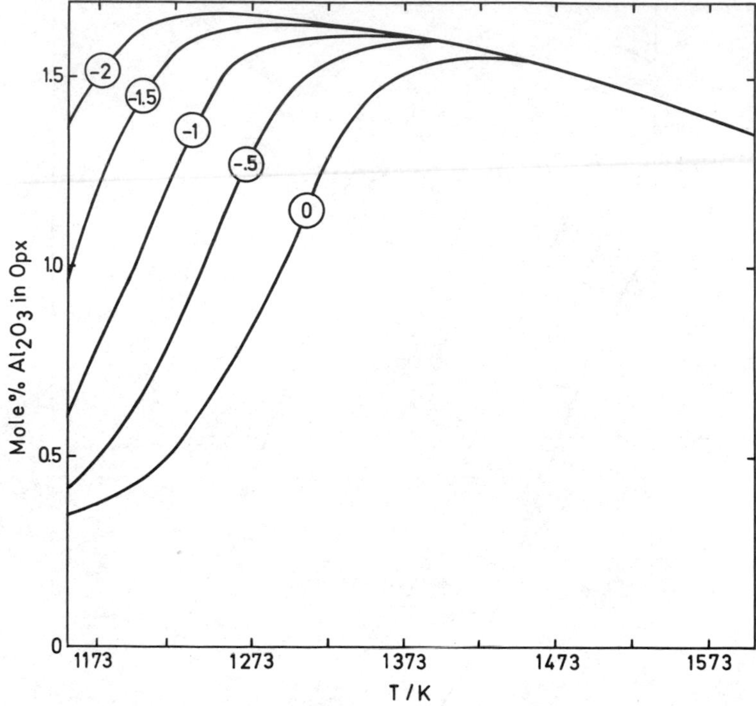

Fig. 18. Isobars of Al_2O_3 mole percentage in orthopyroxene in the high-temperature region. Isobar labels are in log (*P*, in bar).

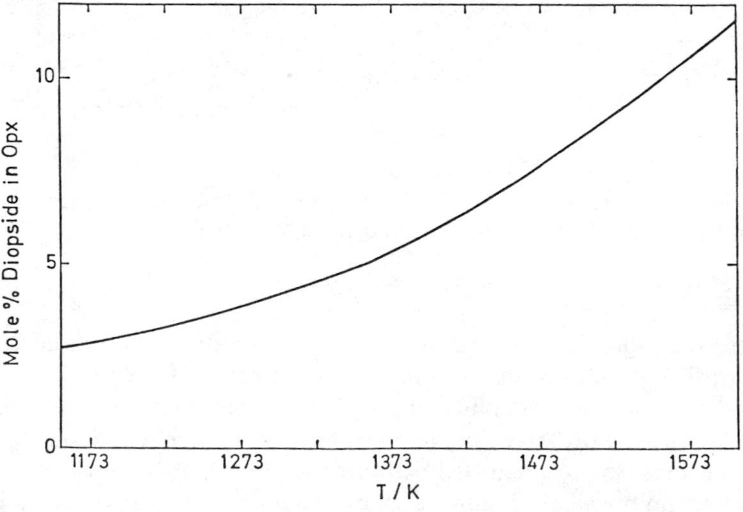

Fig. 19. Mole percentage of orthodiopside in orthopyroxene in the high-temperature region. Unlike in clinopyroxene, pressure (1 bar or below) has no influence on this relationship.

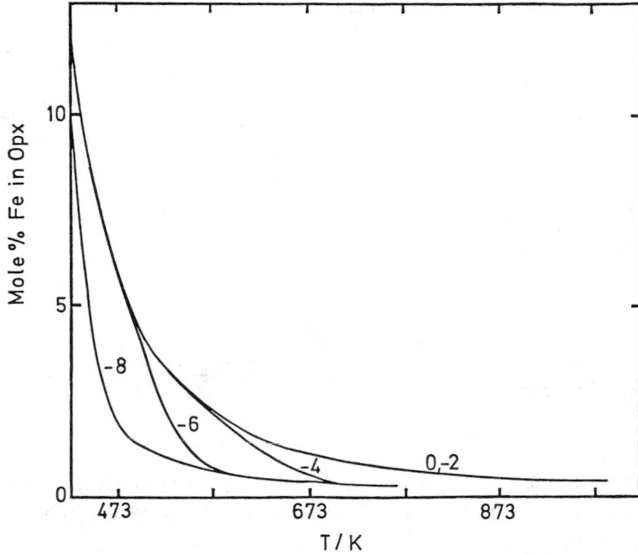

Fig. 20. Isobars of ferrosilite mole percentage in orthopyroxene in the low-temperature region. The numbers on the curves represent log (P, in bar). Below 673K there is a significant effect of pressure variation. The iron content decreases with decreasing pressure.

The iron content of orthopyroxene in equilibrium with the solar gas is small at high temperature and attains significant values only when the temperature lowers to 900K and less (see Fig. 20). It appears that an iron content comparable to that found in the H and L chondrite pyroxenes can be attained only at temperatures lower than 500K, provided the gas composition (H/O) remains at the cosmic abundance of elements. The pressure effect on the ferrosilite content is small. If temperature estimates from coexisting pyroxenes are higher than 500K, it must reflect equilibration under nonsolar H/O gas composition.

Olivine

Olivine was considered a solid solution of fayalite ($FeSi_{0.5}O_2$), forsterite ($MgSi_{0.5}O_2$), and $NiSi_{0.5}O_2$. The nickel content was found not to be significant at any temperature. The mole percentage of fayalite is plotted in a composition–temperature diagram in Fig. 21. At temperatures higher than 900K, the fayalite content is small. Similar to ferrosilite in orthopyroxene, the fayalite content of olivine increases with decreasing temperature. Again, to obtain olivines with compositions as found in ordinary chondrites, the temperature must have been quite low if the chondritic minerals have equilibrated with a gas of solar abundance of H, O, C.

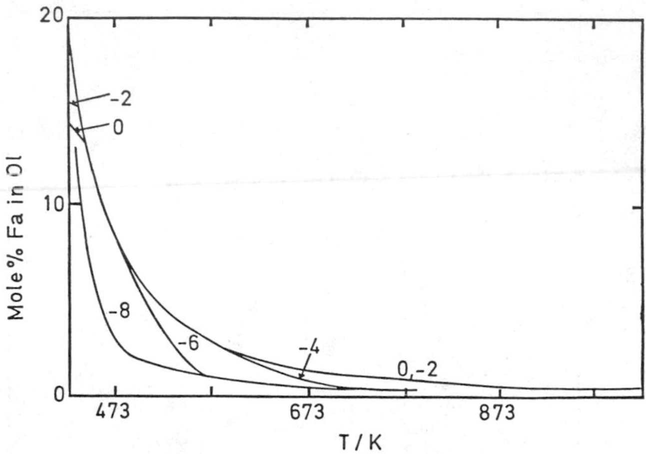

Fig. 21. Isobars of fayalite mole percentage in olivine in the low-temperature region. The numbers on the curves represent log (*P*, in bar). The pressure behavior of olivine composition is similar to that of orthopyroxene in Fig. 20.

Plagioclase Feldspar

A ternary solution model for feldspar with albite ($NaAlSi_3O_8$), anorthite ($CaAl_2Si_2O_8$), and orthoclase ($KAlSi_3O_8$) was used in the study. At high temperatures the orthoclase content of feldspar remains low, under 5 mol %. Figures 22 and 23 show the albite concentration as a function of pressure and temperature in the high- and low-temperature regions, respectively. At all pressures, the albite content of plagioclase increases with decreasing temperature. The slopes of the isobars increase as a function of decreasing pressure. Thus at 1 bar, the albite content increases gradually from a few mole percent at 1400K (not shown) to ~82 mol % at 1100K. At 10^{-8} bar, however, the change from a pure anorthite at 900K to an albite-rich (~82 mol %) plagioclase takes place by a lowering of temperature by only 100K.

Above ~600K, the isobars in the composition–temperature field behave systematically, following a parabolic or a sigmoidal trend and all curves merging with the 1-bar curve at an albite content of ~82 mol %. As temperature decreases, structural complexities in the ternary feldspar increase and the solution model used here, although valid at ~973K, may not rigorously account for the nonideal behavior at the low-temperature end. We note that between 583 and 723K, the albite content of plagioclase is independent of pressure and temperature and remains at about 82 mol %. Beyond 723K, the effect of pressure is quite pronounced and, depending on the pressure, the same albite content may be found over a range of 200K or more.

Fe–Ni Alloy

The metal phase was considered to be a solid solution of Fe, Ni, and Si at high temperatures. Silicon concentration was found to be below 0.015 mol % and

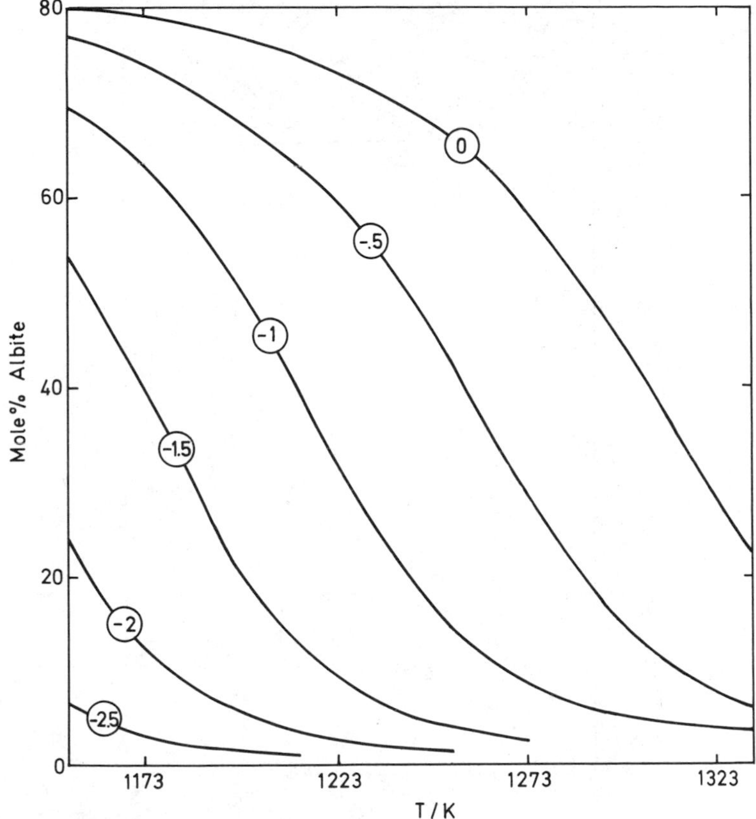

Fig. 22. Isobaric albite content (Mol %) of plagioclase in the high-temperature region. The curves are isobars with values in log (P, in bar). With pressure decreasing below 10^{-3} bar, the mole percentage of albite attains values of less than one.

decreasing with decreasing temperature. The Fe content of metal remains at ~94.5 mol % over a broad range of pressure and temperature. With increasing temperature the isobaric Fe concentrations drop slowly at first and rapidly later (see Fig. 24). The terminations of the isobars occur at the condensation curve (not shown; see Fig. 7). These results are comparable to those of Sears (1978).

Ilmenite and Garnet

We found small quantities of ilmenite and garnet as parts of equilibrium assemblages. Ilmenite occurs over a wide range of pressure and temperature (see Fig. 10). It has a narrow field of existence at low pressures at about 1250K, it condenses again at about 1150K, and then it remains stable down to the lowest temperature investigated except from an enclosed region where rutile is stable.

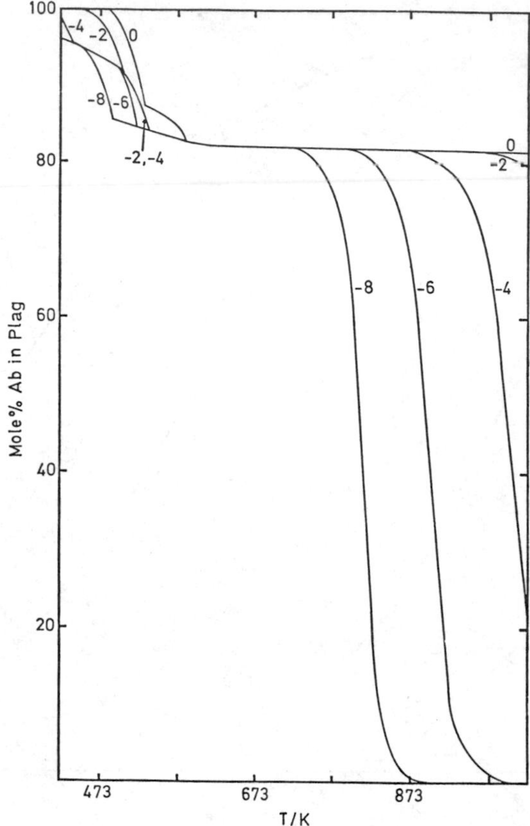

Fig. 23. Isobaric albite content (mol %) of plagioclase in the low-temperature region. The numbers on the curves represent log (P, in bar). The albite content of ~82 mol % is insensitive to any variation in pressure and temperature between 583 to 723K. Above 723K and below 583K compositions change significantly with pressure.

Ilmenite is highly magnesian at high temperatures. The $MgTiO_3$ content decreases with decreasing temperature as shown in Fig. 25. There is a significant effect of pressure on the composition.

To our knowledge, almandine-rich garnet is not noted in meteorites perhaps because it has a small field of stability at low temperatures in the pressure-temperature diagram of the solar gas. We found some effect of pressure on its composition (Fig. 26). It is not unlikely that this result is spurious in view of the extrapolation of the almandine thermochemical data to low temperatures. Grossular has been found in meteorites but not in these equilibrium-computed results. Andradite has been observed in meteorites but lack of thermochemical data did not permit its inclusion in the calculations.

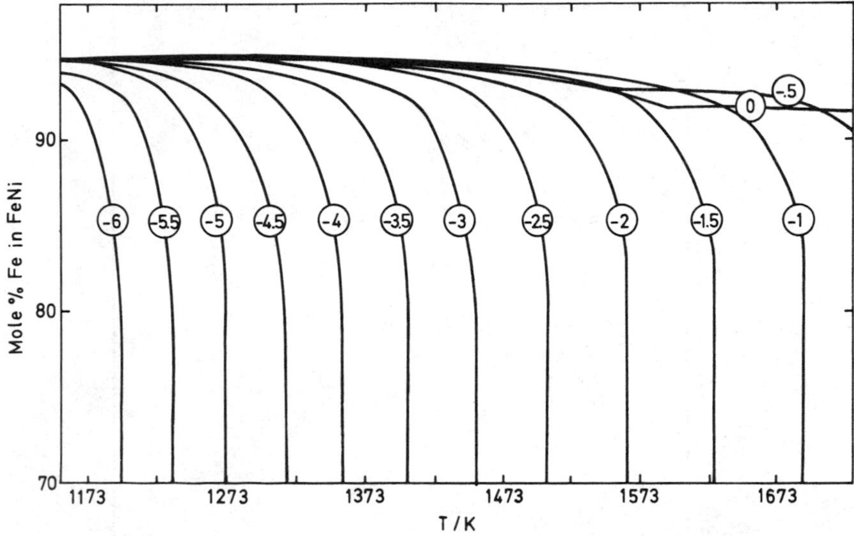

Fig. 24. Isobars of Fe mole percentage in Fe–Ni–Si alloy in the high-temperature region. Isobar labels are in log (*P*, in bar). The Fe curves are complementary to the Ni curves (Si being small) as plotted by Sears (1978). The rapidly decreasing Fe with pressure is a consequence of its greater volatility relative to Ni.

Models of Terrestrial Planets

During the last two decades several models of the formation of terrestrial planets based on equilibrium condensation or nonequilibrium processes have been proposed. Basically, most features required to be explained by any model concern Earth because it is Earth we know most about. Since man's excursion to the moon much new information, particularly on the trace element chemistry of the lunar rocks, has been added to our chemical data. Limited chemical data on Mars and Venus from the Viking and the Russian probes, respectively, are also at hand. Important chemical and physical characteristics of the terrestrial planets have been summarized by Dermott (1978), Ringwood (1979), and the BVSP (1981). In spite of these new data and the many hypotheses presented during the last few decades, the enigma of explaining the genetic relationship between the primitive solar nebula and the meteorites and the planets continues.

As this chapter is mainly based on results that may be obtained only through thermochemical equilibrium calculations, we are limited to considering models that use such data. The models of Larimer and Anders (1967), Lewis (1972a, 1974), Clark *et al.* (1972), and Turekian and Clark (1975) have been based on equilibrium calculations. Other models that use equilibrium theory with kinetic interpretations are by Blander and co-workers (Blander and Katz, 1967; Blander and Abdel-Gawad, 1969; Blander, 1971). Models given by Urey (1952)

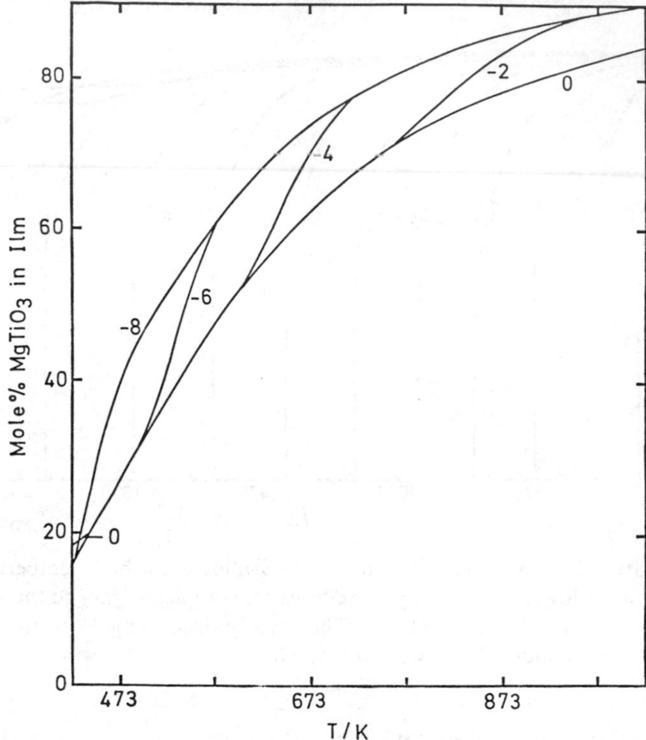

Fig. 25. Isobaric $MgTiO_3$ content (mol %) of ilmenite, a minor phase (<0.3%), plotted against temperature. The $MgTiO_3$–$FeTiO_3$ mixing is assumed ideal. If it is nonideal, the mole percentage of $MgTiO_3$ will be less than shown here. The numbers on the curves represent log (*P*, in bar). The curves are continuous but break at temperatures (not shown) where rutile appears.

and Ringwood (1966) begin with a cold nebula with planetary ices and particles accreting homogeneously. Heating of the primitive planets takes place from radioactive or gravitational accumulation of heat, leading to planetary differentiation into core, mantle, and crust. The latest status of this art has been reviewed by Ringwood (1979) and the BVSP (1981). Our equilibrium calculations do not apply to the cold homogeneous accretion theory. The fact that we do not pursue this further should not be construed as a rejection of these ideas.

We begin by considering the models to which our data are applicable and among these the Lewis model is the simplest.

Homogeneous Accumulation Along a Thermal–Pressure Gradient

Anders (1974) considered that chemical differences among chondrites may be caused by their formation in a cooling solar nebula. Lewis (1972a) calculated

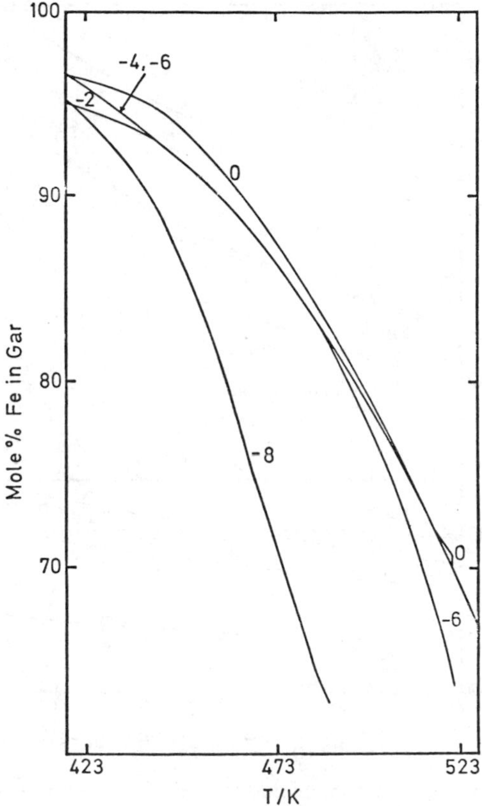

Fig. 26. Isobars of almandine mole percentage in garnet plotted against temperature. The numbers on the curves represent log (*P*, in bar). The pressure effect on the almandine content is quite significant.

the properties of equilibrium condensates at different temperatures and suggested that different planetary densities resulted from accumulation of the condensates at decreasing temperatures away from the heliocenter. Cameron and Pine's (1973) model of a massive nebula with a hot interior seemed to be an appropriate physical model for this hypothesis. Even though Cameron (1978) does not continue with some physical aspects of the 1973 model, the attainment of high temperatures is attested by an Allende refractory inclusion (Grossman, 1972). Lewis (1974) calculated possible thermal–pressure gradients in the nebula. Figure 27 from Barshay and Lewis (1976) explains the Lewis model. Along the adiabat, planets accumulated from equilibrium condensates. Mercury formed at high temperatures, with mostly iron and little silicates. Venus, Earth, and Mars formed at successively lower temperatures and pressures. Thus Mars retained most of the hydrous species and FeS, as well as other silicates, and little or no Fe. Further on, water ice became stable, followed by condensation of other volatiles as ices (Lewis, 1972b).

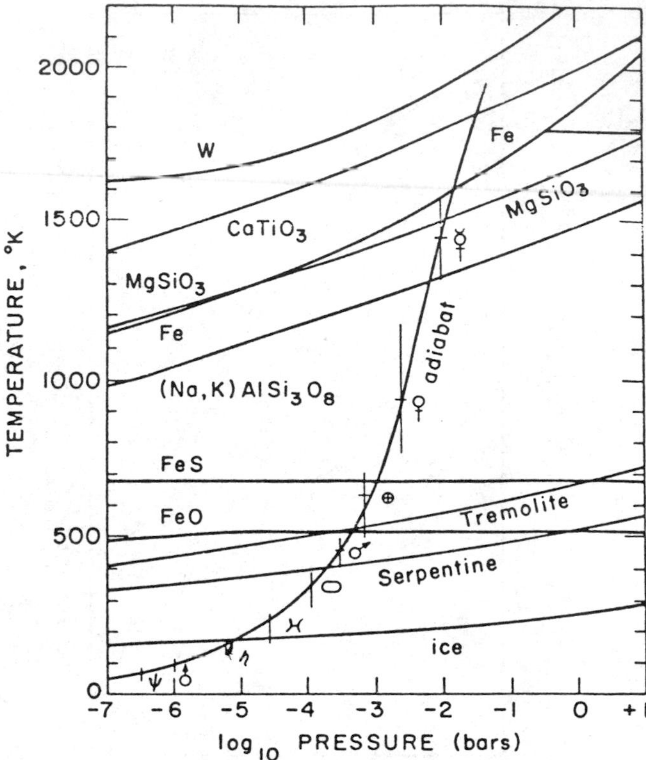

Fig. 27. The equilibrium chemistry of material of solar composition as a function of temperature and pressure. After Barshay and Lewis (1976).

As clearly stated by Lewis (1974) this model was designed to explain certain broad physical–chemical characteristics of the planets, which the model does elegantly. It was not supposed to provide details that would necessarily comply with all possible geochemical characters of the planets and meteorites. Let us consider the consequences of such a model by computing equilibrium assemblages at suitable temperatures simulating Mercury, Venus, Earth, and Mars. Since Lewis (1974) indicates ranges of temperature for each planet over which condensates may accumulate, it is difficult to choose specific temperatures and pressures. As examples we chose temperature–pressure combinations of 1473K and 10^{-2} bar for Mercury, 900K and 10^{-3} bar for Venus, 600K and 10^{-3} bar for Earth, and 423K and 10^{-3} bar for Mars. Tables 7–10 show the calculated equilibrium assemblages. Although it is too much to expect nature to have a strict book-keeping, in modeling we have to have some basic criteria satisfied and in this case we find (a) the average density of the condensates, which is generally higher than the zero-pressure densities of the planets, decreases regularly away from the heliocenter, and (b) the relative fraction condensed (total mass of the con-

Table 7. Equilibrium assemblage at 1473K and 10^{-2} bar in the solar gas.

Phase	Density	Mol %	Composition[a]		
γ-FeNi	7.921	50.78	Fe 0.940	Ni 0.058	Si 0.002
Olivine	3.213	44.00	Fo 1.000		
Clinopyroxene	3.280	2.58	Cen 0.073	Di 0.913	Al_2O_3 0.014
Plagioclase	2.760	2.57	An 1.000		
Ti_2O_3	4.575	0.07			
Average density = 5.59					
Fraction condensed[b] = 0.65					

[a] The element abundances are from Anders and Ebihara (1982).

[b] Fraction condensed is the mass relative to the total mass condensed at 600K (Table 5).

Table 8. Equilibrium assemblage at 900 K and 10^{-3} bar in the solar gas.

Phase	Density	Mol %	Composition			
γ-FeNi	8.069	4.50	Fe 0.812	Ni 0.188		
α-FeNi	7.913	39.55	Fe 0.963	Ni 0.037		
Olivine	3.215	30.62	Fo 0.998	Fa 0.002		
Orthopyroxene	3.209	17.18	En 0.982	Odi 0.013	Fs 0.001	Al_2O_3 0.004
Clinopyroxene	3.276	4.76	Cen 0.040	Di 0.942	Al_2O_3 0.001	Jd 0.016
Plagioclase	2.638	3.27	An 0.149	Ab 0.798	Or 0.054	
TiO_2	4.245	0.11				
Average density = 5.28						
Fraction condensed = 0.95						

Note: See Table 7 for fraction condensed and for elemental abundances.

Table 9. Equilibrium assemblage at 600K and 10^{-3} bar in the solar gas.

Phase	Density	Mol %	Composition				
γ-FeNi	8.425	3.04	Fe 0.470	Ni 0.530			
α-FeNi	7.899	28.59	Fe 0.976	Ni 0.024			
Olivine	3.242	32.66	Fo 0.975	Fa 0.025			
Orthopyroxene	3.227	16.13	En 0.972	Odi 0.004	Fs 0.019	Al_2O_3 0.005	
Clinopyroxene	3.277	4.74	Cen 0.006	Di 0.985	Cfs 0.001	Al_2O_3 0.001	Jd 0.007
Plagioclase	2.638	3.24	An 0.146	Ab 0.812	Or 0.041		
Orthoclase	2.561	0.04	Ab 0.031	Or 0.969			
Ilmenite	4.328	0.22	Fe 0.485	Mg 0.515			
FeS	4.830	11.33					
Average density = 4.89							
Fraction condensed = 1.00							

Note: See Table 7 for fraction condensed and elemental abundances.

Table 10. Equilibrium assemblage at 423 K and 10^{-3} bar in the solar gas.

Phase	Density	Mol %	Composition			
γ-FeNi	8.666	2.99	Fe 0.236	Ni 0.764		
α-FeNi	7.880	2.88	Fe 0.995	Ni 0.005		
Olivine	3.467	57.24	Fo 0.784	Fa 0.216		
Orthopyroxene	3.314	2.16	En 0.868	Odi 0.001	Fs 0.131	
Clinopyroxene	3.227	5.68	Cen 0.001	Di 0.995	Cfs 0.001	Jd 0.003
Garnet	4.301	1.69	Py 0.021	Alm 0.965	Gr 0.014	
Plagioclase	2.620	2.65	Ab 0.999	Or 0.001		
Ilmenite	4.649	0.22	Fe 0.845	Mg 0.155		
Biotite	2.842	0.52	Ann 0.898	Phl 0.102		
FeS	4.830	23.97				
Average density = 4.05						
Fraction condensed = 1.08						

Note: see Table 7 for fraction condensed and elemental abundances.

densed elements/total mass of the condensed elements at 600K and 10^{-3} bar) increases in the same direction. Proto-Mercury and Proto-Mars would have too much material and it would have to be removed by some physical process, e.g., aerodyanamic drag, as a process of removing material from Proto-Mercury to the Sun (Weidenschilling, 1977).

Tables 7–10 also show the compositions of the minerals in the equilibrium assemblages. Except for those forming at 423K, the minerals are all highly reduced. To obtain a compositionally suitable olivine for Earth's mantle (Fe^{2+}/(Fe^{2+} + Mg) ≈ 0.12), we have to find a postaccretionary oxidation mechanism. Explaining the planetary densities, masses, and oxidation states are problems that all models must address and in this respect the equilibrium model brings us close but not close enough to a solution.

Nonhomogeneous Accumulation Model

Although the proposal was evident in the works of Eucken (1944) and Larimer and Anders (1967), the applicability of the nonhomogeneous accumulation model was fully explored first by Turekian and Clark (1969) and then in the works of Anders (1971), Grossman (1972), Clark *et al.* (1972), and Turekian and Clark (1975). Figure 27 shows that planets received condensates from within a narrow pressure–temperature range, each differing according to its distance from the center. Thus the temperature in the part of the nebula where Earth accreted may never have reached beyond the range indicated by a vertical bar in Fig. 27 (~600 ± 100K). The nonhomogeneous model, on the other hand, starts with the assumption that the inner region of the nebula within the orbit of Mars was everywhere heated such that all solids were evaporated. At 10^{-3} bar, this would happen at 1729K. As the gas cooled, the refractories corundum, melilite, perovskite, and spinel condensed and formed the nucleus around which Earth would grow by a layer-by-layer accumulation of condensing material. This, then, is a nonequilibrium model of Earth–the clean separation of the different layers depending on the cooling and settling rate of the condensates. An Fe–Ni core would never be in contact with the lower mantle, with mostly silicates forming at medium temperature. On top of the lower mantle and upper mantle and crust would form, with largely primitive carbonaceous chondrite material. Planetary differentiation processes would take over and, as their work is cut down substantially, only minor adjustments would be necessary to obtain the present-day Earth. As the upper mantle is largely chondritic, it would contain close to the observed abundances of siderophile and volatile elements and the ferrous to ferric ratio. In the cold homogeneous models (e.g., Ringwood, 1966) and the hot homogeneous model of Lewis (1974) discussed above, the upper mantle, lower mantle, and core would all be in equilibrium initially, causing the upper mantle to be depleted in the siderophile elements, such as Ni, Co, and Mn. The heterogeneous or nonhomogeneous accumulation model avoids this difficulty.

Ringwood (1979) discusses the following specific difficulties with the heterogeneous model:

a. Presently the mantle is considered relatively homogeneous. Because of a zoned structure with Fe–Ni already in the core, there would be insufficient gravitational energy to provide heat that could cause a convective mixing of the mantle in a primitive Earth.
b. The mantle silicates would be almost devoid of iron because of their formation under reducing conditions (see Figs. 20 and 21) at medium temperatures. This is contrary to Ringwood's (1979) inference that the $Fe^{2+}/(Fe^{2+} + Mg)$ ratio in the mantle should be 0.12.
c. An upper mantle being formed of a late accreted component would be depleted in high-temperature condensates, such as Ca, Al, Ti, and rare earth elements. This is contrary to the observation on the pyrolite mantle composition.
d. The Earth's core contains substantial quantities of light elements and in this model there is no mechanism for introducing these.
e. An accretion time scale of 10^3 to 10^4 years is too short, posing dynamical problems (Levin, 1972) (see also chapter 1).
f. In the upper mantle many trace elements, such as Ir, Au, Ge, Se, and S, are depleted relative to Ni.

This model for the formation of layered planets suffers from a lack of constraints on the fractionation process. Depending on the choice of cooling rate and effectiveness of separation, very different compositional layering may be found. Therefore, to characterize components, J. A. Wood (1962), Anders (1964), and Larimer and Anders (1967) used the compositional information that formed the basis of classifying chondrites. They recognized that the chondrites were a mixture of high-temperature and low-temperature phase assemblages. The chondrite model was extended by Anders and co-workers (e.g., Larimer and Anders, 1970;

Anders, 1971, 1977; Morgan and Anders, 1980) and Wänke and co-workers (e.g., Wänke *et al.*, 1974; Wänke, 1981) and Ringwood (1977, 1979). Although the general consensus among geochemists seems to agree on such an approach, the nature and number of components and the mechanisms of their mixing have been hotly debated.

Anders (1971) finds that meteorites consist of at least seven separately recognizable components, each forming within a range of temperature in a cooling nebula. Wänke and co-workers favor fewer components, and Ringwood (1979) would work with only two. There is undoubtedly merit in using the large amount of compositional data on meteorites and the assumption that the nebular processes that brought about the distinction between various classes of meteorites also led to the distinction in planetary compositions. However, the planetary model compositions are estimated by mass balancing the components on the basis of some geochemically and geophysically inferred compositional criteria, lending a certain amount of subjectivity to the work. With the presently available data, this seems to be unavoidable.

Formation of Mercury and High-Temperature Planetary Components

In models of planetary formation it is vital that an account be kept of the average mass and the average density of the condensing species. We have noted that in a system of homogeneous composition, the mass and density are functions of the pressure–temperature gradient. Contrary to what may be inferred from the data in Tables 7–10, the relationship between average mass or average density of the condensates and temperature is not linear. As shown in this section, this relationship is complex, which is fortunate because the planetary masses and densities are not linearly related to heliocentric distance. If it could be established that relative masses of suitable densities were present at appropriate heliocentric distances, forming equilibrium assemblages, the success of the equilibrium theory would be insured. Lewis (1972a) adopted this approach but did not pursue it with detailed calculations, which will be presented in this section.

Planetary Densities

To compare the densities of planets with those of equilibrium assemblages calculated at a pressure of 1 bar, we need the 1 bar densities of the planets. However, these quantities cannot be calculated with precision without data on the high-pressure–high-temperature behavior of the planetary material. Calculation of zero-pressure densities is possible through the use of such model-dependent equations as Murnaghan's equation (Stacey, 1977 pp. 172–173), which expresses the bulk modulus–pressure relationship linearly. The compo-

Table 11. Zero-pressure density profile of Earth.

	Stacey	Lyttleton (1982)
Inner core	7.3	—
Outer core	6.3[a]	6.107 ± 0.004
Mantle	4.1	4.057 ± 0.004
Upper mantle Crust	3.3	3.298 ± 0.015

[a] Data from Stacey (1972) for a liquid core. Other data from Figure 6.17 of Stacey (1977).

Table 12. Data on planetary densities.

Planet	Mean density[a]	Mean density at 10 kbar[b]	Zero-pressure density corrected[c]	Mass relative to ~Earth
Mercury	5.42	5.3	~5.3	0.055
Venus	5.25	3.96	4.40	0.815
Earth	5.51	4.07	4.45	1.000
Mars	3.96	3.73	~3.85	0.108

[a] Mean density from Ash *et al.* (1971).
[b] Mean density at 10 kbar from Ringwood (1979).
[c] Zero-pressure density corrected from Lewis (1972a).

sitional and/or phase changes imply that integration of such equations may give zero-pressure densities only at a certain depth or for an entire zone (e.g., lower mantle or upper mantle), not for the entire Earth. Table 11 shows the zero-pressure density distribution in the Earth. The data are from Stacey (1972, 1977) and Lyttleton (1982). Some data on planetary masses and densities are presented in Table 12. Note that the planetary masses and densities do not change regularly as one moves away from the Sun.

One of the difficulties we noted in the model of Lewis (1974) was the fact that Proto-Mercury was too massive. We decided to study in detail the masses and densities of the calculated equilibrium mineral assemblages in order to find a suitable combination that would bring us closer to Mercury's mass and density.

Mass and Density Calculation

Earlier we discussed the method of calculating equilibrium phase relationships. The total Gibbs free energy was minimized, with the mass balances as subsidiary constraints, and the final results therefore contain actual amounts of each condensed species. In Tables 5 and 9 we showed examples of output of the equilitrium mixture at 600K and 10^{-3} bar. This assemblage is chosen as reference when calculating the relative mass of an equilibrium assemblage. The density of each phase and the average density of the entire condensed product at equilibrium are calculated from molar amounts, molar masses, and molar volumes.

Figure 10 showed how several condensation curves (6–16) formed a tight band at pressures of 10^{-2} bar and below. This region marks a rather abrupt change that takes place between the refractory assemblage corundum–melilite–perovskite and the major rock-forming assemblage olivine–plagioclase–pyroxenes. Major

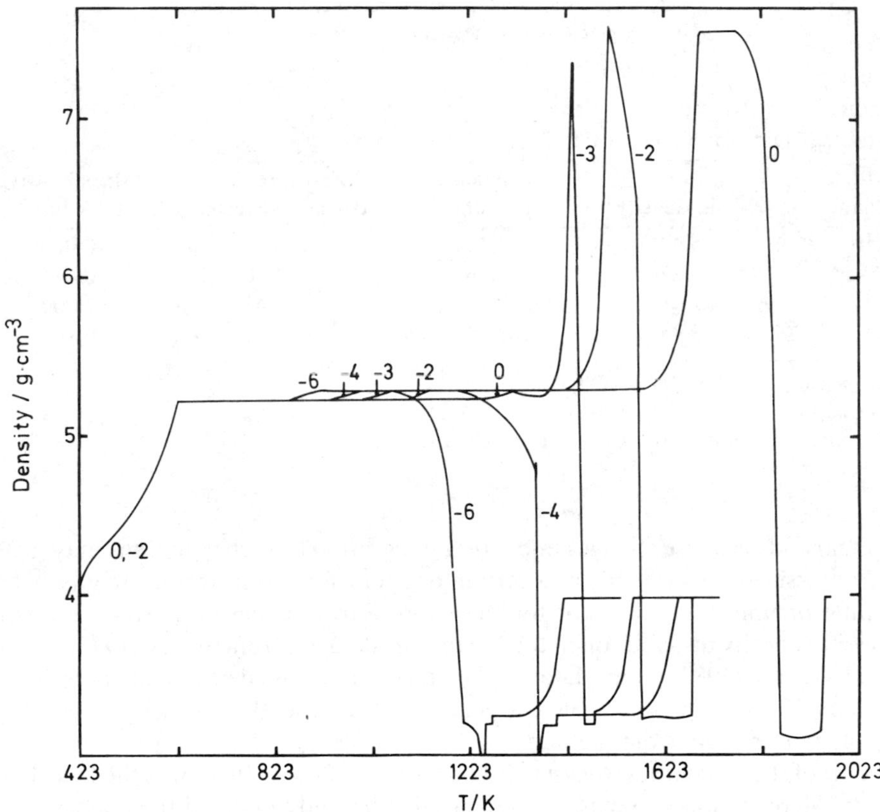

Fig. 28. Average density of the condensates as a function of pressure and temperature in the solar nebula. As the metal Fe–Ni condenses before most major silicates (olivine, plagioclase, and pyroxenes) until the pressure is lowered close to 10^{-4} bar, the average density of the condensing material remains high. This is shown by the density curves between 1 and 10^{-3} bar. At pressures 10^{-4} bar and below, the mixing of condensing plagioclase and clinopyroxene and at lower temperatures of olivine lowers the average density to 5.3 g · cm^{-3} or less. Between ~1000 and 621K, the average density lies between 5.2 and 5.3 g · cm^{-3} at all pressures from 1 to 10^{-6} bar. The effect of pressure continues to be small below 621K, where the density curve bends sharply toward values between 4.1 and 4.2 g · cm^{-3} at 423K. The temperature of 621K corresponds to the condensation of troilite (FeS).

fractionation of elements would take place in this pressure–temperature region of the nebula affecting the amount and the average density of the condensates. Figure 28 shows the variation in average density of the equilibrium assemblages as a function of pressure and temperature. The condensation starts with corundum. As temperature decreases, corundum is joined by low-density melilite (3.0 $g \cdot cm^{-3}$) and medium-density perovskite (4.0 $g \cdot cm^{-3}$) and spinel (3.6 $g \cdot cm^{-3}$). The curves rise abruptly as soon as the Fe–Ni alloy joins the condensates. With decreasing temperatures, the density may reach above 7 $g \cdot cm^{-3}$ at pressures between 1 and 10^{-3} bar. With further decrease in temperature, the average density drops to a value between 5.2 and 5.3 $g \cdot cm^{-3}$ at all pressures and remains so until the condensation temperature of troilite (FeS) is reached; at this point the density of the equilibrium assemblage decreases to between 4.1 and 4.2 $g \cdot cm^{-3}$ at 423K. As could also be deduced from Fig. 10, Fig. 28 shows that any effective separation of metal and silicates (other than melilite) could take place only at pressures of $\sim 10^{-3}$ bar or above.

Variation in the relative mass fraction of the equilibrium assemblages as a function of pressure and temperature is shown in Fig. 29. With decreasing temperature the mass starts accumulating, slowly at first but then faster, and

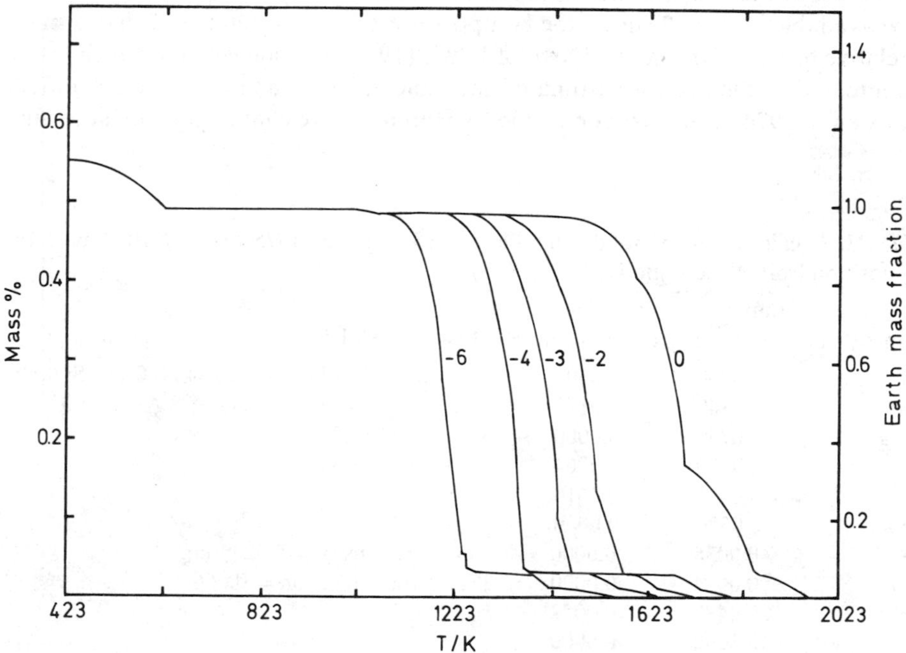

Fig. 29. Relative total mass of condensates (mass of condensed elements × 100/total mass of all elements in the system) as a function of pressure and temperature in the solar nebula. The axis on the right shows the scale in Earth mass units based on the assumption that Earth contains all the material condensed at 600K. See text for application.

within a temperature fall of about 200K at 1 bar nearly 90% of the elements Si, Al, Mg, and Ca are condensed. This range of codensation temperature becomes narrower with decreasing pressure and at 10^{-6} bar a major condensation takes place within 100K.

Formation of Mercury

If we can determine the pressure and temperature conditions at which the accumulating material is of a suitable mass and density for the formation of Mercury, the problem of removing excess material will be eliminated. Relative to Earth, Mercury contains 5.5% material of zero-pressure density of about 5.3 $g \cdot cm^{-3}$ (Table 12). Figures 28 and 29 show that a material amounting to about 6–7% of Earth mass with a suitable density would form in the temperature range of 1570K–1580K at a pressure of 10^{-2} bar. We note in Fig. 28 that density rises very rapidly in this temperature range. Actual computation shows that the closest we can approach the present size and density of Mercury is at 1575.7K and 10^{-2} bar (Table 13). A fraction of a degree Kelvin of temperature is used because the density of the condensates changes very strongly in this region.

The temperature 1575.7K and the pressure 10^{-2} bar are to be regarded as averages of the region of the nebula where suitable material for Mercury would be available. Table 13 shows the composition of Proto-Mercury. It has a mass relative to that of the model Earth of Lewis (1974) of 0.066 which is much lower compared to the mass we estimated in Table 7. Thus the homogeneous model of Lewis (1974) is at least applicable to Mercury if we change the median value

Table 13. Condensation of solids suitable for Mercury at 1575.7K and 10^{-2} bar. Initial composition from Anders and Ebihara (1982).

Element	Initial amount (mol)	Condensed amount (mol)	Phase	Mol %	Composition
Si	1.0000	0.0306	γ-FeNi	42.17	Fe 0.834 Ni 0.165 Si 0.001
Al	0.0849	0.0847	Mel	40.62	Geh 0.998 Åker 0.002
Mg	1.0750	0.0000	Cor	15.87	
Fe	0.9000	0.0264	Ti_2O_3	1.34	
Ca	0.0611	0.0610			
Na	0.0570	0.0000			
K	0.0038	0.0000	Average density = 5.32 $g \cdot cm^{-3}$		
Ti	0.0024	0.0020	Earth mass fraction = 0.066		
Ni	0.0493	0.0052			
C	12.1000	0.0000			
H	27,200.0	0.0000			
O	20.1000	0.2521			
S	0.5150	0.0000			
N	2.4800	0.0000			

of the temperature to 1575.7K, which is still within the error bar from temperatures along the gradient (Fig. 27).

Formation of Other Planets

If we adopt the mass–density relationship approach to the formation of Venus, Earth, and Mars, all of which have a zero-pressure density below 5 $g \cdot cm^{-3}$, we find ourselves working with homogeneous condensation at temperatures below 500K. There would be little difference between such a model and the early models of Urey (1952) and Ringwood (1966, 1979). There would be problems of (a) disposing of gaseous reduction products, (b) forming a core out of chemical equilibrium with the mantle (Grossman, 1972; Ringwood, 1979), and (c) a short time scale required for accretion (Levin, 1972). The necessity to form a core out of equilibrium with the mantle seems to be a stumbling block to any theory that proposes homogeneous accumulation. This problem was presented by Ringwood (1979) as follows.

Core–Mantle Disequilibrium

Because of their greater ability to dissolve in metal solutions than in silicates, such elements as Ni, Pt, and Co would be preferentially partitioned into the metal phase. Therefore, in coexisting Fe and olivine in equilibrium, most Ni would be in solution with Fe, making the distribution coefficient Ni in olivine/Ni in Fe very small. If the Fe–Ni alloy that is now in Earth's core was ever in chemical equilibrium with the mantle silicates, the latter should be practically devoid of Ni and, similarly, of other such siderophile elements as Mn, Pt, and Co.

Ringwood and Kesson (1977) estimated the abundances of the siderophile elements in the mantle from chemical data on peridotites and eclogites. Next they assumed that in a chondritic Earth what is not in the mantle must be in the core. Knowing the mantle/core ratio, it was easy to determine the abundances of the siderophile elements in the core. This approach lead to mantle/core ratios of Ni, Pt, and Co that are several hundred times the experimentally determined ratios, indicating that the mantle is enriched in siderophile elements and cannot have been in equilibrium with a metal.

It would appear that the evidence of core–mantle disequilibrium rules out homogeneous models and favors strongly a heterogeneous accumulation of material to form planets. We note, however, that all data on abundances of siderophile elements in the mantle are for the *upper mantle* only, and no information is available on the siderophile content of the *lower mantle*. Therefore it is possible to consider that whereas the lower mantle may be in chemical equilibrium with the core, the upper mantle may not. This distinction makes it possible to propose the following model for the other three terrestrial planets.

The Homogeneous Model for the Core and Lower Mantle

Many authors (e.g., Grossman and Larimer, 1974; Lewis, 1974) have considered pressures in the range of 10^{-3} to 10^{-4} bar as possible nebular pressures at which condensation has taken place. In this range of pressure, the closely spaced condensation curves for Fe–Ni and silicates (Figs. 10 and 28) show that a clean separation of metal and silicates is difficult. It is, therefore, an attractive proposition to consider that there has been no separation initially and a homogeneous mass consisting of Fe–Ni and silicates has formed the core and lower mantle part of the planets. In other words, the separation of Fe–Ni and the silicates happened not in the nebula but subsequent to their accretion in a primitive Venus, Earth, or Mars. This would cause a core–lower mantle equilibrium and there are no data to argue against this.

Could we find material to form homogeneous mixtures of a core and a primitive lower mantle suitable for Venus, Earth, and Mars, in a cooling nebula? We must review some geophysical data on Earth before proceeding to answer this question. Geophysically Earth's major divisions are as shown in Table 14. Earth contains a core with 32% of the total mass, which is surrounded by a lower mantle with 41% of the mass. The lower mantle is transitional into the upper mantle, with a substantial mass (17%) of mixed densities in the transition zone, followed by the upper mantle and crust with close to 10% of the mass. By mass the core and lower mantle amount to 73%. Figure 29 shows that this much material might form over a large pressure and temperature range. However, there is another constraint, namely, condensing Fe–Ni alloy in sufficient amount to form a major part of the core. In this condensing homogeneous mixture of silicates and Fe–Ni, there should be a certain amount of Fe–Ni that is less than 32%, the rest must be the light element needed to obtain a core with the right density, which is on the order of 6.3 $g \cdot cm^{-3}$ at zero pressure (Lyttleton, 1982). (The density of liquid Fe is 7.015 $g \cdot cm^{-3}$.) According to Ahrens (1985) 8–12% of S by weight would be required to dilute the Fe–Ni alloy to bring the density down. Sulfur was also favored by Murthy and Hall (1970), as it was previously by Ringwood (1966), who now favors oxygen (Ringwood, 1979). If we accept an average value of $\sim$10% S by weight, corresponding to nearly 8% FeS by weight (total Earth percentage), we must look for a pressure and temperature region in the nebula where an initial Fe–Ni core with a mass equal to 24% of that of Earth could condense, thus making a total core with a S content corresponding to 3% of the Earth mass (24% Fe–Ni + 8% FeS = 29% Fe–Ni + 3%S).

In this model, we are committed, at the very outset, to a hypothesis that Earth has formed by mixing of a high-temperature component with a low-temperature assemblage containing FeS. This method should also lead to formation of meteorites, which consist of varying proportions of two (or more) such components. Let us, therefore, use the compositional data both on meteorites and on various equilibrium assemblages at high temperatures simultaneously to find an internally consistent data set. This could be done by plotting Al/Si against Mg/Si (atomic abundance ratios) in meteorites and in the computed high-temperature equilibrium assemblages. Jagoutz *et al.* (1979) plotted the mass ratios of Mg/Si and Al/Si to derive the mantle composition. Figure 30 shows the equilibrium condensation path of the high-temperature mineral assemblages at a pressure of 10^{-2} bar. Other pressures between 10^{-2} and 10^{-3} bar and somewhat lower

Table 14. Geophysical data on model Earth (Dziewonski *et al.*, 1975, Ringwood, 1979).

Region	Radius (km)	Density ($g \cdot cm^{-3}$)[a]	Mass %
Inner core	0–1217.1	$13.01219 - 8.45292R^2$	2
Outer core	1217.1–3485.7	$12.58416 - 1.69929R - 1.94128R^2 - 7.11215R^3$	30
Lower mantle	3485.7–5701	$6.81430 - 1.66273R - 1.18531R^2$	41
Transition zone	5701–5951	$11.11978 - 7.87054R$	17
Upper mantle	5951–6352	$7.15855 - 3.85999R$	10
Crust	6352–6371	2.80–2.90	0.4

[a] R = radius/6371.

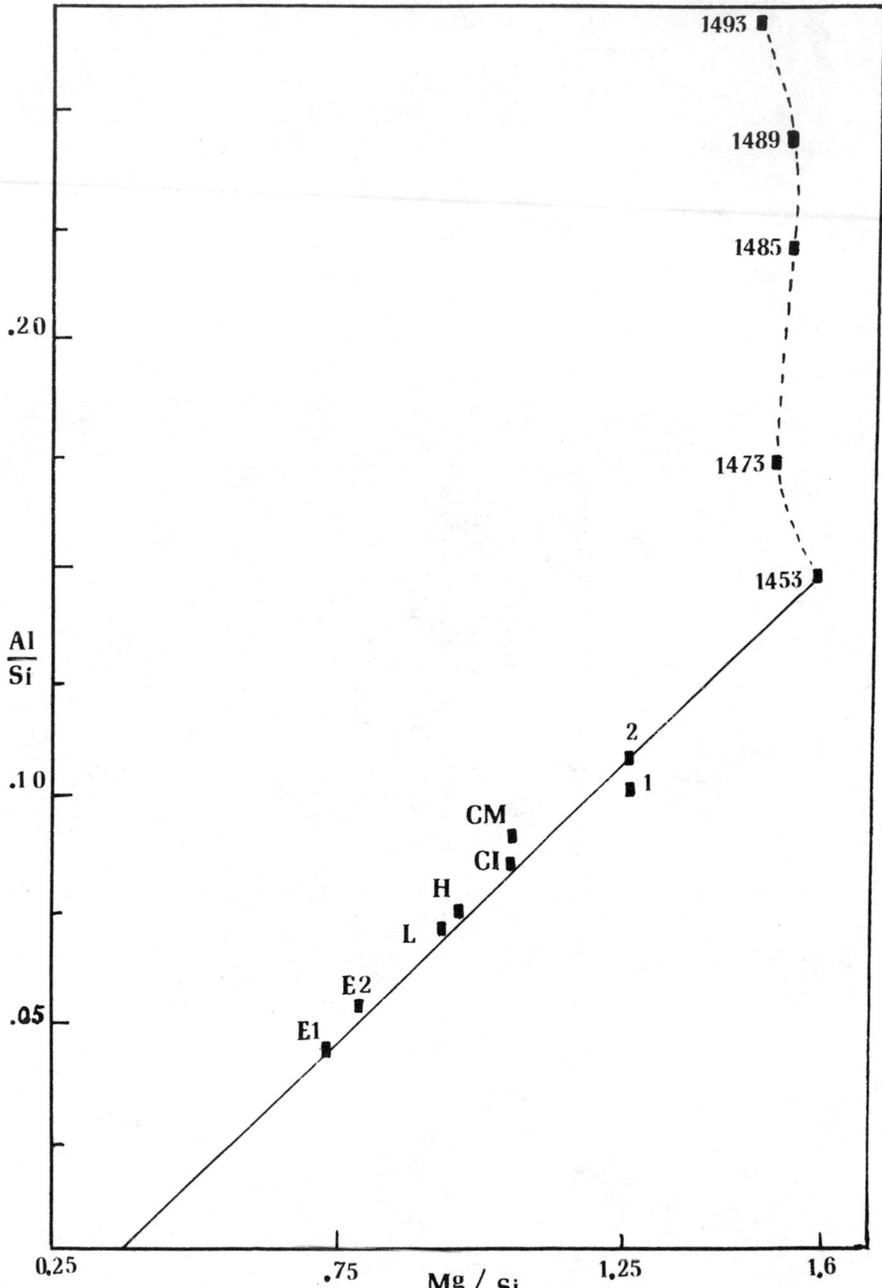

Fig. 30. Atomic ratio Al/Si against Mg/Si for various meteorites and calculated condensates. The compositions of the chondrites CI, CM, E, H, and L are from the compilation of Dodd (1981). For a gas of solar composition (Anders and Ebihara, 1982) the calculated condensate compositions appear on the curve in the plot. With this plot it is possible to consider mixtures of the CI chondrite and the condensates yielding various chondrites and Earth's mantle. See text for other details

temperatures may be chosen for the computation with similar results. The figure shows that a straight line drawn through the chondrites El and CI (data from Anders and Ebihara, 1982) passes close to the points H and L and intersects the equilibrium condensation path at 1453K. The data on the equilibrium assemblage at this temperature are shown in Table 15. This assemblage would form a major part of the core and lower mantle. Because it is linearly related to compositions of meteorites for Mg, Al, and Si, a single mixing process could account for the generation of all these planetary rocks.

The amount of Fe–Ni alloy condensed at 1453K is 0.87 mol (Fe 0.82, Ni 0.05) accounting for 24% of the Earth mass. On this scale, the rest of the 1453K assemblage is calculated to contain 38.9%. Together the high-temperature component accounts for 62.9% of the total Earth mass. This must be mixed eventually with a low-temperature component, which would form the upper mantle and the crust and also add about 8% FeS to the inner core. The total solar mass is very large in comparison to the mass of the terrestrial planets. Therefore a supply of material of CI composition must have been available throughout the thermal history of accretion of the planets and the meteoritic parental bodies.

Formation of the Earth

We have shown that the formation of Mercury could be a purely high-temperature process of condensation. Therefore it was attractive to consider that some parts of other planets might also have formed in the inner region of the nebula. It was suggested that the material that eventually formed the core and part of the mantle of Earth could have condensed homogeneously at a temperature of 1453K and a pressure of 10^{-2} bar. For reasons discussed earlier and the geochemical constraints on the composition of the upper mantle, it is clear that Earth as a whole could not have formed at such a high temperature. In fact, one of the strong arguments in favor of cold accretion theory was an elegant and simple calculation by Ringwood (1966) in distributing chondritic material into core and mantle and finding that the latter matched closely in composition with the peridotites. Therefore the upper mantle cannot be purely a high-temperature condensation product. It is evident then that these results lead us to a two-component model of Earth. We have been able to characterize the high-temperature component of Earth thermochemically, which in turn should constrain the composition of the low-temperature component. Before proceeding with the determination of the low-temperature component, it is important to review Ringwood's (1966) calculations on the distribution of matter in a chondritic Earth.

Ringwood's Chondritic Earth

Ringwood (1966) suggested that Earth was chondritic in composition. Table 16 shows a method of calculating the composition of the mantle from a carbonaceous

Table 15. Phase equilibrium at 1453K and 10^{-2} bar representing parts (Fe–Ni) of Earth's core and lower mantle. Initial composition as in Table 13.

Element	Condensed amount (mol)	Residual amount (mol)	Phase	Mol %	Composition
Si	0.5728	0.4272	γ-FeNi	47.23	Fe 0.942 Ni 0.055 Si 0.003
Al	0.0849	0.0000	Ol	48.11	Fo 1.000
Mg	0.9113	0.1637	Cpx	2.32	Cen 0.094 Di 0.894 Al_2O_3 0.012
Fe	0.8218	0.0782	Plag	2.27	An 1.000
Ca	0.0611	0.0000	Ti_2O_3	0.06	
Na	0.0000	0.0570			
K	0.0000	0.0038	Average density = 5.43 $g \cdot cm^{-3}$		
Ti	0.0024	0.0000	Earth mass fraction = 0.73		
Ni	0.0483	0.0010			
C	0.0000	12.1000			
H	0.0000	27,200.0			
O	2.2446	17.8554			
S	0.0000	0.5150			
N	0.0000	2.4800			

Table 16. Composition of Earth as derived by reduction from the composition of CI carbonaceous chondrites (after Ringwood, 1966, except for column 1).

	(1)[a]	(2)	(3)	(4)	(5)
SiO_2	33.1	33.3	35.9	29.8	43.2
MgO	23.8	23.5	25.2	26.3	38.1
FeO	35.6	35.5	6.1	6.4	9.3
Al_2O_3	2.4	2.4	2.6	2.7	3.9
CaO	1.9	2.3	2.5	2.6	3.7
Na_2O	1.2	1.1	1.2	1.2	1.8
NiO	2.0	1.9	—	—	—
Total	100.0	100.0	73.5	69.0	100.0
Fe			24.9	25.8	
Ni			1.6	1.7	
Si			—	3.5	
Total			26.5	31.0	

[a] Explanation: (1) Oxide weight percent of solar gas on a S-, C-, and H_2O-free basis. (2) Average composition of principal components of CI carbonaceous chondrites on a S-, C-, and H_2O-free basis. (3) Analysis from column 2 with FeO/(FeO + MgO) reduced to be consistent with probable value for Earth's mantle (= 0.12). (4) Analysis from column 3 with sufficient SiO_2 reduced to elemental Si[b] to yield a total silicate-to-metal ratio of 69/31, as in Earth. (5) Model mantle composition: silicate phase from column 4 recalculated to 100%.

[b] This is to satisfy the requirement that not all Earth's core be made up of Fe–Ni. Some of it, perhaps as much as 3–4% by weight, must be a light element, e.g., K, Si, S, or O_2, all of which have been considered as likely candidates in recent years.

Table 17. Weight percent oxide compositions of peridotite and model mantles (data from Ringwood, 1979).

Oxide	Primordial mantle from CI chondrite	Pyrolite	Peridotite (Hutchison *et al.*, 1970)
SiO_2	48.2	45.1	44.4
TiO_2	0.15	0.2	0.04
Al_2O_3	3.5	3.3	1.7
Cr_2O_3	0.7	0.4	0.5
MgO	34.0	38.1	42.3
FeO	8.1	8.0	8.0
MnO	0.5	0.15	0.1
CaO	3.3	3.1	1.6
Na_2O	1.6	0.4	0.1
K_2O	0.15	0.03	0.04

chondritic (CI) composition. In column 3 in the table, part of the FeO is reduced to reach the center and form the core along with Ni and a light element which in this case is assumed to have been Si. In column 4, part of SiO_2 is reduced and added to the core. What remains in column 5 happens to be closely similar in composition to mantle-derived rocks—peridotites. Table 17 shows the average composition of peridotites along with a new calculation of the primordial mantle from CI chondrite and Ringwood's model mantle. The method of calculation of the primordial mantle is the same as shown in Table 16, except Si is not included in the core. Curiously, this leaves us with an excess of Si in the model mantle as compared to pyrolite or peridotite. Other important depletions in the peridotites relative to primordial mantle are Mn, Cr, Na, and K.

Completion of Earth's Core and Mantle

As noted before, 62.9% of the Earth mass could be shown to accrete at high temperature. The Fe–Ni in the core amounted to 24%; the remaining 38.9% created the lower mantle. The average density of this 38.9% silicate mass would be rather low (Table 15) as compared to the estimated zero-pressure density of 4 $g \cdot cm^{-3}$ by Lyttleton (1982) and of 4.15 $g \cdot cm^{-3}$ by Davies and Dziewonski (1975). The remaining 37.1% of the Earth mass is derived from the accretion of CI chondrite. This might yield a suitable composition for the upper mantle and compositionally modify the lower mantle through which FeS melt might pass to contribute to the growing core.

As mentioned before Tozer (1978) argued in favor of a sudden cooling of the condensates that had grown to sizes of a few microns. The condensates, being good radiators of thermal energy, might cool to temperatures as low as a tenth of the surrounding gas. With the appearance of an increasing amount of condensates, therefore, the gas would start cooling at an accelerating rate to temperatures which, depending on the heliocentric distance, might be in the range of 200–800K. Tozer (1978) considered such a cooling scenario as an argument against heterogeneous condensation theory (Turekian and Clark, 1969). This cooling model, however, would be appropriate for a two-component model of planetary accretion. Earth acquired parts of the core and mantle at 1453K and the remaining was added at a low temperature.

Table 18 shows phase compositions in a CI chondritic material at 603K and 1 bar and at a H/O ratio of 50. Details of mineral compositions and modal percentage phase variations would depend on the temperature and H/O ratio. Because our model requires the presence of substantial amounts of FeS, we chose, out of a large amount of computed data, the temperature and H/O ratio that would most closely fulfill the requirement. Although the actual composition of the material arriving on the surface of the growing Earth would be heterogeneous, we use the phase assemblage as shown in Table 18 as an average for the added material.

The high-temperature core and mantle must be topped with material that should accomplish the following:

1. Form a crust and an upper mantle consistent with petrological and geophysical constraints
2. Suitably modify the composition of the mantle formed from the high-temperature component
3. Permit the attainment of the appropriate core size and core composition

Table 18. Phase equilibrium in CI chondrite at 603K and 1 bar.

Element	Initial amount (mol)	Condensed amount (mol)	Phase	Mol %	Composition
Si	1.0000	1.0000	γ-FeNi	3.85	Fe 0.430 Ni 0.570
Al	0.0850	0.0850	Ol	62.20	Fo 0.738 Fa 0.262
Mg	1.0600	1.0600	Cpx	6.51	Cen 0.002 Di 0.970 Cfs 0.003 Jd 0.024
Fe	0.9000	0.9000	Plag	2.93	An 0.071 Ab 0.929
Ca	0.0721	0.0721	Bi	0.49	Ann 0.164 Phl 0.836
Na	0.0600	0.0600	Mt	0.31	
K	0.0035	0.0035	Ilm	0.18	
Ti	0.0024	0.0024	FeS	23.53	
Ni	0.0470	0.0470			
C	12.1000	0.0000			
H	1005.00	0.0070			
O	20.1000	3.6601			
S	0.5150	0.5042			
N	2.4800	0.0000			

In analogy with the formation process of meteorites, we may consider the completion of Earth by mixing of the CI material with the high-temperature (1453K) condensate. The formation of the mantle might have preceded the meteoritic stage of formation of H and L chondrites and should have taken place while the 62.9% homogeneously deposited mass was differentiating into a core of 24% and a lower mantle of 38.9%. High-temperature condensates forming at some distance from Earth's nucleus, which did not quite catch up with the 62.9% material, arrived later mixed with the CI material as the temperature was falling.

Estimation of Earth's Chemical Composition

The composition of the mixture of CI and the high-temperature component would determine the composition of the mantle. A close estimation of such a composition requires data not only on the major elements as presented here but also on minor and trace elements. The principal constraint we now have is from our model. If Earth's initial core is to grow by an additional 8% FeS by weight, this much FeS is the minimum that must be included in the CI assemblage. Thus, in Proto-Earth, whose initial mass might have exceeded by several percent the mass of the present Earth to account for some loss of material later (e.g., by collision to form the moon), there would have to be some sulfur outside the core. If the low-temperature component was chondritic and formed below 603K its composition would be given by the phase assemblage as shown in Table 18. This material when added to the high-temperature (1453K) component would complete the formation of Earth. Table 19 shows several calculated compositions. Columns 1 and 2 show the compositions of the high- and low-temperature components, respectively. Added together they yield the total composition of Earth as shown in column 3. Carbon, nitrogen, and hydrogen do not appear here because they are not included in solid phases or are included only to a minor extent. The chemical composition of the core is obtained from the Fe–Ni phase formed at 1453K which amounts to 24% of the Earth mass and (Fe, Ni)S from the low-temperature component which amounts to 8%. Column 5 shows the residual of the high-temperature component after the removal of Fe–Ni to the core and column 6 shows the residual of the low-temperature component after the removal of (Fe, Ni)S. These columns may be considered to represent primitive lower and upper mantle, respectively. Note that these are merely stages in calculation and the primitive mantles may not have actually existed because of mixing.

The elemental composition of Earth in column 3 is recast into oxide weight percentage to compare it with those of other models. Table 20 shows such a comparison. Columns 2 and 3 show the model compositions from Morgan and Anders (1980) and Ringwood (BVSP, 1981), respectively. The model composition of this study is low in SiO_2 but high in every other oxide except TiO_2. The silica "deficiency" of the present model is resolved by considering the redistribution of material subsequent to accretion, as discussed below.

Table 19. Calculated chemical compositions of Earth's core and mantle (wt %). Solar gas composition is from Anders and Ebihara (1982).

Element	(1)[a]	(2)	(3)	(4)	(5)	(6)
Si	7.92	5.53	13.45	0.00	7.92	5.53
Al	1.13	0.45	1.58	0.00	1.13	0.45
Mg	10.91	5.07	15.98	0.00	10.91	5.07
Fe	22.60	9.90	32.50	27.40	0.00	5.10
Ca	1.21	0.57	1.78	0.00	1.21	0.57
Na	0.00	0.27	0.27	0.00	0.00	0.27
K	0.00	0.03	0.03	0.00	0.00	0.03
Ti	0.06	0.02	0.08	0.00	0.06	0.02
Ni	1.40	0.54	1.94	1.69	0.00	0.25
O	17.68	11.53	29.21	0.00	17.68	11.53
S	0.00	3.18	3.18	2.91	0.00	0.27
Total	62.91	37.09	100.00	32.00	38.91	29.09

[a]Explanation: (1) High-temperature component (1453K, 10^{-2} bar). (2) Low-temperature component (603K, 1 bar). Initial composition CI. (3) Total Earth chemical composition. (4) Earth's core obtained by adding 8% (by weight) of FeS to the high-temperature Fe–Ni phase (24 wt %). Ni in FeS in the low-temperature component has been included in the same proportion as it occurs in Fe–Ni alloy. (5) Primitive lower mantle. High-temperature component without the Fe–Ni alloy. (6) Primitive upper mantle. Low-temperature component without 8% of (Fe–Ni)S.

Table 20. Model chemical compositions of Earth (wt %).

	a	*b*	*c*
Mantle + crust			
SiO_2	42.06	47.9	45.9
TiO_2	0.19	0.20	0.3
Al_2O_3	4.36	3.90	3.9
MgO	38.73	34.10	38.0
FeO	9.60	8.9	8.1
CaO	3.63	3.2	3.2
Na_2O	0.53	0.25	0.1
K_2O	0.05	0.02	0.0129
NiO	0.46	—	—
S	0.39	—	—
Core			
Fe	85.6	84.5	86.2
Ni	5.3	5.6	4.8
S	9.1	9.0	1.0
O	—	—	8.0

[a]This work.
[b]Morgan and Anders (1980).
[c]Ringwood (in BVSP, 1981).

The Mixing and Differentiation Process

At 1453K the condensates arriving from the nebular feeding zone for Earth would form a nucleus and the growth would take place partly with mostly high-temperature component and partly with a mixture of the high- and low-temperature components as discussed in the previous section. Differentiation caused by gravitational heating would be continuous with Fe–Ni melt moving to the core to be later joined by FeS melt from the low-temperature component. Establishment of a thermal gradient in the primitive mantle would result in effective differentiation with 8% FeS loss to the core, some Fe–Ni addition to the lower mantle, and some Na, K, S, and other volatile loss to the crust.

What amount of mixing of high- and low-temperature components is necessary to obtain the present mantle composition? Table 21 shows the elemental compositions of model upper mantle rocks in columns 1 and 2. To obtain similar Si, Al, and Mg for the upper mantle in the present case nearly 40% (by volume) of the primitive lower mantle should mix with the primitive upper mantle material. Column 3 shows the calculated abundances of all elements in the upper mantle (after mixing) according to the above scheme, except for Fe–Ni, which is the same as in the ultramafics in column 1. The remaining Fe–Ni, as well as other elements of the high-temperature component, have been assigned to the lower mantle (column 4, Table 21). In the mixing process the primitive upper mantle gains Al, Mg, Ca, and Ti and loses Fe and Ni to the lower mantle. The composition shown in column 3 represents undepleted upper mantle and therefore

Table 21. Elemental compositions of model mantles.

Element	(1)[a]	(2)	(3)	(4)
Si	1.000	1.000	1.000	1.000
Al	0.104	0.108	0.108	0.148
Mg	1.253	1.254	1.253	1.591
Fe	0.145	0.145	0.145	0.275
Ca	0.083	0.084	0.0847	0.107
Na	0.014	0.016	0.0382	0.000
K	0.0019	0.0006	0.0022	0.000
Ti	0.0036	0.0036	0.0031	0.0042
Ni	0.0050	0.0047	0.0048	0.0161
O	—	—	3.754	3.919
S	—	4.6E-5	0.0272	0.000

[a]Explanation: (1) Ultramafics (Jagoutz *et al.*, 1979), similar to pyrolite (Ringwood, 1979). (2) Mantle (Wänke, 1981). (3) Upper mantle composition obtained by mixing the primitive upper mantle with 40% (by volume) of the primitive lower mantle Fe–Ni is assigned on the basis of the observed Fe^{2+}/Mg ratio in peridotites (column 1). (4) Lower mantle composition as the remainder of the high-temperature component with added Fe–Ni from the primitive upper mantle.

the volatile elements are present in a higher proportion than in the peridotites or in ultramafics. There is little doubt that the upper mantle could not remain chondritic in abundance of many elements because of the postaccretionary modifications from the melting and generation of magmas.

Many recent workers (Verhoogen, 1961; Braginsky, 1963; Loper, 1978; Gubbins *et al.*, 1979) have considered the core dynamo resulting from the crystallization of the inner core in a cooling system. Braginsky (1963) proposed that the differentiation of the components of the outer core mixture contributed importantly to fluid convection. Loper's (1978) quantitative model for the gravitationally powered dynamo shows that a field of 700 Gauss could be maintained if the inner core grew at a constant rate over the age of Earth. Although geophysicists disagree on the magnitude of the effect (Gubbins *et al.*, 1979), this model of the growth of the inner core is generally accepted. We note that the growth of the outer core by addition of FeS as proposed here would be compatible with the gravitational convection model of the core.

What happens when FeS melt arrives at the growing core–mantle interface is of particular interest. To understand the process of mixing FeS with liquid Fe–Ni, let us turn to the phase diagram of the Fe–FeS system as shown in Figure 31 adopted from Usselman (1975). The binary system is characterized by a eutectic at 998°C and 1 atm. For pure Fe, the liquidus starts at 1539°C and occurs at decreasing temperatures with increasing FeS in solution. A eutectic is reached at a composition of 75 wt % FeS and a temperature 998°C. The liguidus rises again, reaching 1230°C, the melting temperature of pure FeS. Usselman (1975) studied the system up to pressures of 100 kbar and found the eutectic system to continue, with the eutectic temperature rising and the eutectic composition becoming richer in Fe with increasing pressure. Usselman's (1975) extrapo-

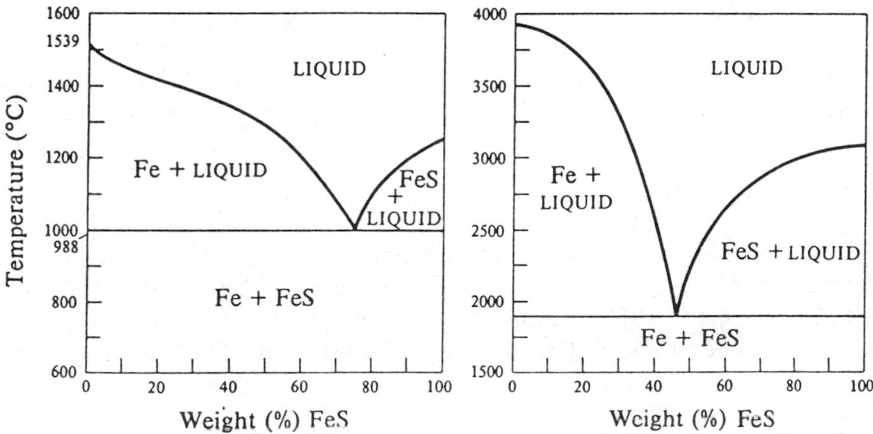

Fig. 31. The Fe–FeS eutectic (data from Usselman, 1975). As FeS dissolves in the melt the composition and temperature of the melt change and on reaching the liquidus, Fe(Ni) starts crystallizing.

lated results showed that what happens between 1 bar and 100 kbar in the binary Fe–FeS system would continue to happen at the core–mantle boundary. Similar conclusion may be drawn from the studies of Brett and Bell (1969) and Ryzhenko and Kennedy (1973), which show that the Fe–FeS eutectic composition is relatively intensitive to pressure. According to Usselman (1975), it remains at 15–17 wt % S at core pressures.

Because the FeS melt would mix with Fe(–Ni) melt, the new solution would find itself in the field of stability for pure Fe(–Ni) crystal and a Fe–FeS liquid. As a result, Fe–Ni crystals would form and settle toward the center, eventually forming a solid inner core. The Fe–Ni liquid moving upward will meet the FeS liquid, and the mixing process will develop a convective nature.

We note that this model for the formation of the Earth in two stages, first by generating the high-temperature component from the nebula and then adding the low-temperature material, which also contributes FeS to the lower mantle and the primitive core, solves two problems simultaneously. The first problem is that of the composition of the outer core, which must have liquid properties with Fe, Ni, and several weight percent of a light element. The second problem is being consistent with a mechanism for providing energy to the core for the dynamo effect, which is believed to be the explanation of geomagnetism. The present model favors solving the energy problem by the convection generated by the crystallization of Fe–Ni in the ternary liquid Fe–Ni–S. This is a variation of the gravitational convection theory of Gubbins (1974) and Loper (1978), who proposed crystallization of Fe–Ni from a melt of Fe–Ni and formation of a layer depleted in Ni outside the inner core, which rises up and produces convection. Such a convection would be energetically more feasible in the Fe–Ni–S system as proposed here. Because the crystallization of Fe–Ni at the top of the outer core results in a direct transportation of material in the Fe–Ni–S core, the process should be efficient in driving the dynamo. There would be a further addition to the energy of the core. As suggested by Verhoogen (1961, 1973), progressive growth of the solid inner core would result in the release of latent heat of crystallization. It is likely that the process of addition of FeS melt to the outer core and the growth of the solid inner core is still continuing.

Acknowledgments

We thank E. Olsen and R. F. Mueller for a critical review of the manuscript. The research of S.K.S. was supported by NSF grants EAR-8206171 and INT-8318029 and by PSC-BHE award 6-64164. The financial support to G.E. from the Swedish Natural Science Research Council (NFR) is gratefully acknowledged.

References

Ahrens, T. V. (1985) Pyrite: Shock compression, isentropic release and composition of the earth's care. *EOS* **66**, 371.

Alfvén, H. (1978) The band structure of the solar system, in *The Origin of the Solar System*, edited by S. F. Dermott, pp. 41–48, John Wiley, New York.

Alfvén, H., and Arrhenius, G. (1976) *Evolution of the Solar System*, NASA SP-345, U.S. Govt. Printing Office, Washington, D.C., 600 pp.

Anders, E. (1964) Origin, age and composition of meteorites, *Space Sci. Rev.* **3**, 583–714.

Anders, E. (1971) Meteorites and the early solar system, *Ann. Rev. Astron. Astrophys.* **9**, 1–34.

Anders, E. (1977) Chemical compositions of the Moon, Earth, and eucrite parent body, *Phil. Trans. Roy. Soc. London*, Ser. A **285**, 23–40.

Anders, E., and Ebihara, M. (1982) Solar-system abundances of the elements, *Geochim. Cosmochim. Acta* **46**, 2363–2380

Arrhenius, G. (1978) Chemical aspects of the formation of the solar system, in *The Origin of the Solar System*, edited by S. F. Dermott, pp. 521–581, John Wiley, New York.

Ash, M. E., Shapiro, I. I., and Smith, W. B. (1971) The system of planetary masses, *Science* **174**, 551–556.

BVSP (1981) *Basaltic Volcanism Study Project 1976–1979*, Pergamon Press, New York, 1286 pp.

Barin, I., and Knacke, O. (1973) *Thermochemical Properties of Inorganic Substances*, Springer-Verlag, Berlin, 921 pp.

Barin, I., Knacke, O., and Kubaschewski, O. (1977) *Thermochemical Properties of Inorganic Substances, Supplement*, Springer-Verlag, Berlin, 861 pp.

Barshay, S. S., and Lewis, J. S. (1976) Chemistry of primitive solar material, *Ann. Rev. Astron. Astrophys.* **14**, 81–94

Black, D. C. (1978) Isotopic anomalies in solar system material—what can they tell us? in *The Origin of the Solar System*, edited by S. F. Dermott, pp. 583–598, John Wiley, New York.

Blander, M. (1971) The constrained equilibrium theory: Sulphide phases in meteorites, *Geochim. Cosmochim. Acta* **35**, 61–76.

Blander, M., and Abdel-Gawad, M. (1969) The origin of meteorites and the constrained equilibrium condensation theory, *Geochim. Cosmochim. Acta* **33**, 701–716.

Blander, M., and Katz, J. L. (1967) Condensation of primordial dust, *Geochim. Cosmochim. Acta* **31**, 1025–1034.

Blencoe, J. G., Merkel, G. A., and Seil, M. K. (1982) Thermodynamics of crystal-fluid equilibria, with applications to the system $NaAlSi_3O_8$–$CaAl_2Si_2O_8$–SiO_2–NaCl–$CaCl_2$–H_2O, *Adv. Phys. Geochem.* **2**, 191–222.

Bohlen, S. R., and Boettcher, A. L. (1982) Experimental investigations and geological applications of orthopyroxene geobarometry, *Amer. Mineral.* **66**, 951–964.

Boynton, W. V. (1975) Fractionation in the solar nebula: condensation of yttrium and the rare earth elements, *Geochim. Cosmochim. Acta* **39**, 569–584.

Braginsky, S. I. (1963) Structure of the F layer and reasons for convection in the Earth's core, *Dokl. Akad. Nauk. SSSR* **149**, 1311–1314.

Brett, R., and Bell, P. M. (1969) Melting relations in the Fe-rich portion of the system Fe–FeS at 30 kb pressure, *Earth Planet. Sci. Lett.* **6**, 479–482.

Cameron, A. G. W. (1971) The early evolution of the solar system, in *Symposium on the Evolutionary and Physical Properties of Meteoroids*, Int. Astron. Union. Colloq., No. 13, Albany, New York.

Cameron, A. G. W. (1973) Abundances of the elements in the solar system, *Space Sci. Rev.* **15**, 121–146.

Cameron, A. G. W. (1978) The primitive solar accretion disk and the formation of the planets, in *The Origin of the Solar System*, edited by S. F. Dermott, pp. 49–74, John Wiley, New York.

Cameron, A. G. W. (1982) Elementary and nuclidic abundances in the solar system, in *Essays in Nuclear Astrophysics*, edited by C. A. Barnes, D. N. Schramm, and D. D. Clayton, 459 pp., Cambridge University Press, Oxford.

Cameron, A. G. W., and Pine, M. R. (1973) Numerical models of the primitive solar nebula, *Icarus* **18**, 377–406.

Charlu, T. V., Newton, R. C. and Kleppa, O. J. (1981)Thermochemistry of synthetic $Ca_2Al_2SiO_1$ (gehlenite) $Ca_2MgSi_2O_7$ (åkermanite) melilites, *Geochim. Cosmochim. Acta* **45**, 1609–1917.

Clark, S. P., Jr., Turekian, K. K., and Grossman, L. (1972) Model for the early history of the Earth, in *Nature of the Solid Earth*, edited by E. C. Robertson, pp. 3–18, McGraw-Hill, New York.

Clayton, R. N., Grossman, L., and Mayeda, T. K. (1973) A component of primitive nuclear composition in carbonaceous meteorites. *Science* **182**, 485–488.

Cressey, G., Schmid, R., and Wood, B. J. (1978) Thermodynamic properties of almandine-grossular garnet solid solutions, *Contrib. Mineral. Petrol.* **67**, 397–404.

Danckwerth, P. A., and Newton, R. C. (1978) Experimental determination of the spinel peridotite to garnet peridotite reaction in the system $MgO–Al_2O_3–SiO_2$ in the range 900°–1,100°C and Al_2O_3 isopleths of enstatite in the spinel field, *Contrib. Mineral. Pertrol.* **66**, 189–201.

Davies, G. F., and Dziewonski, A. M. (1975) Homogeneity and constitution of the earth's lower mantle and core, *Phys. Earth Planet. Int.* **10**, 336–343.

Dermott, S. F. (Ed.) (1978) *NATO Advanced Study Institute: The Origin of the Solar System*, 668 pp., John Wiley, New York.

Dodd, R. T., Jr. (1981) *Meteorites: A Petrologic-Chemical Synthesis*, 368 pp., Cambridge University Press, Cambridge.

Drozd, R. J., and Podosek, F. A. (1976) Primordial ^{129}Xe in meteorites, *Earth Planet. Sci. Lett.* **31**, 15–30.

Dziewonski, A., Hales, A., and Lapwood, E. (1975) Parametrically simple earth models consistent with geophysical data, *Phys. Earth Planet. Int.* **10**, 12–48.

Eriksson, G. (1975) Thermodynamic studies of high temperature equilibria. XII. SOLGASMIX, a computer program for calculation of equilibrium compositions in multiphase systems, *Chem. Scr.* **8**, 100–103.

Eriksson, G., and Rosén, E. (1973) Thermodynamic studies of high temperature equilibria. VIII. General equations for the calculation of equilibria in multiphase systems, *Chem. Scr.* **4**, 193–194.

Eucken, A. (1944) Physikalische-Chemische Betrachtungen über der früheste Entwicklungsgeschichte der Erde, *Nach. Akad. Wiss. Göttingen, Math-Phys.* Kl. **1**, 1–25.

Eugster, H. P., Albee, A. L., Bence, A. E., Thompson, J. B. Jr., and Waldbaum, D. R. (1972) The two-phase region and excess mixing properties of paragonite-muscovite crystalline solutions, *J. Petrol.* **13**, 147–179.

Fegley, B., Jr., and Lewis, J. S. (1980) Volatile element chemistry in the solar nebula: Na, K, F, Cl, Br and P, *Icarus*, **41**, 439–455.

Ferry, J. M., and Spear, F. S. (1978) Experimental calibration of the partitioning of Fe and Mg between biotite and garnet, *Contrib. Mineral. Petrol.* **66**, 113–117.

Fujii, T. (1977) Fe–Mg partitioning between olivine and spinel. *Carnegie Inst. Wash. Yearb.* **76**, 563–569.

Fujii, T., and Scarfe, C. M. (1982) Equilibration experiments on natural peridotite and basalt: recalibration of the olivine-spinel geothermometer. *EOS* **63**, 471.

Ganguly, J. (1979) Garnet and clinopyroxene solid solutions, and geothermometry based on Fe–Mg distribution coefficient, *Geochim. Cosmochim. Acta* **43**, 1021–1029.

Ganguly, J., and Saxena, S. K. (1984) Mixing properties of alumino silicate garnets: constraints from natural and experimental data, and applications to geothermobarometry, *Amer. Mineral.* **69**, 88–97.

Gasparik, T. (1981) Mixing properties of the dioside-jadeite solid solution, *Geol. Soc. Amer. Abstr.* **13**, 456–457.

Goettel, K. A., and Barshay, S. S. (1978) The chemical equilibrium model for condensation in the solar nebula: assumptions, implications, and limitations, in *The Origin of the Sol System*, edited by S. F. Dermott, pp. 611–627, John Wiley, New York.

Grossman, L. (1972) Condensation in the primitive solar nebula, *Geochim. Cosmochim. Acta* **36**, 597–619.

Grossman, L., and Clark, S. P., Jr. (1973) High-temperature condensates in chondrites and the environment in which they formed, *Geochim. Cosmochim. Acta* **37**, 635–649.

Grossman, L., and Larimer, J. W. (1974) Early chemical history of the solar system, *Rev. Geophys. Space Phys.* **12**, 71–101.

Grossman, L., and Olsen, E. (1974) Origin of the high-temperature fraction of C2 chondrites, *Geochim. Cosmochim. Acta* **38**, 173–187.

Gubbins, D. (1974) Theories of the geomagnetic and solar dynamos, *Rev. Geophys. Space Phys.* **12**, 129–137.

Gubbins, D., Masters, T. G., and Jacobs, J. A. (1979) Thermal evolution of the Earth's core, *Geophys. J. Roy. Astron. Soc.* **59**, 57–99.

Guggenheim, E. A. (1967) *Thermodynamics, an Advanced Treatment for Chemists and Physicists*, 390 pp., North-Holland, Amsterdam.

Haselton, H. T., and Newton, R. C. (1980) Thermodynamics of pyrope-grossular garnets and their stabilities at high temperatures and high pressures, *J. Geophys. Res.* **85**, 6973–6982.

Helgeson, H. C., Delany, J. M., Nesbitt, H. W., and Bird, D. K. (1978) Summary and critique of the thermodynamic properties of rock-forming minerals, *Amer. J. Sci.* **278-A**, 1–229.

Herzberg, C. T. (1978) Pyroxene geothermometry and geobarometry: experimental and thermodynamic evaluation of some subsolidus phase relations involving pyroxenes in the system $CaO–MgO–Al_2O_3–SiO_2$, *Geochim. Cosmochim. Acta* **42**, 945–957.

Hoyle, F. (1960) On the origin of the solar nebula, *Q. J. Roy. Astron. Soc.* **1**, 28–55.

Hoyle, F., and Wickramasinghe, N. C. (1968) Condensation of the planets, *Nature (London)* **217**, 415–418.

Hutchison, R., Paul, D. K., and Harris, P. G. (1970) Chemical composition of the upper mantle, *Mineral. Mag.* **37**, 726–729.

Jagoutz, E., Palme, H., Baddenhausen, H., Blum, K., Cendales, M., Dreibus, G., Spettel, B. Lorenz, V., and Wänke, H. (1979) The abundances of major, minor and trace elements in the earth's mantle as derived from primitive ultramafic nodules, *Geochim. Cosmochim. Acta, Suppl.* **11**, 2031–2050.

JANAF Thermochemical Tables with Supplements (1971, 1974, 1975, 1978, 1982) National Bureau of Standards of the U.S. Department of commerce, Washington, D. C.

Jenkins, D. M., and Newton, R. C. (1979) Experimental determination of the spinel peridotite to garnet peridotite inversion at 900°C and 1,000°C in the system $CaO–MgO–Al_2O_3–SiO_2$ and at 900°C with natural garnet and olivine, *Contrib. Mineral. Petrol.* **68**, 407–419.

Kerrick, D. M., and Darken, L. S. (1975) Statistical thermodynamic models for ideal oxide and silicate solid solutions, with application to plagioclase, *Geochim. Cosmochim. Acta* **39**, 1431–1442.

Kubaschewski, O. (1982) *Iron—Binary Phase Diagrams*, 185 pp., Springer-Verlag, Berlin.

Kubaschewski, O., and Alcock, C. B. (1979) *Metallurgical Thermochemistry*, 449 pp., Pergamon Press, Oxford.

Kushiro, I., and Yoder, H. S., Jr. (1966) Anorthite-forsterite and anorthite-enstatite reactions and their bearing on the basalt-eclogite transformation, *J. Petrol.* **7**, 337–362.

Lane, D. L., and Ganguly, J. (1980) A1203 solubility in orthopyroxene in the system $MgO–Al_2O_3–SiO_2$: a reevaluation, and mantle geotherm, *J. Geophys. Res.* **85**, 6963–6972.

Larimer, J. W. (1967) Chemical fractionations in meteorites—I. Condensation of the elements, *Geochim. Cosmochim. Acta* **31**, 1215–1238.

Larimer, J. W. (1973) Chemical fractionations in meteorites—VII. Cosmothermometry and cosmobarometry, *Geochim. Cosmochim. Acta* **37**, 1603–1623.

Larimer, J. W., and Anders, E. (1967) Chemical fractionations in meteorites—II. Abundance patterns and their interpretation, *Geochim. Cosmochim. Acta* **31**, 1239–1270.

Larimer, J. W., and Anders, E. (1970) Chemical fractionations in meteorites—III. Major element fractionations in chondrites, *Geochim.* Cosmocchim. Acta **34**, 367–387.

Larimer, J. W., and Bartholomay, M. (1979) The role of carbon and oxygen in cosmic gases: some applications to the chemistry and mineralogy of enstatite chondrites, *Geochim. Cosmochim. Acta* **43**, 1455–1466.

Levin, B. J. (1972) Origin of the Earth, *Tectonophysics* **13**, 7–29.

Lewis, J. S. (1972a) Metal/silicate fractionation in the solar system, *Earth Planet. Sci. Lett.* **15**, 286–290.

Lewis, J. S. (1972b) Low-temperature condensation from the solar nebula, *Icarus* **16**, 241–252.

Lewis, J. S. (1974) The temperature gradient in the solar nebula, *Science* **186**, 440–443.

Lewis, J. S., and Prinn, R. G. (1983) *Planets and Their Atmospheres: Origin and Evolution*, 470 pp., Academic Press, New York.

Lewis, J. S., Barshay, S. S. and Noyes, B. (1979) Primordial retention of carbon by the terrestrial planets, *Icarus* **37**, 190–206.

Lindsley, D. H. (1980) Phase equilibria of pyroxenes at pressures >1 atmosphere, *Rev. Mineral.* **7**, 289–307.

Lindsley, D. H. (1983) Pyroxene thermometry. *Amer. Mineral.* **68**, 477–493.

Lindsley, D. H., Grover, J. E., and Davidson, P. M. (1981). The thermodynamics of the $Mg_2Si_2O_6–CaMgSi_2O_6$ join: a review and an improved model, *Adv. Phys. Geochem.* **1**, 149–175.

Loper, D. E. (1978) Some thermal consequences of a gravitationally powered dynamo, *J. Geophys. Res.* **83**, 5969–5970.

Lord, H. C., III (1965) Molecular equilibria and condensation in a solar nebula and cool stellar atmospheres, *Icarus* **4**, 279–288.

Lyttleton, R. A. (1982) *The Earth and Its Mountains*, 206 pp., John Wiley, New York.

Mah, A. D., and Pankratz, L. B. (1976) *Contributions to the data on theoretical metallurgy. XVI. Thermodynamic properties of nickel and its inorganic compounds*, Bull 668, 125 pp., U.S., Bur. Mines, Washington, D.C.

Morgan, J. W., and Anders, E. (1980) Chemical composition of Earth, venus and Mercury, *Proc. Natl. Acad. Sci. U.S.A.* **77**, 6973–6977.

Mueller, R. F. (1963) A comparison of oxidative equilibria in meteorites and terrestrial rocks, *Geochim. Cosmochim. Acta* **27**, 273–278.

Mueller, R. F. (1972) Stability of biotite: a discussion, *Amer. Mineral.* **57**, 300–316.

Mueller, R. F., and Saxena, S. K. (1977) *Chemical Petrology with Applications to the Terrestrial Planets and Meteorites*, 394 pp., Springer-Verlag, New York.

Murthy, V. R., and Hall, H. T. (1970) The chemical composition of the earth's core: possibility of sulphur in the core, *Phys. Earth Planet. Int.* **2**, 276–282.

Newton, R. C., Charlu, T. V., and Kleppa, O. J. (1980) Thermochemistry of the high structural state plagioclases, *Geochim. Cosmochim. Acta* **44**, 933–941.

Orville, P. M. (1972) Plagioclase cation exchange equilibria with aqueous chloride solution: results at 700°C and 2000 bars in the presence of quartz, *Amer. J. Sci.* **272**, 234–272.

Perchuk, L. L., and Lavrent'eva, I. V. (1983) Experimental investigation of exchange equilibria in the system cordierite-garnet-biotite, *Adv. Phys. Geochem.* **3**, 199–239.

Perchuk, L. L., Podlesskii, K. K., and Aranovich, L. Ya. (1981) Calculation of thermodynamic properties of end-member minerals from natural parageneses, *Adv. Phys. Geochem.* **1**, 111–129.

Perkins, D., III, and Newton, R. C. (1980) The compositions of coexisting pyroxenes and garnet in the system $CaO–MgO–Al_2O_3–SiO_2$ at 900°–1,100°C and high pressures, *Contrib. Mineral. Petrol.* **75**, 291–300.

Perkins, D., III, Holland, T. J. B., and Newton, R. C. (1981). The Al_2O_3 contents of enstatite in equilibrium with garnet in the system $MgO–Al_2O_3–SiO_2$ at 15–40 kbar and 900°–1,600°C, *Contrib. Mineral. Petrol.* **78**, 99–109.

Reeves, H. (1978) The origin of the solar system, in *The Origin of the Solar System*, edited by S. F. Dermott, pp. 1–18, John Wiley, New York.

Ringwood, A. E. (1966) Chemical evolution of the terrestrial planets, *Geochim. Cosmochim. Acta* **30**, 41–104.

Ringwood, A. E. (1977) Composition of the core and implications for the origin of the earth, *Geochem. J.* **11**, 111–135.

Ringwood, A. E. (1979) *Origin of the Earth and Moon*, 295 pp., Springer-Verlag, New York.

Ringwood, A. E., and Kesson, S. E. (1977) Basaltic magmatism and the bulk composition of the Moon, II. Siderophile and volatile elements in Moon, Earth and chondrites: Implications for lunar origin, *The Moon* **16**, 425–464.

Robie, R. A., Hemingway, B. S., and Fisher J. R. (1978) *Thermodynamic Properties of Minerals and Related Substances at 298.15 K and 1 Bar (105 Pascals) Pressure and at Higher Temperatures*, U.S. Geol. Surv. Bull. 1452, 456 pp.

Ryzhenko, B., and Kennedy, G. C. (1973) The effect of pressure on the eutectic in the system Fe–FeS, *Amer. J. Sci.* **273**, 803–810.

Sack, R. O. (1980) Some constraints on the thermodynamic mixing properties of Fe–Mg orthopyroxenes and olivines, *Contrib. Mineral. Petrol.* **71**, 257–269.

Saxena, S. K. (1968) Chemical study of phase equilibria in charnockites, Varberg, Sweden, *Amer. Mineral.* **53**, 1674–1695.

Saxena, S. K. (1971) $Mg^{2+}–Fe^{2+}$ order–disorder in orthopyroxene and the $Mg^{2+}–Fe^{2+}$ distribution between coexisting minerals, *Lithos* **4**, 345–354.

Saxena, S. K. (1973) *Thermodynamics of Rock-Forming Crystalline Solutions*, 188 pp., Springer-Verlag, New York.

Saxena, S. K. (1981a) The $MgO–Al_2O_3–SiO_2$ system: free energy of pyrope and Al_2O_3–enstatite, *Geochim. Cosmochim. Acta* **45**, 821–825.

Saxena, S. K. (1981b) Fictive component model of pyroxenes and multicomponent phase equilibria,, *Contrib. Mineral. Petrol.* **78**, 345–351.

Saxena, S. K., and Chatterjee, N. (1986) Thermochemical data on mineral phases. I. The system CaO–MgO–Al_2O_3–SiO_2, *J. Petrol.* (In press)

Saxena, S. K., and Ghose, S. (1971) Mg^{2+}–Fe^{2+} order–disorder and the thermodynamics of the orthopyroxene crystalline solution, *Amer. Mineral.* **56**, 532–559.

Saxena, S. K., and Ribbe, P. H. (1972) Activity-composition relations in feldspars, *Contrib. Mineral. Petrol.* **37**, 131–138.

Saxena, S. K., and Eriksson, G. (1983a) High temperature phase equilibria in a solar-composition gas, *Geochim. Cosmochim. Acta* **47**, 1865–1874.

Saxena, S. K., and Eriksson, G. (1983b) Low to medium-temperature phase equilibria in a gas of solar composition, *Earth and Planet Sci. Lett.* **65**, 7–17.

Sears, D. W. (1978) *The Nature and Origin of Meteorites*, 187 pp., Oxford University Press, New York.

Smith, W. R., and Missen, R. W. (1982) *Chemical Reaction Equilibrium Analysis*, 364 pp., Wiley–Interscience, New York.

Spencer, K. J., and Lindsley, D. H. (1981) A solution model for coexisting iron–titanium oxides, *Amer. Mineral.* **66**, 1189–1201.

Stacey, F. D. (1972) Physical properties of the Earth's core, *Geophys. Surv.* **1**, 99.

Stacey, F. D. (1977) *Physics of the Earth*, 414 pp., John Wiley, New York.

Tozer, D. C. (1978) Terrestrial planet evolution and the observation consequences of their formation, in *The Origin of Solar System*, edited by S. F. Dermott, pp. 433–462, John Wiley, New York.

Turekian, K. K., and Clark, S. P. Jr. (1969) Inhomogeneous accretion of the Earth from the primitive solar nebula, *Earth Planet. Sci. Lett.* **6**, 346–348.

Turekian, K. K., and Clark, S. P., Jr. (1975) The non-homogeneous accumulation model for terrestrial planet formation and the consequences for the atmosphere of Venus, *J. Atmos. Sci.* **32**, 1257–1261.

Turnock, A. C., and Lindsley, D. H. (1981) Experimental determination of pyroxene solvi for $\leqslant 1$ kbar at 900 and 1000°C, *Can. Mineral.* **19**, 255–267.

Urey, H. C. (1952) *The Planets: Their Origin and Development*, 245 pp., Yale University Press, New Haven.

Usselman, T. M. (1975) Experimental approach to the state of the core: Part II. Composition and thermal regime, *Amer. J. Sci.* **275**, 291–303.

Verhoogen, J. (1961) Heat balance of the Earth's core. *Geophys. J.* **4**, 276–281.

Verhoogen, J. (1973) Thermal regime of the Earth's core. *Phys. Earth Planet. Int.* **7**, 47–58.

Wänke, H. (1981) Constitution of terrestrial planets, *Phil. Trans. Roy. Soc. London, Ser. A* **303**, 287–302.

Wänke, H., Baddenhausen, H., Palme, H., and Spettel, B. (1974) On the chemistry of the Allende inclusions and their origin as high temperature condensates, *Earth Planet. Sci. Lett.* **23**, 1–7.

Weidenschilling, S. J. (1977) The distribution of mass in the planetary system and solar nebula, *Astrophys. Space Sci.* **51**, 153–158.

Wohl, K. (1953) Thermodynamic evaluation of binary and ternary liquid systems, *Chem. Eng. Prog.* **49**, 218–219.

Wood, B. J., and Kleppa, O. J. (1981) Thermochemistry of forsterite-fayalite olivine solutions, *Geochim. Cosmochim. Acta* **45**, 529–534.

Wood, J. A. (1962) Chondrules and the origin of the terrestrial planets. *Nature (London)* **194**, 127–130.

Woolfson, M. M. (1978) Star formation and interactions between stars, in *The Origin of the Solar System*, edited by S. F. Dermott, pp. 163–178, John Wiley, New York.

Yamada, H., and Takahashi, E. (1982) Subsolidus phase relations between coexisting garnet and pyroxenes at 50 to 100 kbar in the system $CaO–MgO–Al_2O_3–SiO_2$, *Terra Cognita* **2**, 260–261.

Zen, E-An (1972) Gibbs free energy, enthalpy, and entropy of ten rock-forming minerals: calculations, discrepancies, implications, *Amer. Mineral.* **57**, 524–553.

Chapter 3
Thermodynamic Properties and Conditions of Formation of Minerals in Enstatite Meteorite

M. I. Petaev and I. L. Khodakovsky

Introduction

Knowledge of the chemical and mineralogical composition of terrestrial matter and its evolution from the earliest stages of Solar system history to the present is one of the foundations of the modern Earth Sciences. One effective method of studying the evolutionary process of terrestrial matter at the planetary stage of Solar system history is a comparative planetological analysis of data on the structure and composition of the outer shells of the terrestrial planets as obtained by spacecraft missions. However, the early history of the terrestrial planets has been considerably obliterated by the processes of matter differentiation, and only traces of these processes can be seen on the surface of the Earth, Moon, and Venus. Therefore, information on the earliest preplanetary stages of Solar system evolution can be obtained only by the study of its minor bodies: the comets, asteroids, and meteorites, the matter of which may not have experienced extensive planetary differentiation. We witness only the beginning of a serious cosmochemical investigation of comets and asteroids, and reliable data on their chemical and mineralogical compositions is likely to be obtained only in the not too distant future. Hence, reliable information on the composition of the minor bodies of the Solar system may be obtained now from studies of meteorites, numerous samples of which are available in terrestrial meteorite collections. Thus, detailed studies of meteorites of different types and estimation of their formation conditions form the basis for the reconstruction of the physicochemical conditions in the Solar system at the earliest stages of its evolution.

The physicochemical conditions of meteoritic matter formation may be reconstructed using either laboratory experimental modeling or numerical modeling based on the physicochemical laws of the evolution processes of meteoritic matter. The first method, if the experimental procedure is correct, may provide precise values of process parameters. However, this method runs into certain difficulties, such as the long duration of these processes, the cosmic scale of the system, etc. As a result, the use of this method is rather limited and in some cases neglected. The second method is free from these deficiences; nevertheless it also has some constraints, such as (1) the possibility of considering the major factors of the processes only; (2) The possibility of considering the process in terms of equilibrium thermodynamics only; (3) limitations in the accuracy of parameter estimations because of uncertainties in the thermodynamic data used; (4) finally, the availability of thermodynamic data for all of the concerned meteoritic minerals, as well as for other minerals that can be formed in the system under different mineral-forming conditions. Despite these limitations thermodynamical modeling of formation conditions of meteoritic matter appears to be the most practical, if not the only one possible.

Meteorites show considerable variations in mineralogical composition among their respective groups, but the differences in their chemical composition are small. This suggests that meteorites of several groups may have formed from a single parent material, but the redox and P–T conditions of their formation may have been different. The extreme manifestations of differences in formation conditions can be seen in the case of the enstatite and carbonaceous chondrites. The former contain refractory sulfides and nitrides, which are formed under reducing conditions. The latter consist of hydrosilicates, sulfates, and carbonates, which are formed under oxidizing conditions.

The present chapter deals with enstatite meteorite formation under the conditions of the high-temperature reducing zone of the solar nebula. We consider the following parts of the problem below:

a. The peculiarities of the mineralogical composition of enstatite meteorites and the processes of the evolution of this material.
b. The thermodynamic properties of the enstatite meteorite minerals.
c. Some physicochemical aspects of the nebular condensation of enstatite meteorite minerals.
d. Some physicochemical aspects of the thermal metamorphism of enstatite meteorite matter in their parent bodies.

The Peculiarities of the Mineralogical Composition of the Enstatite Meteorites and the Evolution of this Material

Enstatite meteorites comprise a small group of meteorites (about 40) with unusual chemical and mineralogical compositions. The main features of the chemical and mineralogical composition of enstatite meteorites in comparision with meteorites of other chemical groups have been reviewed by Dodd (1981). Compared to most other stoney meteorites the enstatite meteorites are richer in Fe and S, and sulfur enters into the composition of sulfides only, whereas the Fe is distributed between the metal and sulfides. The ferrous content in the silicates usually does not exceed some tenths of a percent, and in unequilibrated meteorites it rarely amounts to a few percent. In the enstatite meteorites the (Mg, Ca, Al, Ti)/Si ratios are lower than in other meteorites. Trace element contents in enstatite chondrites unchanged by metamorphism is about that observed in the CI chondrites. The enstatite meteorites differ mineralogically from other meteorites by the presence of Na, Ca, Mg, Cr, and Ti sulfides and Si and Ti nitrides (Table 1), which are formed under reducing conditions. The majority of these minerals are found in enstatite meteorites only.

Table 1. The mineralogy of the enstatite meteorites.

Minerals	Chemical formulas	Frequencies of occurence[a]
Alabandite	MnS	El (7) Au (6)
Caswellsilverite	$NaCrS_2$	Au (1)
Chromite	$FeCr_2O_4$	Au (2)
Clinoenstatite	$Mg_2Si_2O_6$	EH (9) Au (6)
Cohenite	$(Fe, Ni)_3C$	EH (4) EL (2)
Copper	Cu	EL (2) Au (1)
Cristobalite	SiO_2	EH (7) Au (1)
Daubreelite	$FeCr_2S_4$	EH (6) EL (9) Au (8)
Diopside	$CaMgSi_2O_6$	EL (2) Au (9)
Djerfisherite	$K_3(Cu, Na)(Fe, Ni)_{12}S_{14}$	EH (3) Au (4)
Graphite	C	EH (5) EL (8) Au (2)
Heideite	$FeTi_2S_4$	Au (1)
Kamacite	$\alpha - Fe, Ni$	EH (10) EL (10) Au (10)
Lawrencite	$(Fe, Ni)Cl_2$	EH (1)
Niningerite	$(Mg, Fe)S$	EH (7)
Oldhamite	CaS	EH (8) EL (8) Au (8)
Olivine	$(Mg, Fe)_2SiO_4$	EH (5) Au (10)
Orthoenstatite	$Mg_2Si_2O_6$	EH (5) EL (10) Au (10)
Osbornite	TiN	EH (1) EL (5) Au (4)
Perryite	$(Ni, Fe)_5(Si, P)_2$	EH (6) EL (1) Au (1)
Plagioclases:	$(Na, Ca)(Si, Al)AlSi_2O_8$	
Albite		EH (7) Au (10)
Oligoclase		EL (10)
Richterite	$Na_2CaMg_5Si_8O_{22}(OH, F)_2$	EH (1) Au (1)
Roedderite	$(Na, K)_2Mg_5Si_{12}O_{30}$	EH (1) Au (1)
Schreibersite	$(Fe, Ni)_3P$	EH (6) EL (9) Au (9)
Sinoite	Si_2N_2O	EL (6)
Sphalerite	$\alpha - ZnS$	EH (2) EL (4) Au (1)
Taenite	$\gamma - Fe, Ni$	EH (1) EL (2) Au (5)
Tridimite	SiO_2	EH (3) EL (1)
Troilite	FeS	EH (9) EL (1) Au (10)
Quartz	SiO_2	EH (3) EL (1)

[a] The values in parentheses are written after the indices of meteorite chemical types. In calculating the frequences of occurence 10 samples of each meteorite type are taken into account

The enstatite meteorites are divided now into three types (Yavnel', 1984): EH chondrites, EL chondrites, and enstatite achondrites or aubrites (Au), which differ from one another in chemistry, mineralogy, and structures (Petaev and Skripnik, 1983).

EH chondrites are characterized by the high content of Fe and S (29.15–35.02 and 4.9–6.12 wt. %, respectively). These meteorites contain chondrules, but chondrites are not found in all EH chondrites. The pyroxene of EH chondrites is generally clinoenstatite (0.66–1.8 mol % $FeSiO_3$), but some of these contain orthoenstatite also. The composition of plagioclase corresponds to albite (1.0–2.7 mol % anorthite). The kamacite of EH chondrites contains metallic silicon in solid solution (2.6–3.9 wt %). The content of Ti in troilite is small (0.29–0.48 wt %). EH chondrites contain from 200 to 500 ppm of nitrogen in their bulk composition, but the nitrides osbornite and sinoite are not found in these meteorites, except in Abee. Two of these meteorites include all three polymorphs of silica: tridymite, cristobalite, and quartz. However, the EH chondrites contain niningerite. The coexistence of silica and forsterite in the chondrules of EH chondrites suggests that all the minerals of these meteorites are out of equilibrium with each other.

EL chondrites, bearing 20.71–28.95 wt % of iron and 0.48–4.44 wt % of sulfur are well recrystallized, and contain practically no chondrules. The plagioclase of these meteorites is richer in Ca (13.1–16.8 mol % of anorthite), but the kamacite is poorer in Si (0.92–1.7 wt %). Unlike the EH chondrites, the Ti content in the troilite of EL chondrites is slightly higher, but it also is not high (0.55–0.77 wt %). Only the EL chondrites contain sinoite, which may be regarded as the typomorphic mineral of EL chondrites.

Aubrites are drastically different from chondrites in their structures and textures. All aubrites except Shallowater are breccias, which consist of crystals or crystal fragments of orthoenstatite up to a few centimeters in length. These meteorites are characterized by low bulk contents of Fe and S (0.47–2.8 and 0.16–0.5 wt %, respectively). In contrast to chondrites, the pyroxenes of the aubrites contain practically no iron (0.01–0.12 mol % $FeSiO_3$). The Si content in kamacite varies greatly, not only from meteorite to meteorite (o.12–2.44 wt %), but among the grains of a single meteorite. The aubrites are characterized by high Ti content in troilite on the average, 1–5 wt %), which also varies greatly from grain to grain. For example, in the Bustee aubrite the Ti content is variable over the range 0.2–16.3 wt %. The same meteorite contains heideite, a unique sulfide of iron and titanium. The ordinary accessory minerals of aubrites are diopside and forsterite. Some aubrites include chromite.

The enstatite meteorites contain many unusual minerals. Table 1 lists the minerals of enstatite meteorites and their chemical formulas. This table also includes information on the frequences of occurence of each mineral in the meteorites of various types. The frequency of occurence was calculated from data (Petaev and Skripnik, 1983) on the mineralogical composition for 10 samples of each meteorite type.

Among the more than 30 minerals found in the enstatite meteorites, the accessory sulfides and nitrides are of particular interest because they can be regarded as indicators of the physicochemical parameters of the evolution processes of the enstatite meteorite matter. In this case the types of occurrence of the minerals and their structural relationships with other minerals can be important for estimating the genesis of several minerals as well as the mineral associations as a whole. Some papers (Yudin, 1972; Yudin and Smyshlyaer, 1964; Buseck and Holdsworth, 1972; Keil, 1968; Keil and Andersen, 1965a, Keil and Brett, 1974; Keil and Snetzinger, 1967; Ramdohr, 1973) report detailed data on the types of occurence of accessory minerals in enstatite meteorites and their structural relationships with other minerals.

The occurrences of osbornite and sinoite—two nitrides found in enstatite meteorites—greatly differ from one another. Osbornite is found in the EL chondrites and aubrites as fine, well-developed crystals or crystal fragments from 1 to 8 microns in size. It is uniformly distributed in many meteorites. Some osbornite crystals show intergrowth with troilite and daubreelite. Sinoite is found only in EL chondrites and occurs as rather large grains, up to 200 microns in size, the individual grains are frequently aggregated. Sinoite occurs in the silicates as well as at the boundaries of the metal grains. In the Yilmia chondrite sinoite occurs as platey grains similar to twinned crystals.

The sulfides oldhamite, niningerite, alabandite, daubreelite, and heideite also differ from one another in occurence. Niningerite, alabandite, and daubreelite are usually intergrown with troilite, and they frequently occur as lamellae in the troilite. In constrast to niningerite, alabandite and daubreelite sometimes occur as grains isolated from the troilite. Oldhamite and heideite are usually separated from other sulfides. Heideite is found only as discrete, unshaped masses at the grain boundaries of silicates. It is not in contact with other opaque minerals. Oldhamite, if it is in contact with other sulfides, forms individual grains. These differences in the occurences of the enstatite meteorite minerals suggest that they are formed by several processes.

At present most investigators suggest that the major differences in the redox state of various meteorite groups, as well as some of the features of their chemical and mineralogical compositions, have been determined by a series of complex physicochemical processes, which took place in the gas and dust of the solar nebula and the meteorite parent bodies. In turn, each specific process has taken place at a separate stage of the evolution of the materials of the solar system.

According to modern hypotheses, solar system formation is started by the collapse of an interstellar gas and dust cloud, and this process results in the formation of a proto-Sun and the solar nebula around it. The conditions in the inner zone of the nebula are greatly different from those in the outer zone. As Cameron's and Fegley's (1982) calculations evidence, in the inner zone of the nebula, at a radial distance up to 0.95 from the rotation axis and at heights up to 0.1 from the midplane, the temperature exceeds 1600K, so that the dust must be wholly vaporized. It is apparent, that the proto-matter of the enstatite meteorite and some irons were not condensed as such. In the outer zone of the nebula where the carbonaceous chondrites were formed, the dust here is likely be vaporized incompletely, as indicated by the isotopic anomalies observed in the carbonaceous chondrites (Lavrukhina, 1982). The anomalies seen in these chondrites may be interstellar grains that have survived.

The main mineral-forming process in the nebula, as many scientists suggest, is the condensation of the nebular gas. The condensation can be divided into two types of mineral-forming reactions:

1. Reactions for the formation of submicronsize solids (initial condensates) from a homogeneous gas phase, for example:

$$Ti(g) + 0.5\,N_2(g) \rightleftharpoons TiN(s) \tag{1}$$

2. Reactions between initial condensates and gas with the formation of new minerals, for example:

$$SiC(s) + Mg(g) + 4\,H_20(g) \rightleftharpoons MgSiO_3(s) + CO(g) + 4\,H_2(g) \tag{2}$$

The next process of solar system evolution, the accretion of dust, begins even at the earliest stages of condensation and results in the formation of more coarse grains over a wide range of sizes, from submicrons to submillimeters. The grains continue to react with the nebular gas. However, the behavior of the reactions is varied and is dependent on the grain size. The micron-size grains are equilibrated with the nebular gas, but the millimeter-size grains appear to react with the gas on their surfaces only. Thus, the formation of new minerals on the grain surfaces keeps the earlier condensates within the grains from reacting with the nebular gas.

As a result of such nebular processes as condensation, accretion, and accumulation, bodies with various densities, from friable to compact (Safronov, 1982), are formed in the solar nebula. At this stage the initial meteorite parent bodies (Levin and Simonenko, 1977) are likely to be formed. The interiors of the parent body were heated enough to metamorphose the meteorite matter. The metamorphism can result in the partial melting of the interiors of the meteorite parent bodies. The source of the heating of the parent body interiors is probably the gravitational energy released during the growth of parent body itself. This process evidently produces the regolith on the parent body surface, where the characteristic brecciated structure of many aubrites and some chondrites have been formed. In the interiors of the parent bodies the mineralogical composition of meteoritic matter can be altered by metamorphic reactions, for example:

$$2\,CrS(s) + 2\,FeS(s) \rightleftharpoons FeCr_2S_4(s) + Fe(s) \tag{3}$$

A major factor in the metamorphism is probably the high temperatures. The effect of pressure is likely to be not so significant; at least, high-pressure minerals are not found in enstatite meteorites.

There are two major mineral-forming processes in the enstatite meteorites, —condensation, and thermal metamorphism—which are responsible for their chemical and mineralogical composition. Stated as such we leave out of consideration one of the most important processes of meteoritic matter transformation—the formation of chondrules—and this process can significantly alter the chemical and mineralogical composition of meteoritic matter. Structural and mineralogical features suggest that chondrules are formed by different processes under a very wide range of P–T parameters. Consequently, a special study is needed for the understanding of chondrule formation, and this study is beyond the scope of our report.

Thermodynamic Properties of Enstatite Meteorite Minerals

Knowledge of the thermodynamics of meteoritic minerals is needed in order to analyze the physicochemical processes of meteoritic matter evolution. Many investigators have studied several thermodynamic characteristics of enstatite meteorite minerals as well as some of the physicochemical parameters of the reactions involving these minerals, and corresponding data are now available. Such information is available also for some minerals and phases that are not found in the enstatite meteorites but can be formed in the system under different mineral-forming conditions. The results of thermodynamic investigations for some minerals are summarized in handbooks, but for other minerals only separate experimental data are available. Moreover, some thermodynamic properties of accessory enstatite minerals provided in handbooks (JANAF, 1971–1978; *Termicheskie konstanty veschestv*, 1974; Mills, 1974) are out of date and must be corrected. Although this chapter is not aimed at a detailed consideration of the methods of calculating and estimating thermodynamic properties of minerals, it is useful to outline briefly the methods used. It is useful also to list the sources of the experimental data.

The standard enthalpies of formation of compounds were calculated from available data, such as $\Delta H^\circ(298.15)$ of reactions involving the compounds we are interested in and $\Delta H_f^\circ(298.15)$ for the all reacting substances, on the basis of Hess' law. The values of $\Delta H^\circ(298.15, \text{reaction})$ were calculated from the available experimental values of $\Delta G^\circ(T; \text{reaction})$, on the basis of the third law of thermodynamics.

The coefficients of the Mayer–Kelley equations were calculated from the heat capacity and heat content data using Shomate's (1944) method as modified by Khodakovsky (1975).

Experimental data on the low-temperature heat capacities of some sulfides of Cr and Ti are not available, and hence their $C_p^\circ(298.15)$ and $S^\circ(298.15)$ values as well as the coefficients of the Mayer–Kelley equations were estimated. The values of standard heat capacities were estimated by several methods, such as:

1. Using the Neumann–Kopp rule.
2. Using the suggestion of the equality of the gram-atom heat capacities of compounds that consist of the same elements but in various proportions.

3. Using exchange reactions, for example:

$$Cr(s) + MeS(s) \rightleftharpoons CrS(s) + Me(s) \tag{4}$$

   In this case it is suggested that ΔC_p°(298.15; reactions) is equal to 0.
4. Using reactions for the synthesis of complex sulfides from simple ones on the assumption that ΔC_p°(298.15; reaction) is equal to 0.

The standard entropies were also estimated by several methods:

1. Using Latimer's rule
2. On the basis of a comparison of entropies of sulfides of the same element but characterized by various Me/S ratios
3. Using reactions for the synthesis of complex sulfides from simple ones on the assumption that ΔS°(298.15; reaction) is equal to 0.

Of the values estimated by these methods, the average arithmetical value is recommended for use in thermodynamical calculations.

The coefficients of the Mayer-Kelley equations were estimated in two ways:

1. Using the available values of $C_p^\circ(T)$ assuming the equality of gram-atom heat capacities of compounds that consist of the same elements but in various proportions.
2. Using reactions for the synthesis of sulfides on the assumption that ΔC_p°(298.15; reaction) is a constant over the temperature range considered.

The uncertainties of the estimated values are taken to be equal to 5% of the recommended value of C_p°(298.15) and 10% of the recommended value of S°(298.15). The uncertainties of the recommended ΔH_f°(298.15) values are calculated by the equation:

$$\delta = (\delta_1^2 + \delta_2^2 + \delta_3^2)^{0.5}$$

where δ_1 is the author's experimental error of $\Delta G^\circ(T$; reaction) determination; δ_2 is the uncertainity calculated from the scatter in the ΔH°(298.15; reaction) values; δ_3 is the uncertainity of ΔH_f°(298.15) values included in the calculations.

The uncertainties of the values obtained by experiment as well as the accepted handbook values are taken to be the same as those reported by the authors.

Thermodynamic properties of enstatite meteorite minerals and some related compounds (Table 2) are fitted to the international recommended standard data (CODATA, 1978). In recalculating the literature data, information on the thermodynamical properties of all the reacting substances is taken from the handbooks (*Termodinamicheskie svojstva individual'nykh vechchestv*, 1978–1983; Pankratz, 1982), which, in turn, are fitted to the standard data (CODATA, 1978). The sources of the thermodynamic data for the compounds presented in the Table 2 are listed below.

Clinoenstatite. The thermodynamic properties are taken from (Robinson and Haas, 1982).

Forsterite and diopside. The thermodynamic properties from the handbook by Robie *et al.* (1978) are accepted.

Table 2. Thermodynamic properties of the enstatite meteorite minerals and related substances.

Compounds	$S°(298.15)$ $(J \cdot mol^{-1} \cdot K^{-1})$	$H_f°(298.15)$ $(kJ \cdot mol^{-1})$	$C_p°(T) = a + b \times 10^{-3}T + c \times 10^5 T^{-2}$			Temperature range (K)	$H°$ (tr) $(kJ \cdot mol^{-1})$
			a	b	c		
$Mg_2Si_2O_6$, clinoenstatite	135.54 ± 0.37	−3090.30 ± 8.24	205.43	39.66	−52.55	298–1600	
Mg_2SiO_4, forsterite	94.11 ± 0.10	−2051.1 ± 1.4	149.83	27.363	−8.52	298–2163	
$CaMgSi_2O_6$, diopside	142.7 ± 0.2	−3210.70 ± 9.12	191.82	83.079	−42.795[a]	298–1600	
Fe(s)	27.28 ± 0.13	0	−0.925	50.802	9.5604	298–1042	0.0
—	—	—	183.090	−122.51	0	1042–1185	0.912 ± 0.013
—	—	—	23.033	8.795	4.979	1185–1667	1.11 ± 0.21
—	—	—	−179.88	89.397	2000.7	1667–1811	—
FeS, troilite	60.33 ± 0.28	−100.08 ± 0.84	21.71	110.46	0	298–411	2.38
—	—	—	72.80	0	0	411–598	0.5
—	—	—	51.04	9.96	0	598–1468	—
SiO_2, quartz	41.46 ± 0.2	−910.7 ± 1.0	40.497	44.601	−8.322	298–847	—
—	—	—	58.422	10.032	−1.927	847–1200	—
—	—	—	67.643	4.1924	−23.732	1200–1996	—
$FeCr_2O_4$, chromite	142.0 ± 1.7	−1460.0 ± 0.7	153.63	31.58	−26.09	298–2123	—
Si_2N_2O, sinoite	44.26 ± 0.22	−925.0 ± 7.4	11.95	16.62	−44.32	298–2500	—
TiN, osbornite	30.33 ± 0.18	−337.6 ± 1.3	49.83	3.93	−12.38	298–1738	—
CaS, oldhamite	56.5 ± 1.0	−477.0 ± 8.4	46.29	8.37	−1.188	298–2800	—
MgS, niningerite	50.33 ± 0.40	−348 ± 5	45.544	8.757	−2.307	298–2500	—
$FeCr_2S_4$, daubreelite	207.1 ± 0.3	−566.8 ± 7.6	174.57	12.76	−12.06	298–1335	—
$FeTi_2S_4$, heideite	226.97 ± 0.42	−662 ± 11	188.52	69.576	−1.21	298–1573	—
Cr_2O_3, escolaite	81.1 ± 5.0	−1140.6 ± 1.7	1926.1	−12842[b]	—	298–306	—
—	—	—	55756	−357840[c]	—	306–310	—
—	—	—	2031.7	−11636[d]	—	310–335	—
—	—	—	134.44	−12.62	−28.398[e]	335–2705	—

TiO, honquiite	34.79 ± 0.15	−542.0 ± 4.0	68.496	−40.652	−17.14[f]	298–1265	3.6 ± 1.2
—	—	—	51.328	12.10	−11.641	1265–1810	2.7 ± 0.9
—	—	—	48.187	11.678	17.298	1810–2030	—
TiO_2, rutile	50.62 ± 0.3	−944.00 ± 0.80	63.196	11.82	−10.347	298–2185	—
CrN, carlsbergite	36.8 ± 1.7	−118.0 ± 1.7	46.44	6.61	0	298–1700	—
Cr_2N(s)	64.0 ± 1.7	−131.8 ± 2.5	64.85	27.53	0	298–1700	—
Fe_4N, roaldite	155 ± 16	−11.1 ± 4.2	112.8	34.1	0	298–753	8.4 ± 4.2
—	—	—	112.8	34.1	0	753–953	—
Si_3N_4(s)	64.20 ± 0.32	−787.8 ± 3.0	98.877	77.739	24.199[g]	298–4000	—
CrS(s)	63 ± 6	−150.5 ± 0.6	31.77	49.06	0	298–450	—
—	—	—	51.643	4.904	0	450–1840	—
Cr_2S_3(s)	149 ± 15	−432.3 ± 2.2	90.54	48.6	10.64	298–1800	—
Cr_3S_4, brezinaite	212 ± 21	−599.2 ± 2.0	126.33	68.4	16.25	298–1800	—
TiS(s)	58 ± 6	−282.1 ± 6.0	42.43	8.30	0	298–2200	—
TiS_2(s)	78.21 ± 0.16	−400.8 ± 10.5	70.599	9.387	−5.385	298–1010	—

[a] $C_p^\circ(T) = a + b \times 10^{-3}T + c \times 10^5 T^{-2} + d \times 10^{-6} T^2 + e \times 10^{-9} T^3$; $d = 21.718$.
[b] $d = 22756$.
[c] $d = 575300$.
[d] $d = 17573$.
[e] $d = 8.438$.
[f] $d = 31.626$.
[g] $d = -21.211$; $e = 2.376$.

Kamacite. In consideration of equilibria involving kamacite we suggest that its thermodynamic properties are the same as those of pure iron, which are taken from the handbook by Pankratz (1982).

Troilite. The thermodynamic properties are taken from the handbook by Naumov *et al*. (1974).

Quartz. The thermodynamic properties are taken from the handbook *Termodinamicheskie svojstva individual'nykh vechchestv* (1979).

Chromite. The standard entropy and heat capacity values are taken from the handbook *Termicheske svojstva vechchestv* (1974). The coefficients of $C_p^\circ(T)$ equation are calculated from the data of Naylor (1944). The experimental data are given by an equation with an accuracy of $\pm 1.5\%$. The standard enthalpy of formation is calculated from the data of Rezukhina *et al*. (1965) and Zabejvorota *et al*. (1980).

Sinoite. The values of $C_p^\circ(298.15)$ and $S^\circ(298.15)$ are taken from the report by Kochchenko and Grinberg (1982a). The cofficients of the $C_p^\circ(T)$ equation are calculated from Fegley's (1981) data. The data used are given by an equation with an accuracy of $\pm 0.5\%$. The $\Delta H_f^\circ(298.15)$ value is calculated from Ehlert *et al*. (1980) and Ryall and Muan (1969).

Osbornite. The thermodynamic properties are taken from Khodakovsky and Petaev (1981).

Oldhamite and niningerite. The thermodynamic properties are taken from the handbook *Termodinamicheskie svojstva individual'nykh vechchestv* (1981).

Daubreelite. The thermodynamic properties are taken from Petaev *et al*. (1982).

Heideite. The values of $C_p^\circ(298.15)$, $S^\circ(298.15)$, and $\Delta H_f^\circ(298.15)$ were calculated from experimental data of Petaev *et al*. (1983). The coefficients of the $C_p^\circ(T)$ equation are estimated. However, it is necessary to note that at a temperature of 340K the estimated values differ from the experimental value by 7%. One of the reasons for this discrepancy may be the existence of a phase transition in crystals of $FeTi_2S_4$ (Muranaka, 1973; Keil and Brett, 1974) in the temperature range from 700 to 800K. Thus, more reliable values of $C_p^\circ(T)$ equation coefficients may be obtained only by experimental measurements of the heat capacity of heideite at high temperatures.

Escolaite and titanium oxides. The thermodynamic properties are taken from *Termodinamicheskie svojstva individual'nykh vechchestv* (1982).

Chromium nitrides. The thermodynamic properties are taken from Khodakovsky and Petaev (1981).

Roaldite. The values of $C_p^\circ(298.15)$, $S^\circ(298.15)$, and $\Delta H_f^\circ(298.15)$ are taken from *Termicheskie konstanty vechchestv* (1974). The $C_p^\circ(T)$ equation coefficients are borrowed from Kelley's (1937) summary. The upper limit of roaldite stability is estimated from the phase diagram of the Fe–N system (Parnjipe *et al*., 1950). The parameters of the magnetic transition in crystals of Fe_4N are also taken from *Termicheskie konstanty vechchestv* (1974).

Silicon nitride. The values of the $C_p^\circ(298.15)$ and $S^\circ(298.15)$ are taken from the report by Kochchenko and Grinberg (1982b). The $\Delta H_f^\circ(298.15)$ value and the

$C_f^\circ(T)$ equation coefficients are from *Termodinamicheskie svojstva individual'nykh vechchestv* (1979).

CrS. The $C_p^\circ(298.15)$ and $S^\circ(298.15)$ values are estimated. The Mills's (1974) coefficients of the $C_p^\circ(T)$ equations are fitted to the estimated $C_p^\circ(298.15)$ value. The $\Delta H_f^\circ(298.15)$ is calculated from the data of Moriyama *et al.* (1977) and from the reports of Young *et al.* (1973) and Hager and Elliott (1967).

Cr_2S_3. The $C_p^\circ(298.15)$ and $S^\circ(298.15)$ values are estimated. The $C_p^\circ(T)$ equation coefficients were calculated from the data of Volovik *et al.* (1979). The value of $\Delta H_f^\circ(298.15)$ is calculated from the data of Young *et al.* (1973).

Brezinaite. The $C_p^\circ(298.15)$, $S^\circ(298.15)$, and coefficients of $C_p^\circ(T)$ equation are estimated. The $\Delta H_f^\circ(298.15)$ is calculated from data of Young *et al.* (1973) and Igaki *et al.* (1971).

TiS. The values of $C_p^\circ(298.15)$ and $S^\circ(298.15)$ are estimated. The Mills' (1974) coefficients of the $C_p^\circ(T)$ equation are fitted with an estimated $C_p^\circ(298.15)$ value. The $\Delta H_f^\circ(298.15)$ were calculated from the data of Pelino *et al.* (1979), which are consistent with the earlier data of (Edwards *et al.* (1971), Fransen and Gilles (1965), and Fransen (1963).

TiS_2. The values of $C_p^\circ(298.15)$ and $S^\circ(298.15)$ are taken from the report by Beyer (1983). These values are in good agreement with the data of Todd and Coughlin (1952). The $C_p^\circ(T)$ equation coefficients are calculated from the experimental data of Todd and Coughlin (1952) and Mraw and Naas (1979). The $\Delta H_f^\circ(298.15)$ value is from Whittingham (1978) and this value is consistent with the data of Abendroth and Schlechten (1959).

Some Physicochemical Aspects of the Nebular Condensation of Enstatite Meteorite Minerals

As we noted, at present many scientists consider nebular condensation to be the primary mineral-forming process of meteorite matter evolution. The condensation sequence has been calculated by many investigators, but the greatest progress in the solution of this problem was made by Grossman (1972, 1973; Grossman and Larimer, 1974). Based on one of the models of nebula evolution (Cameron and Pine, 1963) he calculated the condensation sequence of minerals in the system, including the 20 most abundant elements (Cameron, 1973). Among the more than 90 solids taken into consideration, the condensates are represented by such meteoritic minerals as corundum, perovskite, melilite, spinel, nickel–iron, diopside, forsterite, anorthite, enstatite, alabandite, troilite, and magnetite. However, the typomorphic minerals of the enstatite meteorites, such as graphite, cohenite, quartz, and sinoite, are not stable under the conditions considered. Recently Saxena and Eriksson (1983) carried out similar calculations, but they have taken into account the possibility of formation of both silicate and oxide solid solutions. As the authors note, their major results are similar to those of Grossman, but some details of solid solution compositions are of very great interest. It was found that ferrous-free silicates and metal containing more than 0.015 mol % of Si are not condensed from the solar composition gas. Enstatite meteorite minerals, such as oldhamite, niningerite, osbornite, graphite, and cohenite are also not stable under the conditions considered.

Many authors (Blander, 1971; Larimer, 1968, 1975; Baedecker and Wasson, 1975; Herndon and Suess, 1976; Lattimer and Grossman, 1978; Larimer and Bartholomay, 1979; Sears, 1980; Khodakovsky, 1982) have tried to elucidate the formation conditions of enstatite meteorite minerals. Suggested hypotheses differ from one another both in initial postulates and in conclusions.

Blander (1971), Herndon and Suess (1976), and Sears (1980) prove that the enstatite meteorite minerals condensed from the solar composition gas under high pressures of from 10^{-1} to 100 atm. In this case whether some condensation reactions will proceed may be constrained by nucleation kinetics, and some minerals may condense at lower temperatures than their equilibrium ones. However, the suggestion itself, of condensation under high pressures, is not consistent with both modern physical models of the nebula (Fegley and Lewis, 1980; Cameron and Fegley, 1982), as well as the enstatite meteorite data on the primary noble gas content (Hertogen *et al.*, 1983). A conclusive criticism of the papers (Blander, 1971; Herndon and Suess, 1976) is given by Larimer and Bartholomay (1979).

Based on his own calculations of the condensation sequence, Sears (1980) concludes that even at a pressure equal to 1 atm the oldhamite can condense only from gas decreased in oxygen about 20% relative to the solar gas. This indicates conclusively that the enstatite meteorite minerals could not have condensed from the solar gas even at high pressures.

Some investigators (Larimer, 1968, 1975; Baedecker and Wasson, 1975, Lattimer and Grossman, 1978; Larimer and Bartholomay, 1979; Khodakovsky, 1982) consider that enstatite meteorite minerals condensed from a gas under other than solar redox conditions. Larimer (1968) proposed to consider the C/O ratio in the gas as a criterion of gas redox conditions. This parameter was chosen because at the high temperatures and low pressures characteristic of the system considered carbon monoxide is the most stable compound of both oxygen and carbon. Consequently, only the oxygen that is not bonded to carbon can react with other elements, and the higher the C/O ratio is, the greater is the reduction state of the gas. Baedecker and Wasson suggested that the enstatite meteorite minerals condensed from a gas with an H/O ratio increased by factor of 1000 relative to solar composition. However, this suggestion has not yet been studied.

Larimer and Bartholomay (1979) calculated the condensation sequence in the system, including 11 elements over a wide range of temperatures and pressures (300–2500K and 10^{-6} to 1 atm, respectively). The abundances of the elements are taken from Cameron's (1973) table. The C/O ratio varies from 0.56 (solar) to 1.4. Similar calculations in the system with C/O = 1.2 were carried out by Lattimer and Grossman (1978).

The calculations show that the condensation sequence significantly depends on the C/O ratio. At a C/O $<$ 0.85 the initial condensates are oxides of Al, Mg, Ti, Ca, and silicates, and the condensation sequence is similar to that calculated by Grossman (1972). At a C/O $\geqslant$ 0.85, carbides, nitrides, and sulfides of these elements are stable under the conditions considered, and they are formed by reactions between initial condensates (oxides and silicates) and nebular gas. At a C/O $\geqslant$ 1 the initial condensates are enstatite meteorite minerals, such as graphite, oldhamite, osbornite, and cohenite, which is a metastable phase. As the temperature decreases, these minerals must react with the nebular gas to form oxides and silicates and the nitrogen, sulfur, and carbon must return to the gas phase. Under these conditions the condensation of the major enstatite meteorite minerals, such as ferrous-free enstatite, Si-bearing kamacite, and troilite is also possible.

Based on the foregoing results of condensation calculations, Khodakovsky (1982) analyzed the formation conditions of high-temperature accessory minerals of the enstatite meteorites, and refractory minerals of Ca–Al inclusions (CAI's) and concluded that the C/O ratio in the nebular gas was close to 1. Indeed, in the system of interest the dominant gaseous species of carbon is carbon monoxide, and the most stable solid phase is graphite. If we consider the formation of the nebula and take into account the very high volatilization temperature of graphite (more than 4000K) we can infer (Petaev and Khodakovsky, 1981) that the carbon will change from the solid phase to the gas phase by oxidation. In this case the C/O ratio in the gas does not exceed 1. At high temperatures and low pressures the contents of hydrocarbons as well as other gaseous species of carbon are appreciably lower than the carbon monoxide content, so that the nebular gas C/O ratio must be close to 1. Therefore, it is expedient to consider the condensation of minerals from the nebular gas with a C/O $\leqslant$ 1 only. Based on this postulate, and Larimer's and Bartholomay's (1979) results, we estimated the sequence and temperatures of condensation of some enstatite meteorite minerals from a gas with a C/O = 1 (Table 3). Niningerite can

Table 3. The condensation temperatures of some enstatite meteorite minerals and related compounds under C/O = 1 and $P_{tot} = 10^{-4}$ atm.[a]

Compounds	T_{cond}(K)
SiC	1650
TiN	1540
Fe_3C	1360
AlN	1300
CaS	1250
$CaMgSi_2O_6$	1130
C	1100
Mg_2SiO_4	1100
$MgSiO_3$	960

[a] Data from Larimer and Bartholomay (1979).

be stable in the system considered at temperatures below 900K provided methane and graphite are not formed. Investigation of the kinetics of the conversion of CO to CH_4 (Mendybaev *et al.*, 1985) reveals that such conditions may be taking place in the nebula, so that the niningerite may have condensed from the nebular gas.

Recently Petaev *et al.* (1985) considered the possibility of the condensation of some sulfides and nitrides in the nebula with a $C/O = 1$. For estimating the possibility of condensation of any mineral we need to know both the thermodynamical properties of all the considered compounds and some physicochemical parameters of the system, such as temperature, pressure, and the composition of the gas phase. In our calculations we accepted the adiabatic model of the nebula (Fegley and Lewis, 1980), which were earlier used (Dorofeeva *et al.*, 1982; Mendybaev *et al.*, 1985) to calculate the equilibrium gas composition in the H–N–C–O–Si–S–P system by the "Gibbs energy minimization" method (Dorofeeva and Khodakovsky, 1981). The abundances of the elements, exept oxygen, are taken from Cameron (1973); the abundance of oxygen was assumed to be equal to the carbon abundance.

The temperatures of condensation of the Si, Cr, Ti, and Fe compounds are estimated by the method described in detail by Larimer (1967) and Grossman (1972). We have taken into consideration a number of possible condensates, such as SiC, Si_2N_2O, Si_3N_4, TiN, TiO, TiS, $FeTi_2S_4$, Fe, CrN, Cr_2N, CrS, Cr_2O_3, and $FeCr_2S_4$. The results are presented in Fig. 1, where solid lines show the temperature dependencies of the fugacities of nebular gases and the dashed lines show those of the fugacities of the same gases equilibrated with the compounds enclosed in brackets. The activities of the solids are taken to equal to unity. As Fig. 1 shows, the initial condensates are represented by SiC (1775K), TiN (1765K), Fe (1505K), and CrS (1260K).

After the condensation of the above-mentioned compounds, the bulk of the Si, Ti, Cr, and Fe enter into solids, and hence the transfer of the elements from one phase to another is determined by the reactions between initial condensates and the nebular gas. In this case the possibility of any compound formation can be estimated by consideration of reactions giving the compounds in which one is interested. In the calculations the activities of solids (except Si dissolved in kamacite) are taken to be equal to unity. The activity of silicon was calculated based on the available data on the silicon contents in meteoritic kamacite (Petaev and Skripnik, 1983) and the activity coefficients of silicon (Sakao and Elliott, 1975).

The results are presented in Figs. 2–4 in terms of separate systems. Shown here again are the temperature dependencies of the H_2O/H_2 and H_2S/H_2 ratios (dashed and dotted lines). The former ratio is characterized by variation of the redox state of the nebular gas, and the latter serves as a criterion for the possibility of reactions between sulfides and nitrides under nebular conditions.

Figure 2 shows that SiC reacts with the gas at 1190K to give forsterite; at 1150K enstatite is beginning to form. Metal condensing at 1505K contains about 10 atom % of silicon in solid solution, and the silicon content is reduced as the

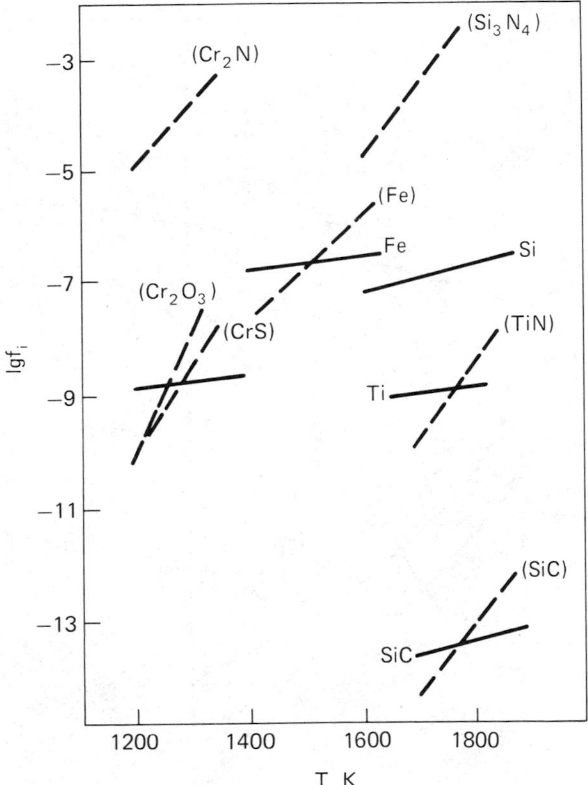

Fig. 1. The condensation temperatures of Si, Ti, Fe, and Cr compounds. Solid lines show the temperature dependence of the fugacities of nebular gases, dashed lines of the same gases equilibrated with the compounds named in brackets.

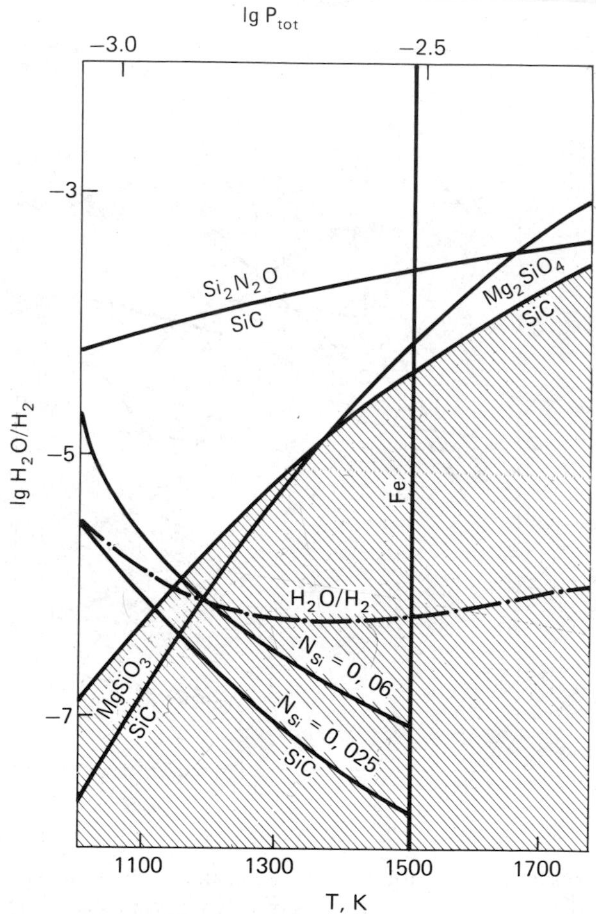

Fig. 2. The equilibria of Si-bearing minerals with nebular gases. The stability field of moissanite is shaded.

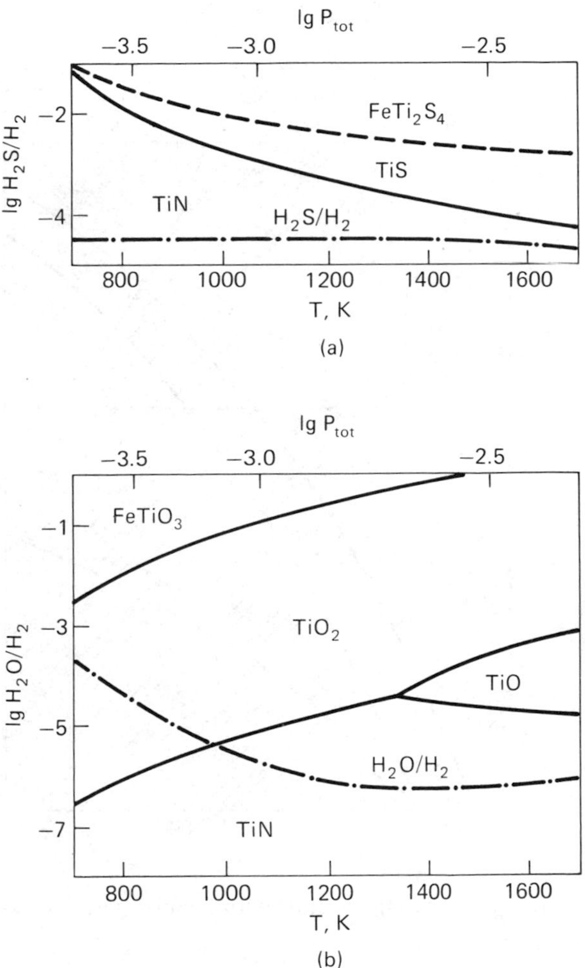

Fig. 3. The equilibria of Ti-bearing minerals with nebular gases.

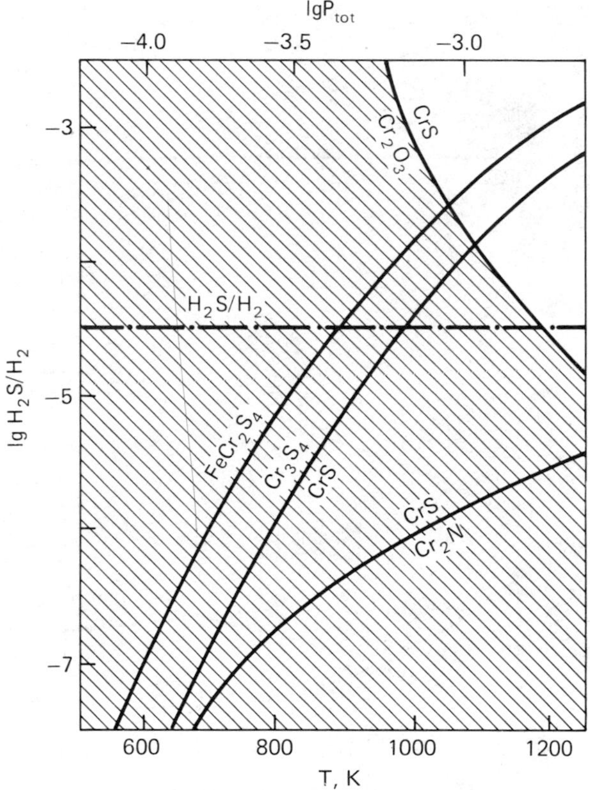

Fig. 4. The equilibria of Cr-bearing minerals with nebular gases. The stability field of escolaite is shaded.

temperature is decreased. The silicon contents characteristic of EH (about 6 atom %) and EL (about 2.5 atom %) chondrites are attained at temperatures of 1200K and 1000K, respectively. Sinoite is not stable under the conditions considered.

The equilibria between Ti-bearing phases are shown in Fig. 3. Under nebular conditions the Ti sulfides are not stable over all of the temperature range considered (Fig. 3a). As the temperature is decreased, osbornite must react with the gas at 975K to form rutile (Fig. 3b).

Chromium monosulfide must react with the gas at 1180K, giving escolaite (Fig. 4); however, this mineral has not been found yet in meteorites. Other sulfides and nitrides of chromium are not stable under nebular conditions over the entire temperature range considered. Chromite is also not stable (the equilibrium between chromite and escolaite is not shown in Fig. 4).

Our results are consistent with those of Larimer and Bartholomay (1979), and the small disagreements in condensation temperatures result from differences in the values of total pressure. In the adiabatic model the pressure increases with raising temperature, and hence our condensation temperatures are larger than those of Larimer and Bartholomay (1979). Therefore, our calculations suggest that sinoite, daubreelite, and heideite can not have condensed from the nebular gas, unlike enstatite, forsterite, kamacite, troilite, osbornite, oldhamite, and niningerite, which the available data (Table 3) show could have condensed from the gas with C/O = 1. However, the mechanism of such enrichment of the gas by carbon is not yet clear. Also, it is not clear whether this enrichment has been local or whether the elevated C/O ratio has been inherent to the nebula as a whole.

Some Physicochemical Aspects of the Thermal Metamorphism of Enstatite Meteorite Matter

As we noted above, condensation, accretion, and accumulation in the nebula have resulted in the formation of meteorite parent bodies, which have experienced the effects of high temperatures and pressures. Many scientists show that it is the process of thermal metamorphism which is responsible for the structural variations observed among enstatite meteorites. This suggestion has been supported recently by the similar trace element trends observed in the enstatite meteorites (Ikramuddin *et al.*, 1976; Biswas *et al.*, 1980; Hertogen *et al.*, 1983; Wolf *et al.*, 1983), on the one hand, and the Abee (EH4) samples heated at temperatures from 400 to 1400°C (Ikramuddin *et al.*, 1976; Biswas *et al.*, 1980), on the other.

Some investigators attempt to estimate the *P–T* parameters of the thermal metamorphism of enstatite meteorite matter. A review of their studies is presented by Dodd (1981). Of these studies the results based on the distribution of Fe, Si, and Ca among enstatite, kamacite, troilite, quartz, and oldhamite (Larimer and Buseck, 1974) are apt to be statistically reliable. Under the assumption that the minerals are equilibrated, the authors calculated equilibration temperatures of 710 ± 150°C (983K) and 820 ± 150°C (1093K) for the EH and EL chondrites, respectively. It is necessary to note that these are only lower limits of metamorphic temperatures, and they are consistent with other estimates (Dodd, 1981). When evaluating the pressures that have acted on enstatite meteorite matter we can place only an upper limit of about 3 kbar, based on the stability of cristobalite (Mineraly Spravochnik, 1974).

Upon high-rank metamorphism enstatite meteorite matter not only is recrystallized but it may experience partial melting followed by the loss of the melt. It is in terms of such a metamorphic–anatexic model that some investigators (Mason, 1966; Biswas *et al.*, 1980; Dodd, 1981) explain the variations of chemical and mineralogical composition among enstatite meteorites of several types. This suggestion is consistent with the observations of the Eu anomaly in aubrites (Wolf *et al.*, 1983) which is one of the indicators of igneous differtiation. When evaluating the temperature of this process, the temperature of the on set of melting (1318K) of the enstatite meteorite silicate component (Dodd, 1981) may be considered as a lower limit. Concurrent with this model, there are other suggestions as to the genesis of enstatite meteorites (Wasson and Wai, 1970; Baedecker and Wasson, 1975; Sears, 1980). These authors show that the major chemical differences among enstatite meteorites of several types result from nebular processes, and that metamorphism is responsible for structural differences only. It is not possible now to decide between these two theories, but the metamorphic–anatexic model, which is consistent with the trace element data, appears to be more conclusive. However, some features of the mineralogical composition of the enstatite meteorites, such as:

a. the occurence of osbornite in the EL chondrites and aubrites and its absence in all EH chondrites except Abee and
b. the occurence of djerfisherite in EH chondrites and aubrites, and its absence in intermediate EL chondrites

are not clear in terms of this model. It appears that EH chondrites are not a parent material for EL chondrites and some of the aubrites. It is possible that EL chondrites had their own proto-matter of type EL 3–4, which was recently found by Ivanov *et al.* (1984) as an inclusion in the Kaidun carbonaceous chondrite. However, the traces of the action of thermal metamorphism on the enstatite meteorite matter are clearly seen. The foregoing estimates evidence that it was a high-temperature process, which apparently brought the minerals of the enstatite meteorites of several types to equilibrium. To test this assumption we calculated the phase quilibria in the mineral systems at 983K, 1093K, and 1318K. The pressure is taken to be equal to 1 atm. Taking into consideration the high content of the kamacite and troilite in the enstatite chondrites, we suggest that the sulfur fugacity was buffered by the interaction of these minerals. For the reaction:

$$Fe(s) + 0.5\,S_2(g) \rightleftharpoons FeS(s) \qquad (5)$$

we calculated sulfur fugacity values of $10^{-10.43}$, $10^{-8.91}$, and $10^{-6.21}$ atm at temperatures of 983K, 1093K, and 1318K, respectively. The results are presented in Figs. 5–8 as diagrams with $\log f_{O_2}$ vs. $\log f_{N_2}$. In analyzing the diagrams we can establish the stable assemblages of enstatite meteorite minerals, and we can also estimate the fugacities of oxygen and nitrogen equilibrated with the mineral assemblage of a distinct meteorite type. These fugacities may be regarded as characteristic parameters of thermal metamorphism. The uncertainties of estimated fugacity values are determined by the uncertainties of calculation of $\log K^\circ(T;\text{ reaction})$ for each mineral equilibrium, and these can be calculated by the equation:

$$\delta(\log K) = (R \ln 10)^{-1} [\delta^2(\Delta S) + T^{-2} \delta^2(\Delta H)]^{0.5} \tag{6}$$

where $\delta(\log K)$ is the uncertainty of the $\log K^\circ(T;\text{ reaction})$, $\delta(\Delta S)$ is the uncertainty of the $\Delta S^\circ(298.15;\text{ reaction})$ calculation, and $\delta(\Delta H)$ is the uncertainty of the $\Delta H^\circ(298.15;\text{ reaction})$ calculation.

This allows us to estimate the parameters of thermal metamorphism of several of the enstatite meteorite types.

EH Chondrites (Fig. 5)

The metamorphism temperature is 710 ± 150°C (983K). The sulfur fugacity is controlled by the kamacite–troilite buffer, and it is equal to $10^{-10.43\pm0.05}$ atm. The oxygen fugacity of $10^{-32.5\pm0.05}$ atm is determined by the equilibrium between quartz (or, more precisely, one of the silica polymorphs found in these

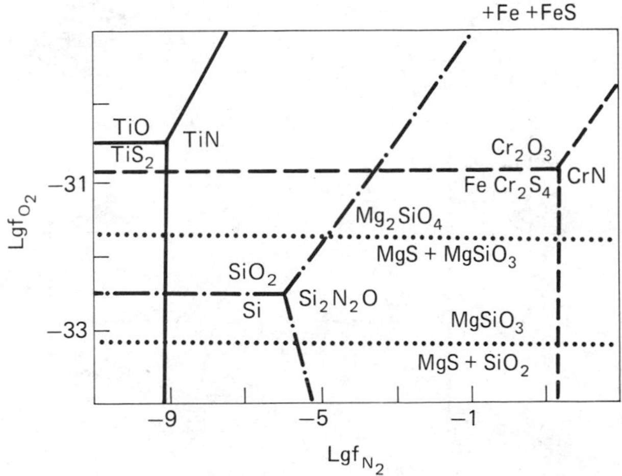

Fig. 5. The phase relations in the mineral systems at 983K, $\log f_{S_2} = -10.43$ and $\log a_{Si} = -6.49$. The legend: solid lines, Fe-Ti–N–S–O system; dashed lines, Fe–Cr–N–S–O system, dashed-and-dotted lines, Si–N–O system; dotted lines, Ca–Mg–Si–S–O system.

meteorites) and silicon dissolved in the kamacite. The upper limit of the nitrogen fugacity of $10^{-8.9\pm0.6}$ atm is estimated by the absence of osbornite in these meteorites. The assemblage of accessory minerals of EH chondrites (quartz + daubreelite + oldhamite) is equilibrated under these conditions. However, the coexistence of enstatite, forsterite, niningerite, and quartz in EH chondrites suggests that metamorphism did not bring all of the minerals of the EH chondrites to equilibrium. The occurrence of forsterite, generally in chondrules, unrecrystallized or in partially recrystallized chondritic structures of meteorites of this type is supported by this inference.

EL Chondrites (Fig. 6)

The metamorphism temperture is 820 ± 150°C (1093K). The sulfur fugacity is controlled by the kamacite–troilite buffer and it is equal to $10^{-8.91\pm0.04}$ atm. The nitrogen and oxygen fugacities are determined by the equilibrium between sinoite and silicon dissolved in the kamacite. These are equal to $10^{(-34.3 \text{ to } -27.5)\pm0.4}$ atm and $10^{(-5.5 \text{ to } -1,6)\pm0.4}$ atm, respectively. The EL chondrite mineral assemblage (enstatite + kamacite + troilite + sinoite + osbornite + daubreelite + oldhamite) is at equilibrium under the estimated conditions of metamorphism. The recrystallized structure of the EL chondrites is supported by this inference.

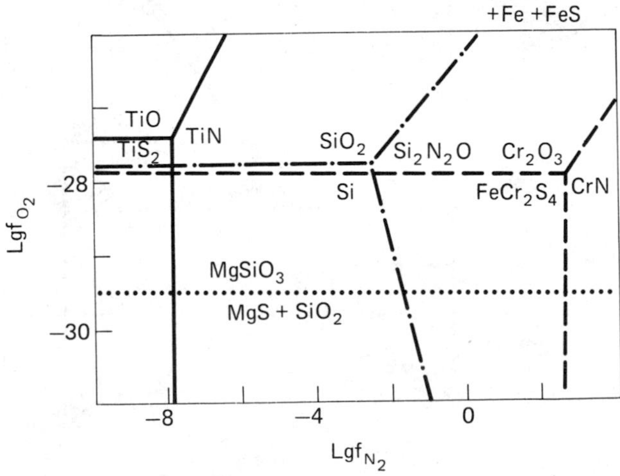

Fig. 6. The phase relations in the mineral systems at 1093K, $\log f_{S_2} = -8.81$ and $\log a_{Si} = -6.68$. The legend is the same as in Fig. 5.

Aubrites (Figs. 7 and 8)

The metamorphism temperture of 1318K is the lower limit of temperatures experienced by aubrites. The upper limit of oxygen fugacity of $10^{-23.3\pm0.8}$ atm is determined by the stability of Cr_2S_3, which must be formed at temperatures above 1280K as a result of daubreelite decomposition according to the reaction:

$$FeCr_2S_4\ (s) = FeS(s) + Cr_2S_3(s) \tag{7}$$

The occurrence of daubreelite rather than Cr_2S_3 in aubrites is possibly the result of the slow cooling of the aubrite material, when Cr_2S_3 reacts with troilite to form daubreelite again. If cooling were fast Cr_2S_3 might be preserved as a quenched phase. The nitrogen fugacity of $10^{-5.7\pm0.4}$ atm is determined by the equilibrium between osbornite and titanium disulfide dissolved in troilite. The sulfur fugacity is controlled by the kamacite–troilite buffer. Under these conditions enstatite, Si-bearing kamacite, Ti-bearing troilite, oldhamite, osbornite, and Cr_2S_3 are at equilibrium. Heideite found in the Bustee aubrite is metastable relative to simple titanium and iron sulfides in the temperature range considered. However, forsterite and diopside associated with enstatite and oldhamite are not stable under the estimated conditions (Fig. 8). Taking this fact into consideration and the small contents of metal and troilite in the aubrites, it may be suggested that the buffering role of the kamacite–troilite interaction is not important in the metamorphism of aubrite materials.

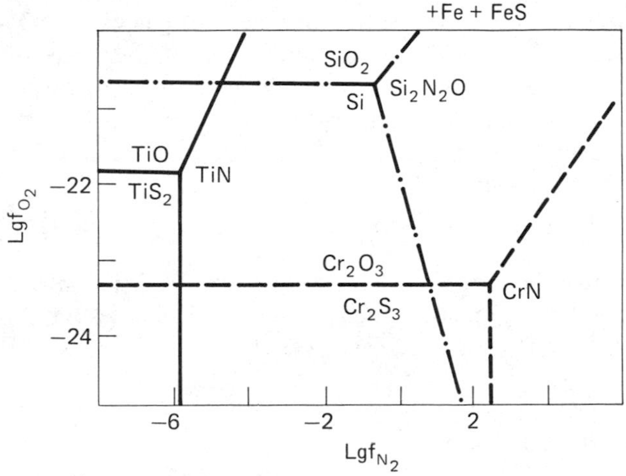

Fig. 7. The phase relations in the mineral systems at 1318K, $\log f_{S_2} = -6.23$ and $\log a_{Si} = -6.09$. The legend is the same as in Fig. 5.

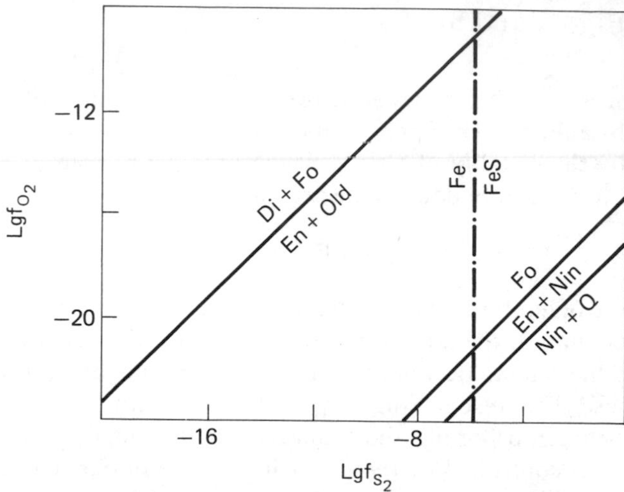

Fig. 8. The phase relations in the Ca–Mg–Si–S–O system at 1318K. The legend: Di diopside; Fo, forsterite; En, enstatite; Old, oldhamite; Nin, niningerite; and Q, quartz.

Conclusions

The studies carried out and the literature data allow us to establish the genesis of some enstatite meteorite minerals.

The major minerals—enstatite, Si-bearing kamacite, and troilite—and some accessory minerals—osbornite, oldhamite, niningerite, and alabandite—could have been formed by the condensation of a nebular gas with C/O = 1. Most of these phases are stable at the *P*–*T* of metamorphism. Other accessory minerals—daubreelite, heideite, and sinoite—cannot have been formed by condensation, but they are likely secondary metamorphogenic minerals.

Daubreelite could have been formed from chromium monosulfide and troilite according to the reaction:

$$2\,CrS(s) + 2\,FeS(s) \rightleftharpoons FeCr_2S_4(s) + Fe(s) \tag{8}$$

Sinoite could have been formed by the interaction of silicon dissolved in the kamacite and O-bearing phases, such as silica and silicates with nitrogen. The mechanism of heideite formation is not yet clear.

Acknowledgments

We are very grateful to Prof. S. K. Saxena for the invitation to take part in this volume.

References

Abendroth, R. P., and Schlechten A. W. (1959) A thermodynamic study of the titanium-sulfur system in the region $TiS_{1.9.3}$ to $TiS_{0.80}$, *Trans. Met. Soc. AIME* **215**, 145–151.

Baedecker, P. A., and Wasson J. T. (1975) Elemental fractionations among enstatite chondrites, *Geochim. Cosmochim. Acta* **39**(5), 735–765.

Beyer, R. P. (1983) Heat capacities of titanium disulfide from 5.87 to 300.7 K, *J. Chem. Eng. Data* **28**(3), 347–348.

Biswas, S., Walsh, T., Bart, G., and Lipschutz, M. E. (1980) Thermal metamorphism of primitive meteorites—XI. The enstatite meteorites: origin and evolution of a parent body, *Geochim. Cosmochim. Acta* **44**(12), 2097–2140.

Blander, M. (1971) The constrained equilibrium theory: sulfide phases in meteorites, *Geochim. Cosmochim. Acta* **35**(1), 61–76.

Buseck, P. R., and Holdsworth, E. F. (1972) Mineralogy and petrology of the Yilmia enstatite chondrite, *Meteoritics* **7**(4), 429–447.

Cameron, A. G. W. (1973) Abundance of the elements in the solar system, *Space Sci. Rev.* **15**(1), 121–146.

Cameron, A. G. W., and Fegley, M. B. (1982) Nucleation and condensation in the primitive solar nebula, *Icarus* **52** (1), 1–14.

Cameron, A. G. W., and Pine M. R. (1973) Numerical models of the primitive solar nebula, *Icarus* **18**(3), 377–406.

Codata (task group on key values for thermodynamics) (1978). *Tentative Set of Key Values for Thermodynamics*, Pt. VII, Spec. Rept. No. 7, 27 pp.

Dodd, R. T. (1981) *Meteorites: A Petrological-Chemical Synthesis*, Cambridge Univ. Press, Cambridge, 368 pp.

Dorofeeva, V. A., and Khodakovsky, I. L. (1981) Raschet ravnovesnogo sostava mhogo-komponenthykh sistem 'metodom minimizatsii' po konstantam ravnovesiya,–*Geokhimiya* **1**, 129–135.

Dorofeeva, V. A., Petaev, M. I., and Khodakovsky, I. L. (1982) On the influence of nebular gas chemistry on the condensate compositions,–Abstracts, *Lunar Planet. Sci. Conf XIII*, Houston, Texas, U.S.A., Pt. 1, pp. 180–181.

Edwards, J. G., Fransen, H. F., and Gilles, P. W. (1971) High-temperature masspectrometry vaporization and thermodynamics of titanium monosulfide, *J. Phys. Chem.* **54**(2), 545–554.

Ehlert, T. S., Dean, T. P., Billy M., and Labbe, J. C. (1980) Thermal decomposition of the oxynitride of silicon, *J. Amer. Ceram. Soc.* **63**(3/4), 235–236.

Fegley, M. B. (1981) The thermodynamic properties of silicon oxynitride, *J. Amer. Ceram. Soc.* **64**(9), C124–C126.

Fegley, M. B., and Lewis, J. S. (1980) Volatile element chemistry in the solar nebula: Na, K, F, Cl, Br, P, *Icarus* **41**(3), 439–455.

Fransen, H. F. (1963) The high-temperature vaporization of some oxides and sulfides of titanium, *Dissert. Abstr.*, **23**, 2713.

Fransen, H. F., and Gilles, P. W. (1965) A thermodynamic study of the vaporisation of titanium monosulfide, *J. Chem. Phys.* **42**(3), 1033–1040.

Grossman, L. (1972) Condensation in primitive solar nebula, *Geochim. Cosmochim. Acta* **36**(5), 597–619.

Grossman, L. (1973) Refractory trace elements in Ca-Al-rich inclusions in the Allende meteorite, *Geochim. Cosmochim. Acta* **37**(6), 1119–1140.

Grossman L., and Larimer, J. W. (1974) Early chemical history of the solar system, *Rev. Geophys. Space Phys.* **12**(1), 71–101.

Hager, P., and Elliott, F. (1967) The free energies of formation of CrS, Mo_2S_3 and WS_2, *Trans. Met. Soc. AIMF* **239**(4), 513–520.

Herndon, J. M., and Suess, H. E. (1976) Can enstatite meteorites form from a nebula or solar composition? *Geochim. Cosmochim Acta* **40**(4), 395–399.

Hertogen, J., Janssens, M., Takahashi, H., Morgan, J. W., and Anders, E. (1983) Enstatite chondrites: trace element clues to their origin, *Geochim. Cosmochim. Acta* **47**(12), 2241–2255.

Igaki, K., Ohashi, N., and Mukami, M. (1971) Phase relation of nonstoichiometric chromium sulfide CrS_x in the range of 1.200 to 1.400, *J. Phys. Soc. Japan* **31**(5), 1424–1430.

Ikramuddin, M., Binz, C. M., and Lipschutz, M. E. (1976) Thermal metamorphysm of primitive meteorites. II. Ten trace elements in Abee enstatite chondrite heated at 400–1000°C, *Geochim. Cosmochim. Acta* **40**(2), 133–142.

Ivanov, A. V. (1984) Uglistyi khondrite Kaidun: intensivnoe peremeshivanie vechchestva pri firmirovanii meteoritnykh roditel'skikh tel, *Tez. Dokl. 27-go Mezhd. Geol. Kongr.*, p. 297 Vol. 5, Nauka, Moskva.

JANAF Thermochemical tables with Supplements (1971, 1974, 1975, 1978). National Bureau of Standards, U.S. Dept. of Commerce, Washington, D.C.

Keil, K. (1968) Mineralogical and chemical relationships among enstatite chondrites, *J. Geophys. Res.* **73**(22), 6945–6976.

Keil, K., and Andersen, C. A. (1965a) Occurence of sinoite, Si_2N_2O in meteorites, Nature (London) **207**(4998), 745.

Keil, K., and Andersen, C. A. (1965b) Electron microprobe study of the Jajh deh Kot Lalu enstatite meteorite, *Geochim. Cosmochim. Acta* **29**(6), 621–632.

Keil, K., and Brett, R. (1974) Heideite, $(Fe, Cr)_{1+x}(Ti, Fe)_2S_4$, a new mineral in the Bustee enstatite achondrite, *Amer. Mineral* **59**(5/6), 465–470.

Keil, K., and Snetzinger, K. G. (1967) Niningerite: a new meteoritic sulfide, Science **155**(3761), 451–454.

Kelley, K. K. (1937) *Contributions to the Data of Theoretical Metallurgy*, Bull. No. 407, U.S. Bur. Mines.

Khodakovsky, I. L. (1975) Nekotorye voprosy termodinamiki vodnykh rastvorov pri vysokikh temperaturakh i davleniyakh, in: *Fisiko-khimicheskie problemy gidrotermal'nykh i magmaticheskikh protsessov*, Moskva pp. 124–150, Nauka.

Khodakovsky, I. L. (1982) On carbon to oxygen ratio in solar nebula, Abstr., *Lunar Planet. Sci. Conf. XIII*, Houston, Texas, U.S.A., Pt. 1, pp. 385–386.

Khodakovsky, I. L., and Petaev, M. I. (1981) Termodinamicheskie svoict va i usloviya obrazovaniya osbornita, sinoita i karlsbergita v meteoritakh, *Geokhimiya*, No 3, 329–340.

Kochchenko, V. I., and Grinberg, Ya. H. (1982a) Termodinamicheskie svoistva Si_2N_2O, *Izv. AN SSSR, Ser. Neorg. Mater.* **18**(6), 1047–1049.

Kochchenko, V. I., and Grinberg, Ya. H. (1982, b) Termodinamicheskie svoistva Si_3N_4, *Izv. AN SSSR, Ser. Neorg. Mater.* **18**(6), 1064–1066.

Larimer, J. W. (1967) Chemical fractionations in meteorites—I. Condensation of elements, *Geochim. Cosmochim. Acta* 31, 1215–1238.

Larimer, J. W. (1968) An experimental investigation of oldhamite CaS; and the petrological significance of oldhamite in meteorites, *Geochim. Cosmochim. Acta* **32**(5), 965–982.

Larimer, J. W. (1975) The effect of C/O ratio on the condensation of planetary material, *Geochim. Cosmochim. Acta* **39**(2), 389–392.

Larimer, J. W., and Bartholomay, M. (1979) The role of carbon and oxygen in cosmic gases: some applications to the chemistry and mineralogy of enstatite chondrites, *Geochim. Cosmochim. Acta* **43**(9), 1455–1466.

Larimer, J. W., and Buseck, P. R. (1974) Equilibrium temperatures in enstatite chondrites, *Geochim. Cosmochim. Acta* **38**(3), 471–477.

Lattimer, J. M., and Grossman, L. (1978) Chemical condensation sequences in supernova ejecta, *Moon Planets* **19**, 169–184.

Lavrukhina, A. K. (1982) O prirode izotopnykh anomalij v meteoritakh, *Meteoritika* No. 41, 78–92.

Levin, B. Yu., and Simonenko, A. N. (1977) Voprosy proischozhdeniya meteoritov, *Meteoritika* No. 36, 3–23.

Mason, B. (1966) Enstatite chondrites, *Geochim. Cosmochim. Acta* **30**(1), 23–39.

Mendybaev, R. A., et al. (1985) O kinetike reaktsij vosstanovleniya CO i N_2 v khimicheskoj evolyutsii doplanetnogo oblaka, *Geokhimiya* **8**, 1206–1217.

Mills, R. S. (1974) *Thermodynamic Data for Inorganic Sulphides, Selenides and Tellurides*, Butterworths, London, 845 pp.

Mineraly Spravochnik (1974) *Diagrammy Fasovykh Sostoyanij*, Moskva vyp. 1, Nauka, 514 pp.

Moriyama, J., et al. (1977) Thermodynamic study of the sulphides of Cr and Ni by EMF measurements, abstr., *Fifth Intern. Conf. Chem. Thermodyram.*, August 1977, Ronneby, Sweden, p. 84

Mraw, S. C., and Naas, D. F. (1979) The heat capacity of stoichiometric titanium disulfide from 100 to 700 K: absence of the previous reported anomaly at 420 K, *J. Chem. Thermodynam.* **11**(6), 585–592.

Muranaka, S. (1973) Order–disorder transition of vacancies in iron titanium sulfide ($FeTi_2S_4$), *Mater. Res. Bull.* **8**(6), 679–686.

Naumov, G. B., et al. (1974) *Handbook of Thermodynamic Data*, Rept. No USGS-WRD-74-001,

Naylor, B. F. (1944) High-temperature heat contents of ferrous and magnesium chromites, *Ind. Ing. Chem* **36**(10), 933–934.

Pankratz, L. B. (1982) *Thermodynamic Properties of Elements and Oxides*, Bull. No 672, U.S. Bur. Mines, 509 pp.

Parnjipe, V. G., Cohen, M., Bever, M. B., and Froc, C. F. (1950) The iron–nitrogen system, *J. Metals* **188**(2), 261–267.

Pelino, M., Viswanadham, P., and Edwards, J. G. (1979) Vaporization chemistry and thermodynamics of titanium monosulfide by the computer-automated simultaneous Knudsen-torsion effusion method, *J. Phys. Chem.* **83**(23), 2964–2969.

Petaev, M. I., and Khodakovsky, I. L. (1981) Ob usloviyakh obrasovaniya karbidov v ehnstatitovykh i zheleznykh meteoritakh, *Tez. Vses. sov. po geokhimii ugleroda*, Moskva. pp. 308–311, Geokhi and SSSR,

Petaev, M. I., and Skripnik, A. Ya. (1983) O mineral'nom sostave ehnstatitovykh meteoritov, *Meteoritika* No. 42, 86–92.

Petaev, M. I., et al. (1982) Termodinamicheskie svojstva dobreelita i usloviya obrasovaniya ego v meteoritakh,–*Geokhimiya* No. 5, 690–703.

Petaev, M. I., et al. (1983) Thermodynamic properties and origin of meteoritic minerals. III. Heideite, $FeTi_2S_4$, Abstr., *Lunar Planet. Sci. Conf. XIV*, Houston, Texas, U.S.A., Pt. 2, pp. 600–601.

Petaev, M. I., et al. (1986) O genesise mineralov ehnstatitovykh meteoritov, *Geokhimiya* (in press).

Ramdohr P. (1973) *The Opaque Minerals in Stony Meteorites*, Academir Verlay, Berlin, 245 pp.

Rezukhina, T. N., et al. (1965) Termodinamicheskie svojstva khromita zheleza iz ehlektrokhimicheskikh izmerenij, *Ehlektrokhimiya* **1**(9), 2014–2020.

Robie, R. A., Hemingway B. S., and Fisher J. R. (1978) Thermodynamic Properties of Minerals and Related Substances at 298.15 K and 1 bar (10^5 Pascals) Pressure and at Higher Temperatures, Bull, No. 1452, U.S. Geol. Soc., Washington, D.C., 456 pp.

Robinson, J. R., and Haas, J. L. (1982) Thermodynamic and thermochemical data for 10 minerals in the magnesia–silica–water system obtained from improved evaluation procedures, Abstr., *XII IUPAC Conf. Chem. Thermodyram*, London, p. 15.

Ryall, W. R., and Muan, A. (1969) Silicon oxynitride stability, *Science* **165**, 1362–1364.

Safronov, V. S. (1982) Sovremennoe sostoyanie teorii proiskhozhdeniya Zemli, *Izv. AN SSSR, Ser. Fizika Zemli* No. 6, 5–25.

Sakao, H., and Elliott, J. F. (1975) Thermodynamics of dilute b.c.c. iron silicon alloys, *Metall. Trans. A*, **6A**(10), 1849–1851.

Saxena, S. K., and Eriksson, G. (1983) High temperature phase equilibria in a solar-composition gas, *Geochim. Cosmochim. Acta* **47**(11), 1865–1874.

Sears, D. (1980) Formation of the E chondrites and aubrites— a thermodynamic model, *Icarus* **43**(1) 184–202.

Shomate, C. H. (1944) High-temperature heat content of magnesium nitrate, calcium nitrate and barium nitrate, *J. Amer. Chem. Soc.* **66**(6), 928–929.

Termicheskie konstanty vechchestv (1974)–Moskva. Vol. 7, VINITI, *Termodinamicheskie svoistva individual'nykh vechchestv* (1978–1983) Moskva. Vols. 1–4, Nauka.

Todd, S. S., and Coughlin, J. T. (1952) Low temperature heat capacity, entropy at 298.16 K and high temperature heat content of titanium disulfide, *J. Amer. Chem. Soc.* **74**(2), 525–526.

Volovik, L. S., et al. (1979) Termodinamicheskie svojstva sul'fidov perekhodnykh metallov, *Izv. AN SSSR, Ser. Neorg. Mater.* **15**(4), 638–642.

Wasson, J. T., and Wai, C. M. (1970) Composition of the metal, schreibersite and perryite of enstatite achondrites and the origin of enstatite chondrites and achondrites, *Geochim. Cosmochim. Acta* **34**(2), 169–184.

Whittingham, M. S. (1978) Chemistry of intercalation compounds metal guests in chalcogenide hosts, *Prog. Solid State Chem.* **12**(1), 41–99.

Wolf, R., Ebihara, M., Richter, G. R., and Anders, E. (1983) Aubrites and diogenites: trace element clues to their origin, *Geochim. Cosmochim. Acta* **47**(12), 2257–2270.

Yavnel', A. A. (1984) Nekotorye voprosy klassifikatsii meteoritov, *Dokl. 27-go Mezhd. Geol. Kongr.*, Moskva. vol. 11, pp. 79–86, Nauka.

Yudin, I. A. (1972) Mineragraficheskoe issledovanie meteorita Pilistvere, *Meteoritika* No. 83–89.

Yudin, I. A., and Smyshlyaev, S. I. (1964) Khimiko-mineragraficheskoe issledovanie neprosrachnykh mineralov meteoritov Norton County i Staroe Pes'yanoe, *Meteoritika* No. 25, 96–128.

Young, D. J., Smeltzer, W. W., Kirkaldy, J. S. (1973) Nonstoichiometry and thermodynamics of chromium sulfides, *J. Electrochem. Soc.* **120**(9), 1221–1224.

Zabejvorota, N. S., et al. (1980) Svobodnaya ehnergiya reaktsii obrasovaniya $FeCr_2O_4$, *Izv. AN SSSR, Ser. Neorg. Mater.* **16**(1), 181–183.

Chapter 4
Lithospheric–Atmospheric Interaction on Venus

V. P. Volkov, M. Yu. Zolotov, and I. L. Khodakovsky

Introduction

Physicochemical methods, especially equilibrium calculations of the phase relations in multicomponent natural systems, inevitably are applied to the investigation of objects that are impractical for experimental treatment.

The exploration of Venus, including the theoretical prediction of the chemical composition of its atmosphere and clouds and the mineral composition of its rocks by thermodynamical methods, has been undertaken for a number of years and is associated with certain advances, which are the subject matter of this chapter.

Venus can be considered a rather appropriate object for the physicochemical modeling of global natural processes. Indeed, this planet is like a specific laboratory in space for promoting the interaction of atmospheric gases with minerals and rocks under conditions approximating chemical equilibrium. The high surface temperature (735K), the absence of seasonal and the negligible latitudinal temperature variations, combined with the presence of a number of chemically active atmospheric gases, are all regarded as favorable factors.

The X-ray fluorescent analyses of Venus' soil obtained by Venera 13 and Venera 14 landers (Surkov *et al.*, 1983), as well as the first radar topographic mapping from Pioneer Venus Orbiter (Masursky *et al.*, 1980), have sharply improved the investigation of lithospheric–atmospheric interactions. In the following text we shall discuss the available data on the physical and chemical properties of Venus' surface rocks together with the tropospheric and cloud particle chemical composition, these being the reservoir of elements outgassed from the planetary interior. This set of data allows us to understand the probable direction of chemical weathering processes as well as to outline a schematic representation of recycling within the outer planetary shells.

The numerical simulation of physicochemical processes in heterogeneous systems enables us to describe the lithospheric–atmospheric interaction in terms of a chemical model. Such a model, consistent with the cosmochemical and geological data, could be regarded as a reasonable base for the solution of the sophisticated problems of the origin and evolution of Venus' atmosphere.

It might be well to point out that the recent data on Venus are summarized in Hunten *et al.* (1983), whereas atmospheric chemistry and lithospheric–atmospheric interaction are specially considered in Volkov (1983) and Lewis and Prinn (1984).

Venus' Planetary Surface: Geology, Physical Properties and the Chemical Composition of Venus' soil

Venus is enveloped by a dense, opaque atmosphere with a permanent cloud cover. The latter prevented all attempts to obtain any data on the character of its surface until the beginning of the 1960s, when radar studies were carried out.

Dramatic breakthroughs in our knowledge of Venus were made in 1978–1979 after the radar mapping of 93% of Venus' surface by the Pioneer Venus Orbiter resulted in the altimetric map at 1 : 50000000 scale, with an arial resolution of 30 km and 200 m of elevation (Masursky *et al.*, 1980) (Fig. 1).

Geological and Physiographic Outline of Venus' Surface

The surface of Venus is much smoother than the surfaces of all the other inner planets: more than 90% of its surface area is defined as having ± 1.5 km elevation in relation to the datum level (the median radius) of 6051.6 km. The hypsometric histogram of Venus has only one mode (Fig. 2) in contrast to that of Earth. Unimodality of the histogram is interpreted as evidence of critical differences between Venusian and terrestrial orogenic processes (McGill *et al.*, 1983).

According to the Pioneer Venus radar map the predominant physiographic province (60% of surface area) is related to so called "rolling plains" (no more than 2 km above the datum level); the lowest regions, termed lowlands, occupy ~27% (below the datum level); highlands comprise about 13% and are defined as areas with more than 2-km elevation. The highest point on Venus' surface is Maxwell Montes, with the height of 11.1 km; the lowest point corresponds to the bottom of Diana Chasma and is 2 km below the datum level.

A radically new stage in our knowledge of Venus was commenced with the Venera 15 and Venera 16 space missions, which orbitted Venus with the aim of making up a radiobrightness image mosaic of the northern hemisphere at a scale of 1 : 4000000 with a resolution of 1–2 km (Barsukov *et al.*, 1984, 1985). On the basis of this mosaic, a physiographic map was drawn involving about 70% of the northern hemisphere surface area and all four types of physiographic provinces distingusihed by Masursky *et al.* (1980). Four types of geological features are

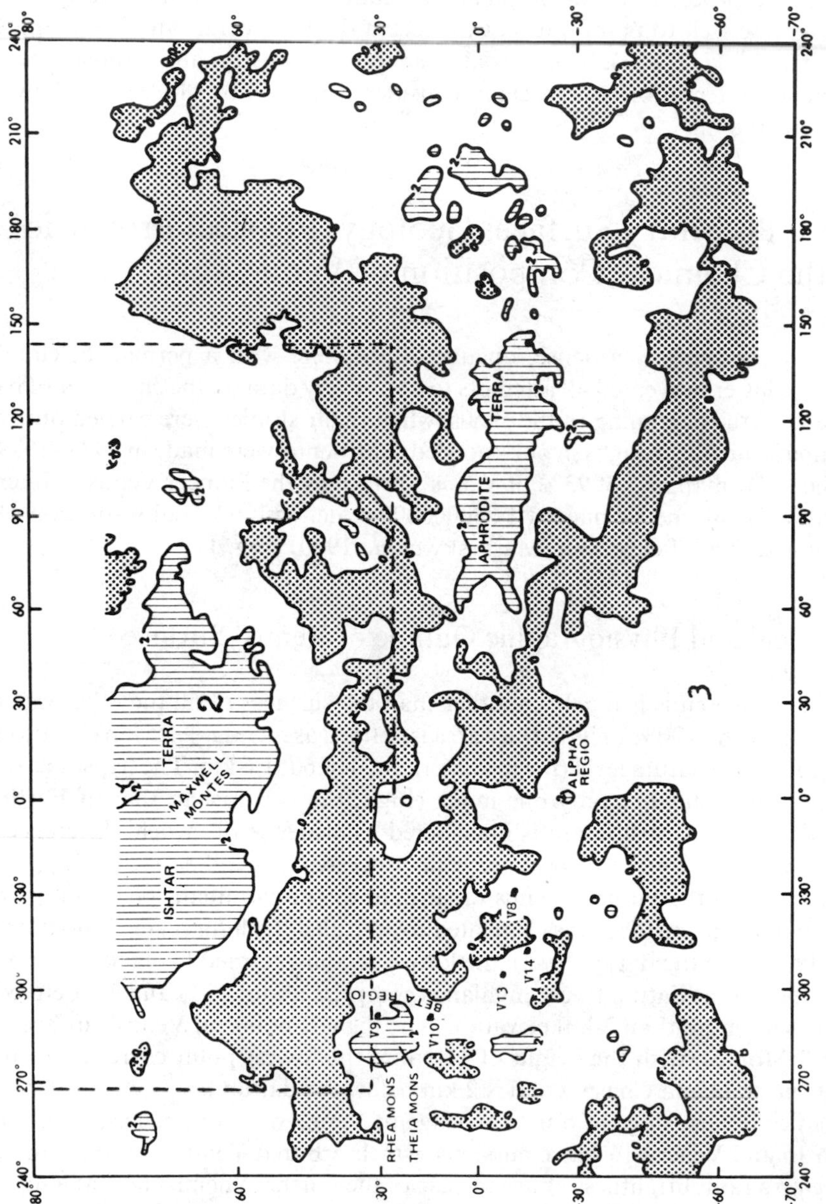

Fig. 1. Venus physiographic provinces (after Masursky *et al.*, 1980). 1, Lowlands; 2, highlands; 3, rolling plains. The area covered by radar mapping of Venera 15 and Venera 16 Orbiters is contoured by dashed lines (Barsukov *et al.*, 1984).

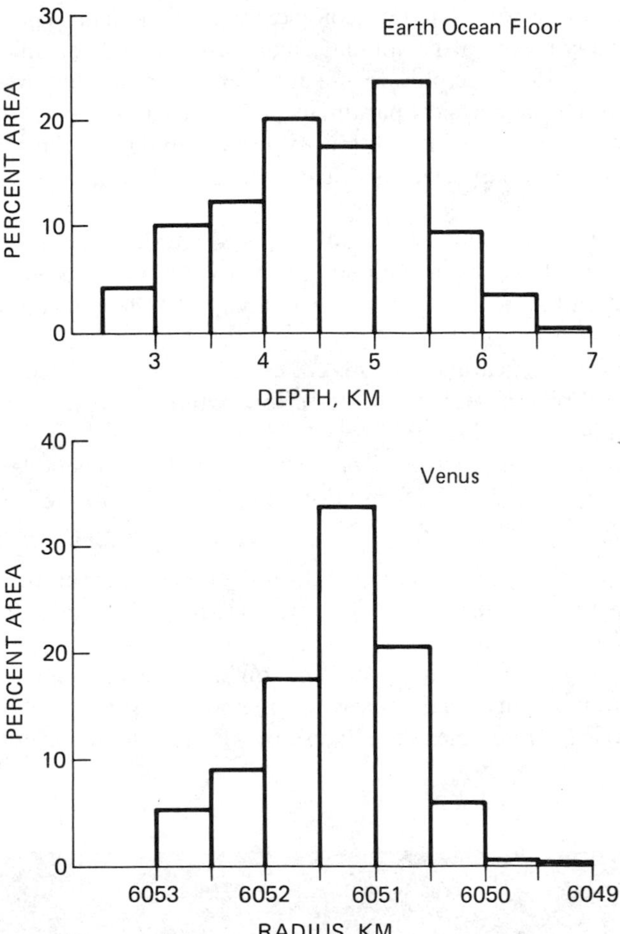

Fig. 2. Hypsograms for Earth's ocean mode and for Venus elevations < 6053 km (Phillips and Malin, 1983). By permission from *Venus*, edited by D. M. Hunten, L. Colin, T. M. Donahue, and V. I. Moroz. University of Arizona Press, Tucson, Copyright 1983.

designated in Barsukov *et al.* (1984): volcanic terrain, especially the basaltic lava sheets; volcanic–tectonic structures, in particular domelike uplifts connected with riftlike features, e.g., Beta Regio (Fig. 1); disjunctive tectonic terrain, having no analogs on any inner planet ("parquette-like" structure); and finally, impact craters.

The absolute geologic age of the volcanic terrain was estimated to be 0.5–1.0 billion years according to crater density. Other physiographic types are so poorly cratered that a statistical treatment is not representative.

In the following a number of conclusions on Venus' geology associated with the problem in question and derived from the radar mapping are presented:

Venus should be considered a geologically active planet. The formation of Venusian relief is a result of a combination of cratering and volcanic and tectonic activity. Venus' global tectonics has a number of unique features despite the similarity between the physical parameters of Venus and Earth.

A heavily cratered ancient continental crust, similar to the known Lunar, Martian, and Mercurian crust, is not observed on Venus within the mapped surface area.

Tectonic–magmatic activity, including global basaltic volcanism, has ceased on Venus, probably not later than 500 million years ago. Recent volcanism, if any, is restricted to a few localities (Masursky *et al.*, 1980; Ksanfomality *et al.*, 1983).

The Unaltered appearance of the craters and associated tectonomagmatic terrain (0.5–1.0 billion years) indicates a low activity of exogenic processes at subsequent stages of geological history.

At present, evidence of fluvial erosion is not distinguishable and aeolian activity has not resulted in the formation of global geomorphic features.

The television panoramas of the landing sites transmitted by the Venera 9, 10, 13, and 14 landers are representative of the planetary surface at the rolling plains and marginal areas of the volcanic Beta Regio (Florensky *et al.*, 1977, 1983a, b) (Figs. 3 and 4).

The Venera 9 landing site is situated on a slope of ~20° covered with flattened, slablike stony fragments with an underlying relatively fine-grained soil. The landscape resembles the talus slope of an arid zone of the Earth. The

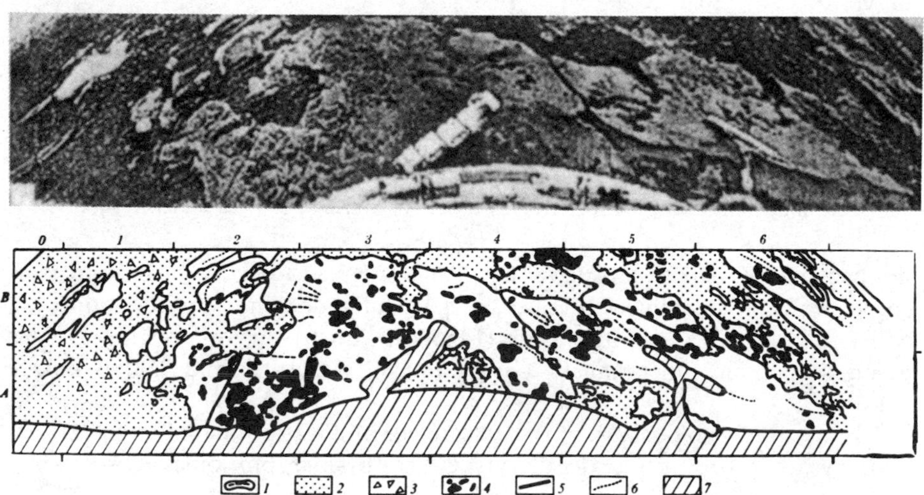

Fig. 3. TV panorama transmitted by Venera 10 and its geomorphic scheme (Florensky *et al.*, 1979). 1, Outcrops of bedrock; 2, soil; 3, undivided talus; 4, cellular voids on slabs; 5, cracks; 6, "honeycomb" weathering; 7, Venera 10 lander details.

Fig. 4. TV panoramas transmitted by Venera 13 (above) and Venera 14 (below) (Florensky *et al.*, 1982).

Venera 10 landing site is located on a plain, locally covered by a thin layer of fine-grained soil with protruding outcrops of bedrock (Fig. 3). Three types of exogenic working are proposed: (a) formation of cracks; (b) honeycomb weathering, similar to that of terrestrial deserts; (c) surface "etching," resembling cellular weathering.

Venera 13 and Venera 14 television panoramas are interpreted as the images of volcanic–sedimentary terrain comprised of layered pyroclastic tephroitic ejecta or lithified products of weathered ancient volcanic rocks. Florensky *et al.* (1983b) prefer a sequence consisting of an aeolian–sedimentary origin of the layered rocks followed by a cyclic process, on Venus' surface, of weathering → transportation → deposition → lithification → weathering. Presumably this cycling, as well as the formation of cells on mottled slabs (Venera 10 panorama), involves the chemical interaction of atmospheric gases with minerals of the surface rocks.

Physical Properties of Venus' Soil

In addition to the data on the physiography of Venus' surface in megascopic and microscopic scale the, physical properties of Venus' soil were investigated by means of Earth-based remote sensing.

The physical conditions at Venus' surface resemble thermostata. Indeed, the Venera and Pioneer Venus space missions revealed the average temperature at the surface to be 734K at 90 $kg \cdot cm^{-2}$ pressure (Avduevskiy *et al.*, 1983). Direct comparison of day and night temperatures at the same hypsometric level indicates that the pressure differential is about 0.2 $kg \cdot cm^{-2}$, corresponding to a temperature differential of 3K.

The temperature and pressure at the planetary surface is in satisfactory agreement with the adiabatic atmospheric model advanced as far back as 1972 (Marov, 1972; Kuz'min and Marov, 1974). The atmospheric pressure above the highlands is lower than that above the lowlands by no less than 50 atm and the maximal temperature differential as a function of hypsometric level is estimated as 85–90K.

The dielectric constant (ε) was derived for the first time from Earth-based radar measurements, the mean value being 4.7 ± 0.8 (Kuz'min and Marov, 1974, see Table 1), corresponding to a mean density for the surface rocks of 2.3 ± 0.4 $g \cdot cm^{-3}$. These estimates indicate the low probability of a wide distribution of loose fines similar to those of the lunar regolith ($\varepsilon \approx 2.7$). The high reliability of these data was corroborated later by radiophysical measurements made by the Venera 8, 9, and 10 missions and the Pioneer Venus Orbiter.

Significant variations of dielectric constant values (Table 1) are considered as evidence of essential differences in the physical properties of surface rocks. A double-layer model of the surface, consisting of a 0.5- to 1.0-m-thick regolith-like soil overlying a dense bedrock, was proposed in by Kroupenio (1972) and Warnock and Dickel (1972). Final conclusions on the distribution of regolith-like soils and their interrelation with dense bedrock would be premature.

The discovery of regions with an extremely high dielectric constant ($\varepsilon = 27 \pm 7$) associated with high radiobrightness areas, e.g., Theia Mons (Fig. 1), was reported after the Pioneer Venus mission. Presumably the high values of the dielectric constant should be interpreted as a result of the presence of highly electroconductive minerals, especially ones with pyrite or iron oxides as main components (Ford and Pettengill, 1983).

Direct measurements of the physicomechanical properties of Venus' soil were carried out at the landing sites of Venera 13 and Venera 14 (Table 1). Bearing capacity was derived from the data obtained by a special penetrator (Kemurdzhian *et al.*, 1983). The tests revealed that the Venera 13 landing site was covered by loose, highly porous soil, whereas the bearing capacity at the Venera 14 landing site corresponded to that of fractured or porous compact rocks. The specific electric resistance of the soil was measured in the same experiment (Table 1) and was found to be one or two orders of magnitude lower in relation to a basalt heated up to 750K. Supposedly, this phenomenon can be interpreted in terms of the high electroconductivity of the surface soil in accordance with the conclusions made from the Pioneer Venus Orbiter radar measurements (Ford and Pettengill, 1983).

The physicomechanical characteristics were also evaluated from measurements of impact loading at the landing sites of the Venera 13 and Venera 14 space probes. The results were compared with Earth-based experiments which consisted of the jettisoning of lander models onto different types of landing surfaces. The bearing capacity at the Venera 13 landing site was found to be in a good agreement with penetrator measurements (Table 1). The Venera 14 landing site

Table 1. Physical properties of Venus surface.

Dielectric constant	Density, ρ, $g \cdot cm^{-3}$	Determination method; space mission	Bearing capacity, $kg \cdot cm^{-2}$	Specific electric resistance, ohm · m	Volume density, $g \cdot cm^{-3}$	Porosity, %	Determination method; space mission
4.7 ± 0.8	2.3 ± 0.4	Earth-based radiolocation (Kuz'min and Marov, 1974)	4–5	—	1.4–1.5	~50	Impact loading, Venera 13 (Avduevskiy *et al.*, 1983)
			2.0 4.5	—	1.15–1.20 1.4–1.5	~60 ~50	Impact loading, Venera 14 (Avduevskiy *et al.*, 1983)
3.2 ± 1.3	1.40 ± 0.55	Radioaltimetry, Venera 8 (Bashmashnikov *et al.*, 1976)	2.6–1.0	89	—	—	Dynamic penetration, Venera 13 (Kemurdjian *et al.*, 1983)
$\varepsilon_1 = 1.5$ $\varepsilon_2 = 8.3$	$\rho_1 = 1.2–1.9$ $\rho_2 = 2.2–2.7$	Earth-based radiolocation (Kroupenio, 1972; Warnock and Dickel, 1972)	65–250	73	—	—	Dynamic penetration, Venera 14 (Kemurdjian *et al.*, 1983)
3.0 ± 9.1	—	Bistatic radiolocation, Veneras 9 and 10 Orbiters (Kolosov *et al.*, 1981)					
4.1 ± 0.9	—	Earth-based radiolocation (Alexandrov *et al.*, 1980)					
4.2 ± 0.8	~2.2	Radiolocation, Pioneer Venus Orbiter [McGill *et al.*, 1983; Ford and Pettengill, 1983]					
—	2.8 ± 0.1	Gamma-ray detector, Venera 10 (Surkov, 1977)					

probably is covered by a loose, regolithic layer with a thickness of about 10 cm. Therefore, the instrumental data are in general agreement with the presentation of Venera 13 and Venera 14 landing sites as a volcanic–sedimentary terrain. What kind of geological agent could be responsible for moving the rock particles released in the process of chemical weathering or other type of exogenic working? Aeolian activity is considered the dominant exogenic factor in the absence of oceans, glaciers, and rivers at Venus' surface.

Aeolian and Volcanic Activity

The theoretical evaluation of near-surface wind velocity was made as early as 1972 using Doppler shift measurements in relation to parachute lander descent (Kerzhanovich and Marov, 1983). The wind velocity was found to be about $0.5\ m \cdot s^{-1}$. Later this was supported by direct anemometric measurements at the Venera 9 landing site ($0.5\ m \cdot s^{-1}$) and Venera 10 landing site (0.8–$1.0\ m \cdot s^{-1}$) (Avduevskiy *et al.*, 1983).

Indirect data derived from the interpretation of Venera 9, 10, and 13 TV panoramas and atmospheric physical property measurements indicate that dust outbursts originated as a result of lander impact (Table 2).

There is no evidence for the existence of significant aeolian features on Venus' surface, as these were not observed even by the 1-km resolution radar mapping (Barsukov *et al.*, 1984); such morphological forms also were not found on the TV

Table 2. Dust characteristics on Venus' surface at lander sites (modified from Garvin 1984).

Lander	Geomorphic characteristics	Dust phenomenon
Venera 9	Blocks on soil (35% cover)	Dust spurt (10–25 s), particle radii 5–50 microns, dust content 0.1–$0.01\ kg \cdot m^{-3}$, thickness of dust layer at the surface 10–100 microns (Ekonomov *et al.*, 1980); no effects in the panorama
Venera 10	Bedrock, soils	Darkening in center of panorama; dust properties similar to Venera 9
Venera 11	High surface albedo	No dust spurt
Venera 12	Low surface albedo	Dust spurt (20–30 s)
Venera 13	Blocks, layered bedrock, soils	Dust spurt (40 s); dust and pebbles on both sides of lander ring, brightness variations in panoramas
Venera 14	Layered bedrock, little soil	No dust spurt; dust on one side of lander ring, bright variations in panoramas
Pioneer Venus, "Day"	?	Dust cloud (210 s) according to nephelometer data, hight of cloud ascension up to 100 m; particle radii 5–50 microns

panoramas. In spite of this, theoretical implication and modeling experiments support the concept of aeolian activity as a dominant factor of modification of the recent planetary surface.

Modeling experiments in carbon dioxide atmospheres under 100 atm of pressure (Greeley *et al.*, 1984) showed that the transport of particles with diameters up to 75 microns could be effected by saltation, providing wind speeds were in the range of 1–3 $m \cdot s^{-1}$. Ripple marks are suggested as probable aeolian features. The particle behavior of aeolian transport on Venus evidently resembles the sorting of materials on the ocean bottom at 1000 m depth. Similar conclusions were derived from the theoretical calculation of optimal particle radii and threshold friction speed (Fig. 5). The fraction with particle diameters lower than 30 microns was predicted to be transported by suspension (designated "dust"). It is just this fraction that is considered the predominant component of aeolian "sedimentary" rocks, because the ripple marks formed by the "sand" fraction (20–1000 microns) are not observed on the TV panoramas (Florensky *et al.*, 1983b).

Low wind speed on the surface of Venus results in a low kinetic energy for transported particles. Hence the deformation of rocks caused by impacting dust particles is estimated as only 0.01% of that on the terrestrial surface. Therefore aeolian abrasion on Venus is proposed to be a subordinate process.

The alternative concept of widespread aeolian sedimentation on Venus' surface is reported in Warner (1983). The highlands are considered erosion areas, and the rolling plains are suggested to be sedimentary deposits of volcanic origin (pyroclastics) as well as products of chemical weathering. It is assumed that

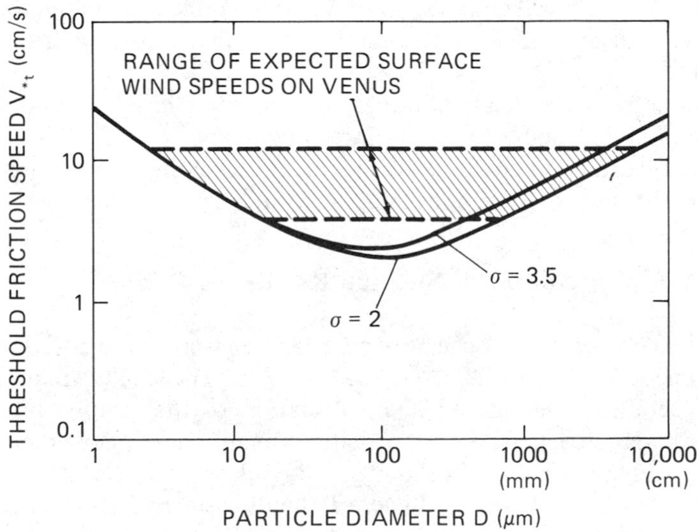

Fig. 5. Threshold fraction speed (V_{*t}) as a function of particle diameter D under Venus' conditions for particle densities σ of 2 $g \cdot cm^{-3}$ and 3.5 $g \cdot cm^{-3}$ (McGill *et al.*, 1983).

recycling has been attained as a result of lithostatic subduction of volcanic material with succeeding remelting in the abyssal zones of Venus' crust.

Unfortunately the origin of the thin layering in rocks at the Venera 13 and Venera 14 sites could not be elucidated adequately. Some authors reported that the pyroclastics on Venus could have originated only if their genesis provided an extremely high volatile content ($CO_2 > 4\%, H_2O > 3\%$), (Garvin *et al.*, 1982; Wilson and Head, 1983). This concept leads to an alternative conclusion: the volatile content of Venus' magmas is substantially higher in relation to terrestrial equivalents or the atmospheric pressure corresponding to these volcanic events was at least one order of magnitude lower than on contemporary Venus. The layered rocks observed on the Venera TV panoramas are considered the products of multistage lava effusion (Wilson *et al.*, 1984). Other investigators (Frenkel' and Zabalueva, 1983) evaluated the rate of solidification of Venusian basalts and concluded it to be markedly higher than on the Earth; therefore thin-layered lava sheets are not plausible.

In the following we summarize the information on the character of exogenic processes on Venus.

Exogenic working on Venus is characterized by extremely low speed as evidenced by the unaltered appearance of impact craters as well as the conservation of the ancient (0.5–1.0 billion years) tectonic–magmatic physiographic features.

Mechanical weathering as an equivalent of terrestrial geological events is absent because there is no liquid water, no biosphere, and negligible climatic contrast.

Data on the physicomechanical properties of Venus' soil indicate that a global cover of loose regolith-like fines is not developed. Wide distribution of porous, thin-layered volcanic rocks, presumably of aeolian–sedimentary origin, is suggested.

Aeolian activity on Venus is rather intensive but has not resulted in the formation of global landforms as on Earth and Mars.

Chemical weathering could be considered as resulting in corrosion and degradation of surface rocks and favoring mechanical weathering.

Chemical Composition of Surface Rocks

The first data on Venus' rock chemical composition were obtained from gamma-ray experiments on Venera landers (Surkov, 1977). The results are presented in Table 3. Detailed investigation and correlation of the relationships among petrogenic and radioactive elements classifies Venera 8 landing site rocks as high-potassium magnesial lavas (shoshonites) (Barsukov *et al.*, 1982a) or alkaline syenites (Florensky *et al.*, 1983a). Venera 10 landing site rocks are related to trap formation basalts (Barsukov, 1985) (Fig. 6). These data indicate that there exist at least two petrochemical rock types, derived from the magmatic and/or metamorphic differentiation of Venus' lithosphere.

Table 3. Radioactive element content in Venus' rocks (Surkov, 1977).

Sample	K, %	U, ppm	Th, ppm
Venus' rocks at Venera 8 landing site	4.0 ± 1.2	2.2 ± 0.7	6.5 ± 0.2
Venus' rocks at Venera 9 landing site	0.5 ± 0.1	0.6 ± 0.2	3.7 ± 0.4
Venus' rocks at Venera 10 landing site	0.3 ± 0.2	0.5 ± 0.3	0.7 ± 0.3
Terrestrial basalts	0.76	0.86	2.1
Terrestrial granites	3.24	9.04	21.9

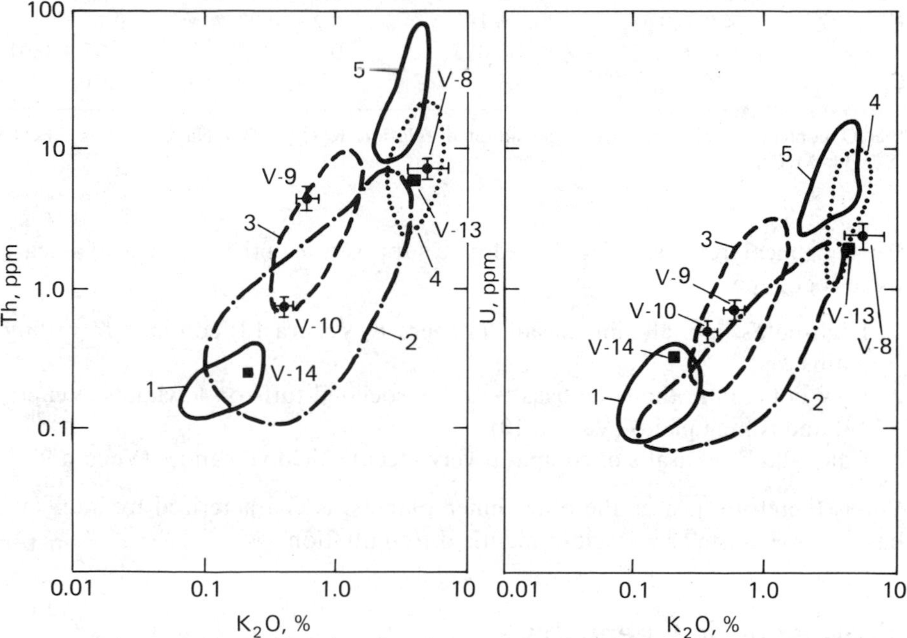

Fig. 6. The Th–K and U–K systematics of Venus' and terrestrial magmatic rocks (Barsukov, 1985). 1. Basalts of midoceanic ridges; 2, tholeiites, calc–alkaline series of active continental margins; 3, traps, Afar tholeiites; 4, nepheline syenites; 5, alkaline basaltoids.

Direct elemental analyses of Venus' rocks were carried out by means of an X-ray fluorescent spectrometer at the Venera 13 and Venera 14 landing sites (Barsukov *et al*., 1982d; Surkov *et al*., 1983). The Venera 13 landing site, corresponding to a rolling plain, is represented by alkaline, high-potassium basalts (phonolitic tephrites). The Venera 14 landing site is confined to an area of a transitional type between lowlands and rolling plains and is represented by rocks basically similar to the terrestrial oceanic tholeiitic basalts (Tables 4 and 5).

Table 4. Chemical composition of Venus' rocks and their terrestrial equivalents at the lander sites of Venera 13 and Venera 14 (Surkov *et al*., 1983; Volkov and Khodakovsky, 1984).

Oxides	Venera 13	Leucitic basalt	Venera 14	Tholeiitic basalt
SiO_2	45.1 ± 3.0	46.18	48.7 ± 3.6	50.6
TiO_2	1.59 ± 0.45	2.13	1.25 ± 0.41	1.2
Al_2O_3	15.8 ± 3.0	12.74	17.9 ± 2.6	16.3
FeO	9.3 ± 2.2	9.86	8.8 ± 1.8	8.8
MnO	0.2 ± 0.1	0.19	0.16 ± 0.08	0.2
MgO	11.4 ± 6.2	8.36	8.1 ± 3.3	8.5
CaO	7.1 ± 0.96	8.16	10.3 ± 1.2	12.0
Na_2O[a]	2.0 ± 0.5	2.36	2.4 ± 0.4	2.4
K_2O	4.0 ± 0.63	6.18	0.2 ± 0.07	0.1
S	0.65 ± 0.40	0.036	0.35 ± 0.28	0.07 ± 0.01
Cl	<0.3	—	<0.4	0.01

[a] Na_2O content is calculated from the adopted relation $K_2O/(K_2O + Na_2O)$ versus $\Sigma FeO/(\Sigma FeO + MgO)$.

Petrochemical treatment (Fig. 7) leads to a supposition of the existence of at least three rock types:

1. High-potassium, alkaline basalts (Venera 8, Venera 13) confined to rolling plains
2. Low-potassium, tholeiitic basalts and associated tuffs on lowlands (Venera 14) and rolling plains (Venera 10)
3. Calc–alkaline basalts of comparatively recent shield volcanoes (Venera 9)

Venus therefore, just as the other inner planets, is characterized by intensive basaltic volcanism as a result of mantle differentiation.

Tropospheric Chemistry

After the Venera 4 space mission in 1967 it became clear that carbon dioxide was the predominant component of Venus' troposphere (97%); molecular nitrogen comprises about 3%, and other components in total make up about 0.1% (H_2O, SO_2, Ar, etc.). The chemical composition data are presented in Table 6 and Fig. 8.

Venus' troposphere is characterized by a uniform mixing of carbon dioxide and nitrogen, resulting in their equac concentration at any level. At the same time, differences in the rate of chemical and physical atmospheric processes have resulted in concentration gradients of microcomponents. It is well established that water vapor, sulfur dioxide, and carbon monoxide contents vary in the vertical profile of the troposphere (Fig. 8). Unfortunately data on chemical

Table 5. Normative composition of Venus' rocks and their terrestrial equivalents (Barsukov *et al.*, 1982a).

C.I.P.W.[a] norms	Venera 13	Leucitic basalt	Shoshonite	Teschenite	Missourite	Venera 14	Tholeiitic basalt
Hypersthene	—	—	4.7	—	—	18.2	14.2
Olivine	26.6	16.6	9.4	13.7	22.2	9.1	8.1
Diopside	10.2	29.4	8.3	32.8	31.4	9.9	21.2
Anorthite	24.2	6.2	23.1	20.1	11.9	38.6	33.6
Albite	3.0	—	29.9	—	—	20.7	20.3
Orthoclase	25.0	11.8	23.0	—	—	1.2	0.3
Nepheline	8.0	11.2	—	11.2	5.1	—	—
Leucite	—	20.6	—	18.4	21.6	—	—
Ilmenite	3.0	4.2	1.6	2.5	2.7	2.3	2.3
Monticellite	—	—	—	1.3	5.1	—	—

[a] Cross, Iddings, Pirsson and Washington normative method of calculation may be found in many petrological texts.

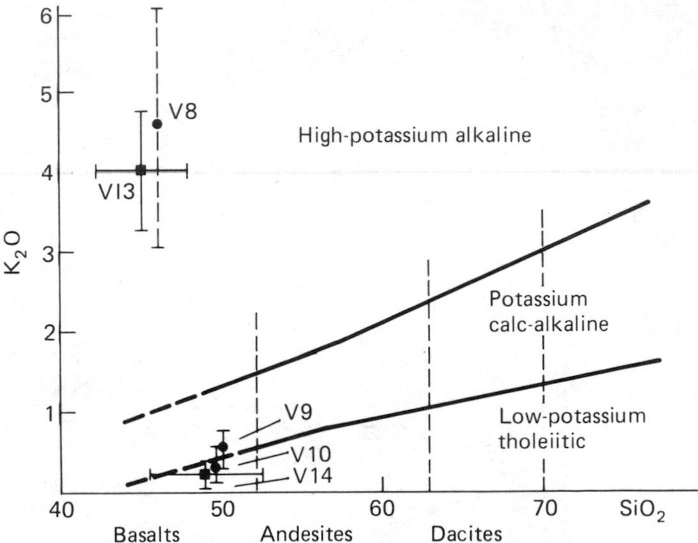

Fig. 7. Petrochemical systematics of Venus' and terrestrial magmatic rocks on a SiO_2–K_2O diagram (Barsukov *et al.*, 1982a). The compositions of Venus' rocks are shown with error bars.

Table 6. Chemically active microcomponents in Venus' troposphere (instrumental data).[a]

Gas	Altitude of measurement, km	Mixing ratio, ppm	Instrumental technique, space mission
COS	37–29	40 ± 20	Gas chromatography, Venera 13, 14
	22	<2	Gas chromatography, Pioneer Venus
	<20	<3	Mass spectrometry, Pioneer Venus
H_2S	37–29	80 ± 40	Gas chromatography, Venera 13, 14
	22	<2	Gas chromatography, Pioneer Venus
	<20	3 ± 2	Mass spectrometry, Pioneer Venus
S_2	<52	0.02–0.8	Spectrophotometry, Venera 11, 12
HCl	~ 70	0.42–0.61	Earth-based spectroscopy
	<50	<10	Mass spectrometry, Pioneer Venus
HF	~ 70	<1	Earth-based spectroscopy
O_2	~ 70	<1	Earth-based spectroscopy
	<60	<50	Spectrophotometry, Venera 11, 12
	52	43.6 ± 25.2	Gas chromatography, Pioneer Venus
	42	16.0 ± 7.4	Gas chromatography, Pioneer Venus
O_2	<42	<20	Gas chromatography, Venera 11, 12
	58–35	18 ± 4	Gas chromatography, Venera 13, 14
H_2	58–49	25 ± 10	Gas chromatography, Venera 13, 14

[a] Data on H_2O, SO_2, and CO measurements are shown in Fig. 8.

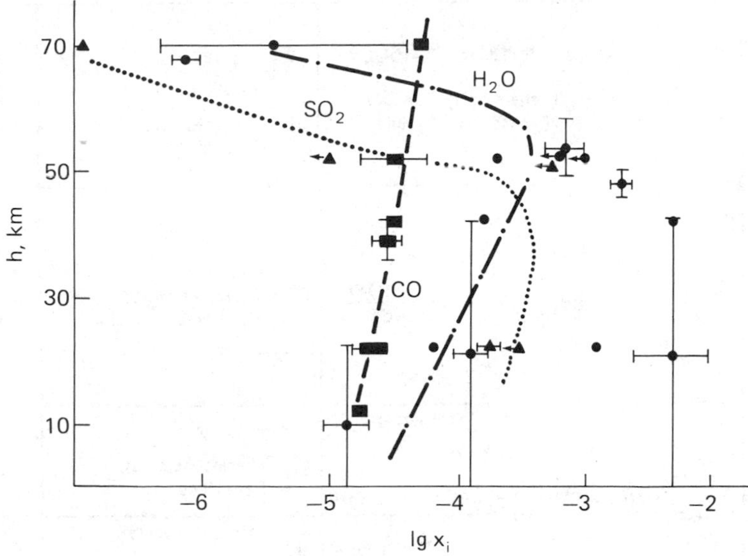

Fig. 8. Distribution of water vapor, sulfur dioxide, and carbon monoxide on Venus' low troposphere (instrumental data)

composition below 20 km are not available; therefore assumptions and extrapolations are used in relation to a number of microcomponents.

As early as 1964, R. Mueller proposed a threefold structure of Venus' atmosphere in terms of atmospheric chemistry (Fig. 9).

1. A zone of thermochemical reactions, where the atmospheric composition is buffered by minerals of surface rocks
2. A zone of 'frozen" chemical equilibrium, where gas concentrations correspond to the equilibrium chemical composition of the zone of thermochemical reactions
3. A zone of photochemical reactions in the upper atmosphere

The assumption of mineral buffering of tropospheric chemical composition was also used by Lewis (1970) to calculate near-surface tropospheric compositions (Table 7) based on carbon dioxide and water vapor determinations from the Venera 4, 5, and 6 space missions and spectroscopic measurements of CO, HC1, and HF content.

A chemical model of Venus' troposphere was proposed by Florensky *et al.*, (1978). The troposphere was subdivided into three zones in relation to the prevalence of different chemical processes:

1. An upper zone of photochemical processes predominately above the cloud deck and within the upper cloud layers
2. An intermediate zone of concurring photochemical and thermochemical

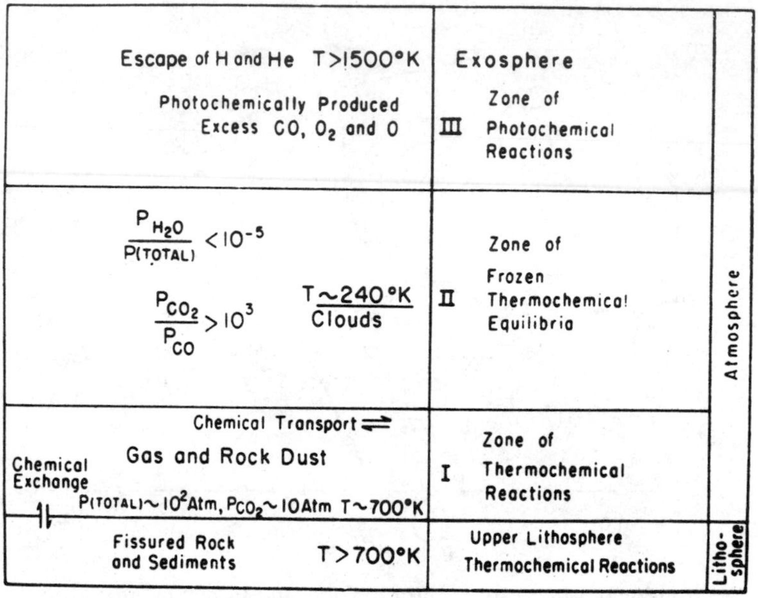

Fig. 9. Vertical zoning of Venus' atmosphere (Mueller, 1964).

Table 7. Calculated mixing ratios of gases in near-surface troposphere of Venus (microcomponents).

	Source			
Component	Lewis (1970)	Khodakovsky (1982)	Krasnopolsky and Parshev (1981)	Zolotov (1985)
CO	2×10^{-4}	1.7×10^{-5}	1.5×10^{-5}	7.2×10^{-6}
H_2O	5×10^{-4}	2×10^{-5}	2×10^{-4}	2×10^{-5}
SO_2	3×10^{-7}	1.3×10^{-4}	1.3×10^{-4}	1.3×10^{-4}
H_2S	5×10^{-6}	5.2×10^{-8}	3×10^{-7}	8×10^{-9}
COS	5×10^{-5}	2.3×10^{-5}	2×10^{-5}	3×10^{-6}
S_2	2×10^{-8}	1.8×10^{-7}	10^{-7}	1.3×10^{-8}
H_2	7×10^{-7}	2.4×10^{-9}	2×10^{-8}	10^{-9}
O_2	8×10^{-26}	10^{-23}	—	10^{-23}

reactions involving the troposphere below the cloud base and lower cloud layers and characterized by nonequilibrium conditions

3. A lower zone of the troposphere assumed to represent thermochemical equilibrium conditions.

According to this model the variations of microcomponent concentrations were predicted to be a function of the temperature and density of the atmosphere in vertical profile (Fig. 10).

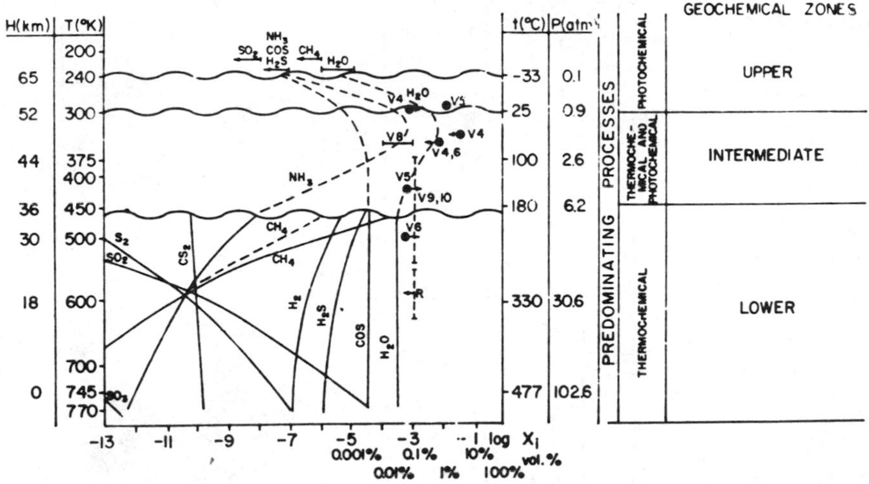

Fig. 10. Chemical model of vertical zoning in Venus' atmosphere (Florensky *et al.*, 1978). Uniform distribution of CO_2, N_2, H_2O, and CO is adopted; the relative distribution of sulfur-bearing gases is calculated at $x_{COS} = 3 \times 10^{-5}$. Measurements of gas concentrations are shown according to Venera missions and earth-based data.

A comparison of instrumental and calculated data on atmospheric chemistry became available for the first time after the completion of the Venera 11 and 12 and the Pioneer Venus space missions (Barsukov *et al.*, 1980a). The calculated vertical tropospheric profile of carbonylsulfide content is shown on Fig. 11. The mole fraction of COS on the logarithmic scale is plotted versus a temperature interval of 670–750K, corresponding 0–10 km altitude and derived from the equilibrium constant of the reaction:

$$3\,CO + SO_2 \leftrightarrows COS + 2\,CO_2,$$

$$K_p = \frac{x_{COS} \cdot x_{CO_2}^2}{x_{CO}^3 \cdot x_{SO_2}} \cdot P_{tot}^{-1}. \tag{1}$$

The values of CO_2, CO, and SO_2 mole fractions were taken from Venera 12 experimental determinations (Gel'man *et al.*, 1979). The equilibrium value of COS is substantially higher in relation to its experimental value ($\leqslant 10$ ppm) (Table 6) and even at 10-km altitude attains an implausible content exceeding 10,000 ppm. Thus the chemical equilibrium, if any, could be predicted only at the atmosphere–surface interface (Fig. 11). In addition to this conclusion the equilibrium composition of the troposphere allows for a steep vertical gradient of sulfur dioxide (Fig. 10), which is not observed in experiments.

The interpretation of instrumental data obtained by the Venera 11 and 12 and Pioneer Venus space missions formed the basis for the chemical models of Venus' troposphere (Barsukov *et al.*, 1980a; Krasnopolsky and Parshev, 1981). Both models use the initial assumption of general nonequilibrium conditions in Venus'

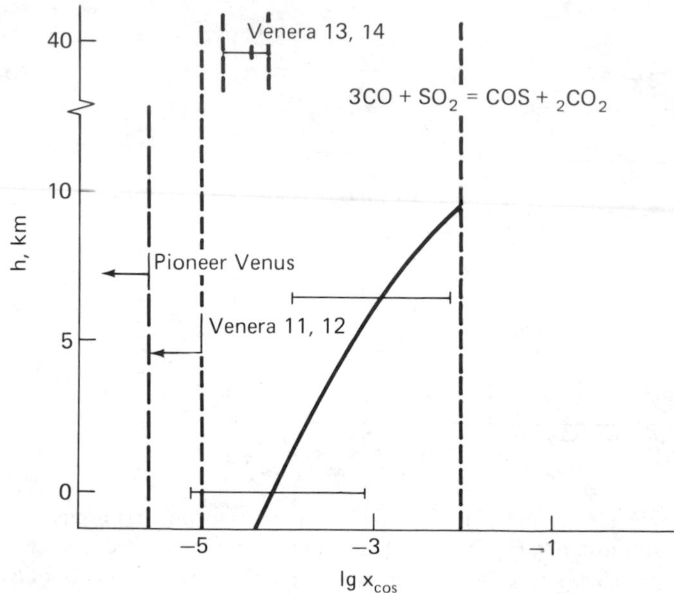

Fig. 11. Carbonyl sulfide mixing ratios (x_{COS}) in Venus' troposphere according to equilibriam calculations and instrumental data (Khodakovsky, 1982, modified). Error bars are related to equilibrium calculations in Eq. (1).

troposphere, with the lower equilibrium layer near the surface. The chemical equilibrium in this layer could be attained as a result of basically constant high temperature at the surface as well as an acceleration of the gas-phase reaction rate owing to catalytic properties of the soil. The tropospheric chemical composition was assumed to be constant and considered to reflect the "frozen" equilibrium composition corresponding to near-surface *P–T* conditions.

The calculations of near-surface tropospheric chemical composition carried out using the extrapolation of instrumental data obtained as space missions for levels above 20 km indicate the prevalence of oxidized forms of sulfur-bearing gases ($SO_2 > H_2S + COS$).

For the qualitative evaluation of redox conditions at Venus' surface the chemical indicators "Contrast" were mounted on Venera 13 and Venera 14 landers (Florensky *et al.*, 1983c). The indicator involves the observation of the color reaction of sodium pyrovanadate reduction with the formation of vanadium oxides:

$$Na_4V_2O_7 + CO_2 + CO \rightleftharpoons V_2O_4 + 2\,Na_2CO_3, \qquad (2)$$

$$Na_4V_2O_7 + 2\,CO \rightleftharpoons V_2O_3 + 2\,Na_2CO_3. \qquad (3)$$

These reactions proceed at 750K (Venus' surface) at oxygen partial pressures lower than 10^{-21} atm (Fig. 12). Examination of the TV image of the landing

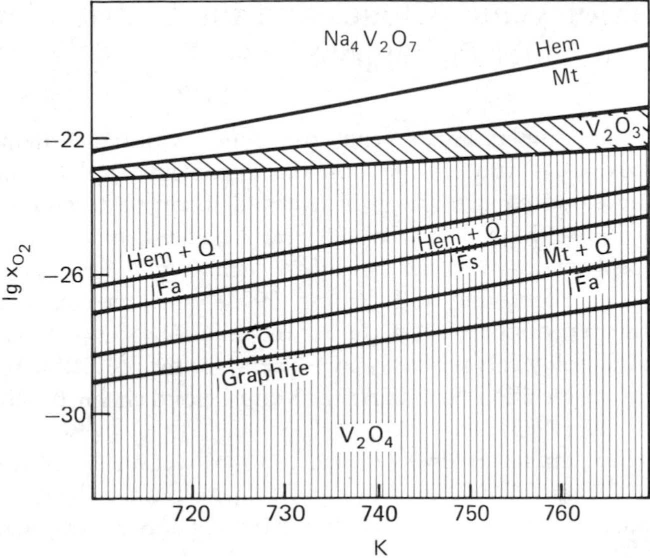

Fig. 12. Oxygen mole fraction in Venus' troposphere as a function of temperature in terms of chemical indicator "Contrast" data. Modified after Florensky *et al.*, (1983c). The white field corresponds to sodium pyrovanadate stability field; the heavily hatched region is the V_2O_4 stability field; the thinly hatched region is the V_2O_3 stability field. Hem, hematite; Mt, magnetite; Q, quartz; Fa, fayalite; Fs, ferrosilite.

torus at the Venera 13 and Venera 14 panoramas revealed that the albedo of the indication label was extremely high, corresponding to the process of reduction. Presumably the reduction is a result of the presence of carbon monoxide. Its calculated concentration from the equilibrium constants of Eqs. (2) and (3) gives the lower limit of $X_{CO} \geqslant 10^{-5}$ ($X_{O_2} \leqslant 10^{-23}$, being in agreement with the equilibrium calculations based on the instrumental data of Venera 11 and 12 and Pioneer Venus (Table 7).

Unfortunately the determination of tropospheric chemical composition by the Venera 13 and Venera 14 landers is not satisfactory with respect to the problem discussed because it is difficult to explain the coexistence of molecular oxygen (~ 18 ppm) and reduced sulfur-bearing gases (80 ppm of H_2S; 40 ppm of COS) (Table 6), even providing the kinetic constraints (Andreeva *et al.*, 1983).

Thus the theoretical treatment of Venus' near-surface tropospheric chemical composition leads to alternative cases, one characterized by the predominance of oxidized sulfur-bearing gases ($SO_2 > H_2S + COS$) (Barsukov *et al.*, 1980a, 1982b; Khodakovsky, 1982) and one where $H_2S + COS > SO_2$ (Lewis, 1970; Lewis and Kreimendahl, 1980). A final coclusion is not possible because of the lack of direct measurements of tropospheric composition at the planetary surface, as well as discreparcies in chemical composition data obtained by different space missions.

Chemistry of Venus' Clouds and the Distribution of Volatiles in Outer Planetary Shells

The high surface temperature of Venus, associated with the permanent 20-km-thick cloud cover, has attracted attention to the problem of volatiles distribution, especially the cloud condensates, which could incorporate components outgassed from Venus' lithosphere. By 1967 HCl and HF had been indicated in Venus' atmosphere, and in 1972 the sulfuric acid hypothesis was put forward (Sill, 1972; L. D. Young and A. T. Young, 1973). Thus the chemistry of sulfur and halogens was considered as a clue to the solution of a number of cosmochemical and petrological problems and as useful for the prediction of cloud particle chemical composition. Summaries of Venus' cloud chemistry are presented by Esposito *et al.* (1983) and Volkov (1983).

According to data from Venera 12, 13, and 14 direct measurements the main components of Venus' clouds are sulfur and chlorine but the chemical forms and their distribution are not known as the experimental data are extremely variable (Table 8).

Sulfur

Presumably sulfur could be considered the dominant component of the clouds. The sulfuric acid hypothesis (Sill, 1972; A. T. Young, 1973) is supported by measurements of the optical and microphysical properties of the cloud particles, the character of the water vapor and sulfur dioxide concentration vertical profiles within the troposphere, and the correspondence of cloud base level to the physicochemical properties of sulfuric acid solutions.

The cyclic processes in Venus' clouds are schematically depicted in Fig. 13 (Knollenberg and Hunten, 1980). Gaseous sulfuric acid is generated above the cloud deck at a level exceeding 60 km within the zone of photochemical reactions in the process of sulfur trioxide hydration, the latter being the product of sulfur dioxide photochemical oxidation. The oxidation is performed by the atomic oxygen produced in the photolysis of carbon dioxide and or sulfur dioxide. The existence of liquid particles of sulfuric acid is presumably governed by chemical condensation. The lifetime of sulfuric acid aerosols varies from several hours in lower cloud layers up to several months at the cloud deck (Knollenberg and Hunten, 1980). The regeneration of sulfur dioxide is provided by thermochemical reactions in the troposphere below the cloud base (45–47 km).

Elemental sulfur condensates are also suggested as plausible components of Venus' clouds. The theoretical treatment of condensation from the dimeric sulfur vapor with a mole fraction of 0.02–0.8 ppm (San'ko, 1980) is presented in Barsukov *et al.* (1982c). It was demonstrated that solid and liquid supercooled sulfur particles could exist within Venus' cloud cover, maximal concentration of cloud particles being 0.6 $mg \cdot m^{-3}$ ($r = 1$ μm) (Table 8).

Table 8. Proposed cloud condensates on Venus.

Component	Altitude interval of measurement, km	Condensation level, km	Aerosol mass loading per unit volume $mg \cdot m^{-3}$	Reference
H_2SO_4(sol)	70.0–56.5	70.0–48.0	1.0	Knollenberg and Hunten (1980)
	56.5–50.5		5.0	
S(c, l)		~48	0.6	Barsukov *et al.* (1982c)
Sulfur-bearing compounds	54.0–47.0		0.10 ± 0.03	Surkov *et al.* (1981)
			1.10 ± 0.13	Surkov *et al.* (1982)
Chlorine-bearing compounds	54.0–47.0		2.10 ± 0.06	Surkov *et al.* (1981)
			0.16 ± 0.04	Surkov *et al.* (1982)

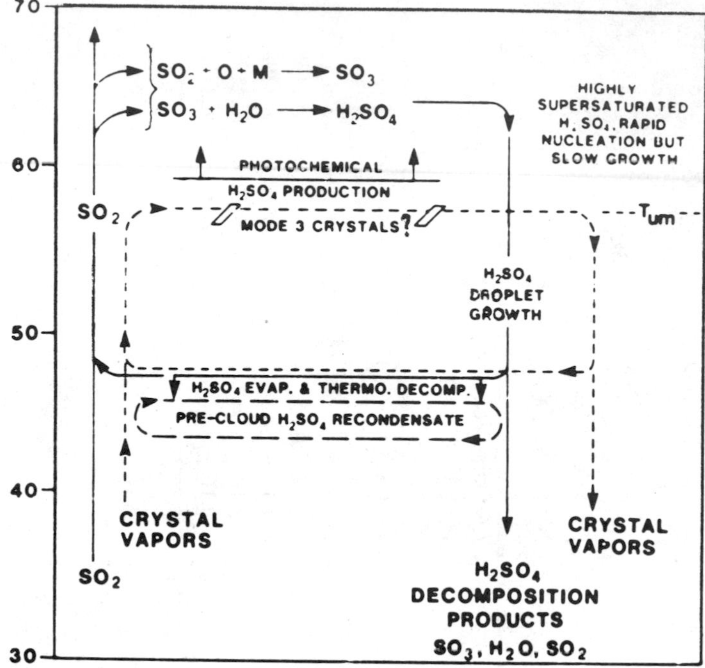

Fig. 13. Cycling processed in Venus' clouds (Knollenberg and Hunten, 1980).

Chlorine

Spectroscopic detection of HC1 at the cloud deck level (Connes *et al.*, 1967) was considered as an indication of chlorine cloud condensates. Identification of chlorine in X-ray fluorescence analyses of aerosol particles by the Venera 12, 13, and 14 space probes stimulated the theoretical treatment of the problem. Ammonia chlorides (Surkov and Andreichikov, 1973), mercury (Lewis, 1969), iron (Kuiper, 1969), and alumina chlorides (Krasnopolsky and Parshev, 1981) were suggested as probable candidates. Barsukov *et al.*, (1982c) and Volkov (1983) give detailed thermodynamical predictions of probable chlorine compounds in cloud particles. The Following assumptions were used in the calculations:

Equilibrium condensation of a corresponding aerosol proceeds at the appropriate tropospheric level, provided the chemical composition of the gaseous parcel is similar to the "frozen" equilibrium composition at the troposphere–surface interface.

Elemental inventories in the outer shells of the Earth and Venus are the same; furthermore, the complete outgassing of the Venusian crust with respect to a given element is assumed.

Elemental content in the troposphere is sufficient to insure the concentration of condensate exceeding 0.1 $mg \cdot m^{-3}$ corresponding to a mole fraction of gas within the 10^{-7} to 10^{-6} interval.

These points were used as the starting assumptions for evaluating the probable condensation of chlorine compounds involving 47 elements. The stability of condensable compounds at the surface was revealed, mole fractions of predominant gaseous phases in the near-surface troposphere were calculated, and condensation levels of predominant vapors were derived from the saturation curves (Tables 9).

The calculations lead to the following conclusions: the equilibrium condensation of chloride vapors 46 elements is thermodynamically forbidden, or the concentration of condensed particles is negligible ($x_{Me_mCl_n} \leqslant 10^{-10}$) (Fig. 14). This is also true for ammonia, iron, aluminum, and mercury chlorides, suggested previously as possible candidates. The only halide to which this thermodynamical constraint is not extended is antimony chloride vapor ($SbCl_3$), which form hydrated condensates (SbOCl).

Lewis and Fegley (1982) used new thermodynamic data for SbS. This resulted in a change in the thermodynamical prediction: gaseous antimony sulfide rather than antimony chloride was regarded as the predominant antimony-bearing gas in Venus' troposphere. Antimony sulfides were suggested to be stable crystalline phases at the planetary surface (Table 9). This conclusion is in conflict with the still more stringent upper limits of antimony-bearing gas content and this excludes antimony compounds as possible candidates. It should be pointed out that the existence of crystalline cloud particles themselves has recently been considered open to question (Toon *et al.*, 1984).

A. T. Young (1977) and Volkov *et al.*, (1979) Have shown the thermodynamic instability of cloud particles of liquid hydrochloric acid as well as liquid compounds of $SOCl_2$, SO_2Cl_2, and H_2S_3 at *P–T* conditions corresponding to Venus' cloud layer. Thus the significant number of thermodynamically forbidden chlorine compounds in Venus' clouds is indicative of the improbability of chlorine compound equilibrium condensation as cloud particles.

Mercury

As early as 1969, when it became evident that high temperatures prevailed at Venus surface, the hypothesis of mercury-bearing clouds was advanced (Lewis, 1969). Complete outgassing of mercury followed by equilibrium condensation of mercury-bearing vapors was assumed. This could have resulted in the formation of cloud particles consisting of liquid mercury, solid HgS, $HgCl_2$; etc. The theoretical prediction by Barsukov *et al.*, (1982c) indicated that the predominant mercury-bearing gaseous form was gaseous elemental mercury. Liquid and solid mercury condensates should be stable in Venus' clouds above 52 km (Fig. 15) under the assumption of complete outgassing of mercury from Venus' crust and the equality of mercury inventories on Venus and Earth providing as atmospheric mercury content corresponding to $x_{Hg} \approx 10^{-5}$. However, the Venera 12

Table 9. Results of the thermodynamic calculation of volatile compounds in Venus' troposphere and at the planetary surface (data from Barsukov *et al*., 1982c; Lewis and Fegley, 1982).

Gas[a]	Maximal mixing ratio,[b] x_i	Condensation level, km	Condensed phase at the planetary surface
As_4, As_4O_6, *AsS*, As_2, $AsCl_3$, AsO, As	$\leqslant 10^{-7}$	—	As_2S_3(l), As_4S_4(l)
SbS, $SbCl_3$, Sb_4O_6 (italic), SbCl, Sb_4, Sb_2, Sb, SbO, $SbCl_5$	$\leqslant 3.10^{-8}$	—	Sb_2S_3(c), Sb_4O_6 (senarmontite)
SO_2, COS, H_2S, S_n (italic)	$x_{S_2} \approx 1.8 \times 10^{-7}$	$\geqslant$48[S(c, l)]	FeS_2 (pyrite), $CaSO_4$ (anhydrite)
Se_2 (italic), Se, $SeCl_2$, SeO_2, SeO	$\leqslant 10^{-5}$	$\geqslant$18[Se(l)]	Se(l)
Te_2, $TeOCl_2$, Te, $TeCl_2$, TeO_2, TeO, $TeCl_4$	$\leqslant 10^{-7}$	$\geqslant$18[Te(c)]	Te(l)
Hg, $HgCl_2$, HgCl	$\leqslant 10^{-8}$	$\geqslant$62[Hg(c)]	—
$PbCl_2$, PbCl, Pb, PbO	$\leqslant 1.2 \times 10^{-8}$	$\geqslant$16[$PbCl_2$(c)]	PbS (galena)
$FeCl_2$ (italic), $FeCl_3$	$\leqslant 10^{-12}$	—	—
Al_2Cl_6 (italic)	$\leqslant 10^{-39}$	—	—

[a] The predominant gas is shown at the head of the sequence; condensed-phase formula is italic.

[b] Recent instrumental technique is not valid for the determination of atmospheric particle content lower than 0.1 ppm ($x_i < 10^{-7}$).

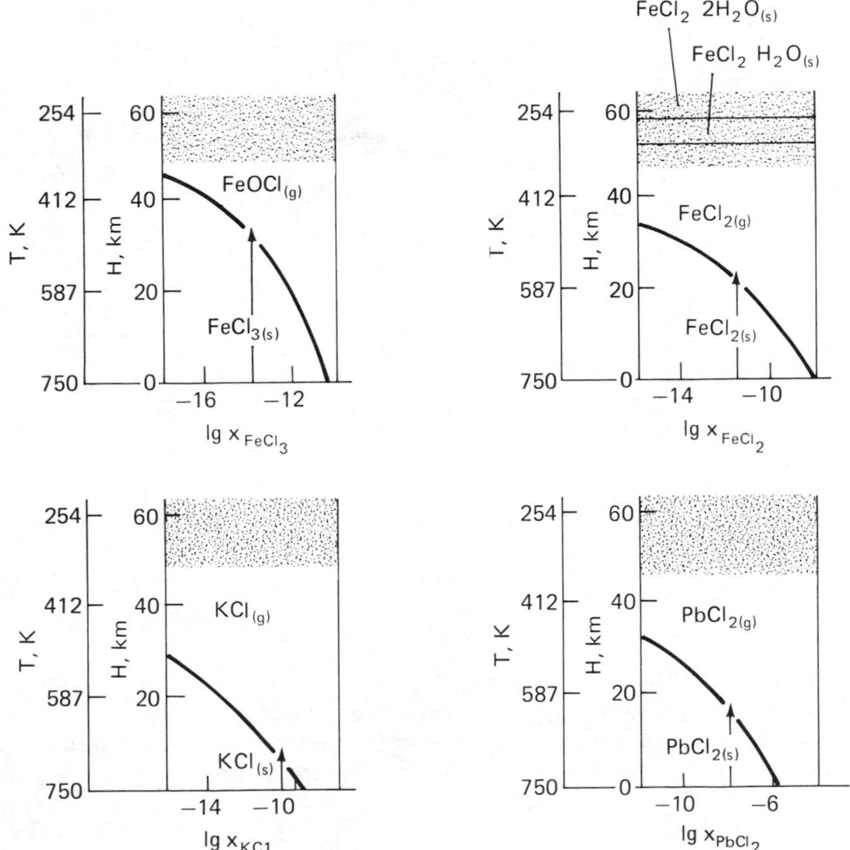

Fig. 14. Saturation curves of metal chloride vapors in *P–T* conditions of Venus' troposphere (Volkov, 1983). The altitude level of condensation corresponds to the intersection of the vertical arrow with the saturation curve. The dotted region corresponds to the cloud layer.

experiment on cloud particle composition measurements (Surkov *et al.*, 1981) revealed that the mercury content of the aerosol was lower than 0.05 $mg \cdot m^{-3}$, corresponding to $x_{Hg} < 10^{-8}$. Radiooccultation experiments on Mariner 10 (Kliore *et al.*, 1979) also demonstrated the incompatibility of the presence of liquid mercury aerosols with the experimental data.

Therefore, the estimated maximal mercury content in Venus' clouds is about 3.5 orders of magnitude lower in relation to its calculated tropospheric content based on complete outgassing of mercury from the crust. This may be explained either in terms of relatively moderate mercury outgassing as a result of its incorporation in isomorphic form in certain minerals and/or by the primary depletion of mercury as a consequence of the position of Venus' orbit with respect to the Sun.

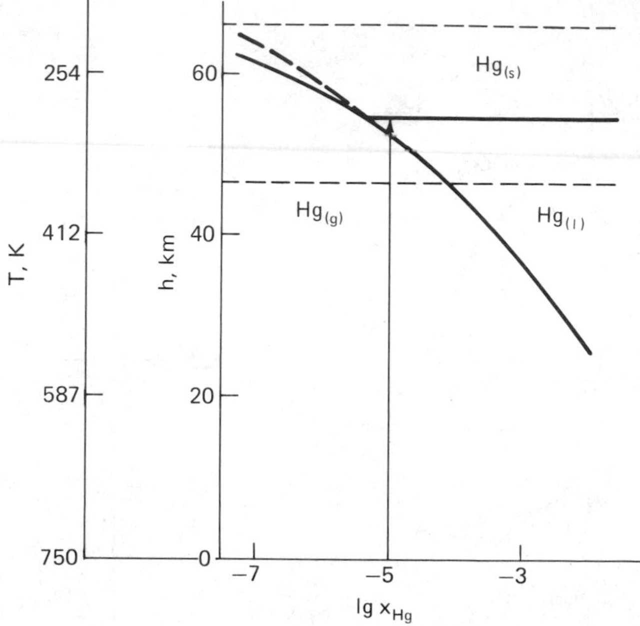

Fig. 15. Mercury phase diagram applied to the *P–T* conditions of Venus' troposphere (Barsukov *et al.*, 1982c). The metastable liquid mercury stability field is shown by a dashed curve. Cloud base and cloud deck are shown by dashed lines. The altitude level of condensation corresponds to the intersection of the vertical arrow with the saturation curve ($x_{Hg} = 10^{-5}$).

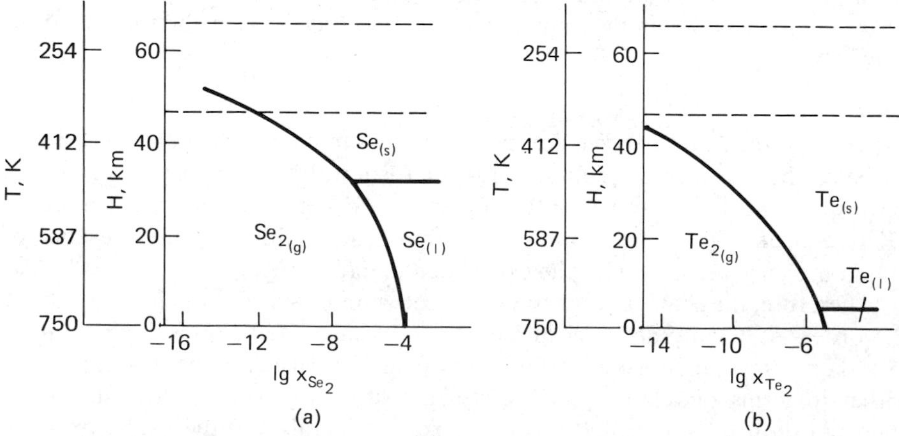

Fig. 16. Selenium (a) and tellurium (b) phase diagrams applied to the *P–T* conditions of Venus' troposphere (Barsukov *et al.*, 1982c). Cloud base and cloud deck are shown by dashed lines.

Selenium and Tellurium

According to thermodynamic predictions liquid selenium and tellurium are stable at Venus' surface (Fig. 16). The pressure of saturated selenium and tellurium vapors above Se(l) and Te(l) is about 1.5–2 orders of magnitude higher in relation to calculated values obtained under similar assumptions, than for that of mercury. The altitude levels of Se(1) and Te(c) condensation in this case do not exceed 18 km (Fig. 16), with the mass loading of hypothetical aerosol particles not lower than 200 cm^{-3} for particles with 1-micron radius. However, only thin hazes are observed below 47 km, the mass loading being $\leqslant 20\ cm^{-3}$ (Marov *et al.*, 1980). Hence, selenium (Se_2) and tellurium (Te_2) mole fractions are several orders of magnitude lower than calculated values. This may be interpreted as a consequence of incomplete outgassing with respect to these volatiles.

Arsenic

Analogous thermodynamic treatment was carried out with regard to arsenic (Barsukov *et al.*, 1982c). The upper limit of the predominant arsenic-bearing gas in Venus' troposphere (As_4O_6) was established as $x_{As_4O_6} \leqslant 10^{-4}$, i.e., exceeding the average arsenic content in the Earth's crust. More complete calculations were published by Lewis and Fegley (1982) and it was shown that the liquid arsenic sulfides could be stable on Venus' surface, resulting in a very low upper limit for predominant arsenic-bearing gas in Venus' troposphere (As_4) determined as $x_{As_4} \leqslant 10^{-7}$. This limit excludes the possibility of equilibrium condensation of arsenic-bearing compounds in Venus' clouds and a related suggestion of the predominant storage of arsenic in the troposphere.

Experimental and theoretical investigations of Venus' cloud chemistry indicated that sulfur could be considered the main component of cloud particles present in the form of liquid droplets of concentrated sulfuric acid and liquid and solid particles of elemental sulfur.

The calculations of equilibrium condensation of a number of compounds in the troposphere allows us to draw the following conclusions:

Equilibrium condensation of chlorine-bearing compounds in Venus' troposphere is thermodynamically forbidden.

The mercury inventory in the outer shells of Venus may be significantly lower in relation to the Earth as a result of the primary depletion of Venus' protomatter, because Venus is located in a zone of the solar system relatively close to the Sun.

Thermodynamic Prediction of the Mineral Composition of Venus' Surface Rocks

The high temperature of the planetary surface and the absence of significant latitudinal and seasonal temperature variations are the basis for assumptions on chemical quasi-equilibrium at the atmosphere–surface interface, allowing the possibility of the applying thermodynamic methods to the prediction of mineral composition.

Theoretical Models of Venus' Rock-Mineral Composition

The first such approach was applied by Mueller (1963, 1964, 1969) to the development of a chemical model of Venus' atmosphere and the estimation of possible mineral assemblages in Venus' rocks. Some of his conclusions, as follows are still valid:

The temperature and pressure at Venus' surface are in good agreement with P–T parameters of the silicatecarbonate mineral equilibria, especially with respect to the so-called "wollastonite" equilibrium:

$$\underset{\text{wollastonite}}{CaSiO_3} + CO_2 \rightleftharpoons \underset{\text{calcite}}{CaCO_3} + \underset{\text{quartz}}{SiO_2}. \tag{4}$$

Calcite is considered to be the only stable carbonate mineral in Venus' rocks.

Iron-bearing minerals are considered buffers in relation to oxygen partial pressure, the latter being in turn a buffering factor with respect to sulfide–sulfate mineral equilibria in rocks.

Graphite and metals in the elemental form are suggested to be unstable at the planetary surface.

Nitrogen is not incorporated in any mineral at Venus' surface.

Chlorine- and fluorine-bearing minerals are predicted to be stable mineral phases.

Interesting implications were expressed in thermodynamical modeling of lithospheric–atmospheric interactions on Venus by Lewis (Lewis, 1969, 1970). The main aim of this work was to reveal plausible values of temperature and pressure at Venus' surface corresponding to certain mineral buffers, provided that the tropospheric composition be determined by spectroscopic measurements of CO_2, CO, HCl, and HF and water vapor content within the wide range of 1–7000 ppm (Venera 4 and Venera 5 data). The complete chemical equilibrium was assumed to be attained in the global troposphere–planetary surface system up to the cloud deck. Sixty-four mineral equilibria were calculated; selected reactions are presented in Table 10. Three sets of temperature and pressure values were predicted as plausible and one of these was found to be in good agreement with the direct instrumental measurements Venera 7 obtained at a later time ($P = 120 \pm 20$ bar; $T = 747 \pm 20$K).

The calculations indicated that the hydrated minerals could be represented by tremolite; carbon, fluorine, and chlorine were predicted to be incorporated in calcite, fluorite, and halite, respective. Carbonyl sulfide was proposed as a dominant, sulfur-bearing tropospheric gaseous component, buffered by iron oxides and sulfides (Table 10). The carbonyl sulfide mixing ratio was calculated to be 5.10^{-5} (Table 7), well above the corresponding spectroscopic upper limit of

Table 10. Buffering mineral assemblages of Venus' rocks (Lewis, 1970).

Mineral assemblage	Gaseous phase	Partial pressure, atm
$CaCO_3$ + SiO_2 + $CaSiO_3$ Calcite quartz wollastonite	CO_2	118.3
$Ca_2Mg_5Si_8O_{22}(OH)_2$ + $MgSiO_3$ + $CaMgSi_2O_6$ + SiO_2 tremolite enstatite diopside quartz	H_2O	6×10^{-2}
$FeMgSiO_4$ + $MgSiO_3$ + Fe_3O_4 olivine enstatite magnetite	$CO + CO_2$	$P_{CO} = 2.4 \times 10^{-2}$, $P_{CO_2} = 118.3$
$NaCl$ + Al_2SiO_5 + SiO_2 + $NaAlSi_2O_6$ halite andalusite quartz jadeite	HCl	1.2×10^{-4}
CaF_2 + SiO_2 + $MgSiO_3$ + $Ca_2MgSi_2O_7$ fluorite quartz enstatite akermanite	HF	1.2×10^{-4}
FeS + Fe_3O_4 troilite magnetite	COS $CO + CO_2$	5×10^{-3}

($x_{COS} < 10^{-7}$). This estimate was developed in the framework of a hypothesis for a possible sink of sulfur in cloud particles in the form of mercury sulfides (Lewis, 1969). At present we know that the actual concentration gradient of sulfur-bearing gases (SO_2) in the tropospheric vertical profile is a result of cyclic generation of sulfuric acid liquid condensates.

The mineral composition of Venus' soil was calculated by Karpov *et al.* (1971). For the first time the multicomponent open system was used with the atmospheric composition measured on Venera 5 and Venera 6 as starting values. The upper limit of oxygen content (1000 ppm) was taken, so the predicted mineral composition corresponded to extremely oxidized conditions.

The next step in the theoretical modeling of the calculation of phase relations in multicomponent systems was accomplished by Khodakovsky *et al.* (1979) and Barsukov *et al.* (1980a, b, 1982b, 1983). In several models of Venus' soil calculated before the direct chemical determinations performed by the Venera 13 and Venera 14 landers, complete chemical equilibrium in the lithosphere–atmosphere interaction was assumed and the average chemical compositions of terrestrial rock types were used as starting values. The volatile component content corresponded to the tropospheric chemical composition obtained by the Venera 11 and Venera 12 landers and the Pioneer Venus at different altitude levels. Chlorine and fluorine content was taken from early spectroscopic measurements above the cloud deck (L. D. Young, 1972). All calculations were carried out using the Gibbs free-energy minimization method by a special numerical procedure (Schvarov, 1978). Phase relations in the 16-component system involving more than 150 substances with thermodynamic values from the data bank of the Laboratory, or from the thermodynamics of natural processes were calculated. A priori unstable minerals, such as clay minerals, zeolites, and hydroxides, were excluded. Selected results of calculations are presented in Table 11.

It should be noted that certain constraints on our models are connected with the virtual absence of kinetic data related to heterogeneous chemical reactions. For example, the intensity and scale of rock alteration in atmospheric–lithospheric interactions are beyond thermodynamic prediction. Actually it is possible to establish limiting values of mineral composition given the pressure and temperature data and related tropospheric composition. These limiting values allow us to describe the character of rock weathering as well as to estimate the lower limits of gas partial pressure, providing that these volatiles are incorporated in minerals of the surface rocks (Table 12). A number of conclusions were arrived at, as follows:

Sulfur could be involved in the crust composition in the form of sulfides (pyrite) and sulfates (anhydrite), with the concentration well above that of primary magmatic rocks.

Hydrated minerals are unstable; their formation is possible provided the tropospheric water vapor content exceeds 300 ppm, this being an order of magnitude higher than found by the instrumental measurements.

Calcite and dolomite are unstable if oxidized sulfur (SO_2) is the predominant form of sulfur-bearing gas in the near-surface troposphere.

Table 11. Mineral composition of Venus' rocks (calculated data).

Mineral	Basalt[a]	Soil-13[b]	Soil-14[c]
Diopside	—	1.1	5.0
Enstatite	13.7	28.0	18.0
Anorthite	25.4	22.7	37.9
Albite	28.6	—	19.8
Orthoclase	10.0	18.7	1.2
Nepheline	—	12.2	—
Quartz	5.7	—	3.4
Magnetite	6.6	10.1	9.5
Sphene	—	4.0	3.1
Rutile	0.9	—	—
Tephroite	0.2	0.2	0.2
Fluorapatite	1.0	0.1	0.1
Anhydrite	7.9	2.8	1.3
Marialite	—	—	0.5
Sylvine	—	0.1	—

[a] Average basalt after Daly (Khodakovsky *et al.*, 1979).
[b] Venus' soil at Venera 13 landing site; the multisystem is close with respect to sulfur and chlorine (Volkov and Khodakovsky, 1984).
[c] Venus' soil at Venera 14 landing site; the multisystem is close with respect to sulfur and chlorine (Volkov and Khodakovsky, 1984).

Table 12. Partial pressure of gas phase as a constraint of mineral formation on Venus' surface ($T = 750$ K; $P = 96.1$ atm) (Volkov, 1983).

Mineral	Partial pressure, atm
Pyrite, FeS_2	$P_{SO_2} > 7.0 \times 10^{-3}$
	$P_{CO} > 2.8 \times 10^{-3}$
Anhydrite, $CaSO_4$	$P_{SO_2} > 1.3 \times 10^{-2}$
	$P_{CO} < 4.2 \times 10^{-3}$
Fluorflogopite, $KMg_3[AlSi_3O_{10}]F_2$	$P_{HF} > 2.5 \times 10^{-5}$
Marialite, $3\,NaAlSi_3O_8 \cdot NaCl$	$P_{HCl} > 1.4 \times 10^{-4}$
Tremolite, $Ca_2Mg_5[Si_4O_{11}]_2(OH)_2$	$P_{H_2O} > 0.03$

Fluorine and chlorine presumably are incorporated in weathered rocks in the form of fluorine–apatite, marialite, etc.

The stability of minerals relatively rare under terrestrial conditions in Venus' rocks was evaluated as a byproduct of the thermodynamic treatment of the condensation of chlorine-bearing compounds in Venus' troposphere (Barsukov *et al.*, 1982c). The silicates of Li, Be, Zr, Th, and Sn; barium and strontium sulfates; sulfides of Pb, Zn, Cu, Bi, As, Sb, Ag, Cd, Ni, and Co; and oxides of Cr and Sb were predicted to be thermodynamically stable phases. Gold and elements of the platinum group should also be predicted. It should be noted that many of these elements are presumably represented by isomorphic forms involved in the rock-forming minerals.

According to our theoretical modeling the significant sink of sulfur in the crust is in the form of sulfide and/or sulfate formation in weathered rocks. This is considered a main factor in the chemical interaction of the troposphere with Venus' rocks (Khodakovsky *et al.*, 1979; Khodakovsky, 1982; Barsukov *et al.*, 1982b; Volkov, 1983).

After the Venera 13 and Venera 14 space missions the elemental composition of Venus' soil at two landing sites became available and new calculations of mineral composition were performed (Barsukov *et al.*, 1983). The starting values involved known tropospheric measurements of CO_2, CO, SO_2, and H_2O content. It was revealed that in the multicomponent system open with respect to all volatiles the sulfur abundance in rocks was far above actual instrumental data (Table 4). Therefore, either the chemical equilibrium with relation to sulfur dioxide is not attained or the sulfur dioxide content in the near-surface troposphere at the landing site is well below 130 ppm, i.e., the value adopted from Venera 12 measurements.

Results of calculations carried out with the assumption of an inert behavior of sulfur and chlorine are presented in Table 11. The mineral composition of the soil is characterized by the presence of 1–3% anhydrite, about 10% of magnetite, and traces (lower than 1%) of sylvine and marialite.

The absence of chemical equilibrium with respect to sulfur dioxide in soils at the Venera 13 and Venera 14 landing sites could be interpreted as a result of a low reaction rate for interaction of sulfur dioxide with minerals and/or of small exposure age of mineral particles in relation to the troposphere. In general terms the sulfur content could be considered a function of the exposure age. The maximal enrichment of weathered rocks with sulfur is suggested to correspond to the chemical composition of the rock closely approximating complete chemical equilibrium with tropospheric gases. So our calculations indicated that the composition of weathered basalts in equilibrium with the troposphere could have a sulfur content of up to 11% (Barsukov *et al.*, 1980b, 1982b).

The problem of valent forms of iron and sulfur as a function of redox conditions at the planetary surface is discussed by Lewis and Kreimendahl (1980). According to the assumption of chemical equilibrium between the atmosphere and Venus' crust and using the data on tropospheric chemical composition (concentration of CO, NH_3, CH_4, H_2), the lower limit of oxygen partial pressure at the surface was estimated as $P_{O_2} \geqslant 10^{-22.8}$ bar. The upper limit of P_{O_2} can be calculated if we take into account the following constraints: $x_{\Sigma s} > 10^{-4}$ (instrumental data) and $x_{H_2S} \geqslant x_{SO_2}$. Then the upper limit of P_{O_2} is $\leqslant 10^{-22.0}$. Such an extremely narrow interval of predicted oxygen partial pressure ($10^{-22.8}$ bar $< P_{O_2} < 10^{-22.0}$ bar) was interpreted in terms of a sulfide–sulfate mineral buffer (Fig. 17):

$$\underset{\text{pyrite}}{FeS_2} + 2\,\underset{\text{calcite}}{CaCO_3} + 7/2\,O_2 \rightleftharpoons 2\,\underset{\text{anhydrite}}{CaSO_4} + \underset{\text{wüstite}}{FeO} + 2\,CO_2. \qquad (5)$$

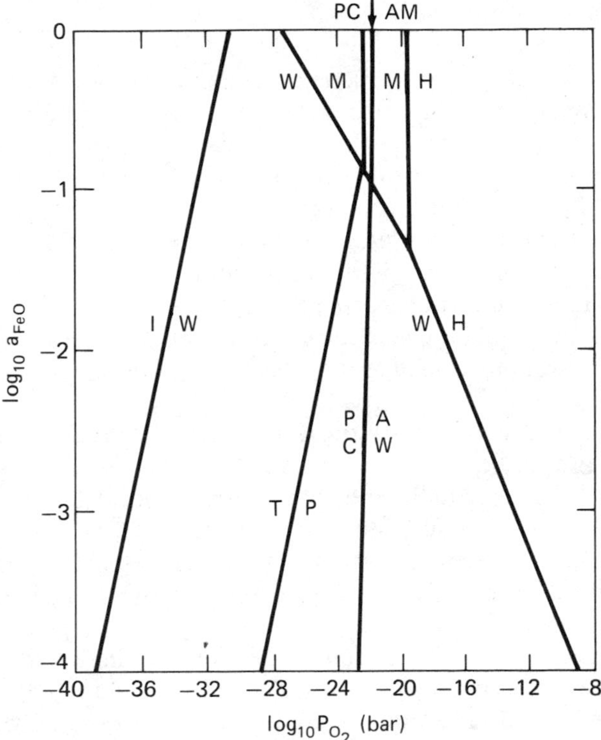

Fig. 17. Stability fields of minerals in Venus' surface rocks ($T = 750K$) in terms of a_{FeO} vs. log P_{O_2} diagram (Lewis and Kreimendahl, 1980). I, Native iron; W, wüstite, FeO; T, troilite, FeS; P, pyrite, FeS_2; C, calcite, $CaCO_3$; A, anhydrite; M, magnetite, Fe_3O_4; H, hematite, Fe_2O_3.

Lewis and Kreimendahl concluded that the relative contents of different sulfur-bearing gases defined the oxidation extent of Venus' troposphere, whereas the absolute concentrations reflected the Fe^{3+}/Fe^{2+} ratio in Venus' crust. It was inferred that Venus' crust was extremely depleted in ferric iron in comparison with the Earth: the Fe^{3+}/Fe^{2+} ratio was estimated to be one or two orders of magnitude below that of the Earth' crust.

This conclusion evidently is valid only if complete chemical equilibrium is attained in the global system of atmosphere–Venus' crust. Our calculations of the mineral composition of Venus' soil at the Venera 13 and Venera 14 landing sites indicated that the chemical equilibrium with respect to sulfur dioxide was not achieved, this being inconsistent with the conclusion on the only slightly oxidized crust. The final inference evidently will be possible only when highly reliable measurements of near-surface tropospheric chemical composition are available.

Carbonates in Venus' Soil

The problem of the stability of carbonates on Venus' surface is actively discussed in the literature because the carbonates can be considered products of the interaction of the carbon dioxide atmosphere with silicate minerals. Urey (1951) proposed that the carbon dioxide content in the terrestrial atmosphere could be controlled by silicate–carbonate equilibria in the presence of liquid water. Adamčik and Draper (1963) suggested that these equilibria on Venus could be attained in an anhydrous environment, although over an extended period of time. In the above-mentioned paper and in Mueller (1964) a number of silicate–carbonate equilibria were calculated to compare deduced carbon dioxide partial pressures with available estimates of P_{CO_2} values in Venus' atmosphere (Fig. 18).

According to thermodynamic estimates, siderite, magnesite, and dolomite are unstable under the present P–T conditions of Venus' surface. Yet the equilibrium P_{CO_2} value corresponding to the mineral assemblage of calcite, wollastonite, and quartz ("wollastonite" equilibrium, Eq. 4) is surprisingly near the actual atmospheric pressure of Venus' surface at 750K. The application of refined thermodynamic data (Lewis, 1970; Vinogradov and Volkov, 1971) supports the early inferences.

The special treatment of reaction rate in Eq. (4) indicated that chemical equilibrium could be attained in Venus' conditions during the geologically negligible time period of several hundred years, i.e., instantaneously (Mueller

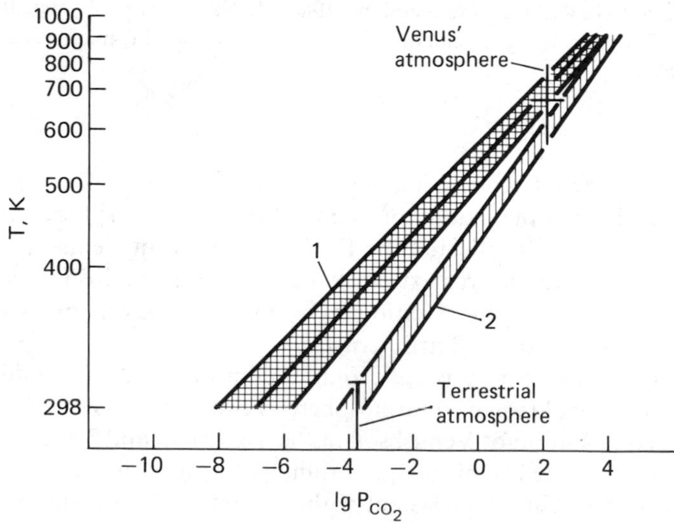

Fig. 18. Curves for two types of carbonation reactions applicable to Venus (Mueller and Saxena, 1977). 1, Reaction (4); 2, $MgSiO_3 + CaCO_3 + SiO_2 = CaMgSi_2O_6 + CO_2$ Hatched areas correspond to the uncertainty intervals of thermodynamic data.

and Kridelbaugh, 1973). Lewis (1971) proposed that the "wollastonite" equilibrium could be considered a buffering factor in relation to surface temperature. Indeed, the temperature difference between day and night hemispheres on Venus may be reflected in the diurnal shift of the "wollastonite" equilibrium with the endothermic release of carbon dioxide from the soil at the day side. This idea was supported by Dobrovolskis (1983) in terms of atmospheric circulation: diurnal adsorption and desorption of carbon dioxide by surface soils was considered to govern atmospheric tides on Venus. The discussion was continued by Jakosky (1984), who pointed out that insolation could not heat the soil to the needed depth because of the low heat of conductivity of the surface material. Thus, the thermal decomposition of carbonates, if any, is confined to an extremely thin layer, which is far below the estimated value.

However, the influence of gases other than carbon dioxide on the stability of carbonates was not considered by Lewis (1970) and Lewis and Kreimendahl (1980). It is evident that calcite is stable only if the reduced forms of sulfur-bearing gases predominate in the near-surface troposphere ($COS + H_2S > SO_2$). The equilibrium calculations of mineral composition carried out using the complex interrelations of sulfur-bearing gases (Barsukov *et al.*, 1980b, 1982b) (see above) indicated that carbonates should be unstable in weathered basalts and granites.

The special thermodynamic treatment of the interaction of carbonates with sulfur-bearing gases was carried out by Zolotov (1985), who inferred the instability of calcite and dolomite at Venus' surface.

The interaction of calcite with sulfur dioxide in the absence of free oxygen is expressed by the reaction (Terradellas and Bonnetain, 1973):

$$CaCO_3 + 1.5\,SO_2 \rightleftharpoons CaSO_4 + CO_2 + \tfrac{1}{4}S_2. \qquad (6)$$

The reaction proceeds at Venus' surface temperature without any kinetic constraints. In accord with Eq. (6) the thermodynamic stability of a number of minerals has been depicted as a function of P_{SO_2} and P_{S_2} (Fig. 19). From Fig. 19 it is seen that the stability of carbonates increases with increasing temperature and pressure, corresponding to lower hypsometric levels. Calcite is stable, provided the sulfur dioxide mole fraction is below 10^{-6}. Dolomite is evidently unstable, for a neglible content of sulfur dioxide ($\sim$4 ppb) is sufficient for the chemical decomposition of this mineral. Once magnesite ($MgCO_3$) is formed as a product of the interaction of magnesium silicates with atmospheric carbon dioxide, it is likely to be stable on practically the whole surface, excluding Montes Maxwell, because the temperature of conversion of magnesite to magnesium sulfate corresponds to the hypsometric level above 10 km.

The presence of sulfur dioxide in the near-surface troposphere should therefore be regarded as a severe constraint preventing the formation of carbonates at Venus' surface. Burried sea sediments consisting of carbonates could perhaps be found on Venus provided these rocks were preserved at certain depths in the crust. This speculation is evidently considered one of the possible deductions from the hypothesis of an ancient ocean on Venus (Donahue *et al.*, 1982).

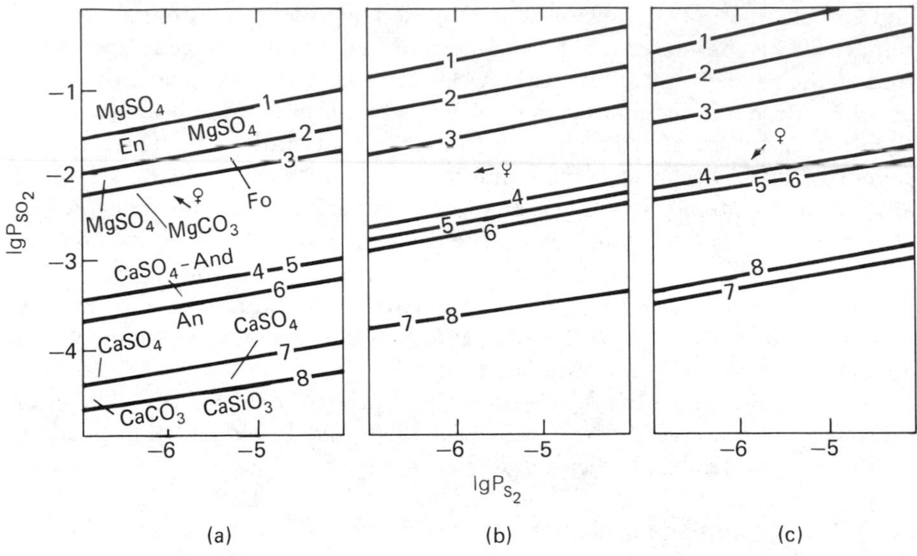

Fig. 19. Stability of certain minerals at the Venusian surface as a function of SO_2 and S_2 partial pressure (Zolotov, 1985). A, 675K, 54 atm, +7.8 km; B, 735K, 90 atm, O km; C, 765K, 115.3 atm, −4 km. En, Enstatite; Fo, forsterite; An, anorthite; And, andalusite; Di, diopside. ♀ shows Venus data for SO_2 and S_2. The lines (n = 1–8) correspond to equilibrium conditions in the reactions:

1) $MgSiO_3 + 1.5\,SO_2 = MgSO_4 + SiO_2 + 1/4\,S_2$

2) $Mg_2SiO_4 + 3\,SO_2 = 2\,MgSO_4 + SiO_2 + 1/2\,S_2$

3) $MgCO_3 + 1.5\,SO_2 = MgSO_4 + CO_2 + 1/4\,S_2$

4) $CaAl_2Si_2O_8 + 1.5\,SO_2 = CaSO_4 + Al_2O_3 + 2\,SiO_2 + 1/4\,S_2$

5) $CaMg[SiO_3]_2 + 3\,SO_2 = CaSO_4 + MgSiO_3 + SiO_2 + 1/2\,S_2$

6) $CaAl_2Si_2O_8 + 1.5\,SO_2 = CaSO_4 + Al_2SiO_5 + SiO_2 + 1/4\,S_2$

7) $CaCO_3 + 1.5\,SO_2 = CaSO_4 + CO_2 + 1/4\,S_2$

8) $CaSiO_3 + 1.5\,SO_2 = CaSO_4 + SiO_2 + 1/4\,S_2$

Hypsometric Control of Mineral Stability

The significant differences of temperature and pressure between highlands and lowlands (up to 100K and 65 atm) can be contemplated as factors of mineral equilibria shift resulting in the dependence of mineral composition upon the hypsometric level. This idea was reported for the first time by Walker (1975) and later expressed more particularly by Florensky *et al.* (1977) and Nozette and Lewis (1982). A number of gas–mineral buffer equilibria were calculated (Fig. 20) and it was demonstrated that the regions of maximal temperatures and pressures (lowlands) corresponded to the conditions of the dehydration decom-

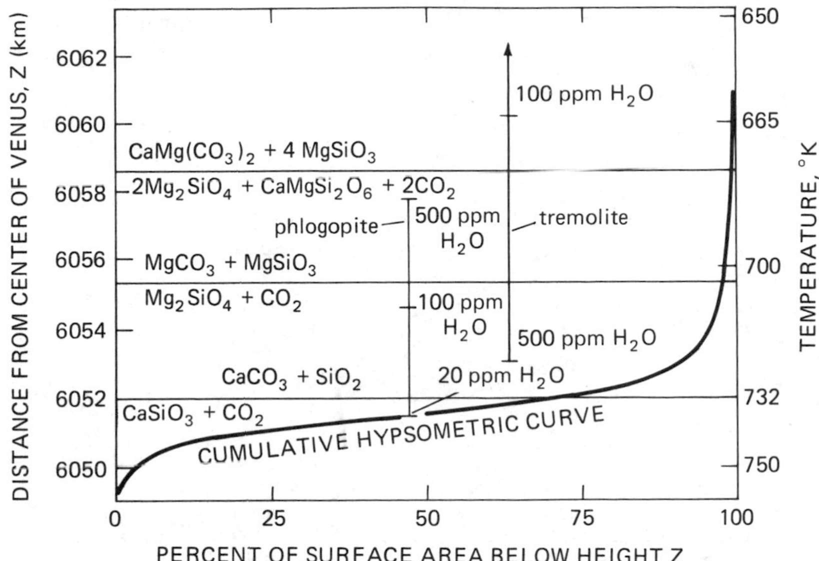

Fig. 20. Equilibrium altitudes for selected weathering reactions on Venus, plotted with cumulative hypsometric curve. Modified from Nozette and Lewis, (1982). The equilibrium altitudes of reactions:

$$\underset{\text{diopside}}{2\,CaMgSi_2O_6} + \underset{\text{enstatite}}{3\,MgSiO_3} + \underset{\text{quartz}}{SiO_2} + H_2O \rightleftharpoons \underset{\text{tremolite}}{Ca_2Mg_5Si_8O_2{}^2(OH)_2}$$

and

$$\underset{\text{microcline}}{KAlSi_3O_8} + \underset{\text{enstatite}}{3\,MgSiO_3} + HF \rightleftharpoons \underset{\text{phlogopite}}{KMg_3AlSi_3O_{10}F_2} + H_2O + \underset{\text{quartz}}{3\,SiO_2}$$

are plotted for a range of x_{H_2O}.

position of tremolite and carbonate (Fig. 21). At the same time the highlands were considered to be regions for the stable existence of these minerals. It was therefore suggested that recycling could involve the aeolian transport of weathered material of different mineral compositions as well as buffering of water vapor and carbon dioxide partial pressures. However, the most reliable measurements of water vapor content in the troposphere (Moroz *et al.*, 1983) are far below the lower limit of H_2O concentration needed to allow the stability of tremolite. Also, the stability of carbonates discussed above is doubtful. Therefore, the whole mechanism of chemical weathering in terms of hypsometric control of hydration and carbonatization presents a number of problems.

According to our predictions of Venus' soil mineral composition (Khodakovsky, 1982; Barsukov *et al.*, 1980b, 1982b; Volkov, 1983), hypsometric control is proposed as a permissible factor in relation to sulfur-bearing minerals: pyrite could be stable on highlands, whereas anhydrite would be a characteristic mineral of lowlands. The indirect support of this speculation is found in Ford

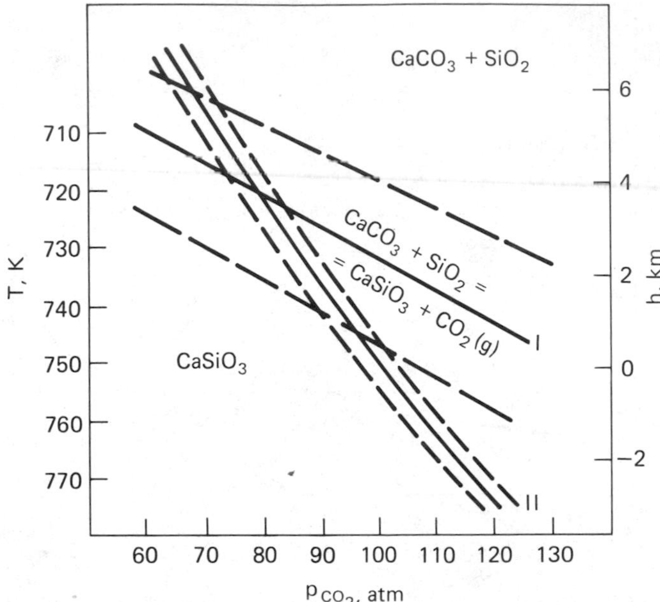

Fig. 21. The temperature dependence of "wollastonite" equilibrium reaction as a factor of weathering on Venus (Florensky *et al.*, 1977). I, *P–T* curve of reaction (4) (from Vinogradov and Volkov, 1971); II, carbon dioxide adiabatic curve for Venus' troposphere for $x_{CO_2} = 0.97^{+0.3}_{-0.4}$ plotted according to Kuz'min and Marov (1974).

and Pettengill (1983). Estimates of the dielectric constant and derived values of electroconductivity of the surface soil were obtained from thermal radio emission observations on the Pioneer Venus Orbiter. The areas with extremely high electroconductivity (pyrite?) were proved to be confined only to highlands.

Finally, we shall formulate conclusions about the mineral composition of Venus' surface rocks as inferred from the interpretation of Venera 13 and Venera 14 X-ray fluorescent analyses of the soil and related theoretical predictions.

Interaction of Venus' basaltic rocks, the only reliably established rock type, with the troposphere has resulted in the alteration of their chemical and mineral composition, i.e., in chemical weathering.

The main form of weathering is expressed in the incorporation of sulfur in pyrite and/or anhydrite, these are considered to be the dominant secondary minerals.

Redox conditions at Venus' surface presumably correspond to the low oxygen partial pressure ($P_{O_2} \approx 10^{-21}$ atm), determining the stability of magnetite. Oxidation of ferrous iron by atmospheric water vapor cannot be excluded.

Hydration processes at Venus' surface are thermodynamically forbidden because the lower limit of water vapor content needed to provide the hydration is two orders of magnitude below the instrumental data.

Carbonatization of Venus' rocks in the form of calcite and dolomite formation is not plausible. According to theoretical estimates, the relatively high tropospheric content of sulfur dioxide ($\sim$130 ppm) prevents the preservation of carbonates. Graphite is also thermodynamically unstable.

Chlorine and fluorine are predicted products of lithospheric–atmospheric interaction in the form of marialite, fluorine-bearing apatite, etc.

Nitrogen-bearing minerals at Venus' surface are thermodynamically unstable.

The adiabatic temperature and pressure gradient in the tropospheric vertical profile could permit the variation of mineral assemblages as a function of hypsometrical level. It is suggested that pyrite is stable in the highlands while anhydrite may be confined to lowlands.

The Inventories of Chemically Active Volatiles and Their Cycles on Venus

The modeling of the mineral composition of Venus' rocks, as well as great advances in our knowledge of atmospheric and cloud chemistry, is considered a starting point for the general discussion of the volatile distribution in the outer shells of Venus. The estimates of volatile inventories compared with the known terrestrial data are presented in Table 13.

Nitrogen

It was mentioned above that nitrogen-bearing minerals are unstable at Venus' surface (Mueller, 1963). Therefore, the total nitrogen inventory is confined to the atmosphere, as is the completion of the nitrogen cycle.

The predominant nitrogen-bearing atmospheric gas is represented as molecular nitrogen, N_2, with the concentration about 3 vol. %. Venera 8 measure-

Table 13. Volatiles inventory[a] in the outer shells of Earth and Venus

Component	Venus' atmosphere	Earth			
		Atmosphere	Hydrosphere	Crust	Total
CO_2	4.8×10^{23}	1.5×10^{18}	1.4×10^{20}	4.1×10^{23}	4.1×10^{23}
N_2	6.1×10^{21}	1.95×10^{21}	7×10^{17}	0.85×10^{21}	2.8×10^{21}
H_2O	2.1×10^{19}	5×10^{19}	1.9×10^{24}	0.4×10^{24}	2.3×10^{24}
S	$4–6 \times 10^{19}$	1.8×10^{12}	0.13×10^{22}	2.6×10^{22}	2.7×10^{22}
Cl	2.5×10^{17}	—	2.6×10^{22}	0.95×10^{22}	3.55×10^{22}
F	5.0×10^{15}	—	1.8×10^{18}	0.86×10^{22}	0.86×10^{22}

[a] The inventory of CO_2, N_2, H_2O, and S in Venus' atmosphere is estimated using the chemical analyses of the atmosphere from the Venera 11, Venera 12, and Pioneer Venus space missions, the inventory of Cl and F is estimated using the spectroscopic data. The estimates in relation to Earth are carried out using the corresponding values from Barsukov (1981) and Volkov (1983).

ments revealed that ammonia content in the troposphere below the cloud base was quite high, in the interval of 100–1000 ppm (Surkov *et al.*, 1973). However, succeeding experiments using mass-spectrometric and gas-chromatographic methods were unable to detect ammonia. A. T. Young (1977) presented some arguments supporting the suspicion of experimental artifact in the Venera 8 measurements. It is likely that the colored characteristic reaction registered on Venera 8 is related to the presence of sulfuric acid rather than ammonia.

Small quantities of nitrogen oxides NO_x($\sim$30 ppb) were detected in Venus' stratosphere above the cloud deck. These compounds may have been generated as a result of electrical discharges (lightning) providing such activity is comparable to that in the terrestrial atmosphere (Borucki *et al.*, 1981). The nitrogen cycle on Venus is governed mainly by photochemical reactions in the stratosphere, with the lifetime of nitrogen oxide molecules estimated as 10^4 years (Yung and DeMore, 1982).

Carbon

Atmospheric inventories of carbon are comprised of carbon dioxide which is the predominant component of Venus' atmosphere. The carbon dioxide content is five orders of magnitude above that of CO and COS, with other carbon-bearing gases being negligible. Graphite is thermodynamically unstable at the P–T conditions on the planetary surface and the formation of calcite and dolomite is highly improbable (see the previous section).

Thus the carbon cycle is presumably confined to atmospheric processes because there is no evidence of a carbon sink in Venus' crust (Volkov, 1983).

Among these atmospheric processes the photochemical reactions are suggested to play a dominant role, especially those which make possible carbon dioxide regeneration from the products of its photodissociation. According to Krasnopolsky and Parshev (1983), photochemical production of carbon monoxide as a result of carbon dioxide photolysis is estimated as 9.10^3 g$\cdot^{-1}$. At the same time the carbon dioxide recombination, expressed by

$$CO + O + M \rightarrow CO_2 + M, \qquad (7)$$

is considerably less effective (McEwan and Phillips, 1975). The conservation of carbon dioxide as a prevailing atmospheric component is presumably provided by fast catalytic oxidation reactions of carbon monoxide conversion to CO_2 involving ClO_x radicals (Prinn, 1971) and HO_x radicals (McElroy *et al.*, 1973). Hydrogen chloride is proposed as a source of these radicals, being produced by the photolytic dissociation of HC1. In compliance with the theories of Krasnopolsky and Parshev (1983), about 98% of carbon monoxide produced by photolysis of carbon dioxide is oxidized with Venus' stratosphere. The residual fraction is removed to the lower troposphere and thermochemically oxidized with conversion to carbon dioxide:

$$CO + SO_3 \rightleftharpoons CO_2 + SO_2. \qquad (8)$$

In the following we shall demonstrate the significant role of this reaction in the sulfur cycle. Here we conclude that the carbon cycle, like the nitrogen cycle, is governed by photochemical processes in Venus' stratosphere and its completion is confined to the atmosphere.

The approximate equality of carbon inventories on Earth and Venus (Table 13) is considered evidence of an early accumulation of equivalent masses of carbonaceous chondrites on both planets. Subsequent oxidation of carbon by water vapor resulted in the formation of carbon monoxide and carbon dioxide outgassed at the planetary differentiation stage (Khodakovsky, 1982).

Sulfur

The sulfur cycle is regarded as one of the main global cycles in Venus' outer shells. Sulfur content in Venus' atmosphere is seven orders of magnitude above the terrestrial value (Table 13); however, the contribution of sulfur in Venus' crust is difficult to evaluate because of the deficiency of information with relation to Venus' rocks. According to chemical analyses of Venus' soil at the Venera 13 and 14 landing sites and related thermodynamic predictions, sulfur-bearing minerals should be important components of surface rocks. Hence the sulfur cycle, in contrast to the carbon and nitrogen cycles, is determined by sulfur behavior in the atmosphere as well as in the crust. Three different parts of the sulfur cycle are distinguished by their relation to the rate of chemical reactions involved in the cycle (von Zahn *et al.*, 1983; Volkov, 1983; Lewis and Prinn, 1984).

The fast atmospheric cycle is described by a set of photochemical reactions in Venus' stratosphere and clouds and are expressed by the following alternative schemes:

$$CO_2 + h\nu \rightarrow CO + O, \tag{9}$$

$$SO_2 + O + M \rightarrow SO_3 + M, \tag{10}$$

$$SO_3 + H_2O \rightarrow H_2SO_4(\text{sol}), \tag{11}$$

$$SO_2 + h\nu \rightarrow SO + O, \tag{12}$$

$$SO + O + M \rightarrow SO_2 + M, \tag{13}$$

$$SO_2 + O + M \rightarrow SO_3 + M, \tag{14}$$

$$SO_3 + H_2O \rightarrow H_2SO_4(\text{sol}). \tag{15}$$

The oxygen production in the scheme (9)–(11) (Winick and Stewart, 1980) is a result of carbon dioxide photodissociation (9), succeeded by catalytic oxidation of sulfur dioxide to sulfur trioxide. The catalytic agents of reaction (10) are presumably OH and HO_2 radicals, being produced as a result of hydrogen chloride photodissociation. In distinction to this Winick and Stewart (1980) and Krasnopolsky and Parshev (1983) suggest stratospheric photolysis of sulfur dioxide as the dominating mechanism of oxygen production without any de-

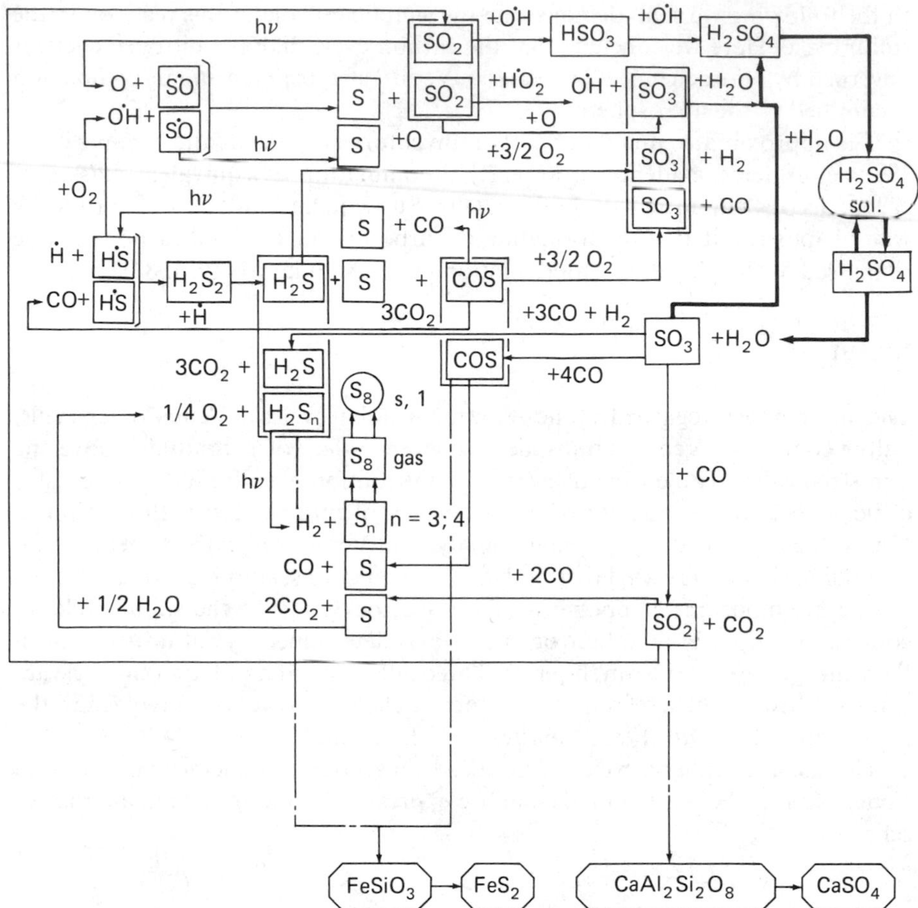

Fig. 22. Generalized scheme of sulfur cycle on Venus: 1, fast atmospheric cycle; 2, slow atmospheric cycle; 3, sink of sulfur-bearing gases in Venus' crust.

pendence on carbon photochemistry (reactions 12 and 13). Elemental sulfur is deduced as a byproduct of such a photochemical process. Both schemes of photochemical reactions are depicted on Fig. 22 and are found to be in good agreement with the instrumental data, indicating a significant decrease in SO_2 and H_2O content in Venus' stratosphere in relation to the troposphere (Fig. 8). The final selection from two photochemical schemes of sulfur dioxide oxidation could be carried out only after the experimental determination of reaction rates of corresponding oxidation processes. The hydration of sulfur trioxide has resulted in the formation of sulfuric acid condensates comprising a 15-km-thick cloud cover. These sulfuric acid aerosols are thermally decomposed at 365–380K, corresponding to a 47–49 km altitude level in accordance with the position of the global cloud base. The completeness of the fast atmospheric cycle is governed by the thermochemical reduction of sulfur trioxide to sulfur dioxide in the low troposphere (reaction 8). The maximal lifetime of sulfur dioxide molecules in the fast atmospheric cycle is estimated as several years (Rossow, 1978).

The slow atmospheric sulfur cycle is confined to the low atmosphere and involves the photochemical and thermochemical processes related to the existence of reduced sulfur-bearing gases (H_2S, COS) as well as elemental sulfur in gaseous and condensed forms (Fig. 22). The stratosphere and cloud layers are regarded as sink regions with respect to H_2S and COS, these being oxidized by molecular oxygen with sulfur trioxide production. The alternative path of sink is a photodissociation process. Elemental sulfur is assumed to be formed as a result of carbonyl sulfide and/or hydrogen sulfide photodissociation (Prinn, 1975), both as hydrogen sulfide photooxidation in the presence of CO and SO_2 and with HS and H radicals as catalyst (Volkov, 1983) (Fig. 22). The completeness of the slow atmospheric cycle is assumed owing to thermochemical reactions in the low troposphere (Lewis and Prinn, 1984) as follows:

$$SO_3 + 4\,CO \rightleftharpoons COS + 3\,CO_2, \tag{16}$$

$$SO_3 + H_2 + 3\,CO \rightleftharpoons H_2S + 3\,CO_2, \tag{17}$$

$$\tfrac{1}{n}\,S_n + H_2 \rightleftharpoons H_2S, \tag{18}$$

$$\tfrac{1}{n}\,S_n + CO \rightleftharpoons COS. \tag{19}$$

Surface rocks are considered catalyzing agents, but it is necessary to emphasize that a free hydrogen content in the troposphere is regarded as among the poorly identified components. The lifetime of sulfur molecules in the slow atmospheric cycle has been evaluated by Lewis and Prinn (1984) as several dozens of years; the latter value should be inferred as the limiting estimate derived from the maximal lifetime of sulfuric acid particles in a fast atmospheric cycle.

The third "geological" sulfur cycle governs the sulfur behavior in both outer shells of Venus, i.e., the atmosphere and the crust. The cycle begins with the production of sulfur-bearing gases of different oxidation states (SO_2, H_2S, COS), resulting from volcanic and tectonic processes and also from lithospheric–atmospheric interaction of possible sulfur-bearing minerals, such as pyrite, with such atmospheric components as CO_2, H_2O, and CO.

The gases are repeatedly cycled in photochemical and thermochemical reactions in Venus' atmosphere. The fast atmospheric cycle has resulted in the long-term existence of condensate particles in the form of sulfuric acid aerosols. The concurrent photochemical and thermochemical reactions in the low atmosphere are suggested to provide the predominance of sulfur dioxide as the main form of tropospheric sulfur-bearing components. The excess of sulfur dioxide in relation to its equilibrium content in the atmospheric–lithospheric system is consumed by the sink of SO_2 in minerals of surface rocks (Fig. 22).

The outgassing of Venus' crust proceeds evidently paroxysmally and at different rates. Thus the scale of sulfur sink in minerals depends on the reaction rates of heterogeneous reactions in the atmospheric–crust interaction as well as on the residence time of mineral particles in contact with Venus' atmosphere, i.e., the rejuvenation rate of Venus' surface. The deficiency of reliable data on the chemical composition of the near-surface atmosphere prevents derivation of a final conclusion on the extent of disequilibrium in atmosphere–crust global system.

The completeness of the slow "geological" cycle is presumed in the conversion of anhydrite to pyrite in Venus' crust (Lewis and Prinn, 1984) but the mechanism of such a process is rather doubtful. Prinn (1985) pointed out that the existence of a permanent sulfuric acid cloud cover by itself could be inferred as evidence of volcanic eruptions on Venus up to the present day, particularly, if we take into account the report (Esposito, 1984) of the periodic variations of SO_2 content in Venus' stratosphere during the last 20 years (these variations are interpreted as the consequence of enormous volcanic events). Indeed, the long-term existence of sulfuric acid clouds providing the source for the sulfur cycling of the geologically active planet should be explained in terms of the dynamic balance of the sulfur outgassing rate and the velocity of heterogeneous reactions between sulfur-bearing gases and surface minerals, i.e., the crustal sulfur sink. An alternative interpretation involves the completeness of the sulfur cycles in the atmosphere providing the chemical equilibrium between the troposphere and the surface rocks with respect to sulfur dioxide.

Water

The absence of liquid water on Venus and the relatively high D/H ratio in the lower atmosphere (Donahue *et al.*, 1982), interpreted in terms of the presumed existence of an ancient ocean on Venus, are both basic facts that have stimulated an intense discussion of the water cycle on Venus.

Water vapor content in the lower atmosphere of Venus is estimated as 100 ppm (von Zahn *et al.*, 1983) corresponding to 2.1×10^{19} g and approximately similar to that of the terrestrial atmosphere (Table 13). However, the total water inventory on Venus is five orders of magnitude below the terrestrial value, so the question arises: is there any water incorporated in minerals of the Venusian crust?

The search for a mechanism of water sink in the near-surface atmosphere or crust is of great interest with regard to the explanation of a significant increase of water vapor content in a vertical atmospheric profile with altitudes in the 0–50 km range. In previous work (Barsukov *et al.*, 1980a) an attempt was made to find some correlation of tropospheric water vapor content variations from different space missions (100–1000 ppm) with large-scale meteorological processes. The hydration of silicates was predicted as a mechanism for providing a high water vapor tropospheric content. The upper limit of water incorporated in the crust up to a depth of 5 km, and involving the dehydration level of the Venusian crust thermal gradient of 7–12 $K \cdot km^{-1}$, was estimated as 7×10^{22} g (Khodakovsky, 1982). The recent available spectrophotometric measurements of water vapor in the lower atmosphere indicate that the H_2O content near the surface does not exceed 20 ppm. This value represents a severe constraint on the possibility of hydration (Barsukov *et al.*, 1982b; Volkov and Khodakovsky, 1984). The water cycle is therefore suggested to be completed within the Venusian atmosphere.

Atmospheric–Lithospheric Interaction: The Evolutionary Approach

The main processes governing Venus' atmospheric evolution may be considered to be the following: outgassing, the greenhouse effect, hydrogen dissipation, and atmospheric–lithospheric interactions. Each component of the total evolutionary development is connected with every other by feedback.

Buffering gas–mineral equilibria may be effective in the evolution of the atmospheric composition. These equilibria evidently are dependent on geological activity, especially the volcanic and tectonic processes determining the recycling of volatiles in Venus' outer shells. Equilibria shifts lead to the changes in the character of the elemental sink in Venus' crust in relation to different volatiles and may be considered factors of chemical evolution of the atmosphere and crust.

Though our knowledge of Venus' evolution is rather schematic, the relatively intensive tectonic and magmatic activity, outgassing, and the high rate of surface rejuvenation can be suggested reliably as events at the early stages of its geologic history (Barsukov *et al.*, 1984).

In accord with a number of data concerned with the history of the Earth, Moon, and Mars, intensive outgassing is discussed as a plausible event during the first billion years and associated with intensive meteoric impacts (Walker, 1977; Khodakovsky, 1982; Barsukov, 1981; Donahue and Pollack, 1983). In particular, the deficiency of radiogenic ^{40}Ar, which is the product of ^{40}K radioactive decay, in Venus' atmosphere when compared to the terrestrial atmosphere could be regarded as the result of early outgassing, providing there were equal potassium inventories on the Earth and on Venus (Donachue and Pollack, 1983). Indeed, shorter duration of magmatic and tectonic activity on Venus may serve as one of the explanations for the radiogenic ^{40}Ar deficiency on Venus.

The assumption of the primary carbon dioxide composition for the ancient Venusian atmosphere with subordinate fractions of nitrogen, inert gases, and water vapor is not a very improbable hypothesis (Walker, 1975; Pollack and Yung, 1980; Donahue and Pollack, 1983). According to D/H measurements from Pioneer Venus (Donahue *et al.*, 1982) the water inventory on ancient Venus is estimated as at least 0.3% in relation to the terrestrial oceans. This figure is two orders of magnitude above the contempory atmospheric inventory and is considered the most probable value (Donahue and Pollack, 1983).

The original carbon dioxide abundance in the ancient Venusian atmosphere is difficult to estimate, because the rate of outgassing is unknown. The convincing evidence of the absence of carbonate rocks on Venus' surface could serve as a clue to the inference that the current abundance of atmospheric CO_2 was never exceeded.

The outgassing of water vapor and carbon dioxide must be considered a factor in the greenhouse processes that have been critical for Venus' evolution. The presence of these infrared absorbers in the atmosphere has resulted in the heating of the atmosphere and the planetary surface as the reflected solar heat energy is lost in the atmosphere. The relative proximity of Venus' orbit to the Sun is suggested to be the dominant factor in the "runaway" greenhouse effect on Venus: because the condensation temperature of water vapor is always below the surface temperature, the production of liquid water is impossible. According to calculations by Watson *et al.* (1984), the surface temperature of ancient Venus could have attained 950–1300K, this value being a function of carbon dioxide and water vapor content and planetary albedo, provided there has been a carbon dioxide pressure of 90 bar and a water vapor partial pressure in the range of 10–90 bar. The cooling of the planetary surface to the present value of 735K is a result of depletion of the water consumed in photochemical dissociation of water vapor and the water sink in Venus' crust.

The problem of water loss and the related oxygen sink is regarded as one of the most important in the evolutionary history of Venus. Water loss was presumably a result of water vapor photolysis associated with hydrogen escape. A certain fraction of water vapor could be consumed in the thermochemical reaction with the reduced gases as carbon monoxide (Pollack and Yung, 1980; Richardson *et al.*, 1984) in the troposphere. A mechanism of water sink in the crust could be proposed in terms of oxidation of iron minerals in the cycling processes of final hydrogen escape (Khodakovsky, 1982). The sorption of water vapor at the surface of volcanic lava effusions was proposed in Fricker and Reynolds (1968).

Free oxygen as a product of water vapor photodissociation should be consumed in the oxidation of atmospheric gases as well as in the oxidation of the crust. The rate of rejuvenation of crust as a factor of recycling is regarded as an important constraint of the oxygen sink. The estimates of recycling rate, given a consumption of enormous masses of oxygen comparable to that of the terrestrial ocean indicate that the hypothesis that the water inventory of Earth is similar to that of Venus is not plausible (Walker, 1975; Pollack and Yung, 1980; Lewis and Kreimendahl, 1980; Donahue and Pollack, 1983).

If the ancient troposphere of Venus was wet, the formation of hydrated minerals in Venus' crust could be inferred with great probability. The water vapor depletion could be reflected in the changes of mineral assemblages consisting of hydrated phases in a stepwise manner, corresponding to a sequence of different water vapor buffers. Present conditions on Venus are presumably characterized by complete dehydration of the surface rocks.

The formation of significant masses of carbonate rocks is reasonable only in the presence of liquid water at Venus' surface. However, providing there was a high water vapor content in the primary, predominantly carbon dioxide atmosphere, the surface temperature would be well above 750K, excluding the coexistence of most carbonates with silicates. Gradual depletion of the water vapor in Venus' atmosphere and the related surface temperature decrease favor carbonate formation, particularly in silicate–carbonate mineral assemblages. The stability of carbonates and the related role of the buffering of the carbon dioxide in Venus' atmosphere is a function of sulfur-bearing gas tropospheric content as discussed in the section, Carbonates in Venus' Soil.

As was indicated in accordance with preliminary results of the high-resolution radar mapping by Venera 15 and Venera 16 orbiters, the geological activity on Venus during the last billion years has practically ceased; likewise, aeolian agents are not significant in the transformation of the Venusian surface on a planetary scale. It was shown above that carbon, nitrogen, and water vapor are not involved in recycling processes on Venus at present. Therefore, sulfur, chlorine, and fluorine are the only volatile elements providing the presumed chemical interactions between Venus' troposhere and crust.

The sulfur cycle was discussed in detail in the section, sulfur; where it was pointed out that the probable disequilibrium with respect to sulfur dioxide between Venus' surface rocks and its troposphere could be explained either in terms of kinetic constraints and/or as owing to the low rate of rejuvenation of Venus' surface.

Variations in sulfur-bearing components of Venus' atmosphere, connected with lithospheric–atmospheric interactions as well as with geological activity, are perhaps the peculiar manifestation of atmospheric evolution after the termination of processes resulting in the establishment of runaway greenhouse conditions and attainment of the present surface temperature. If the preliminary results of the Venera 15 and Venera 16 radar mapping interpretation are supported by succeeding exploration, the assessment that the main events of atmospheric evolution were associated with the period preceeding 0.5–1.0 billion years ago may be confirmed.

Are there any signs of recent volcanism on Venus? Unfortunately Beta Regio and other regions of presumed volcanic origin are out of the area mapped by Venera 15 and Venera 16 orbiters, so direct geological evidence is not available. However, three groups of indirect data are involved in speculations on recent volcanic activity: temporal variations of SO_2 content and submicron aerosols in Venus' stratosphere (Esposito, 1984); the uncompensated character of Venus' crust at Beta Regio, Atla Regio, etc. (Phillips and Malin, 1983); and the clustering of electrical discharges (Ksanfomality *et al.*, 1983) and the heterogeneity of the thermal emission field (Ksanfomality, 1984) above "volcanic" regions.

Theoretical estimates of Venus' volcanic gaseous composition were made by Richardson *et al.* (1984) and Zolotov and Khodakovsky (1985). It was inferred that one could predict sulfur dioxide to be the dominant sulfur-bearing volcanic emanation. Provided the Venusian magmatic melts have a low water content several unusual sulfur-bearing components, such as SF_4, SF_6, F_2, and Cl_2, are suggested (Fig. 23). The inert chemical properties of SF_6 may have resulted in the conservation of this gas in Venus' atmosphere for a long period on the geological time scale. The detection of SF_6 by the Venera 13 and Venera 14 space probes

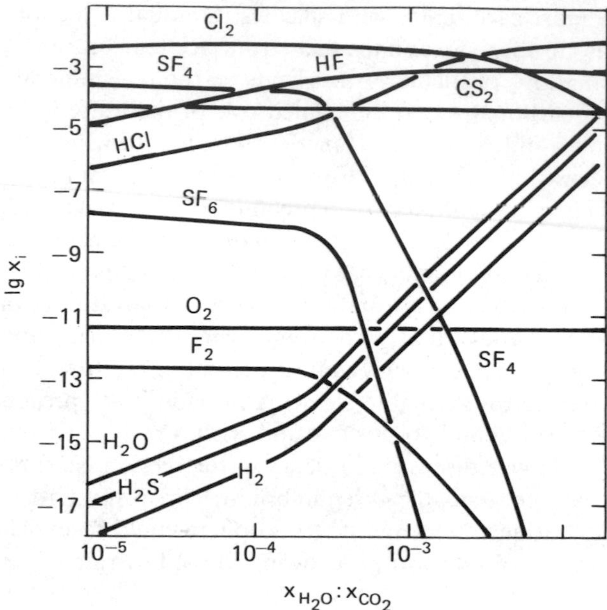

Fig. 23. The microcomponent content in hypothetical Venusian volcanic gases as a function of $H_2O : CO_2$ ratio (1400K, 100 atm) (Zolotov and Khodakovsky, 1985).

(Moroz, 1983) is an indirect evidence of relatively recent volcanic processes on Venus.

The succeeding evolution of Venus' atmosphere involving hydrogen dissipation, the consumption of oxygen derived from water vapor photolysis in the oxidation of atmospheric gases and surface minerals, and the sulfur sink in the crust should result in the decrease of cloud cover thickness, an albedo decrease, and finally the cooling of the atmosphere. If the solar luminosity should increase this scenario would be complicated.

Acknowledgments

The authors are grateful to Prof. V. L. Barsukov, whose efforts were so fruitful in the organization and fulfilment of the Venus exploration program. The authors acknowledge their colleagues in the United States, especially Prof. J. Lewis, for remarks and attention to manuscripts previously submitted for publication in international journals.

References

Adamčik, J. A., and Draper, A. L. (1963). The temperature dependence of the Urey equilibrium and the problem of the CO_2 content of the atmosphere of Venus, *Planet. Space Sci.* **11**, 1303–1307.

Alexandrov, Yu. N., Golovkov, V. K., Dubrovin, V. M., *et al.*, (1980) Radiolocation research of relief and reflectance properties of Venus' surface at wave-length of 39 cm, *Astr. Zhurn.* **57**, 237–249 (in Russian).

Andreeva, N. E., Volkov, V. P., Sidorov, Yu. I., and Khodakovsky, I. L. (1983). On the concentration of minor components in the near-surface Venus' atmosphere: physico-chemical implication, in *Lunar and Planetary Science*, Vol. XIV, pp. 7–8. Lunar and Planetary Institute, Houston, Texas.

Avduevskiy, V. S., Marov, M. Ya., Kulikov, Yu. N., Shari, V. P., Gorbachevskiy, A. Ya., Uspenskiy, G. R., and Cheremukhina, Z. P. (1983) Structure and parameters of the Venus atmosphere according to Venera probe data, in *Venus*, pp. 280–298, edited by D. M. Hunten, L. Colin, T. M. Donahue, and V. I. Moroz. Univ. Arizona Press, Tucson.

Barsukov, V. L. (Ed.) (1981) *Outlines of Comparative Planetology*. Izd-vo Nauka, Moscow (in Russian).

Barsukov, V. L. (1985) Comparative planetology and the early history of the Earth *Geokhimiya* **1985**(1), 3–19.

Barsukov, V. L., Basilevsky, A. T. Burba, G. A., Bobina, N. N., Kryuchkov, V. P., Kuz'min, R. O., Nikolaeva, O. V., Pronin, A. A., Ronca L. B., *et al.*, (1985). The geology and geomorphology of the Venus surface as revealed by the radar images obtained by Venera 15 and 16, in *Proc. 16th, Lunar Planet. Sci. Conf.* (in press).

Barsukov, V. L., Basilevsky, A. T., Kuz'min, R. O., Pronin, A. A., Kryuchkov, V. P., Nikolaeva, O. V., Chernaya, I. M., Burba, G. A., Bobina, N. N., Shashkina, V. P., Markov, M. S., and Sukhanov, A. L. (1984) Geology of Venus as a result of radal image analysis obtained by Venera 15 and Venera 16 automatic interplanetary stations (preliminary data). *Geokhimiya* **1984**(12), 1811–1820 (in Russian).

Barsukov, V. L., Khodakovsky, I. L., Volkov, V. P., and Florensky, K. P. (1980a) The geochemical model of the troposphere and lithosphere of Venus based on new data, in: *COSPAR Space Res.*, Vol. XX, pp. 197–208, 22nd Plen. Meeting, Bangalore, India.

Barsukov, V. L. Volkov, V. P., and Khodakovsky, I. L. (1980b). The mineral composition of Venus surface rocks: a preliminary prediction, in *Proc. 11th Lunar Planet. Sci. Conf.*, pp. 765–773.

Barsukov, V. L., Khodakovsky, I. L., Volkov, V. P., Sidorov, Yu. I., and Borisov, M. V. (1983). The mineral composition of rocks at the landing sites of Venera 13 and Venera 14 descending modules: thermodynamic calculations, in *Lunar and Planetary Science*, Vol. XIV, pp. 19–20. Lunar and Planetary Institute, Houston, Texas.

Barsukov, V. L., Khodakovsky, I. L. Volkov, V. P., Sidorov, Yu. I., Dorofeeva, V. A., and Andreeva, N. E. (1982c). The metal chloride and elemental sulfur condensates in the Venusian troposphere—is it possible? in *Proc. 12th Lunar Planet. Sci. Conf.*, pp. 1517–1532.

Barsukov, V. L., Surkov, Yu. A., Moskalyeva, L. P., Sheglov, O. P., Kharyukova, V. P., Manvelyan, O. S., and Perminov, V. G. (1982a) Geochemical investigations of Venus' surface by Venera 13 and Venera 14. *Geokhimiya* **1982**(7), 899–919.

Barsukov, V. L., Volkov, V. P., and Khodakovsky, I. L. (1982b) The crust of Venus: theoretical models of chemical and mineral composition, in *Proc. 8th Lunar Planet. Sci. Conf., J. Geophys. Res.* **87** Suppl., Part 1, A3–A9.

Bashmashnikov, M. V., Gusljakov, V. T. Yezhkov, V. M., Kerzhanovich, V. V. Natalovich, M. L., and Zeitlin, E. E. (1976) Characteristics of Venus' surface based on radioaltimetric data at Venera 8 landing site, *Kosm. Issled.* **14**, 111–114 (in Russian).

Borucki, W. J., Dyer, J. W., Thomas, G. Z., Jordan, J. C. and Comstock, D. A. (1981). Optical search for lightning on Venus, *Geophys. Res. Lett.* **8**, 233–236.

Connes, P., Connes, J., Benedict, W. S., and Kaplan, L. D. (1967) Traces of HCl and HF in the atmosphere of Venus, *Astrophys. J.* **147**, 1230–1237.

Dobrovolskis, A. R. (1983) Does Venus Breathe? *Icarus* **54**, 539–542.

Donahue, T. M., and Pollack, J. B. (1983) Origin and evolution of the atmosphere of Venus, in *Venus*, pp. 1003–1036, edited by D. M. Hunten, L. Colin, T. M. Donahue, and V. I. Moroz. Univ. Arizona Press, Tucson.

Donahue, T. M., Hoffman, J. H., Hodges, R. R., Jr., and Watson, A. J. (1982) Venus was wet: a measurement of the ratio of D to H, *Science* **216**, 630–633.

Ekonomov, A. P., Golovin, Yu. M., and Moshkin, B. E. (1980) Visible radiation observed near the surface of Venus: results and their interpretation, *Icarus* **41**, 65–75.

Esposito, L. W. (1984) Sulfur dioxide: episodic injection shows evidence for active volcanism, *Science* **223**, 1072–1074.

Esposito, L. W., Knollenberg, R. G., Marov, M. Ya., Toon, O. B., and Turco, R. P. (1983) The clouds and hazes of Venus, in *Venus*, pp. 484–564, edited by D. M. Hunten, L. Colin, T. M. Donahue, and V. I. Moroz. Univ. Arizona Press, Tucson.

Florensky, C. P., Basilevsky, A. T., Burba. G. A., Nikolaeva, O. V., Pronin, A. A., Volkov, V. P., and Ronca, L. B. (1977a) First panoramas of the Venusian surface. in *8th Proc. Lunar Planet. Sci. Conf., Geochim. Cosm. Acta* Suppl. 8, 2655–2664.

Florensky, C. P., Basilevsky, A. T., Burba, G. A., Nikolaeva O. V., Pronin A. A., Selivanov, A. S., Narayeva, M. K., Panfilov, A. S., and Chemodanov, V. P. (1983a) Panorama of Venera 9 and 10 landing sites, in *Venera*, pp. 137–153, edited by D. M. Hunten, L. Colin, T. M. Donahue, and V. I. Moroz. Univ. Arizona Press, Tucson.

Florensky, C. P., Basilevsky, A. T., Kruychkov, V. P., Kuz'min, R. O., Nikolaeva, O. V., Pronin, A. A., Chernaya, I. M., Tyuflin, Yu. S., Selivanov, A. S., Naraeva, M. K., and Ronca, L. B. (1983b) Venera 13 and 14: sedimentary rocks on Venus? *Science*, **221**, 57–58.

Florensky, C. P., Nikolaeva, O. V., Volkov, V. P., Kudryashova, A. F., Pronin. A. A., Gektin, Yu. M., Tchaikina, E. A., and Bashkirova, A. S. (1983c). Redox indicator "Contrast" on the surface of Venus, in *Lunar and Planetary Science*, Vol. XIV, pp. 203–204. Lunar and Planetary Institute, Houston, Texas.

Florensky, C. P., Basilevsky, A. T., Pronin, A. A., and Burba, G. A. (1979). Results of geomorphic interpretation of Venus' panoramas, in *First Panoramas of Venus' Surface*, pp. 107–127. Izd-vo Nauka, Moscow (in Russian).

Florensky, C. P., Basilevsky, A. T., and Selivanov, A. S. (1982) Panoramas of Venera 13 and 14 landing sites (preliminary data), *Astron Vestnik* **3**, 131–138 (in Russian).

Florensky, C. P., Ronca, L. B., Basilevsky, A. T., Burba, G. A., Nikolaeva, O. V., Pronin, A. A., Trakhtman, A. M., Volkov, V. P., and Zasetsky, V. V. (1977b). The surface of Venus as revealed by Soviet Venera 9 and 10, *Geol. Soc. Amer. Bull.* **88**, 1537–1545.

Florensky, C. P., Volkov, V. P., and Nikolaeva, O. V. (1978) A geochemical model of the Venus troposphere, *Icarus* **33**, 537–553.

Ford, P. G., and Pettengill, G. H. (1983) Venus: global surface radio emissivity, *Science* **220**, 1379–1381.

Frenkel', M. Ya., and Zabalueva, E. V. (1983) On the solidification of effusive melt on Venus, *Geokhimiya*, **1983**(9), 1275–1279 (in Russian).

Fricker, P. E., and Reynolds, R. T. (1968) Development of the atmosphere of Venus, *Icarus* **9**, 221–230.

Garvin, J. B. (1984) Dust on Venus: geological implications, in *Lunar and Planetary Science*, Vol. XV, pp. 286–287. Lunar and Planetary Institute, Houston, Texas.

Garvin, J. B., Head, J. W., and Wilson, L. (1982) Magma vesiculation and pyroclastic volcanism on Venus, *Icarus* **52**, 365–372.

Gel'man, B. G., Zolotukhin, V. G., Lamonov, N. I., Levchuk, B. V., Lipatov, A. N., Mukhin, L. M., Nikanorov, D. F., Rotin, V. A., and Okhotnikov, B. P. (1979) Chemical composition of Venus' atmosphere measurements by gas chromatograph on Venera 12, *Kosm. Issled.* **17**, 708–713 (in Russian).

Greeley, R., Marshall, J. R., and Leach, R. N. (1984) Microdunes and other aeolian bedforms on Venus: wind tunnel simulations *Icarus* **60**, 152–160.

Hunten, D. M., Colin, L., Donahue, T. M., and Moroz, V. I. (Eds.) (1983) *Venus*, Univ. Arizona Press, Tucson.

Jakosky, B. M. (1984) Buffering of diurnal temperature variations on Venus, *Icarus* **59**, 478–480.

Karpov, I. K., Kiselev, A. I., and Letnikov, F. A. (1971) *Chemical Thermodynamics in Geochemistry and Petrology*. Irkutsk (in Russian).

Kemurdzhian, A. L., Brodsky, P. N., Gromov, V. V., Grushin, V. P., Kiselev, I. E., Kozlov, G. V., Mitskevich, A. V., Perminov, V. G., Sologub, P. S., Stepanov, A. D., Turobinsky, A. V., Turchaninov, V. N., and Yudkin, E. N. (1983) Preliminary results of physicomechanical properties of soil measurements at Soviet interplanetary automatic station Venera 13 and Venera 14, *Kosm. Issled.* **21**, 323–330 (in Russian).

Kerzhanovich, V. V., and Marov, M. Ya. (1983) The atmospheric dynamics of Venus according to Doppler measurements by the Venera entry probes, in *Venus*, pp. 766–778, edited by D. M. Hunten, L. Colin, T. M. Donahue, and V. I. Moroz. Univ. Arizona Press, Tucson.

Khodakovsky, I. L. (1982) Atmosphere-surface interactions on Venus and implications for atmospheric evolution, *Planet Space Sci.* **30**, 803–817.

Khodakovsky, I. L., Volkov, V. P., Sidorov, Yu. I., and Borisov, M. V. (1979) Venus: preliminary prediction of the mineral composition of surface rocks *Icarus* **39**, 352–363.

Kliore, A. J., Elachi, C., Patel, I. R., and Cimino, J. B. (1979) Liquid content of the lower clouds of Venus as determined from Mariner 10 radio occultation, *Icarus* **37**, 51–72.

Knollenberg, R. G., and Hunten, D. M. (1980) The microphysics of the clouds of Venus: results of the Pioneer Venus particle size spectrometer experiment, *J. Geophys. Res.* **85**(A13), 8039–8058.

Kolosov, M. A., Yakovlev, O. I., Pavelyev, A. G., Kucherjavenkov, A. I., and Milekhin, O. E. (1981) Characteristics of the surface and features of the propagation of radio waves in the atmosphere of Venus from data bistatic radiolocation experiments using Venera 9 and 10 satellites, *Icarus* **48**, 188–200.

Krasnopolsky, V. A., and Parshev, V. A. (1981) Chemical composition of the atmosphere of Venus, *Nature* (*London*) **292**, 610–613.

Krasnopolsky, V. A., and Parshev, V. A. (1983) Photochemistry of the Venus atmosphere, in *Venus*, pp. 431–458, edited by D. M. Hunten, L. Colin, T. M. Donahue, and V. I. Moroz. Univ. Arizona Press, Tucson.

Kroupenio, N. N. (1972) Some characteristics of the Venus surface. *Icarus* **17**, 692–698.

Ksanfomality, L. V. (1984) Infrared thermal emission of Venus, *Kosm. Issled.*, **22**, 252–256 (in Russian).

Ksanfomality, L. V., Scarf, F. L., and Taylor, W. W. (1983) The electrical activity of the atmosphere of Venus, in *Venus*, pp. 565–603, edited by D. M. Hunten, L. Colin, T. M. Donachue, and V. I. Moroz. Univ. Arizona Press, Tucson.

Kuiper, G. P. (1969) On the nature of the Venus clouds, in *Planetary atmospheres*, *IAU Symp. No. 40*, edited by C. Sagan, T. Owen, and H. Smith. D. Reidel, Dordrecht.

Kuz'min, A. D., and Marov, M. Ya. (1974) *Physics of Planet Venus*. Izd-vo Nauka, Moscow (in Russian).

Lewis, J. S. (1969) Geochemistry of the volatile elements on Venus, *Icarus* **11**, 367–386.

Lewis, J. S., (1970) Venus: atmospheric and lithospheric composition, *Earth Planet. Sci. Lett.* **10**, 73–80.

Lewis, J. S., (1971) Venus: surface temperature variations, *J. Atmos. Sci.* **30**, 1218–1220.

Lewis, J. S., and Fegley, B. (1982) Venus: halide cloud condensation and volatile element inventories, *Science* **216**, 1223–1224.

Lewis, J. S., and Kreimendahl, F. A., (1980) Oxidation state of the atmosphere and crust of Venus from Pioneer Venus results, *Icarus* **42**, 330–337.

Lewis, J. S., and Prinn, R. G. (1984) *Planets and Their Atmospheres: Origin and Evolution.* Academic Press, Orlando, Florida.

Marov, M. Ya. (1972) Venus: a perspective of the beginning of planetary explorations *Icarus* **16**, 415–461.

Marov, M. Ya., Lystsev, V. E., Lebedev, V. N., Lukashevich, N. L., and Shari, V. P. (1980) The structure and microphysical properties of the Venus clouds: Venera 9, 10 and 11 data, *Icarus* **44**, 608–639.

Masursky, H., Eliason, E., Ford, P. G., McGill, G. E., Pettengill, G. H., Schaber, G. G., and Schubert, G. (1980) Pioneer Venus radar results: geology from images and altimetry, *J. Geophys. Res* **85**(A13), 8232–8260.

McElroy, M. B., Sze, N. D., and Yung Y. L. (1973) Photochemistry of the Venus atmosphere, *J. Atmos. Sci.* **30**, 1437–1447.

McEwan, M. J., and Phillips, F. L. (1975) *Chemistry of the atmosphere.* Edward Arnold, New Zealand.

McGill, G. E., Warner, J. L., Malin, M. C., Arvidson, R. E., Eliason, E., Nozette, S., and Reasenberg, R. D. (1983) Topography, surface properties and tectonic evolution, in *Venus*, pp. 69–130, edited by D. M. Hunten, L. Colin, T. M. Donahue, and V. I. Moroz. Univ. Arizona Press, Tucson.

Moroz, V. I. (1983) Summary of preliminary results of the Venera 13 and 14 missions in *Venus*, pp. 45–68, edited by D. M. Hunter, L. Colin, T. M. Donahue, and V. I. Moroz. Univ. Arizona Press, Tucson.

Moroz, V. I., Ekonomov, A. P., Golovin, Yu. M., Moshkin, B. E., and San'ko, N. F. (1983) Solar radiation scattering in the Venus atmosphere: the Venera 11, 12 data, *Icarus* **53**, 509–537.

Mueller, R. F. (1963) Chemistry and petrology of Venus: preliminary deductions, *Science* **41**, 1046–1047.

Mueller, R. F. (1964) A chemical model for the lower atmosphere of Venus, *Icarus* **3**, 285–298.

Mueller, R. F. (1969) Planetary probe: origin of the atmosphere of Venus, *Science* **163**, 1322–1324.

Mueller, R. F., and Kridelbaugh, S. J. (1973) Kinetics of CO_2 production on Venus, *Icarus* **19**, 531–542.

Mueller, R. F., and Saxena, S. K. (1977) *Chemical Petrology.* Springer-Verlag, New York.

Nozette, S., and Lewis, J. S. (1982) Venus: chemical weathering of igneous rocks and buffering of atmospheric composition. *Science* **216**, 181–183.

Phillips, R. J., and Malin, M. C. (1983) The interior of Venus and tectonic implications, in *Venus*, pp. 159–214, edited by D. M. Hunten, L. Colin, T. M. Donahue, and V. I. Moroz. Univ. Arizona Press, Tucson.

Pollack, J. B., and Yung, Y. L. (1980) Origin and evolution of Planetary atmospheres, *Ann. Rev. Earth Planet. Sci.* **8**, 425–487.

Prinn. R. G. (1971) Photochemistry of HCl and other minor constituents in the atmosphere of Venus, *J. Atmos. Sci.* **28**, 1058–1068.

Prinn, R. G. (1975) Venus: chemical and dynamical processes in the stratosphere and mesosphere, *J. Atmos. Sci.* **32**, 1237–1247.

Prinn, R. G. (1985) Volcanoes and clouds of Venus, *Sci. Amer* **252**(3), 36–43.

Richardson, S. M., Pollack, J. B., and Reynolds, R. T. (1984) Water loss on Venus: the role of carbon monoxide, *Icarus* **60**, 307–316.

Rossow, W. B. (1978) Cloud microphysics: analysis of the clouds of Earth, Venus, Mars and Jupiter, *Icarus* **36**, 1–50.

San'ko, N. F. (1980) Gaseous sulfur in the Venus atmosphere, *Kosm. Issled.* **18**, 600–608 (in Russian).

Sill, G. T. (1972) Sulfuric acid in the Venus clouds, *Commun. Lunar Planet. Lab.* **9**, 191–198.

Surkov Yu. A. (1977) Geochemical studies of Venus by Venera 9 and 10 automatic interplanetary stations, in *Proc. 8th Lunar Planet. Sci. Conf.*, pp. 2665–2689.

Surkov, Yu. A. (1983) Studies of Venus rocks by Veneras 8, 9 and 10, in *Venus*, pp. 154–158, edited by D. M. Hunten, L. Colin, T. M. Donahue, and V. I. Moroz. Univ. Arizona Press, Tucson.

Surkov, Yu. A., and Andreichikov, B. M. (1973) Composition and structure of the Venus cloud cover, *Geokhimiya* **1973**(10), 1435–1440 (in Russian).

Surkov, Yu. A., Andreichikov, B. M., and Kalinkina, O. M. (1973) On the ammonia content in Venus' atmosphere according to automatic station Venera 8 data, *Dokl. Akad. Nauk SSSR*, **213**, 296–298 (in Russian).

Surkov Yu. A., Kirnozov, F. F., Gurjanov, V. I., Glazov, V. N., Dunchenko, A. G., Kurochkin, S. S., Rasputny, V. N., Kharitonova, E. G., Tatsiy, L. P., and Gimadov, V. L. (1981) Investigation of Venus' cloud aerosol on the automatic interplanetary station Venera 12 (preliminary data), *Geokhimiya* **1981**(1), 3–9 (in Russian).

Surkov, Yu. A., Kirnozov, F. F., Glazov, V. N., Dunchenko, A. G., and Atrashkevich, V. V. (1982) New data on Venus' cloud aerosol (preliminary results on Venera 14 investigations), *Pis'ma v Astron. Zhurn.* **1982**(11), 700–704 (in Russian).

Surkov, Yu. A., Moskalyeva, L. P., Shcheglov, O. P., Kharyukova, V. P., Manvelyan, O. S., and Kiritchenko, V. S. (1983) Determination of elemental composition of rocks on Venus by Venera 13 and Venera 14 (preliminary results), in *Proc. 13th Lunar Planet. Sci. Conf., J. Geophys. Res.* **88**, Suppl., A481–A494.

Shvarov, Yu. V. (1978) On the minimization of the thermodynamical potential of the open chemical system, *Geokhimiya* **1978**(12), 1892–1895 (in Russian).

Terradellas, J., and Bonnetain, L. (1973) Nature des reactions chimiques lors de e'action du dioxyde de soubre sur le carbonate de calcium, *Bull. Soc. Chim. France*, No. 6, P. 1, 1903–1908.

Toon, O. B., Ragent, B., Colburn, D., Blamont, J., and Cot, C. (1984) Large, solid particles in the clouds of Venus: do they exist? *Icarus* **57**, 143–160.

Urey, H. C. (1951) The origin and development of the Earth and other terrestrial planets, *Geochim. Cosmochim. Acta* **1**, 209–277.

Vinogradov, A. P., and Volkov, V. P. (1971) On the wollastonite equilibrium as a mechanism determining Venus' atmospheric composition, *Geochimiya* **1971**(7), 755–759 (in Russian).

Volkov, V. P. (1983) *Chemistry of Atmosphere and Surface of Venus*. Izd-vo Nauka, Moscow (in Russian).

Volkov V. P., and Khodakovsky, I. L. (1984) Physicochemical modelling of the mineral composition of Venus' surface rocks, in *27th Int. Geol. Congr., Repts., Geochemistry and Cosmochemistry*, Vol. 11, pp. 32–38. Izd-vo Nauka, Moscow (in Russian).

Volkov V. P., Khodakovsky, I. L., Dorofeeva, V. A., and Barsukov, V. L. (1979) Main physico-chemical factors controlling chemical composition of Venus' cloud cover, *Geokhimiya* **1979**(12), 1759–1766 (in Russian).

von Zahn, U., Kumar, S., Niemann, H., and Prinn, R. (1983) Composition of the Venus atmosphere, in *Venus*, pp. 299–430, edited by D. M. Hunten, L. Colin, T. M. Donahue, and V. I. Moroz. Univ. Arizona Press, Tucson.

Walker, J. C. G. (1975) Evolution of the atmosphere of Venus. *J. Atmos. Sci.* **32**, 1248–1256.

Walker, J. C. G. (1977) *Evolution of the Atmosphere*. Macmillan, New York.

Warner, J. L. (1983) Sedimentary processes and crustal cycling on Venus, *Proc. 13th Lunar Planet. Sci. Conf., J. Geophys. Res.* **88**, Suppl., A495–A500.

Warnock, W. W., and Dickel, J. R. (1972) Venus: measurements of brightness temperatures in the 7–15 cm wavelength range and theoretical radio and radar spectra for a two-layer sub-surface model, *Icarus* **17**, 682–691.

Watson, A. J., Donahue, T. M., and Kuhn, W. R. (1984) Temperatures in a runaway greenhouse on the evolving Venus: implications for water loss, *Earth Planet. Sci. Lett.* **68**, 1–6.

Wilson, L., and Head J. W. (1983) A comparison of volcanic eruption processes on Earth, Moon, Mars, Io and Venus, *Nature (London)* **302**, 633–636.

Wilson, L., Garvin, J. B., and Head, J. W. (1984) Characteristics of basaltic lava flows on Venus, in *Lunar and Planetary Science*, Vol. XV, pp. 922–923. Lunar and Planetary Institute, Houston, Texas.

Winick, J. R., and Stewart, A. I. F. (1980) Photochemistry of SO_2 in Venus' upper cloud layers, *J. Geophys. Res.* **85**(A13), 7849–7860.

Young, A. T. (1973) Are the clouds of Venus sulfuric acid? *Icarus* **18**, 564–582.

Young, A. T. (1977) An improved Venus cloud model, *Icarus* **32**, 1–26.

Young, L. D., and Young, A. T. (1973) Comment on "The composition of the Venus cloud tops in light of recent spectroscopic data," *Astrophys. J.* **179**, L39–L43.

Young, L. D. (1972) High resolution spectra of Venus, *Icarus* **17**, 632–658.

Yung Y. L., and DeMore, W. B. (1982) Photochemistry of the stratosphere of Venus: implications for atmospheric evolution, *Icarus* **51**, 199–247.

Zoltov, M. Yu. (1985) Sulfur-containing gases in the Venus atmosphere and stability of carbonates, in *Lunar and Planetary Science*, Vol. XVI. Lunar and Planetary Insitute, Houston, Texas.

Zolotov, M. Yu., and Khodakovsky, I. L. (1985) Composition of volcanic gases on Venus, in Lunar and Planetary Science, Vol. XVI. Lunar and Planetary Institute, Houston, Texas.

Chapter 5
Weathering of Martian Surface Rocks

Yu. I. Sidorov and M. Yu. Zolotov

Introduction

The early 1960s marked the beginning of an extremely important trend in Martian investigation, i.e., exploration by space missions. From 1962 to 1976 more than a dozen spacecraft in the framework Mars, Mariner, and Viking space missions were launched, providing new and important data on the composition and structure of the Martian atmosphere and soil, as well as data on the planet's physical characteristics.

The temperature at the planetary surface depends to a great extent on latitude, as well as on diurnal and seasonal variations, but even in Martian summer in the equatorial regions the temperature does not exceed 300K. The mean temperature of the planet is about 210K.

The martian surface is substantially asymmetric. The northern hemisphere is dominated by low, cratered plains of volcanic origin located below the datum level; the southern hemisphere is comprised mainly of heavily cratered terrain similar to the lunar highlands. The immense volcanic structures are situated on the Tharsis Plateau. Among them is the greatest volcano of the solar system–Olympus Mons, 25 km in height. In addition, there are large braided valley systems hundreds of miles long, reminiscent in morphology of the dried channels of terrestrial rivers, and other peculiar features reflecting fluvial and glacial erosion. Tectonic activity is evidenced by numerous faults, grabens, extensive crevasses with a system of anastomosing chasmata. The fault systems are several kilometers in depth, dozens of kilometers in width, and hundreds and even thousands of kilometers in length.

Chemistry and Mineralogy of Martian Soil

Data on the properties of Martian soil provided by Mars, Mariner, and Viking space missions have confirmed the suggestion on the presence of weathering products on the planetary surface (Adams and McCord, 1969), in particular, ferric oxides and hydroxides. The devices mounted on the Viking landers allowed us to obtain the complete characteristic of the surface materials at two points on the planet. However, there are only indirect data available on the composition of the bedrock of the northern volcanic plains of Mars, the Tharsis Plateau, and some continental regions.

Interpretation of data obtained by the Mars-5 mission suggests that in uranium and thorium contents, the rocks of the continental regions analyzed are similar to rocks of the anorthite–norite–troctolite series of lunar materials, whereas young volcanic rocks in the Tharsis region are similar to basalts (Basilevsky *et al.*, 1981). The morphology of young lava flows in the Tharsis and Elysium regions corresponds to basic rocks and the flows are assumed to be ultrabasic lavas (Moore, 1979).

Many authors have attempted to estimate the composition of the surface material by means of comparative analysis of the reflectance spectra of different planetary areas, mainly in the visible and infrared regions, with the spectra of rocks, minerals, and mineral mixtures. The results of the majority of this work are reviewed in detail by Singer *et al.* (1979). Some new data are available in McCord *et al.* (1982a) and Singer (1982), which suggest that there are definite absorption bands, indicating Fe^{3+}, in the reflectance spectra of high-albedo areas. The surface material in these regions is interpreted to be the oxidized weathering product characterized by a high content of ferric iron compounds. The concentration of ferric iron oxides and hydroxides in high-albedo areas is estimated as 6–8 vol %. The presence of hydrated minerals is plausible. The spectra of low-albedo areas, dominating the southern tropical Martian regions, are characterised by diffuse ferrous iron absorption bands as well as solid ferric iron absorption band, presumably because of the presence of pyroxene in the soils.

Investigation of the thermal inertia of the Martian surface implies that low-albedo areas are characterized by high values of thermal inertia coefficients and vice versa (Palluconi and Kieffer, 1981; Christensen, 1982; Kieffer *et al.*, 1977). The material with low thermal inertia can be interpolated as fines with a particle size interval of 17–400 microns (Kieffer *et al.*, 1977). The comparison of data on soil thermal inertia with the results of radar reflectivity measurements suggest the existence of regions of predominantly weathering as well as areas with abundant fine-grained and, possibly, aeolian depositions (Jakosky and Muhleman, 1981; Christensen, 1982).

The surface material of most low-albedo areas presumably is mainly fresh or slightly altered basic and ultrabasic rocks and/or the debris of these rocks. The amount of weathered material must be lower than in areas of high albedo.

The spectral differences of low-albedo areas can be explained by variations in the rock to sand ratio, as well as by certain variability of igneous rock types (McCord *et al.*, 1978, 1982a; Singer *et al.*, 1979).

Data obtained from Viking landers (Baird *et al.*, 1976; Clark *et al.*, 1976; Toulmin *et al.*, 1977), the character of reflectance spectra of the surface material and atmospheric dust (Singer *et al.*, 1979; Singer, 1982; McCord *et al.*, 1982a), and the investigation of soil thermal inertia (Kieffer *et al.*, 1977; Jakosky and Muhleman, 1981; Palluconi and Kieffer, 1981) indicate that the high-albedo surface material is represented by fine-grained oxidized weathering products, homogenized by planetary-scale aeolian processes.

We next consider in detail the results of elemental analyses of soils carried out by the Viking 1 and 2 landers.

The landing sites of the Viking 1 and 2 landers are two volcanic regions of the northern hemisphere—on Chryse and Utopia Plains, respectively—the two sites being separated by about 6500 km. The planetary surface at the landing sites is similar to terrestrial rocky deserts and is represented by a predominantly sandy material with evidence of aeolian activity (Moore *et al.*, 1977). The size of debris is estimated to vary from several to dozens of centimeters (Moore *et al.*, 1977; Garvin *et al.*, 1981). The fines were estimated to contain a great number of particles of the 10–100 micron size. The material of the regolith duricrust partly consists of cemented fragments of fine-grained material (Baird *et al.*, 1976; B. C. Clark *et al.*, 1976). The regolith has a yellowish-brown color (Huck *et al.*, 1977). Several authors (Huck *et al.*, 1977; Evans and Adams, 1979; Strickland, 1979; Evans and Adams, 1981) have compared the reflectance spectra of surface material (in the visible and infrared regions) with corresponding data on different terrestrial rocks and minerals. A distinct absorption line of ferric iron can be observed. The spectra as a whole correspond to weathering products of basalts, especially to altered basaltic pumice (Evans and Adams, 1979, 1981). Some fragments are presumably covered with weathering crust (Strickland, 1979).

Each Viking lander carried an energy-dispersive X-ray fluorescence spectrometer (XRFS) for elemental analysis of samples of the Martian soil (B. C. Clark *et al.*, 1976, 1982). Most of the samples were taken from a depth of 6 cm. The composition of both the fine soil fraction (up to 2 mm in diameter) and larger fragments (up to 10 mm in diameter) was analyzed (Baird *et al.*, 1976). All the samples are greatly similar in chemical composition (B. C. Clark *et al.*, 1982); however, considerable variations can be observed in sulfur and chlorine content: 60 and 100 %, respectively. The pebble-size regolithic fragments of the duricrust (V-1) are characterized by relatively high sulfur and chlorine concentrations (B. C. Clark *et al.*, 1976, 1982). The fragments contain 9.5% SO_3 and 0.9% Cl. Iron content in samples taken beneath the large fragments is lower than in those taken directly from the surface (B. C. Clark *et al.*, 1982). A correlation of sulfur and chlorine concentration and possibly magnesium content can be observed in the samples (B. C. Clark *et al.*, 1982). Similar trends can be observed for Si, Al, and Ca. In a number of samples Fe and Ti contents are well correlated. Some samples indicated the presence of bromine, with a Br/Cl ratio estimated to be about 100 (B. C. Clark and Baird, 1979a). Chemical composition of a typical regolith sample is indicated in Table 1.

The chemical composition of gases emitted from the regolith fragments after heating to 500°C was determined by the gas-chromatograph mass spectrometer (GCMS) on the Viking 1 and 2 landers. The products of pyrolysis were estimated to contain substantial amounts of water, about 1–3% most of which was evapo-

Table 1. Chemical composition of a typical regolith sample from Chryse Planitia (SI) (Baired *et al.*, 1976, B. C. Clark *et al.*, 1976).

Element	SI	Oxides	SI
Si	20.9 ± 2.5	SiO_2	44.7
Al	3.0 ± 0.9	Al_2O_3	5.7
Fe	12.7 ± 2.0	Fe_2O_3	18.2
Mg	5.0 ± 2.5	MgO	8.3
Ca	4.0 ± 0.8	CaO	5.6
Ti	0.51 ± 0.2	TiO_2	0.8
K	<0.25	K_2O	0.3
S	3.1 ± 0.5	SO_3	7.7
Cl	0.7 ± 0.3		
O[a]	50.1 ± 4.3		
X[b]	8.4 ± 7.8		

[a] "Oxygen" is the sum of all elements not directly determined.
[b] X is the sum of nondetected components, including H_2O Na_2O, CO_2, and NO_x.

rated in the range of 200–350°C. The carbon dioxide content was also determined (Biemann *et al.*, 1977; Anderson and Tice, 1979).

The experiments have also been carried out on the presence of magnetic particles in the regolith. Their content is estimated at 1–2 vol % (Hargraves *et al.*, 1977, 1979).

The aeolian homogenization of regolith could be responsible for the similarity of the chemical compositions of material from the two landing sites (Toulmin *et al.*, 1977; B. C. Clark *et al.*, 1982). The bulk of the regolith fines cannot be considered weathering products of rocks at the landing sites but are likely to represent the weathering product of a certain "average" Martian rock. The low content of Si, Al, K, Rb, Zn, and Sr is indicative of the basic and ultrabasic composition of the primary rock, containing labradorite, pyroxenes, and magnetite (Baird *et al.*, 1976; Toulmin *et al.*, 1977; B. C. Clark *et al.*, 1982). This is consistent with the interpretation of data from the Mars 5 orbiter Gamma-ray spectrometry of the surface (Surkov *et al.*, 1980; Basilevsky *et al.*, 1981), as well as with the data from the low-albedo areas of the planet (Singer *et al.*, 1979; McCord *et al.*, 1982a).

If the regolith composition at the Viking 1 and 2 landing sites is assumed to contain a considerable fraction of bedrock meterial, the similarity of chemical composition of the samples analyzed is indicative of the similarity of the igneous rocks on Chryse and Utopia Plains. The morphological analysis has shown that in these regions the igneous rocks are most likely basalts or felsites (Garvin *et al.*, 1981).

Assuming the isochemical character of weathering some authors have attempted to predict the composition of a primary rock (Maderazzo and Huguenin, 1977; McGetchin and Smyth, 1978; Bussod and McGetchin, 1979; Morgan and Anders, 1979). The relatively high iron content of Martian primary rocks may result from a high content of FeO in the mantle, evidenced by its high density (McGetchin and Smyth 1978; Goettel, 1981). The Fe-bearing picrites and komatiites are considered terrestrial analogs of Martian rocks (Maderazzo and Huguenin, 1977; Bussod and McGetchin, 1979; Baird and Clark, 1981).

The composition of Martian primary rocks has also been estimated in terms of the nonisochemical weathering of rocks, minerals, and volcanic glasses (Allen *et al.*, 1980; Gooding, 1980). The results are in agreement with the assumption of the predominance of basic and ultrabasic rocks on the Martian surface.

The alternative model suggests that the presence of a great amount of meteoritic matter be considered as the source of formation for the final products of weathering. The similarity of regolith chemical composition with some types of meteorites is noted by B. C. Clark *et al.* (1979a), B. C. Clark and Baird (1979d), and Baird and Clark (1981); in particular, it corresponds to a mixture of tholeiitic basalts with Cl-type carbonaceous chondrites (B. C. Clark *et al.*, 1979a).

Detailed investigation of four samples of rare meteorites—shergottites, nakhlites (three samples), and Chassigny achondrite—assigned the SNC abbreviation have led to a conclusion that they all may represent analogs of Martian rocks. Determination of crystallization age indicated that all these meteorites are close to 1.36 billion years old. The results of investigations (Burghele *et al.*, 1983) as well as data on the oxygen isotopes (Clayton and Mayeda, 1983) indicate that SNC meteorites may be considered part of one parent body. The isotopic composition of captured rare gases (Bogard and Johson, 1983) and nitrogen (Becker and Pepin, 1984) in shergottite is consistent with that of the Martian atmosphere. Hence, inferences about Martian primary rocks can be drawn from the elemental and mineral compositions of SNC meteorites.

Martian Regolith Mineral Composition

Martian regolith mineral composition can be estimated both from the data on soil chemical composition, magnetic particle content, spectrum characteristics of surface material, and atmospheric dust, and by the theoretical modeling of weathering mechanaisms and rates, as well as calculations of stabilities of certain minerals, etc.

Presumably, the Martian regolith is a mixture of mineral particles represented by silicates, salts, and oxides (Baird *et al.*, 1976; Toulmin *et al.*, 1977). According to the results of mathematical modeling and laboratory experiments (Baird *et al.*, 1976; Toulmin *et al.*, 1977) the Martian regolith mineral composition can be ascribed as following: 80% smectite clays (nontronite, saponite, montmorillonite), 10–13% sulfates, and 5–7% carbonates. The presence of small (up to 5%) amounts of iron and titanium oxides as well as chlorides and nitrates is plausible.

The thermodynamic stability of rock-forming minerals of igneous rocks and the formation of secondary minerals under Martian conditions have been reviewed by Gooding (1978). According to the results of thermodynamic calculations, such primary minerals as quartz, albite, K-feldspar, apatite, and corundum are considered stable mineral phases. Secondary minerals, such as kaolinite, Ca-beidellite, calcium and magnesium carbonates, anhydrite, kieserite, hematite, and maghemite, are stable in the present Martian environment. However, because of low reaction rates the attainment of chemical equilibrium in the global atmosphere–surface rocks system with respect to a number of gaseous components is not real under present conditions. Therefore the presence of metastable phases on the planetary surface is plausible.

The mineral composition of the Martian regolith is a function of the absolute and relative rates gas–solid weathering reactions.

Minerals of Igneous Rocks in the Regolith

The Martian regolith appears to be composed not only of weathering products but also of the debris of rocks and separate fragments of primary minerals (Toulmin *et al.*, 1977). The Martian soil is estimated to contain up to 5–10% unweathered basaltic fragments.

The Martian regolith may also contain a number of thermodynamically stable minerals of igneous rocks, e.g., quartz, apatite, rutile and metastable magnetite, ilmenite, and augite (Gooding, 1978, 1980).

In fact, the infrared regolith spectra at the Viking 1 and 2 landing sites suggest the presence of ferruginous pyroxenes (Huck *et al.*, 1977). The correlation of Fe and Ti content in regolith samples could be interpreted in terms of the presence of ilmenite or titanomagnetite (B. C. Clark and Baird, 1979b). Data on the UV spectrum of atmospheric dust (Pang and Ajello, 1977) presumably indicate the presence of several percent of anatase. We consider the possible presence of magnetite and sulfides in Martian soil in the following two subsections.

The presence of substantial amounts of volcanic glass, olivines, and calcium plagioclases in the regolith is of low probability as these minerals are unstable to mechanical and chemical weathering (Gooding, 1980).

Clay Minerals

Mariner 9 orbiter infrared spectroscopic data provided information on the composition of atmospheric dust. The dominant constituents are estimated to be silica (SiO_2, up to $60 \pm 10\%$) (Hanel *et al.*, 1972; Toon *et al.*, 1977) and montmorillonite (Hunt *et al.*, 1973). The same infrared spectra were interpreted by Aronson and Emslie (1975) as evidence of a different mineralogical composition of the atmospheric dust. It was described as a mixture of feldspatic, quartz, and mica mineral particles, the clay minerals being absent. The mineralogical model of the regolith (Baird *et al.*, 1976; Toulmin *et al.*, 1977) assuming the soil to contain up to 80% smectite clays (47% nontronite, 17% montmorillonite, and 15% saponite) is based on the prediction of montmorillonite in the atmospheric dust. Biological gas-exchange experimental results provide indirect evidence for the presence of clay minerals in the regolith (Oyama and Berdahl, 1977). The laboratory experiments carried out by Banin and Rishpon (1979) and Banin and Margulies (1983) revealed that smectites, in particular montmorrilonites, are important and active constituents of the Martian soil, responsible for such properties as ion exchange, molecular adsorption, and catalysis.

Comparison of visible and infrared montmorillonite and nontronite spectrum data with regolith spectrum data at Viking 1 and 2 landing sites with those of high- and low-albedo Martian regions has not revealed a consistency (Singer, 1981). This conclusion was supported later by other investigators (Singer, 1982; McCord *et al.*, 1982a). However, the new infrared spectrum data on Fe- and Ca-montmorillonites and their mixtures with a small amount of nontronite indicate that hydrated Fe-bearing montmorillonite may be an important constituent of the Martian soil (Banin and Margulies, 1984). The interpretation of the recent infrared spectrum data on Martian regions by McCord *et al.* (1982a) has ruled out montmorillonite as the main soil constituent; however, they predict the presence of Mg-bearing phyllosilicates (talc, serpentine, and Mg-smectite). The color and magnetic variations in nontronite after heating up to 700–900°C closely approach those of the Martian soil (Moskowitz and Hargraves, 1982); when subjected to a shock effect (180–300 kbar) the nontronite becomes darker and redder (Weldon *et al.*, 1982). Nontronites of impact origin may account for some color variations in the soils.

According to thermodynamic calculations, the kaolinite as well as calcium-, sodium-, and potassium- bearing beidellites could be stable, whereas the montmorillonite, saponite, and nontronite are considered metastable minerals under present conditions on the Martian surface (Gooding, 1978).

Laboratory experiments have shown the possibility of the formation of clay minerals as a result of the photostimulated oxidation of ferruginous silicates (Huguenin, 1974); however, the formation of great amounts of clays in this way is not plausible (Gooding, 1978). Volcanic glasses are considered one of the possible sources of clay minerals (Gooding and Keil, 1978; Evans *et al.*, 1980; Allen *et al.*, 1981b). This assumption is based on the estimates of glass decomposition rates in interaction with atmospheric gases. These values are presumably higher in relation to the decomposition rates of silicates. According to the thermodynamic calculations (Gooding and Keil, 1978) the beidellites composed of calcium, sodium, and potassium are predicted to be products of the chemical weathering of feldspathic volcanic glasses under present Martian conditions. The weathering of basic and ultrabasic glasses may lead to the formation of metastable smectites. The rate of decomposition of rocks containing 10% glasses is estimated as 0.4 cm for 10^9 years (Gooding and Keil, 1978).

In the weathering process silicates and glasses as a result of gas–solid reactions, smectite clays could accumulate as intermediate products, decomposition rate, of which is lower than their rate of formation (Gooding, 1978; Gooding and Keil, 1978). However, as pointed out by Toulmin *et al.* (1977) and Gooding (1978), great amounts of clays can have been formed only in the presence of water, probably in the process of chemical sedimentation in ancient seas. Clays may also have been formed in the process of palagonitization, i.e., as a result magma interaction with permafrost rocks (Toulmin *et al.*, 1977; Allen, 1979; Allen *et al.*, 1981b) and under hydrothermal alteration of impactites (Allen *et al.*, 1982).

If ancient weathering processes resulted in the formation of acidic water (B. C. Clark, 1979, 1980; Gooding, 1980) the only stable clay mineral phase would be kaolinite according to thermodynamic predictions (Gooding, 1978); however, if smectite clays were formed at ancient stages of Martian history their interaction with atmospheric CO_2 under present conditions should result in their dehydration and decomposition (Gooding, 1978). The decomposition rate of nontronite and saponite is suggested to be higher than that of relatively stable montmorillonite. The decomposition of montmorillonite occurs in several stages, including the formation of talc (Gooding, 1978). The final products of decomposition of nontronite, saponite, and montmorillonite under contemporary Martian conditions are thought to consist of quartz, hematite or maghemite, Ca- and Mg carbonates, Ca-beidellite, and kaolinite (Gooding, 1978). We emphasize that the amount of kaolinite and beidellites should not be too large, because aluminum content in the regolith is rather low.

Palagonite has been considered by many authors (Toulmin *et al.*, 1977; Soderblom and Wenner, 1978; Evans and Adams, 1979; Evans *et al.*, 1980; Allen *et al.*, 1980, 1981b; Singer, 1981) as a possible analog of a salt-free fraction of the Martian soil. Palagonite is an amorphous-crystalline aggregate of variable chemical composition, formed as a result of the weathering of basaltic glasses in the presence of H_2O in the form of liquid or ice (Geptner, 1977). Palagonitization can occur over a wide range of low-temperature weathering conditions, as well as vid magma interaction with water or ice (Geptner, 1977). The clay minerals, zeolites, carbonates, ferric iron hydroxides, and opal may be present in palagonite besides the amorphous component.

The numerous samples of palagonites appeared to be fairly similar in their chemical composition (except the content of Al, Fe, S, and Cl) as well as in particle size to Martian soil (Evans *et al.*, 1980; Allen *et al.*, 1980, 1981b). The visible and infrared spectra of many palagonite samples almost coincide with regolith spectra at both Viking 1 and 2 landing sites and spectra of Martian regions (Soderblom and Wenner, 1978; Singer, 1981, 1982; Evans and Adams, 1981; Allen *et al.*, 1981b). Palagonite spectra are consistent with reflectance spectra of high-albedo areas with respect to clays.

Large amounts of palagonites on Mars, as on the Earth, could have been formed only in the presence of liquid water. It is possible that since a cold and dry climate set in on Mars, palagonites could have formed as a result of magma interaction with permafrost or in areas of local permafrost ice melting (Toulmin *et al.*, 1977; Soderblom and Wenner, 1978; Allen *et al.*, 1981b) or beneath moving glaciers (Soderblom and Wenner, 1978). The palagonitization process could proceed in the absence of liquid water provided atmospheric water vapor partial pressure exceeds that observed order present conditions. By analogy with dry weathering in Antarctica, zeolites could be considered the dominant constituents of palagonites (Berkley and Brake, 1981).

A smaller quantity of palagonite-like dry material may possibly be formed under present Martian conditions by glass weathering as result of interaction with atmospheric gases (Gooding and Keil, 1978) or by the decomposition of metastable clay minerals (Gooding, 1978).

Under present Martian conditions palagonite is expected to be dehydrated, the amorphous substance to be decrystallized, and metastable minerals (zeolite, smectites, ferric iron hydroxides) to be decomposed (Gooding, 1978; Soderblom and Wenner, 1978). The terrestrial palagonites usually contain up to 10–25% H_2O, evaporated at 160°C (Geptner, 1977; Allen *et al.*, 1981b). The high temperature range of regolith water evaporation (Biemann *et al.*, 1977; Anderson and Tice, 1979) may be suggestive of either completed palagonite dehydration or its absence in the regolith.

Iron Oxides and Hydroxides

The reddish-brown color of the planetary surface, the distinct adsorption lines of ferric and ferrous iron in the spectra of surface materials, the presence of magnetic particles in soil, and a high content of iron in regolith samples, all indicate the presence of iron oxides in the Martian regolith (Sherman *et al.*, 1981).

Despite the thermodynamic instability under current conditions (O'Connor, 1968a; Gooding, 1978) magnetite may be present in the regolith because of its stability to chemical and mechanical weathering (Maderazzo and Huguenin, 1977; Gooding, 1978). Inference of a possibility of intensive weathering of Martian magnetite, drawn from the analysis of its photooxidation under UV radiation (Huguenin, 1973a, Huguenin *et al.*, 1980), was open to question after the publication of newer experimental data (Morris and Lauer, 1980).

The amount of magnetite in the atmospheric aerosol was estimated to be $10 \pm 5\%$ on the basis of interpretation of visible and infrared spectra of the Martian sky, obtained by the Viking 1 and 2 Landers (Pollack *et al.*, 1977). However, the high density of magnetite makes this estimate doubtful.

Magnetite along with maghemite is suggested to be the basic magnetic mineral of the Martian regolith (Hargraves *et al.*, 1977, 1979)

Several investigators of the reflectance spectra of Mars and/or its separate regions (Adams and McCord, 1969; Binder and Jones, 1972; McCord *et al.*, 1978; 1982a; Singer *et al.*, 1979; Morris and Neely, 1981), as well as the spectra from Viking landing sites (Huck *et al.*, 1977; Evans and Adams, 1979; Strickland, 1979) conclude that limonite as a mixture of hydrated and anhydrous forms of Fe_2O_3 may be present in the Martian regolith. Goethite (α-FeOOH) and ferroxyhyte (δ-FeOOH) spectra bear a resemblence to the spectra of the Martian surface more than any of the FeOOH modifications. Besides, ferroxyhyte, owing to its magnetic properties, could be one of the magnetic particle fractions of the regolith

(Burns, 1980; Sherman *et al.*, 1981). It is also possible that some part of the water vapor emitted from soil samples in the Viking experiments was incorporated in iron hydroxides (Biemann *et al.*, 1977).

Some investigators assume that a considerable portion of Fe^{3+} ions is incorporated into the amorphous component, particularly in palagonite (McCord *et al.*, 1982; Sherman *et al.*, 1982; Singer, 1982). For example, magnetic disordered material (amorphous gels, iron sulfates, etc.) is expected to be the basic form of ferric iron in the areas of high albedo (Sherman *et al.*, 1982), as the existence of amorphous Si–Fe gels (McCord *et al.*, 1982b) and gels of variable composition of $(Fe_xAl_{1-x})_2O_3$ (Morris and Lauer, 1984) is plausible. Substance of this kind are formed on the Earth in the process of weathering and their spectra are very near to those of the high-albedo areas of Mars. It is not possible to obtain a reliable estimate of the composition and abundance of ferric iron oxides and hydroxides from the spectra data, because the spectra of oxides, hydroxides, amorphous gels, and some iron-bearing silicates are fairly identical (Sherman, 1984).

The thermodynamic analysis indicates that various FeOOH modifications are unstable under present Martian conditions and are dehydrated to form such stable modifications as hematite (Fish, 1966; O'Connor, 1968a; Gooding, 1978). The modifications of FeOOH may have been formed at the ancient Martian surface at $PH_2O > 10^{-0.6}$ bar or in the presence of liquid water (Gooding, 1978) and not yet completely decomposed (O'Connor, 1968b; Pollack *et al.*, 1970a; Fuller and Hargraves, 1978).

Experimental investigation of goethite dehydration (Pollack *et al.*, 1970a, b) imply that goethite exists at the Martian surface only if its dehydration is kinetically hampered. The investigation of goethite and lepidocrocite photodehydration (Morris and Lauer, 1981) by UV radiation confirms either the suppression or the low rate of FeOOH dehydration in the present Martian environment. The possibility of goethite transport to the surface from the low-temperature regolith layers where this mineral is a stable phase was discussed by Pollack *et al.* (1970b). This transport could be stimulated by aeolian mixing of regolith material. However, as noted in the same report (Pollack *et al.*, 1970b) the rate of goethite dehydration most probably exceeds the rate of such mixing.

In terms of thermodynamics, hematite and maghemite are predicted to be the stable phases of the regolith (Gooding, 1978). Maghemite has been suggested to be the dominant mineral in the soil magnetic fraction (Hargraves *et al.*, 1977, 1979). The presence of nonstoichiometric maghemite for which the infrared spectra are consistent with Martian spectra in the regolith was discussed by Sherman (1982). The gas-exchange experimental results obtained from Viking biological investigations may indirectly indicate the presence of maghemite. The interpretation of the results suggests, that oxygen emanation in the course of experiment can be attributed to H_2O oxidation by hydrogen peroxide in the presence of γ-Fe_2O_3 as catalyst (Oyama and Berdahl, 1977). However, H_2O oxidation by penta- and hexavalent manganese compounds, which have been formed in the process of pyrolusite (β-MnO_2) photooxidation, seems more reasonable (Blackburn *et al.*, 1979). According to thermodynamic calculations, pyrolusite appears to be the most stable manganese mineral at the Martian surface (Blackburn *et al.*, 1979). From the Mn : Fe ratio in SNC meteorites, the Martian soil has been estimated to contain up to 0.5–0.6% of pyrolusite (Blackburn *et al.*, 1979).

Under present conditions hematite and maghemite formation may result from the oxidation of glasses (Gooding and Keil, 1978), iron-bearing silicates (Huguenin, 1974; Gooding, 1978), iron sulfides (Gooding, 1978; Houstley, 1981), and possibly magnetite as well (Huguenin 1973a, b, 1974; Huguenin *et al.*, 1980). Iron oxides may also be formed in the process of nontronite (MacKenzie and Rogers, 1977) or siderite decomposition (Gooding, 1978). The real contribution of each reaction to the formation of the Fe_2O_3 varieties depends on their rates.

Salts

The Martian regolith is believed to contain 8–25% salts. The presence of sulfates, chlorides, bromides, carbonates, and nitrates is most likely (B. C. Clark and Van Hart, 1981).

The lack of correlation of sulfur and iron concentration in regolith samples (B. C. Clark *et al.*, 1982) and the instability of sulfides because of interaction with the Martian atmosphere (Gooding, 1978, Houstley, 1981) may indicate a low content of iron sulfides in soil and, probably their complete absence, although pyrrhotite is assumed to occur in regolith magnetic particles (Hargraves *et al.*, 1977, 1979).

According to the thermodynamic predictions the most stable sulfur-bearing compounds of the Martian regolith are: $CaSO_4$, $MgSO_4 \cdot H_2O$, $FeSO_4$, $FeSO_4 \cdot H_2O$, etc. (Gooding, 1978). Sulfates are estimated to constitute 8–15% of the regolith (B. C. Clark and Van Hart, 1981). The Mariner 9 infrared spectroscopic data provide evidence for the presence of sulfates in the atmospheric dust (Logan *et al.*, 1975). It is also quite possible that particles contained in the regolith surface crust are cemented by sulfates (Baird *et al.*, 1976; Toulmin *et al.*, 1977).

The fact that sulfur-bearing gases were not detected in the course of experimental determination of regolith volatile components (Biemann *et al.*, 1977) can be accounted for by thermal stability of sulfates at temperatures up to 500°C (Kotra *et al.*, 1982).

Under present conditions iron sulfates are assumed to form as a product of sulfide oxidation in the presence of atmospheric gases. This process may also be accompanied by the formation of hematite and elemental sulfur (Gooding, 1978; Houstley, 1981). Besides, there are experimental data suggesting the formation of sulfates as a result the interaction of sulfur-bearing gases and atmospheric aerosols with carbonates and nitrates (B. C. Clark *et al.*, 1979b). The rate of sulfate formation will be higher if carbonates and nitrates react with atmospheric sulfuric acid aerosols, originated as a result of volcanic events and uniformly distributed in the atmosphere (Settle, 1979).

The formation of hydrated calcium, magnesium, and sodium sulfates is most probable in the presence of liquid water (B. C. Clark, 1978; Brass, 1980; Houstley, 1981). If the enrichment of the regolith by salts has been a result of upward migration and subsequent evaporation of solutions at the surface, the magnesium sulfates should be expected to accumulate at surface in association with chlorides (Gooding, 1978; B. C. Clark and Van Hart, 1981).

Because in the present environment epsomite must be dehydrated, kiserite is predicted to be the dominant sulfate, cementing the soil particles in the surface crust (Baird *et al.*, 1976).

A high S/Cl ratio in the soil samples can be attributed to salt precipitation by relatively high-temperature solutions (Brass, 1980). More reasonably, the S/Cl ratio value reflects the same ratio as in the outgassing products of the planetary interior, which were later incorporated in the regolith (B. C. Clark and Baird, 1979c; Clark, 1980; Clark and Van Hart, 1981; Taylor and Hodges, 1981).

High concentrations of sulfur in the Martian regolith in relation to those in terrestrial and lunar rocks may be indicative of the corresponding higher sulfur content in the Martian crust (Clark and Baird, 1979c). The weathering of materials enriched with sulfur from primordial accretion and probably delivered by carbonaceous chondrites, could have resulted in the formation of products similar to the Martian regolith in their chemical composition (B. C. Clark and Baird, 1979a; B. C. Clark *et al.*, 1979a). The incorporation of sulfur in the regolith as a result of the accumulation of outgassing products seems to be most plausible in relation to sulfur enrichment through the weathering of sulfides in primary rocks. The latter is less preferable because the aggregation of weathering products is impossible in an anhydrous environment, whereas the incorporation of outgassed sulfur-bearing compounds is permissible in gas–solid interaction as well (B. C. Clark *et al.*, 1979b).

A comparatively high content of chlorine in the regolith could be attributed to the presence of chlorides, estimated to constitute 0.5–1.5% (B. C. Clark and Van Hart, 1981). Chlorides, as well as sulfates, may cement the particles of the regolith in the surface crust (Toulmin *et al.*, 1977).

Two main sources of chlorine are suggested: volcanic gases and apatite from igneous rocks. Weathering in the absence of water is insufficient for chlorine to be released and accumulated in the surface crust because of the stability of apatite under present conditions (Gooding, 1978). However the dissolution of apatite in aqueous solutions and the upward migration of chlorine is possible providing it is mobilizated from the suite of underlaying primary rocks, several kilometers thick (Clark and Van Hart, 1981). Hence, volcanic gases could be the most plausible source of chlorine in the regolith (B. C. Clark and Baird, 1979a, b).

The concentration of chlorine-bearing compounds in the regolith may result from gas–solid reactions; however, it is only the presence of water that insures the intensive incorporation of chlorine-bearing gases in regolith minerals. In fact, the solutions are assumed to be enriched with chlorine owing to the high solubility of chlorides within the temperature interval of -20 to $-60°C$. Thus the upward migration of solutions under present Martian conditions is considered favorable for chloride accumulation in the regolith as well as in its surface crust (Brass, 1980; B. C. Clark, 1980; B. C. Clark and Van Hart, 1981). Although calcium, magnesium, and iron chlorides are related salts, it is suggested that halite must be the dominant chloride deposited in the regolith as a result of evaporation of solutions, because iron must have been incorporated earlier in hydroxides and clays, and sodium and magnesium should form sulfates and carbonates (B. C. Clark, 1980; B. C. Clark and Van Hart, 1981). Besides, sodium and magnesium chlorides appear to be unstable in the presence of CO_2 and sulfur-bearing gases (B. C. Clark and Van Hart, 1981). Thus the detection of halite in the regolith is more probable in relation to other chlorides; however, the halite content could be limited by the low sodium content of the Martian soil (B. C. Clark and Van Hart, 1981).

The detection of bromine in some soil samples allows us to suggest the presence of bromides in the regolith (B. C. Clark and Van Hart, 1981).

Carbonates and clay minerals are known as typical weathering products. It is assumed that under the high atmospheric carbon dioxide partial pressure relative to that of the terrestrial atmosphere carbonates could exist in the Martian regolith (Baird *et al.*, 1976; Toulmin *et al.*, 1977). The infrared spectra of the regolith and atmospheric dust indicate low concentrations of carbonates; the upper limit of calcium carbonate content is estimated at 5% (Toon *et al.*, 1977, 1979). Some part of the carbon dioxide emanated from the regolith by heating it up to 500°C in the Viking experiments may probably be attributed to partial decomposition of carbonates (Biemann *et al.*, 1977; Toulmin *et al.*, 1977; Kotra *et al.*, 1982).

Sodium and magnesium carbonates are expected to be stable at present on the Martian surface. The formation of dolomite should take place with at a rate exceeding that of calcite and magnesite (O'Connor, 1968a; Gooding, 1978), while Siderite appears to be thermodynamically unstable (Gooding, 1978).

Experiments conducted under conditions approximating those to of the Martian surface, have shown that submicron-sized carbonate films could be formed on silicates under present conditions (Booth and Kieffer, 1978). Carbonates presumably could be considered products of the decomposition of metastable smectite clays (Gooding, 1978) as well as glasses (Gooding and Keil, 1978), although substantial amounts of carbonates may be produced only in the presence of liquid water (Gooding, 1978; B. C. Clark, 1980; B. C. Clark and Van Hart, 1981).

The total content of carbonates in the regolith probably does not exceed 5%, dolomite and calcite being the predominant minerals (B. C. Clark and Van Hart, 1981).

On the basis of estimates of nitrogen content in the ancient Martian atmosphere (McElroy *et al.*, 1976) nitrates have been proposed as nitrogen concentrators in the regolith. It should be emphasized that the interpretation of infrared spectra of Martian surface materials has revealed that the upper limit of nitrate content in the regolith is comparatively low (Toon *et al.*, 1977, 1979).

The interpretation of chemical analyses of Martian soil suggests that the existence of nitrogen-bearing minerals cannot be excluded. Many nitrates are stable at temperatures up to 500°C, although the ammonia salts are probably absent in the regolith or ammonia would have been detected by the Viking chromatomass spectrometer measurements (Biemann *et al.*, 1977).

The content of nitrogen-bearing gases in the atmosphere and the precipitation rate of nitrates and nitrites at the surface have been estimated from the design model of nitrogen photochemistry (Yung *et al.*, 1977). The results show that for a period of 4.5 billion year a 3-mm layer of calcium nitrate could have been accumulated at the surface if the nitrogen content of the atmosphere has been constant.

However, nitrates, like carbonates, appear to be unstable in the presence of sulfur-bearing gases in the Martian atmosphere (B. C. Clark *et al.*, 1979b); also, they may be decomposed in an atmosphere enriched with carbon dioxide (B. C. Clark and Van Hart, 1981). The total content of nitrogen-bearing minerals in the regolith probably should not exceed 0.5%, and possibly they are completely absent (B. C. Clark and Van Hart, 1981).

Water

A number of morphological features suggest the existence of a global cryolithosphere on Mars (Rossbacher and Sheldon, 1981; Kuz'min, 1983). The depth of the permafrost may be as great as 1 km at the equator and within the interval of 2.6–3 km at the poles, respectively (Rossbacher and Sheldon, 1981). The upper level of the permafrost layer, estimated from the morphology of impact craters is located at depths < 350 m at the equator and about 100 m below the surface at a latitude of 40–50°, respectively (Kuz'min, 1980). On the basis of seasonal variations of water vapor partial pressure in the atmosphere, it has been suggested that at latitudes above 40° the regolith contains water ice at the depth of 0.1–1 m below the surface (Farmer and Doms, 1979). In the equatorial regions the ice layer is in disequilibrium with the atmosphere, however the rate of its evaporation is moderate and is largely dependent on temperature and the porosity of the overlying regolith (Clifford and Hillel, 1983). The existence of an ice layer in the equatorial regions evidenced by morphological features suggests the presence of a mechanism for permanent supplement of the ice inventory (Clifford and Hillel, 1983). Interpretation of infrared spectra indicates the presence of adsorption bands of water vapor (McCord *et al.*, 1982a). In low-albedo areas the amount of ice is estimated to be one quarter that of high-albedo areas. The complete absence of ice in low-albedo areas is not excluded as the latter are known to contain only restricted amounts of high-albedo material. The ice may be found only as a component of high-albedo fine-dispersed weathering products, which are abundant in areas of high albedo. The ice may sporadically appear in the form of condensates, in particular, those precipitated during the winter season at the Viking Lander 2 landing site (Stephen, 1981). Three areas, characterized by stable seasonal condensation, were defined in the Martian equatorial region. These areas are expected to represent the main sources of water vapor in the process of condensation (Huguenin and Clifford, 1982; Huguenin *et al.*, 1983). The albedo of these areas abruptly increases with the transition from spring to the summer season in the southern hemisphere. This phenomenon could be attributed to the appearance of water in the regolith (Zisk and Monginis-Mark, 1980).

The permanent (summer) northern and southern polar caps are composed of H_2O and CO_2, respectively, whereas in winter CO_2 may plausibly be condensed on both caps (Farmer and Doms, 1979).

Estimates of water content in the regolith derived from interpretation of infrared spectra of the planetary surface are comprehensively reviewed in (Singer *et al.*, 1979). It is suggested that Martian soil may contain 1–3% bound water. New infrared spectrometry data (McCord *et al.*, 1982a) are consistent with the assumption of the existence of hydrated minerals enriched in magnesium, such as talc, serpentine, and anthophyllite. The presence of hydrated minerals in the northern permanent polar cap cannot be excluded either (R. N. Clark and McCord, 1982). The release of the predominant fraction of H_2O (1–1.9%) from the regolith within the temperature range of 200–350°C (Biemann *et al.*, 1977; Anderson and Tice, 1979) may indicate the absence of abundant ice at the Viking landing sites. The water, released by heating, could have been contained in hydrates or partically adsorbed by soil (B. C. Clark, 1978, 1979). In the temperature range discussed water could be released from montmorillonite, gibbsite, epsomite, some other minerals (Kotra *et al.*, 1982). However, a reliable estimate of the forms of H_2O in existence in the regolith is rather difficult, because most hydrates evolve water over a wide range of temperatures, yet the kinetics may depend on particle size, pressure, rate of heating, etc. (Kotra *et al.*, 1982). Most hydrated minerals in the present-day regolith could have formed during "moist" periods of Martian history (O'Connor, 1968b; Fuller and Hargraves, 1978). The presence of iron hydroxides and other metastable hydrated phases may be explained by their low rates of dehydration (Morris and Lauer, 1981) and be indicative of the absence of a regolith–atmosphere thermodynamic equilibrium with respect to H_2O.

Ancient and Present-Day Weathering

The traces of fluvial erosion on Mars (Masursky *et al.*, 1977; Carr and Clow, 1981; Barner, 1982) are indicative of the existence of a fairly dense ancient atmosphere and related high surface temperature. Similar conclusions can be derived from the interpretation of data on the isotopic composition of atmospheric gases (McElroy *et al.*, 1977; Cess *et al.*, 1980; Fox and Dolgarno, 1983). The relatively high surface temperature and atmospheric pressure could be attributed to a greater heat flux and outgassing, to meteoritic falls, and to volcanic activity. The seasonal warmings occurring against the background of general cooling of the planet could have been caused by variations of the orbital parameters of Mars (Toon *et al.*, 1980), by increase of atmospheric dust content (Pollack *et al.*, 1979), and by other phenomena.

The period of intensive meteoritic bombardment (up to 3.9 billion years ago) was accompanied by fluvial activity and resulted in the formation of numerous valleys on current landforms. The next phase of fluvial erosion was synchronous with the formation of the majority of the "Lava flows" of the northern hemisphere (3.0–3.7 billion years) (Greeley and Spudis, 1981; Neukum and Hiller, 1981). The subsequent uplift of Tharsis Plateau (3.0–3.3 billion years ago) was also accompanied by fluvial activity (Greeley and Spudis, 1981; Neukum and Hiller, 1981).

Detailed geomorphic investigations of Martian valleys indicate that most valleys originate from discrete sources, with water coming from subsurface reservoirs (Carr and Clow, 1981). With this assumtion some authors completely exclude the possibility of the formation of valleys by rainwater. (Pieri, 1980). The valleys must have been developed sporadically; however, the periods of intensive fluvial activity are considered to have been short term and may be dated to the early stages of Martian evolution (Masursky *et al.*, 1977; Carr and Clow, 1981).

The formation of different sediments, particularly iron-rich eluvial material similar to terrestrial basic or ultrabasic weathering crust, was possible during the intensive fluvial periods, presumably accompanied by rainfall and erosion of present-day continental areas (Masursky *et al.*, 1977). Although geomorphological analysis leads to the conclusion of the absence of large and long-lived seas in Martian geological history (Masursky *et al.*, 1977) the clays, carbonates, and evaporites could have accumulated in relatively small reservoirs.

Active volcanism ceased on Mars about 1.5 billion years ago (except in the Tharsis Plateau), and liquid water could have disappeared from the planetary surface about 1 billion years ago, or somewhat earlier (Neukum and Hiller, 1981). The planetary climate must have become similar to the present-day situation (Lucchitta, 1984). Since that period, liquid water may have formed at the surface sporadically, in particular from the melting of the permafrost layer in areas of local heating, especially ones associated with intrusions or the falls of large meteorites.

Following the formation of a cryolithosphere, the weathering may have continued in the sporadically melted upper layer of the regolith. Experiments in so-called "dry" Antarctic valleys showed that common weathering in and under frozen ground may result in the formation of surface materials similar to the Martian regolith. In particular, the enrichment of surface crusts in salts may be observed (Berkley and Brake, 1981; Gibson and Ransom, 1981). Palagonite composed of iron hydroxides, clay minerals, zeolites, as well as different salts—halite, hydrohalite, calcite, mirabilite, astrakhanite, and antarcticite—can usually be formed under these conditions (Berkley and Brake, 1981; Gibson and Ransom, 1981; Gibson *et al.*, 1982, 1984; Wentworth and McKay, 1982).

The salts may be concentrated on the surface by capillary upward seepage of solutions and their subsequent evaporation (Ugolini and Anderson, 1972; Gibson *et al.*, 1982). The seepage of solutions to the surface could be related to the dominant process of enrichment of the Martian regolith and its surface crust in salts (Baird *et al.*, 1976; Toulmin *et al.*, 1977). A high Br/Cl ratio in the regolith (B. C. Clark and Baird, 1979a), characteristic of residual salt systems, is consistent with this interpretation. The oxidation of ferrous iron, carried by ground water to the surface, may have resulted in the enrichment of the regolith in iron hydroxides (Fuller and Hargraves, 1978). Concentrations of sulfur, chlorine, bromine, and iron have arisen by infiltration into the regolith of aqueous solutions both before the formation of the permafrost layer and in frozen ground (Fuller and Hargraves, 1978).

The secondary minerals may have been formed as a result of local melting of permafrost or melting of ice condensed at the surface (Soderblom and Wenner, 1978). The interaction of magma with frozen rocks may have resulted in the formation of palagonites (Toulmin *et al.*, 1977; Allen, 1979). Hazardous floods resulting from rapid ice melting are suggested as a cause for intensive erosion of the surface, weathering of rocks, and subsequent formation of salt crusts at the surface (B. C. Clark, 1978).

The usual weathering process in an aqueous environment in the presence of an oxidative atmosphere must have led to the formation of iron hydroxides, clay minerals, carbonates, sulfates, and chlorides. A high S/Cl ratio in the regolith samples may be accounted by either salt deposition at relatively high temperatures (B. C. Clark, 1980) or more possibly by direct incorporation of volcanic gases in surface material (Settle, 1979).

At the sites of large meteoritic falls the impactites and wallrocks may have undergone hydrothermal alteration, resulting in the formation of clay minerals, iron hydroxides, etc. (Allen *et al.*, 1981a; Clagton and Mayeda, 1983). This kind of hydrothermal activity in the process of meteoritic bombardment presumably predetermined the composition of the regolith, in particular a high S/Cl ratio (Newson, 1980; Berkley and Brake, 1981).

In the following we consider the conditions of contemporary weathering on the planet. The temperature of the Martian surface, depending on latitude, season, and of day, varies from 145 to 245K, rising during the spring–summer season up to 273K. In some equatorial regions the mean atmospheric pressure is estimated to be 600 Pa (6 mbar) (Leovy, 1979). The atmosphere at the surface is composed of: CO_2, 95.32%; N_2, 2.7%; Ar, 1.6%; O_2, 0.13%; CO, 0.07%; and H_2O, 0.13% (Owen *et al.*, 1977). There is a suggestion, derived from the biological gas-exchange experiment, that a certain amount of H_2O_2 and other oxygen–dydrogen oxidants (Oyama and Berdahl, 1977, 1979) are present in the atmosphere and act on the surface materials as oxidants (Huguenin *et al.*, 1979; Hunten, 1979; Huguenin, 1982). The hydrogen peroxide formed photochemically from atmospheric water vapor has been suggested to condense in the regolith. The subsequent decomposition of H_2O_2, resulting in free oxygen release, leads to the oxidation of soil materials (Huguenin, 1982).

Despite the fact that eutectic chloride and sulfate brines may exist at the temperatures of the Martian surface, these types of solutions probably appear only locally and for short periods of time (B. C. Clark, 1979; Brass, 1980; B. C. Clark and Van Hart, 1981). The microcapillary upward migration of solutions to the surface, leading to enrichment of the regolith and its surface crust in salts, is assumed to occur in local areas of the planet (B. C. Clark, 1980; B. C. Clark and Van Hart, 1981).

The main chemical processes taking place at the Martian surface are the following: weathering of glasses and certain minerals in interactions with atmospheric gases (Huguenin, 1974; Gooding, 1978; Gooding and Keil, 1978; B. C. Clark *et al.*, 1979c) and the decomposition and dehydration of metastable minerals (iron hydroxides, smectite clays, hydrated salts) formed in the presence of liquid water (Gooding, 1978). The extremely low rates for these reactions may account for the presence of numerous metastable phases in the soil, indicative of the probable absence of atmosphere–regolith equilibrium with respect to O_2, CO_2, and H_2O. Therefore, the "aqueous" weathering products conceivably are the prevailing constituents of the regolith in relation to the products of gas–solid reactions; likewise, the products of continental weathering must be abundant in relation to the weathering products of the young volcanic plains (Berkley and Brake, 1981).

Thermodynamic Calculations of the Mineral Composition of the Martian Regolith

In the previous sections we have reviewed a great number of reports with a bearing on the mineral composition of Martian soil and have seen how different authors have approached the problem. Estimates of mineral composition using a physicochemical approach are few. Nevertheless, before new data on Martian soil chemical composition are obtained, physicochemical modeling, combined with the available data and theoretical concepts, can be used to yield real estimates of mineral composition of soil as well as some peculiarities of its formation.

We assume that in the absence of liquid water on the Martian surface the main mechanism of chemical weathering of surface materials consists of the interaction of atmospheric gases with surface materials. We also assume that for the period of 0.1–1.0 billion years complete or partial chemical equilibrium between the surface materials and the Martian atmosphere may have been attained with respect to several volatile components. Following these assumptions investigators, especially in the case of Venus (Mueller, 1963; Lewis, 1970; Khodakovsky *et al.*, 1979), came up with a thermodynamical treatment for the purpose of a preliminary prediction of Martian regolith mineral composition.

Unlike Venus, which has a hot dense atmosphere ($T = 735$K and $P_{total} = 9 \times$

10^6 Pa at the surface), Mars has an extremely tenuous one with a mean temperature estimated to be 210K (Moroz, 1978), maximum at the equator being 305K. However, because it is quite possible that in the geologic past the Martian atmosphere has had a high density and temperature, our assumption of chemical equilibrium in the surface–atmospheric interaction is quite plausible. Metastable phases may be present at the planetary surface along with thermodynamically stable minerals. In this case the methods of chemical thermodynamics cannot give a real estimate of mineral composition; however, they reveal the trend of chemical reactions and final (equilibrium) result of these interactions. The soil mineral composition, when unequilibrated with the atmosphere, may be largely dependent on relative and absolute rates of gas–solid reactions and their mechanisms. We consider some kinetic aspects in the example of the estimation of sulfur-bearing atmospheric incorporation into regolith minerals.

The calculations were carried out by Gibbs free-energy minimization method in isobaric–isothermal multicomponent systems (Shvarov, 1978), the mobile components being the main gaseous components of the Martian atmosphere: CO_2, N_2, O_2, H_2O, SO_2, HCL, and HF. The calculations were accomplished under the assumption that incorporation of volatile components in mineral phases may be effective until the attainment of regolith–atmosphere equilibrium with respect to each mobile component. The multicomponent system at $T =$ 240K—including 17 components: H, O, C, Cl, F, P, S, N, Si, Ti, Al, Fe, Mn, Mg, Ca, Na, K—was calculated. The following data were used as primary information: the chemical composition of the planetary atmosphere (Horn *et al.*, 1972; Owen and Sagan, 1972; Owen *et al.*, 1977); the bulk chemical composition of the Martian regolith (Baird *et al.*, 1976; B. C. Clark *et al.*, 1976), terrestrial komatiite (Ringwood, 1975), basalt (R. Daly), and for minerals: anorthitic plagioclase, augite, olivine, magnetite, and pyrite (Deer *et al.*, 1962); and Gibbs free energies for more than 150 substances. The thermodynamic properties were borrowed mainly from Naumov *et al.*, (1971), Report CODATA (1969–1979), *Thermal Constants of Substances* (Glushko, 1965–1982), and *Thermodynamic Properties of Individual Substances* Glushkov, 1979–1983).

The modeling of Martian soil compositions included numerical experiments with various sets of volatile components divided into five models:

1. O_2, CO_2, H_2O, SO_2, HCl, HF, N_2
2. O_2, CO_2, H_2O, HCl, HF, N_2
3. O_2, CO_2, H_2O, HCl, N_2
4. O_2, CO_2, H_2O, HF, N_2
5. O_2, CO_2, H_2O

The results of the equilibrium calculations showed that under the conditions for an open system (model 1) the amount of bound sulfur in the weathering crust of basalts and ultrabasic rocks on Mars is expected to vary within the interval 7–20%, which is rather far from the instrumental value of $3.1 \pm 0.5\%$ (B. C. Clark *et al.*, 1982). The exclusion of SO_2(g) from the system in models 2, 3, and 4 allows us to analyze the situation in which gas–solid reaction with SO_2 do not proceed so that other volatile components can be incorporated into soil minerals.

However, in models 2, 3, and 4 the amount of bound chlorine, under equilibrium conditions, is estimated to be within the interval of 10–25%, markedly exceeding the real chlorine content in the soil, 0.7 ± 0.3% (B. C. Clark *et al.*, 1982).

The comparison of the results of thermodynamic calculation with Martian soil analyses implies that the assumption of equilibrium in the Martian atmosphere–surface rocks system (with respect to sulfur, chlorine, and fluorine) cannot be justified. Undersaturation of the regolith in sulfur, chlorine, and fluorine in relation to equilibrium contents may be attributed to the limited yield of "acid" volcanic gases into the Martian atmosphere over geologic time.

Table 2. Theoretical mineral composition of the Martian regolith (vol %).

Mineral	Model[a]	Chemical[b]	Basalt[c]	Komatiite[d]
Quartz	—	20.8	8.7	—
Microcline	—	—	7.8	0.8
Acmite	—	—	9.3	0.8
Montmorillonite	17.0	23.9	45.9	11.1
Nontronite	47.0	—	—	—
Saponite	15.0	—	—	—
Talc	—	27.9	0.3	51.9
Hematite	—	10.5	3.4	4.9
Rutile	1.0	0.6	0.8	0.1
Pyrolusite	—	0.2	0.2	0.1
Anhydrite	—	13.4	—	—
Kiserite	13.0	0.2	—	—
Calcite	7.0	—	—	—
Dolomite	—	—	22.7	13.8
Magnesite	—	—	—	15.3
Halite	—	0.3	—	—
Sylvine	—	0.5	—	—
Bischofite	—	1.1	—	—
Apatite and other phosphates	—	0.4	0.8	1.1
Carbon[e]	(1.2)	—	3.1	4.0
Chlorine	(0.7)	0.5	0.01	0.01
Fluorine	—	0.04	0.04	0.03
Sulfur	(3.1)	3.2	0.04	0.02
H_2O	(3.2)	2.5	1.6	1.6

[a] Model composition of the regolith corresponding to the chemical analysis of samples S-l (B. C. Clark *et al.*, 1976).
[b] Regolith composition, corresponding to chemical equilibrium in model 5.
[c] Regolith composition, corresponding to a basalt weathering crust (model 5).
[d] Regolith composition, corresponding to a komatiite weathering crust (model 5).
[e] Elemental composition (volatiles) in mass %.

The mineral composition of the soil calculated from model 5 (with the sulfur content fixed in the solid phases to correspond to the elemental composition of the Viking samples) best fits the model composition of the regolith (Baird *et al.*, 1976). Table 2 presents the mineral composition of the Martian soil as well as the altered basalt and komatiite in contrast to the model composition from (Baird *et al.*, 1976). As shown in Table 2, the equilibrium weathering products of basalts may contain up to 45% montmorillonite, 20% carbonates, and 3% hematite. The model weathering products of ultrabasic rocks (komatiite) are characterized by the presence of talc in association with carbonates, montmorillonite, and hematite. These assemblages are consistent with available data on the mineral composition of the soil, although the abundance of carbonates in it is probably considerably lower. Note that nontronite and saponite, suggested as possible components of the clay fraction by Baird *et al.*, (1976), are thermodynamically unstable in all the calculated models.

As suggested by Gooding (1978) clay minerals may be stable in the presence of liquid water at the Martian surface. Our calculations show that in gas–solid systems the only stable clay mineral is montmorillonite.

The water content in the regolith minerals equilibrated with the atmosphere, as estimated from our calculations (model 5), is in good agreement with experimental data, providing indirect evidence for the attainment of equilibrium with respect to H_2O.

The mutual supercession of volatiles appears to be an interesting peculiarity of their geochemical behavior. For example, the formation of sulfates by the interaction of sulfur-bearing compounds with carbonates and chlorides must lead to an increase of the CO_2 and HCl content of the atmosphere.

Table 3 illustrates the mineral associations that are to be expected in the selective weathering of some rock-forming minerals in the present Martian

Table 3. Weathering products of rock-forming minerals under conditions at the Martian surface.

Mineral	Weathering products[a] (Gooding, 1978)	Our data
Olivine	Magnesite, hematite, quartz, *goethite*, *clay minerals*, *talc*	Magnesite, dolomite, hematite, quartz, rutile
Augite	Magnesite, calcite, hematite, quartz, *clay minerals*	Dolomite, hematite, quartz, montmorillonite, talc
Anorthitic plagioclase	Calcite, quartz, corundum, *beidellite*	Calcite, quartz, albite, montmorillonite
Magnetite	Hematite, *goethite*	Hematite
Pyrrhotite	$FeSO_{4(c)}$, $FeSO_4 \cdot H_2O_{(c)}$	—
Pyrite	—	Hematite, $Fe_2(SO_4)_{3(c)}$

[a] Italic names are metastable minerals, formed in the presence of liquid water.

environment at 240K. The set of minerals predicted in our calculations and that predicted by Gooding (1978) are similar in many respects.

The calculation of the equilibrium association of minerals under present-day conditions of interaction in the Martian atmosphere and a matrix of the elemental composition corresponding to Martian soil chemical composition (B. C. Clark *et al.*, 1976) are shown in Table 2, column 2. In this system, with mobile O_2, CO_2, and H_2O and inert behavior of SO_2, HCl, HF, and N_2, water can be incorporated in phylloaluminosilicates (clay minerals, talc), and sulfur—mainly in the form of calcium sulfate, but not magnesium sulfate; chlorine may be present in the form of magnesium, potassium, and sodium chlorides. Also, nitrogen may be fixed in sodium and potassium nitrates.

In the weathering of a crust of basic and ultrabasic rocks thermodynamically stable minerals are represented by montmorillonite, talc, hematite, calcium and magnesium carbonates, calcium and magnesium sulfates, and fluorine–apatite. The model calculations are quite consistent with all data on the mineral composition of the Martian soil.

Unfortunately, the accepted thermodynamic treatment cannot give constraints on the prediction of Martian soil composition; however, it helps to reveal the peculiarities of volatile behavior in the atmosphere–planetary surface system. Further investigation, however, needs to be carried out for a more complete understanding of the chemical forms of sulfur, chlorine, and fluorine in the Martian regolith.

Gas–solid reactions appear to be the only mechanisms for attaining chemical equilibrium of both the atmosphere and surface rocks under present Martian conditions. The regolith composition inevitably depends on the absolute and relative rates of gas–mineral interactions.

The equilibrium calculations lead us to a conclusion that oxidation, hydration, and carbonation represent the main types of chemical processes on the Martian surface. The correlation of rates of these process estimated by Gooding (1978) may be represented in the following way: oxidation → hydration → carbonation. The role of incorporation of microcomponents (SO_2, HCl, HF, etc.) into solid phases is presumably subordinate. Experiments show that the formation of great amounts of carbonates as a result of carbon dioxide interaction under present Martian conditions is not plausible (Booth and Kieffer, 1978). Because H_2O content in the regolith equilibrated with the atmosphere is similar to the water content as determined in soil samples, we may assume that further hydration of soil particles is unlikely. Therefore the oxidation of surface materials by atmospheric oxygen may be the most important border chemical process at the Martian surface.

We next consider the kinetics of sulfur-bearing gas–mineral interactions under present Martian conditions, using sulfur dioxide interactions as an example.

Sulfur dioxide, as one of the major components of volcanic emanations, may be incorporate into mineral phases by interaction with silicates, iron and manganese oxides, chlorides, and carbonates. Thermodynamic calculations of sulfur dioxide equilibrium partial pressures in reactions with minerals suggest that the stability of compounds with respect to $SO_2(g)$ will increase in hierarchy according to the following sequence: $FeO(OH)(c)$, $FeCl_3(c)$, $MgCO_3(c)$, $CaCO_3(c)$, $CaCl_2(c)$, $MnCO_3(c)$, $MnO_2(c)$, $MnCl_2(c)$, $NaCl(c)$, $FeCO_3(c)$, $Fe_2O_3(c)$. It seems that rates of sulfur dioxide interaction with the first members of the sequence are probably higher; however, the interaction rates may be estimated only by using kinetic data.

In order to estimate the rates of sulfur dioxide interaction with calcite we used the kinetic data for this reaction at 300–600°C. It follows from experiments (Van Hote and Dolman, 1979a) that the reaction:

$$CaCO_3(c) + SO_2(g) + 0.5\,O_2(g) = CaSO_4(c) + CO_2(g) \tag{1}$$

proceeds in two stages (Eqs. 2, 3, or 4):

$$CaCO_3(c) + SO_2(g) = CaSO_3(c) + CO_2(g) \tag{2}$$

$$CaSO_3(c) + 0.5\,O_2(g) = CaSO_4(c) \tag{3}$$

$$CaSO_3(c) + 0.5\,SO_2(g) = CaSO_4(c) + 0.5\,S(c) \tag{4}$$

Stage (2) is the slowest and, consequently, it determines the rate of total reaction (1). With free oxygen available reaction (1) proceeds to reaction (3) (Van Hote and Dolman, 1978). Because the energy of the active stage (2) evidently exceeds that of stages (3) and (4) (Van Hote and Dolman, 1979a), stage (2) will remain limiting even at 240K. Thus, calcium sulfate (reaction 3) and elemental sulfur (reaction 4) is (not likely to accumulate at the Martian surface. The quantitative kinetic parameters of stage (2) are difficult to obtain because stage (2) cannot be easily separated from stage (4) in the course of the experiment (Van Hote and Dolman, 1979a).

Reaction (1) was treated in the presence of 2% $CaCl_2(c)$, accelerating stage (2) and impeding stage (4) (Van Hote and Dolman, 1979b). Using data from (Van Hote and Dolman, 1979b) and the Arrhenius equation:

$$K = K_0 e \frac{-E_a}{RT}$$

where E_a, the activation energy of stage (2) $= 154.8\ \text{kJ} \cdot \text{mol}^{-1}$, $K =$ rate constant, we estimated the effective rate constant of stage (2) at 240K to be $K = 10^{-20}\ V(\% \cdot \text{min}^{-1})$ for $CaCO_3(c)$. Considering the experimentally derived dependence of the reaction rate on the sulfur dioxide partial pressure (Van Hote and Dolman, 1979b) we estimated the rate of calcite interaction with sulfur dioxide at $T = 240$K and $P_{SO_2} = 8.5 \cdot 10^{-5}$ Pa to be $\sim 10^{-28}\ V(\% \cdot \text{min}^{-1})$ for $CaCO_3$. Assuming the estimated rate of interaction, we find that only 10^{-15} of calcite may react with sulfur dioxide in a period of 10^{15} min (4×10^9 years). Even considering the great uncertainty of this type of estimate (up to five orders of magnitude), we may infer that under current Martian conditions sulfur dioxide is unlikely to be fixed in the process of interaction with calcite. The reactions of sulfur dioxide with calcium-bearing silicates are probably not effective either.

The interaction of calcium and sodium chlorides with sulfur dioxide is known to proceed only under more ridgid constraints. Experiments show that the reaction:

$$CaCl_2(c) + SO_2(g) + O_2(g) = CaSO_4(c) + Cl_2(g) \tag{5}$$

cannot proceed without the presence of catalysts within the temperature interval 20–850°C (Derlyukova *et al.*, 1973) and that sodium chloride interaction with sulfur dioxide proceeds extremely slowly even at 800–900°C (Heuriksson and Warngvist, 1979). The limiting stage in these processes appears to be sulfur dioxide oxidation, which will proceed only in the presence of catalysts. The anomalously high activation energy for this reaction ($E_a = 336$ kJ $\cdot$ mol^{-1}) suggests that this reaction does not proceed at the temperatures of the Martian surface. We therefore conclude that sulfur dioxide is not chemically active with respect to carbonates or chlorides under the low temperature conditions of the Martian surface. The negligible concentration of sulfur-bearing atmospheric gases and related anomalously high sulfur content in the regolith may be interpreted in terms of a volcanic emanation sink, resulting in the interaction of sulfuric acid aerosols with sulfur bearing rocks (B. C. Clark and Baird, 1979a; Settle, 1979).

A complete understanding of oxidation at the Martian surface will have to wait until additionl data are obtained; otherwise it is impossible to evaluate the effect of oxidation on the composition of the regolith and atmosphere. The low oxidation rates may be considered a factor governing the disequilibrium of the regolith with respect to oxygen calculated from the thermodynamic treatment.

The majority of the regolith minerals (clays, iron hydroxides, sulfates) may have formed under conditions differing from the present environment at the Martian surface, i.e., at higher temperatures and in the presence of liquid water. The possibility that certain amounts of weathering products may have been formed by falls of large meteorites, local melting of the permafrost layer, etc., cannot be excluded. However, when Martian conditions became close to those of the present environment, they probably would not lead to substantial changes in the mineral composition of the regolith-forming components (Zolotov *et al.*, 1983).

Summary

All of the experimental data and theoretical implication lead to the following conclusions about the Martian soil mineral composition and weathering processes.

The martian regolith is considered to be composed of a mixture of chemical and mechanical weathering products. The primary material is probably represented by magmatic rocks of basic and/or ultrabasic composition. The main regolith minerals are predicted to be as follows: quartz, clay minerals, sulfates, iron oxides, phyllosilicates, and titanium and manganese oxides. Subordinate quantities of chlorides, phosphates, and probably nitrates are assumed. The carbonates are also suggested, especially dolomite and/or magnesite. Palagonites as well as hydrated minerals containing iron, sulfur, and chlorine may be present provided the extremely low dehydration rate is not excluded.

Physicochemical treatment indicates the absence of chemical equilibrium with respect to SO_2, HCl, HF, and presumably CO_2. The system is not equilibrated with respect to molecular oxygen providing magnetite exists as a soil component. Martian soil is expected to be in equilibrium only with respect to water vapor. The incorporation of sulfur into regolith minerals under present-day conditions is extremely unlikely according to kinetic estimates.

After the cessation of temperate climatic conditions and with the transition to a completely arid evironment, mechanical weathering should be considered the main geological agent at low temperatures and with a low effectiveness of gas–solid interactions. Nevertheless it is the final products of chemical weathering that are predicted to comprise the predominant fraction of the recent regolith.

It is suggested that the alteration of Martian surface rocks proceeds mainly in terms of mechanical weathering, with a succeeding aeolian homogenization on a planetary scale. Chemical weathering should be considered a competitive process in relation to mechanical weathering only in a comparatevely humid and warm climates.

Acknowledgments

The authors sincerely appreciate the assistance and useful advice of Prof. I. Khodakovsky, N. Andreeva, and Dr. R. Kuz'min. The authors thank Dr. M. Borisov for computer calculations and Prof. S. K. Saxena for his kind invitation to contribute a paper to the present volume.

References

Adams, J. B., and McCord, T. B. (1969) Mars: interpretation of spectral reflectivity of light and dark regions, *J. Geophys. Res.* **74**, 4851–4856.

Allen, C. C. (1979) Volcano-ice interactions on Mars, *J. Geophys. Res.* **84**, 8048–8059.

Allen, C. C., Gooding, J. L., and Keil, K. (1980) Partially weathered basaltic glass—a martian soil analog, in *Lunar and Planetary Science*, Vol. XI, pp. 12–14. Lunar and Planetary Institute, Houston, Texas.

Allen, C. C., Gooding, J. L., and Keil, K. (1981a) Hydrothermally altered impact melt from Brent and Ries craters, in *Lunar and Planetary Science*, Vol. XII, pp. 16–18. Lunar and Planetary Institute, Houston, Texas.

Allen, C. C., Gooding, J. L., Jercinovic, M., and Keil, K. (1981b) Altered basaltic glasses: a terrestrial analog to the soil of Mars, *Icarus* **45**, 347–369.

Allen, C. C., Gooding, J. L., and Keil, K. (1982) Hydrothermally altered impact melt rock and breccia: contributions to the soil of Mars, *J. Geophys. Res.* **87**, 10083–10102.

Anderson, D. M., and Tice, A. R. (1979) The analysis of water in the martian regolith, *J. Mol. Evol.* **14**, 33–38.

Aronson, J. R., and Emslie, A. G. (1975) Composition of the martian dust as derived by infrared spectroscopy from Mariner 9, *J. Geophys. Res.* **80**, 4925–4931.

Baird, A. K., Toulmin, P., Clark, B. C., Rose, H. J., Keil, K., Jr. Christian, R. P., and

Gooding, J. L. (1976) Mineralodic and petrologic implications of Viking geochemical results from Mars: Inter report, *Science* **194**, 1288–1293.

Baird, A. K., Castro, A. J., Clark, B. C., Toulmin, P., Rose, H. J., Keil, K., Jr., and Gooding, J. L. (1977) Viking X-ray fluorescence experiments: sampling strategies and laboratory simulations, *J. Geophys. Res.*, **82**, 4595–4624.

Baird, A. K., and Clark, B. C. (1981) On the original igneous source of martian fines, *Icarus* **45**, 113–123.

Banin, A., and Rishpon, J. (1979) Smectite clays in Mars soil: evidence for their presence and role in Viking biology experimental results, *J. Mol. Evol.* **14**, 133–152.

Banin, A., and Margulies, L. (1983) Simulation of Viking biology experiments suggest smectites not palagonites as martian soil analogues, *Nature* (*London*) **305**, 523–525.

Banin, A., and Margulies, L. (1984) Iron montmorillonite: spectral analogy to Mars soil, in *Lunar and Planetary Science*, Vol. XV, pp. 31–32. Lunar and Planetary Institute, Houston, Texas.

Barner, V. R. (1982) *Channels on Mars*. Univ. Texas Press, 1982.

Basilevsky, A. T., Moskaleva, L. P., Manvelyan, O. S., and Surkov, Yu. A. (1981) Estimation of thorium and uranium contents in the material of Martian surface: new interpretation of gamma-spectrometric measurements by Mars-5 probe, *Geochimistry* **1**, 10–16.

Becker, R. H., and Repin, R. O. (1984) The case for a martian origin of the shergottites: nitrogen and noble gases in EETA 79001, *Earth and Planet. Sci. Lett.* **69**, 225–242.

Berkley, J. L., and Brake, M. J. (1981) Weathering of of Mars: Antarctic analog studies, *Icarus* **45**, 231–249.

Biemann, K. J., Oro, P., Toulmin, P., Orgel, L. E., Nier, A. O., Anderson, D. H., Simmonds, P. G., Flory, D., Diaz, A. V., Rushneck, D. R., Biller, J. E., and Lafleur, A. L. (1977) The search for organic substances and inorganic compounds in the surface Mars, *J. Geophys. Res.* **82**, 4641–4658.

Binder, A. B., and Jones, J. C. (1972) Spectrophotometric studies of the photometric functions, composition and distribution of the surface materials on Mars, *J. Geophys. Res.* **77**, 3005–3020.

Blackburn, T. R., Holland, H. D., and Ceasak, G. P. (1979) Viking gas exchange reaction: simulation on UV-irradiated manganese dioxide substrate, *J. Geophys. Res.* **84**, 8391–8394.

Blackburn, T. R. (1984) Manganese oxides as high-pε redox buffers on Mars, *Icarus* **57**, 307–312.

Bogard, D. D., and Johnson, P. (1983) Martian atmospheric gases trapped in the EETA 79001 Shergottite?, in *Lunar and Planetary Science*, Vol. XIV, pp. 53–54. Lunar and Planetary Institute, Houston, Texas.

Booth, M. C., and Kieffer, H. H. (1978) Carbonate formation in Marslike environments, *J. Geophys. Res.* **83**, 1809–1815.

Brass, G. W. (1980) Stability of brines on Mars, *Icarus* **42**, 20–28.

Burghele, A., Dreibus, G., Palme, H., Rammensee, W., Spettel, B., Weckwerth, G., and Wänke H. (1983) Chemistry of shergottites and the shergotty parent body (SPB): further evidence for the two component of formation, in *Lunar and Planetary Science*, Vol. XIV, pp. 80–81. Lunar and Planetary Institute, Houston, Texas.

Burns, R. G. (1980) Does ferroxyhyte occur on the surface of Mars?, *Nature* (*London*) **285**, 647.

Busod, G., and McGetchin, T. R. (1979) Martian lavas-Reconnaissance experiments on a model ferro-picrite composition, in *Lunar and Planetary Science*, Vol. X, pp. 172–174.

Lunar and Planetary Institute, Houston, Texas.
Carr, M. H., and Clow G. D. (1981) Martian channels and valleys: their characteristics, distribution and age, *Icarus* **48**, 91–117.
Cess, R. D., Ramanathan V., and Owen T. (1980) The martian paleoclimate and enhanced atmospheric carbon dioxide, *Icarus* **41**, 159–165.
Christensen, P. R. (1982) Martian dust manting and surface composition: interpretation of thermophysical properties, *J. Geophys. Res.* **87**, 9985–9998.
Christensen, P. R. (1984) Thermal emissivity of the martian surface: evidence for compositional variations, in *Lunar and Planetary Science*, Vol. XV, pp. 150–151. Lunar and Planetary Institute, Houston, Texas.
Clagton, R. N., and Mayeda, T. K. (1983) Oxygen isotopes in eucrites, shergottites nakhlites and chassignites, *Earth Planet. Sci. Lett* **62**, 1–6.
Clark, B. C., Baird, A. K., Rose, H. J., Klaas, K., Castro, A. J., Kelliher, W. C., Rowe, C. D., and Evans, P. H. (1976) Inorganic analyses of Martian surface samples at the Viking landing sites, *Science* **194**, 1283–1288.
Clark, B. C. (1978) Implications of abundant hydroscopic minerals in the martian regolith, *Icarus* **34**, 645–665.
Clark, B. C. (1979) Chemical and physical micro-environments at the Viking landing sites, *J. Mol. Evol.* **14**, 13–31.
Clark, B. C., and Baird, A. K. (1979b) Chemical analysis of martian surface materials: status report, in *Lunar and Planetary Science*, Vol. X, pp. 215–217. Lunar and Planetary Institute, Houston, Texas.
Clark, B. C., and Baird, A. K. (1979c) Volatiles in the martian regolith, *Geophys. Res. Lett.* **6**, 811–814.
Clark, B. C., and Baird, A. K. (1979a) Is the martian lithosphere sulfur rich? *J. Geophys. Res.* **84**, 8395–8403.
Clark, B. C., Baird, A. K., and Keil K. (1979a) Composition and cosmochemical context of the surface fines of Mars, *Meteoritics* **14**, 367.
Clark, B. C., Kenley, S. L., O'Brien, D. L., Huss, G. R., Mack, R., and Baird, A. K. (1979b) Heterogeneous phase reactions of martian volatiles with putative regolith minerals, *J. Mol. Evol.* **14**, 91–102.
Clark, B. C. (1980) Aqueous transport of salts on Mars, in *Lunar and Planetary Science*, Vol. XI, pp. 152–154. Lunar and Planetary Institute, Houston, Texas.
Clark, B. C., and Van Hart, D. C. (1981) The salts of Mars, *Icarus* **45**, 370–378.
Clark, B. C., Baird, A. K., Weldon, R. J., Tsusaki, D. M., Schnabee, L., and Candelaria, M. P. (1982) Chemical composition of Martian fines, *J. Geophys. Res.* **87**, 10059–10067.
Clark, R. N., and McCord, T. B. (1982) Mars residual north polar cap, Earth-based spectroscopic confirmation of water ice as a major constituent and evidence for hydrated minerals, *J. Geophys. Res.* **87**, 367–370.
Clifford, S. M., and Hillel, D. (1983) The stabilities of ground ice in the equatorial region on Mars, *J. Geophys. Res.* **88**, 2456–2474.
Deer, W. A., Howie, R. A., and Zussman, J. (1962). *Rock-Forming Minerals.* Longmans, London.
Derlyukova, A. E., Tarakanov, B. M., Bunin, V. M., and Evdokimov, V. I. (1973) On the interaction of calcium chloride with oxygen and sulfur oxides, *Zhurn. Neorgan. Khim.* **18**, 2341–2345.
Evans, D. L., and Adams, J. B. (1979) Comparison of Viking lander multispectral images and laboratory reflectance spectra of terrestrial samples, *Proc. 10th Lunar Sci. Conf.*, 1829–1834.

Evans, D. L., Adams, J. B., and Wenner, D. B. (1980) Amorphous gels as possible analogs to martian weathering products, in *Lunar and Planetary Science*, Vol. 11, pp. 271–272. Lunar and Planetary Institute, Houston, Texas.

Evans, D. L., and Adams, J. B. (1981) Comparison of spectral reflectance propertes of terestrial and martian surface determinated from landsat and Viking orbiter multispectral images, in *Lunar and Planetary Science*, Vol. 12, pp. 271–273. Lunar and Planetary Institute, Houston, Texas.

Farmer, C. B., and Doms, P. E. (1979) Global seasonal variation of water vapor on Mars and the implications for permafrost, *J. Geophys. Res.* **84**, 2881–2888.

Fish, F. F. (1966) The stability of goethite on Mars, *J. Geophys. Res.* **71**, 3063–3068.

Fox, J. L., and Dolgarno, A. (1983) Nitrogen escape from Mars, *J. Geophys. Res.* **88**, 9027–9032.

Fuller, A. O., and Hargraves, R. B. (1978) Some consequences of liquid water saturated regolith in early martian history, *Icarus* **34**, 614–621.

Garvin, J. B., Mouginis-Maric, P. J., and Head, J. W. (1981) Characterization of rock population of planetary surface, techniques and priliminary analysis of Mars and Venus, *Moon Planets* **24**, 355–387.

Geptner, A. R. (1977) Palagonite in the process of palagonitization, *Litho. Miner. Resour.* **5**, 113–139.

Gibson, E. K., and Ransom, B. (1981) Soils and weathering processes in the dry valleys in Antarctica: analogs of the martian regolith, *Proc. 12th Lunar Sci. Conf.* 342–344.

Gibson, E. K., Bustin, R., and Wentworth, S. (1982) Development of regoliths in Mars: line environments, in *Lunar and Planetary Science*, Vol. XIII, pp. 259–260. Lunar and Planetary Institute, Houston, Texas.

Gibson, E. K., Presley, B. J., and Hatfield, J. (1984) Salts in the dry valleys of Antarctica, in *Lunar and Planetary Science*, Vol. XV, pp. 302–303. Lunar and Planetary Institute, Houston, Texas.

Goettel, K. A. (1981) Density of the mantle of Mars, *Geophys. Res. Lett.* **8**, 497–500.

Gooding, J. L. (1978) Chemical weathering on Mars: Thermodynamic stabilities of primary minerals (and their weathering products) from mafic igneous rocks, *Icarus* **33**, 483–513.

Gooding, J. L. (1980) Geochemical fractionations during the evolution of martian soils, in *Lunar and Planetary Science*, Vol. XI, pp. 342–345. Lunar and Planetary Institute, Houston, Texas.

Gooding, J. L., and Keil, K. (1978) Alteration of glass as a possible source of clay minerals on Mars, *Geophys. Res. Lett.* **5**, 727–730.

Greeley, R., and Spudis P. D. (1981) Volcanism on Mars, *Rev. Geophys. Space Phys.* **19**, 13–41.

Hanel, R. B., Conrath, W., Hovis, V., Kunde, V., Lowman, P., Maguire, W., Pearl, J., Pirraglia, J., Prabhakara, C., Schlachman, B., Levin, G., Straat, P., and Burke, T. (1972) Investigation of the Martian environment by infrared spectroscopy on Mariner 9, *Icarus* **17**, 423–442.

Hargraves, R. B., Collinson, D. W., Arvidson, R. E., and Spitzer, C. R. (1977) The Viking magnetic properties experiment: primary mission results, *J. Geophys. Res.* **82**, 4547–4558.

Hargraves, R. B., Collinson, D. W., Arvidson, R. E., and Cates, P. M. (1979) Viking magnetic properties experiment: extended mission results, *J. Geophys. Res.* **84**, 8379–8384.

Heuriksson, M., and Warngvist, B. (1979) Kinetics of formation of HCl(g) by the reaction

between NaCl(s) and SO_2, O_2, H_2O, *Ind. Eng. Chim. Proc. Des. Devel.* **18**, 249–254.
Horn, D., McAfee, J. M., Winer, A. M., Herr, K. C., and Pimental, G. C. (1972) The composition of martian atmosphere: minor constituent, *Icarus* **16**, 543–556.
Houstley, R. M. (1981) Considerations concerning the weathering history and sulfur content the martian regolith, in *Lunar and Planetary Science*, Vol. XII, pp. 374–376. Lunar and Planetary Institute, Houston, Texas.
Huck, F. O., Jobson, D. J., Park, S. K., Wall, S. D., Arvidson, R. E., Patterson, W. R., and Benton, W. D. (1977) Spectrophometric and color estimates of the Viking lander sites, *J. Geophys. Res.* **82**, 4401–4411.
Huguenin, R. L. (1973a) Photostimulated oxidation of magnetite 1. Kinetics and alteration phase indentification, *J. Geophys. Res.* **78**, 8481–8493.
Huguenin, R. L. (1973b) Photostimulated oxidation of magnetite 2. Mechanism, *J. Geophys. Res.* **78**, 8495–8506.
Huguenin, R. L. (1974) The formation of goethite and hydrated clay minerals on Mars, *J. Geophys. Res.* **79**, 3895–3905.
Huguenin, R. L., Miller, K. L., and Harwood, W. S. (1979) Frost-weathering on Mars: experimental evidence for peroxide formation, *J. Mol. Evol.* **14**, 103–132.
Huguenin, R. L., Denielson, J., and Clifford, S. (1980) Additional experimental evidence for the photostimulated oxidation of magnetite on Mars, in *Reports Planetary Geology Program 1978–1979*, pp. 147–148. TM-81776 NASA.
Huguenin, R. L. (1982) Chemical weathering and the Viking biology experiments on Mars, *J. Geophys. Res.* **87**, 10069–10082.
Huguenin, R. L., and Clifford, S. M. (1982) Remote sensing evidence for regolith water vapor sources on Mars, *J. Geophys. Res.* **87**, 10227–10252.
Huguenin, R. L., Miller, K. J., and Leschine, S. B. (1983) Mars: a contamination potential? *Advan. Space Res.* **8**, 35–38.
Hunt, G. R., Logan, L. M., and Salisbury, J. H. (1973) Mars: components of infrared spectra and the composition of dust cloud, *Icarus* **18**, 459–469.
Hunten, D. M. (1979) Possible oxidate sources in atmosphere and surface of Mars, *J. Molec. Evol.* **14**, 71–78.
Jakosky, B. M., and Muhleman, D. O. (1981) A comparison of the thermal and radar characteristics of Mars, *Icsrus* **45**, 25–38.
Khodakovsky, I. L., Volkov, V. P., Sidorov, Yu. I., and Borisov, M. V. (1979) Venus: Preliminary prediction of mineral composition of surface rocks, *Icarus* **39**, 352–363.
Kieffer, H. H., Martin T. Z., Peterfreund, A. R., Jakosky, B. M., Miner, E. D., and Palluconi, F. D. (1977) Thermal and albedo mapping of Mars during the Viking primary missions, *J. Geophys. Res.* **82**, 4249–4291.
Kotra, R. K., Gibson, E. K., and Urbanicic, M. A. (1982) Relise of volatiles from possible martian analogs, *Icarus* **51**, 593–605.
Kuz'min, R. O. (1980) Determination of ice-bearing rock-roof depth on Mars from morphology of fresh craters, *Dokl. Akad. Nauk USSR*, **252**, 1445–1448.
Kuz'min, R. O. (1983) *Martian Cryolithosphere*, Izd-vo Nauka, Moscow.
Leovy, C. B. (1979) Martian meteorology, *Ann. Rev. Astron. Astrophys.* **17**, 387–414.
Lewis, J. S. (1970) Venus: atmospheric and lithospheric composition, *Earth Planet. Sci. Lett.* **10**, 73–80.
Logan, L. M., Hunt, G. R., and Salisbury, J. J. (1975) The use of mid-infrared spectroscopy in remote sensing of space targets, in *Infrared and Roman Spectra of Lunar and Terrestrial minerals*, pp. 117–142 edited by C. Karr, Academic Press, New York.
Lucchitta, B. K.194) A late climatic change on Mars, in *Lunar and Planetary Science*,

Vol. XV, pp. 493–494. Lunar and Planetary Institute, Houston, Texas.

Mackenzie, K. J. D., and Rogers, D. E. (1977) Thermal and Mössbauer studies of iron containing hydrous silicate. I-Nontronite, *Thermochim. Acta* **18**, 177–196.

Maderazzo, M., and Huguenin, R. L. (1977) Petrologic interpretation of Viking XFF analysis based on reflection spectra and the photochemical weathering model, *Bull. Amer. Astron. Soc.* **9**, 527–528.

Masursky, H., Boyce, J. M., Dial, A. L., Schaber, G. G., and Strobell, M. E. (1977) Classification and time of formation of martian channels based on Viking data, *J. Geophys. Res.* **82**, 4016–4038.

McCord, T. B., Clark, R. N., and Huguenin, R. L. (1978) Mars: near-infrared spectral reflectances and compositional implications, *J. Geophys. Res.* **83**, 5433–5441.

McCord, T. B., Clark, R. N., and Singer, R. B. (1982a) Mars: near-infrared spectra reflectance of surface regions and compositional implications, *J. Geophys. Res.* **87**, 3021–3032.

McCord, T. B., Singer, R. B., Hawke, B. R., Adams, J. B., Evans, D. L., Head, F. W., Monginis-Mark, P. J., Pieters, C. M., Huguenin, R. L., and Zick, S. H. (1982b) Mars: definition and characterization of global surface units with emphasis on composition, *J. Geophys. Res.* **87**, 10129–10148.

McElroy, M. B., Yung, Y. L., and Nier, A. O. (1976) Isotopic composition of nitrogen: implications for the past history of Mars atmosphere, *Science* **194**, 70–72.

McElroy, M. B., Kong, T. Y., and Yung, Y. L. (1977) Photochemistry and evolution of Mars atmosphere: a Viking perspective, *J. Geophys. Res.* **82**, 4379–4388.

McGetchin, T. R., and Smyth, J. R. (1978) The mantle of Mars: some possible geological implications of its high density, *Icarus* **34**, 512–536.

Michael, W. H., Maoy, A. P., Blackshear, W. T., Tolson, R. H., Kelly, G. M., Brenkle, J. P., Cain, D. L., Fjeldbo, G., Sweetnam, D. N., Coldstecn, R. B., MacNell, P. E., Reasesenberg, R. D., Shapiro, I. I., Boak, T. I. S., Grossi, M. D., and Tany, C. H. (1976) Mars dinamics atmospheric and surface properties: determination from Viking tracking data, *Science* **194**, 1337–1338.

Moore, H. J., Hutton, R. E., Scott, R. F., Spitzer, C. R., and Shorthill, R. W. (1977) Surface materials of the Viking landing sites, *J. Geophys. Res.* **82**, 4497–4523.

Moore, H. J. (1979) Yield strengths of diverse flows on the flanks of Elysium, Ascracus and Arsia Montes, Mars, in *Reports Planetary Geology Program 1978–1979*, pp. 63–64. TM-80339, NASA.

Morgan, J. W., and Anders, E. (1979) Chemical composition of Mars, *Geochim. Cosmochim. Acta* **43**, 1601–1610.

Moroz, V. I. (1978) *Physics of Planet Mars*. Izd-vo Nauka, Koscow.

Morris, R. V., and Lauer, H. V., Jr. 1980. The case against UV photostimulated oxidation of magnetite, *Geophys. Res. Lett.* **7**, 605–608.

Morris, R. V., and Lauer, H. V., Jr. (1981) Stability of goetite (α-FeOOH) and lepidocrocite (δ-FeOOH) to dehydration by UV-radiation: implications for their occurrence on martian surface, *J. Geophys. Res.* **86**, 10893–10899.

Morris, R. V., and Neely, S. C. (1981) Diffuse reflectance spectra of pigmentary-sized iron oxides, iron oxihydroxides and their mixtures: implications for reflectance spectra of Mars, in *Lunar and Planetary Science*, Vol. XII, pp. 723–725. Lunar and Planetary Institute, Houston, Texas.

Morris, R. V., and Lauer, H. V., Jr. (1984) Spectral properties of annealed $(Fe_xAl_{1-x})_2O_3$ gels with applications to Mars, in *Lunar and Planetary Science*, Vol. XV, pp. 571–572. Lunar and Planetary Institute, Houston, Texas.

Moskowitz, B. M., and Hargraves, R. B. (1982) Magnetic changes accompanying the thermal decomposition of nontronite (in air) and its relevance to martian mineralogy, *J. Geophys. Res.* **87**, 10115–10128.

Mueller, R. F. (1963) Chemistry and petrology of Venus preliminary deductions, *Science* **141**, 1046–1047.

Naumov, G. B., Ryzhenko, B. N., and Khodakovsky, I. L. (1971) *Handbook of Thermodynamic Values*. Izd-vo, Atomizdat, Moscow.

Neukum, G., and Hiller, R. (1981) Martian ages, *J. Geophys. Res.* **86**, 3097–3121.

Newson, H. E. (1980) Hydrothermal alteration of impact melt sheets with implication for Mars, *Icarus* **44**, 207–216.

O'Connor, J. T. (1968a) Mineral stability at the martian surface, *J. Geophys. Res.* **73**, 5301–5311.

O'Connor, J. T. (1968b) "Fossil" martian weathering, *Icarus* **8**, 513–517.

Owen, T., and Sagan C. (1972) Minor costituents in planetary atmosphere ultraviolet spectroscopy from the Orbiting Astronomical Observatory, *Icarus* **16**, 557–568.

Owen, T., Biemann, K., Rushueck, D. R., Bieller, J. E., Howath, D. W., and Lableur, A. L. (1977) The composition of the atmosphere at the surface of Mars, *J. Geophys. Res.* **82**, 4635–4639.

Oyama, V. I., and Berdahl, B. J. (1977) The Viking gas exchange experiment results from Chryse and Utopia surface samples, *J. Geophys. Res.* **82**, 4669–4676.

Oyama, V. I., and Berdahl, B. I. (1979) A model of martian surface chemistry, *J. Mol. Evol.* **14**, 199–210.

Palluconi, F. D., and Kieffer, H. H. (1981) Thermal inertia mapping from 60°S to 60°N, *Icarus* **45**, 415–426.

Pang, K. D., and Ajello, J. M. (1977) Complex refractive index of martian dust: wavelength dependence and composition, *Icarus* **30**, 63–74.

Pieri, D. C. (1980) Martian valleys: morphology, distribution, age and origin, *Science* **210**, 895–897.

Pollack, J. B., Pitman, D., Khare, B. N., and Sagan, C. (1970a) Goetite on Mars: A laboratory study of physically bound water in ferric oxides, *J. Geophys. Res.* **75**, 7480–7490.

Pollack, J. B., Wilson, R. N., Goles, G. G. (1970b) A re-examination of the stability of geotite on Mars, *J. Geophys. Res.* **75**, 7491–7500.

Pollack, J. B., Colburn, D., Kahn, R., Hunter, J., and Van Camp, W. (1977) Properties of aerosols in the Martian atmosphere, as infrafed from Viking lander imaging data, *J. Geophys. Res.* **82**, 4479–4496.

Pollack, J. B., Colburn, D. S., Frasar, F. M., Kahn, R., Carlston, C. E., and Pidek, D. (1979) Properties and effects of dust particles suspended in martian atmosphere, *J. Geophys. Res.* **84**, 2929–2945.

Report of the CODATA Task Group on Key Velues for Thermodynamics (1969–1979).

Ringwood, A. E. (1975) *Composition and Petrology of the Earth's Mantle*. McGraw-Hill Book Company, New-York.

Robie, R. A., Hemingway, B. S., Fisher, J. R. (1978) Thermodynamic properties of mineral and related substances at 298, 15K and 1bar (10^5 Pascals) pressure and at higher temperatures. U.S. Geol. Surv. Bull. N1452, Washington, D.C.

Rossbacher, L. A., and Sheldon, J. (1981) Ground ice on Mars: Inventory, distribution and resulting landform, *Icarus* **45**, 39–59.

Settle, M. (1979) Formation and deposition of volcanic sulfate aerozols on Mars, *J. Geophys. Res.* **84**, 8343–8354.

Shvarov Yu. V. (1978) On minimization of thermodynamic potential of open chemical system, *Geochemistry* **12**, 1892–1895.

Sherman, D. M., Burns, R. G., and Burns, V. M. (1981) Assessment of ferric iron oxide minerals likely to occur on Mars, in *Lunar and Planetary Science*, Vol. XII, pp. 970–972. Lunar and Planetary Institute, Houston, Texas.

Sherman, D. M. (1982) Non-stoichiometric maghemite on Mars, in *Lunar and Planetary Science*, Vol. XIII, pp. 720–721. Lunar and Planetary Institute Houston, Texas.

Sherman, D. M., Burns, R. G., and Burns, V. M. (1982) Spectral characteristics of the oxides with application to the Martian bright region mineralogy, *J. Geophys. Res.* **87**, 10169–10180.

Sherman, D. M. (1984) Reassigment of the iron (III) absorption bands in the spectra of Mars, in *Lunar and Planetary Science*, Vol. XV, pp. 764–765. Lunar and Planetary Institute Houston, Texas.

Singer, R. B., McCord, T. B., Clark, R. N., Adams, J. B., and Huguenin, R. L. (1979) Mars surface composition from reflectance spectroscopy: a summary, *J. Geophys. Res.* **84**, 8415–8426.

Singer, R. B. (1981) Spectral constraints on iron-rich smectites as abundant constituents of martian soil, in *Lunar and Planetary Science*, Vol. XII, pp. 996–998. Lunar and Planetary Institute Houston, Texas.

Singer, R. B. (1982) Spectral evidence for the mineralogy of high-albedo soils and dust on Mars, *J. Geophys. Res.* **87**, 10159–10168.

Soderblom, L. A., and Wenner, D. B. (1978) Possible fossil H_2O liquid-ice interfaces in the Martian crust, *Icarus* **34**, 622–637.

Stephen, D. M. (1981) Analysis of condensates formed at the Viking 2 lander site: the first winter, *Icarus* **47**, 173–183.

Strickland, E. L. (1979) Martian soil stratigraphy and rock coatings observed in color-enhanced Viking lander images, *Proc. 10th Lunar Planet. Sci. Conf.*, pp. 3055–3077.

Surkov, Yu. A., Moskaleva, L. P., Manvelyan, O. S., and Kharyukova, V. P. (1980) The analysis of Martian rock gamma-radiation based on Mars 5 space probe, *Kosm. Issled.* **15**, 623–631.

Taylor S. R., and Hodges R. R. (1981) Chlorine and sulfur abundances in Mars and the Moon: implications for bulk composition, in *Lunar and Planetary Science*, Vol. XII, pp. 1082–1083. Lunar and Planetary Institute Houston, Texas.

Glushko, V. P. (Ed.) (1965–1982) *Thermal Constants of Substances*, Vols. I–X. Izd-vo, VINITI, Moscow.

Glushko V. P. (Ed.) (1979–1983) *Thermodynamic Properties of Individual Substances* Vols. I–IV. Izd-vo, Nauka, Moscow.

Toon, O. B., Pollack, J. B., and Sagan, C. (1977) Physical properties of the particles composing the martian dust storm of 1971–1972, *Icarus* **30**, 663–696.

Toon, O. B., Khare, B. N., Pollack, J. B., and Sagan C. (1979) Martian surface composition: comparison of remote spectral studies and in situ X-ray fluorescence analysis, in *Reports Planetary Geology Program*. TM-79729, NASA.

Toon, O. B., Pollack, J. B., Ward, W., Burns, J. A., and Bilski, K. (1980) The astronomical theory of climatic change on Mars, *Icarus* **44**, 552–607.

Toulmin, P., Baird, A. K., Clark, B. C., Keil, K., Rose, H. S., Jr., Christain, R. P., Evans, P. H., and Kellihez, W. C. (1977) Geochemical and mineralogical interpretation of the Viking inorganic chemical results, *J. Geophys. Res.* **82**, 4625–4634.

Ugolini, F. C., and Anderson, D. M. (1972) Ionic migration in frozen Antarctic soil, *Antarct. J.* **7**, 112–113.

Van Hote, G., and Dolman, B. (1978) Cinetiqie de la réaction du sulfite de Calcium avec ë oxygene, *Bull. Soc. Chim. France* **11–12**, 413–418.

Van Hote, G., and Dolman, B. (1979a) Kinetics of reaction of $CaCO_3$ with SO_2 and O_2 below 650°C, *J. Chem. Soc. Faraday Trans.* **75**, 1593–1605.

Van Hote, G., and Dolman B. (1979b) Kinetics of the reaction of $CaSO_3$ and $CaCO_3$ with SO_2 in presente of $CaCl_2$, *Bull. Soc. Chim. Belg.* **88**, 205–213.

Weldon, R. J., Warren, M., Boslough, M. B., and Ahrens, T. J. (1982) Shock-induced color changes in nontronite: implications for the Mars fines *J. Geophys. Res.* **87**, 10102–10114.

Wentworth, S. J., and McKay, D. S. (1982) Silicate weathering and diagenesis in on Antarctic soil-Mars analog, in *Lunar and Planetary Science*, Vol. XIII, pp. 853–855. Lunar and Planetary Institute Houston, Texas.

Yung, Y. L., Strobel, D. F., Kong, T. Y., and McElroy, M. B. (1977) Photochemistry of nitrogen in the martian atmosphere, *Icarus* **30**, 26–41.

Zisk, S. H., and Monginis-Mark, P. J. (1980) Anomalous region on Mars, implication for near-surface liquid water, *Nature* (*London*) **288**, 126–129.

Zolotov, M. Yu., Sidorov, Yu. I., Volkov, V. P., Borisov, M. V., and Khodakovsky, I. L. (1983) Mineral composition of martian regolith: thermodynamic assessment, in *Lunar and Planetary Science*, Vol. XIV, pp. 883–884. Lunar and Planetary Institute Houston, Texas.

Chapter 6
Theoretical Computation of Physical Properties of Mantle Minerals

Michele Catti

Introduction

Methods of direct sampling, as used for crustal rocks, unfortunately cannot be applied to the study of chemical and mineralogical composition of the Earth's mantle; only in some cases for the uppermost mantle layers, which interact with the crust by exchange of matter, are fragments or relics inserted in crustal material subject to petrographic examination. However, in the last two decades much experimental work has been performed in laboratory simulations of pressure–temperature conditions in systems of reasonably probable mantle chemical compositions (cf., e.g., Ringwood, 1975; Liu, 1977; Yagi *et al.*, 1979; Ito *et al.*, 1984). On the basis of these results a model of the mantle can be proposed such that its physical properties are in accordance with those observed by seismological and other measurements. This scheme requires that the relevant physical data be known for all phases that are to be included in the model. In some cases an experimental determination is possible, and the relative account of methods and results is found in Chapter 7 of this book (Weidner, this volume). However, very often it is not, either because synthetic samples are not available in the physical conditions required by the measurement (adequate dimensions or amount, single-crystal state, etc.), or because data are desired for pressure–temperature fields that cannot be attained. In these cases theoretical methods of calculating or sometimes even just estimating the required data can be very useful. The methods that are presently available and the main results obtained so far are considered and reviewed here.

Taking into account the laboratory studies on phase equilibria quoted above, the geochemical data on probable abundance of chemical elements in the mantle and a number of crystal–chemical arguments, the main crystalline phases in these calculations are:

1. Magnesium compounds: periclase MgO; the three polymorphs of Mg_2SiO_4, α (forsterite), β (wadsleyite), γ (ringwoodite) (Akaogi *et al.*, 1984); hypothetical

polymorphs of Mg_2SiO_4 isostructural with K_2MgF_4, Sr_2PbO_4, and Fe_2CaO_4, respectively; the three polymorphs of $MgSiO_3$ (enstatite pyroxene, ilmenite type, perovskite type) (Ito and Matsui, 1977; Ito *et al.*, 1984). For some of these phases, data of crystal symmetry, molar volume and pressure limit of stability are reported in Table 1.

2. A number of analoguous compounds in which Mg is replaced by Fe, Co, Ni, Mn.
3. The polymorphs of silica SiO_2 coesite and stishovite (Akaogi and Navrotsky, 1984).
4. Aluminum silicates with garnet-type structures, $(Mg, Fe)_3Al_2(SiO_4)_3$, pyrope–almandine, and $Ca_3Al_2(SiO_4)_3$, grossular (Liu, 1979).
5. With less importance, the silicates of Ca and CaMg such as $CaMgSiO_4$ (monticellite), $CaMg(SiO_3)_2$ (diopside), $Ca_3Mg(SiO_4)_2$ (merwinite), Ca_2SiO_4 (K_2MgF_4 type), $CaSiO_3$ (perovskite type) (Liu, 1979).

Modern ideas on the mantle constitution assume that, as the depth increases, the three polymorphs, α, β, and γ of Mg_2SiO_4 succeed each other; then a transformation to $MgSiO_3$ (ilmenite type) probably occurs, and finally in the lower layer the perovskite-type $MgSiO_3$ should be present as the dominant phase, possibly associated with stishovite. Besides, a replacement of Mg by about 10–20% of Fe is probable, and a greater mineralogical complexity should be expected in the upper layers, with the presence of garnet phases in addition to Mg_2SiO_4 (Anderson and Bass, 1984).

I shall summarize here the most important physical properties for the modeling of Earth's mantle, which can be simulated by theoretical calculations. First are the atomic structure and density of crystalline phases involved. A knowledge of the structural configuration is necessary to estimate the stability range (in terms of pressure–depth) of the mineral examined: the greater the atomic coordination number (with shorter chemical bonds on the average), the larger is the density and the deeper is the mantle region where the phase is likely to be stable. By means of the density values, the changes of free energy $-p\Delta V$ arising from phase transitions in the mantle can be determined; besides, density data are required for computing the velocities of seismic waves, as is shown below.

Second are the elastic constants c_{ijhk} and the elastic bulk modulus. These quantities are very important, together with the density ρ, in evaluating the velocity v of acoustic waves traveling through the crystalline medium, by the well-known relation:

$$\rho v^2 U_i = \sum_{1^h}^{3} \left(\sum_{1^j}^{3} \sum_{1^k}^{3} c_{ijhk} l_j l_k \right) U_h \tag{1}$$

where U_i is a Cartesian component of the wave polarization vector and l_j is a director cosine of the wave propagation direction. Using this general formula with the complete set of elastic constants, one can calculate the velocity of seismic waves and their polarization vectors for different crystallographic directions. This not only allows one to check the compositional model of the mantle against the observed average seismic data, but also to study finer effects, such as the

Table 1. Some physical and structural data of Mg_2SiO_4 and $MgSiO_3$ crystal phases of the earth's mantle. The coordination numbers CN of Si and Mg atoms, molar volume, calculated density, elastic modulus and upper pressure limit of stability at 1500 K are reported.[a]

	α-Mg_2SiO_4, forsterite	β-Mg_2SiO_4, wadsleyite	γ-Mg_2SiO_4, ringwoodite	$MgSiO_3$ (ilmenite type)	$MgSiO_3$ (perowskite type)
Lattice	Orthorhombic	Orthorhombic	Cubic	Trigonal	Orthorhombic
Space group	Pbnm	Ibmm	Fd3m	$R\bar{3}$	Pbnm
CN of Si	4	4	4	6	6
CN of Mg	6	6	6	6	8
V_{mol}, $cm^3 \cdot mol^{-1}$	43.63[1]	40.47[2]	39.49[3]	26.35[4]	24.44[5]
D_c, $g \cdot cm^{-3}$	3.225[1]	3.477[2]	3.563[3]	3.810[4]	4.108[5]
K, Mbar	1.27[6]	1.66[7]	1.84[8]	?	2.6[9]
p_{limit}, kbar (~1500 K)	150[10]	180[10]	~300[11]	~300[11]	≫300[11]

[a] Sources: (1) Hazen (1976); (2) Horiuchi and Sawamoto (1981); (3) Sasaki *et al.* (1982); (4) Horiuchi *et al.* (1982); (5) Y. Matsui (1982); (6) Graham and Barsch (1969); (7) Mizukami *et al.* (1975); (8) Weidner *et al.* (1984); (9) Yagi *et al.* (1982); (10) Akaogi *et al.* (1984); (11) Ito and Matsui (1977).

seismic anisotropy, whose relevance has been stressed by recent analyses of seismological measurements (Bamford, 1976; Kern and Richter, 1981). If an isotropic approximation is used, the knowledge of the bulk modulus or of the compressibility may be sufficient, instead of the whole set of elastic constants, to estimate seismic velocities.

Third, the lattice vibrational frequencies are required to calculate the thermodynamic poperties depending on temperature (heat capacity, entropy, free energy) for the mantle minerals considered (Kieffer, 1979); in principle, this would lead to a theoretical prediction of pressure–temperature stability fields.

Fourth are the dielectric properties. Although the dielectric behavior of the mantle has been little investigated, knowing the dielectric tensor of the present minerals may be interesting for two reasons: for a characterization of electromagnetic fields inside the earth; for a study of possible ferroelectric phase transitions as sources of energy release in the mantle, according to the assumed abundance of the perovskite-type $MgSiO_3$ phase in the lower region, and to the importance of the perovskite structure in ferroelectric and paraelectric phenomena (Schloessin and Timco, 1977).

All the above physical quantities depend in different ways on the overall interaction energy between atoms forming the microscopic structure of the crystalline phase considered, and from that energy they can be calculated in principle. Thus the central problem is to devise a physically reliable model to express interatomic forces; a large part of this work is devoted to surveying the several potentials introduced for mantle minerals and their optimization procedures on the basis of known physical properties. Then some applications of these potentials to computations relevant for mantle phases are examined, keeping in mind that at present this kind of theoretical analysis on minerals is still at an early stage, so that important progress can be expected in the near future.

Interatomic Potentials

From a chemical point of view, the minerals thought to occur in the mantle are mainly silicates and oxides of group I, II, and III metals and of transition metals. The chemical bond between atoms in these crystalline phases can be considered to be prevalently but not fully ionic; covalent effects are dominant in the silicatic part (Si—O bonds) but should decrease as the coordination number of silicon increases from four (e.g., in olivine) to six (ilmentile- and perovskite-type $MgSiO_3$). Other clear evidences of covalency are observed in the bonds between transition-metal atoms and oxygen. Let us concentrate for the moment fully ionic interactions: in this case, a two-body central potential is quite adequate and the lattice energy can be expressed by the sum of atom–atom V_{ij} terms depending only on the interatomic distances r_{ij} and not on the angular orientation of bonds,

$$E_L = \sum_{ij} V_{ij}(r_{ij}) \tag{2}$$

In the case of partially ionic bonding, such expression can be retained as a first approximation. The $V_{ij}(r_{ij})$ potential generally takes the Born-Mayer form as sum of three contributions:

$$V_{ij}(r_{ij}) = e^2 z_i z_j / r_{ij} + b_{ij} \exp(-r_{ij}/\rho_{ij}) - c_{ij}/r_{ij}^6 \tag{3}$$

The first term represents the electrostatic interaction energy between ions i and j with z_i and z_j electric charges (measured in units of electron charge, e); the second gives the short-range repulsion energy from Pauli's exclusion principle, which applies when electrons with equal spins belonging to different ions approach so as to overlap their clouds; the third term expresses the dispersive (or van der Waals) energy, which is attractive and is caused by a correlation effect between induced polarization charges of electron clouds of neighboring ions (Tosi, 1964).

Whereas the electrostatic energy is given by an "exact" formula, at least within the validity of the ionic model, the expression of repulsive energy (referred to as V_{ij}^R) shown in Eq. (3) is approximate and contains two parameters, b_{ij} and ρ_{ij} (hardness parameter), which have to be optimized empirically so as to reproduce some measured physical property. Several slightly different formulas derived from the general expression $b_{ij} \exp(-r_{ij}/\rho_{ij})$ are found for the repulsive term in the literature. Most of them assume the principle of transferability of parameters, by changing the b_{ij} and ρ_{ij} values for atomic pairs into the single-atom parameters r_i and ρ_i:

$$V_{ij}^R = b \exp[(r_i + r_j - r_{ij})/(\rho_i + \rho_j)]$$

$$b_{ij} = b \exp[(r_i + r_j)/(\rho_i + \rho_j)], \; \rho_{ij} = \rho_i + \rho_j \tag{4}$$

$$V_{ij}^R = b\beta_{ij} \exp[(r_i + r_j - r_{ij})/(\rho_i + \rho_j)], \; b_{ij} = b\beta_{ij} \exp[(r_i + r_j)/(\rho_i + \rho_j)] \tag{5}$$

$$V_{ij}^R = b(\rho_i + \rho_j) \exp[(r_i + r_j - r_{ij})/(\rho_i + \rho_j)]$$

$$b_{ij} = b(\rho_i + \rho_j) \exp[(r_i + r_j)/(\rho_i + \rho_j)] \tag{6}$$

The r_i quantities are the repulsion radii, or basic radii, of atoms; it should be noticed that differences between repulsive radii are very close to differences between corresponding ionic radii, so that the two kinds of radii differ approximately by an additive constant. The b value is a scale factor with the dimension of energy; only two of the three quantities b, r_i, r_j can be varied independently. Expression (5) is a generalization of the original Born–Mayer formula, where a single hardness parameter ρ appears for all atom pairs; the Pauling coefficients $\beta_{ij} = 1 + z_i/N_i + z_j/N_j$ (where N_i, N_j are the numbers of outer electrons in the ith and jth ions) should account roughly for the effect of electronic structure on repulsion. By suppressing these quantities, which have a questionable physical meaning, the simplified expression (4) is obtained. Finally, the last formula (6) (Gilbert, 1968; Busing, 1970) gives the hardness parameter ρ_i a slightly different significance than in the former cases.

A complete expression of the dispersion energy should include not only the reported term $-c_{ij}/r_{ij}^6$, which results from the dipole–dipole interaction between polarized electron clouds, but also higher order multipole expansion terms, such as $-d_{ij}/r_{ij}^8$ (dipole–quadrupole), $-e_{ij}/r_{ij}^{10}$ (quadrupole–quadrupole), and so on. However components higher than dipole–quadrupole are always dropped, and in most cases just the single $-c_{ij}/r_{ij}^6$ term is retained. In principle, the c_{ij} coefficients can be calculated by the London formula (Tosi, 1964), which relates them to experimental atomic polarizabilities and average excitation energies; nevertheless, because such data are not always available, the C_{ij} values are often considered as empirical parameters in the same way as the repulsive b_{ij} and ρ_{ij} quantities. In the latter case the dispersive energy has an ambiguous physical meaning, as it can mask partial covalency effects and simply represent an additional empirical term in the potential; thus the usefulness of the dispersive contribution in the Born–Mayer potential, Eq. (3), for minerals is controversial (Catlow *et al.*, 1982).

Rigorously speaking, the Born–Mayer potential is valid for pure ionic bonds, where the z_i values represent formal ionic charges of closed-shell electron configurations. However, as discussed above, the crystalline phases thought to be present in the earth mantle show chemical bonds with variable covalency/ionicity ratios, which can be divided into three classes for convenience: (1) M—O bonds, where the M metal belongs to groups IA, IIA, and IIIA of the periodic table (e.g., Na, K, Mg, Ca, Al); (2) TM—O bonds, where TM is a transition metal (e.g., Fe, Co, Ni, Mn, Ti); and (3) Si—O bonds. In case (1), the fraction of covalency is minor but generally not quite negligible (except perhaps for alkali metals); its effects can be assumed to be reasonably isotropic, so as to account for them by an electrostatic Born potential with reduced atomic charges, whose modulus is less than that achieved in an ideal closed-shell configuration. In this scheme the charges are additional semiempirical parameters to be optimized, with the constraint of electroneutrality which sets the sum of all atomic charges in the unit cell equal to zero. In bonds of type (2), the *d* orbitals of the TM atom show a partially covalent behavior with directional character, which is analyzed by the crystal field theory: the relative stabilization energy can be calculated as a function of the geometry of the coordination surrounding of the transition metal ion (Gaffney, 1972) and then is added to the Born energy Eq. (3), with fractional charges as in case (1). For Si—O bonds, case (3), with Si in tetrahedral coordination, the covalent effect of the stabilization of sp^3 hybrid orbitals on silicon makes the bond much stronger than those of type M—O or TM—O, as is proved by two kinds of experimental data. The first is given by structural studies of silicates at high pressure, showing that the SiO_4 tetrahedra are practically uncompressible, while nearly all volume decrease of the structure results from shortening of M—O and TM—O bonds and to shrinking of structural voids (Hazen and Prewitt, 1977). The second is represented by vibrational (infrared and Raman) spectra of silicates, where the frequencies of absorption relative to Si—O stretching modes are much higher than for M—O and TM—O bonds.

The most obvious approximation to account for Si—O bonds is a rigid body scheme for the whole SiO_4 tetrahedron, which is treated as undeformable (Catti, 1982): its internal energy is not considered in the expression of the crystal energy, and its geometry is kept fixed in the calculation of energy derivatives with respect to a crystal strain or to atomic displacements. A different way can be followed by treating the Si—O interaction by the usual Born potential with fractional charges, with the addition of another term to "stiffen" the bond. This may be a Morse potential (Price and Parker, 1984):

$$V_{ij} = D_{ij}\{\exp[2\lambda_{ij}(r_{ij}^{\circ} - r_{ij})] - 2\exp[\lambda_{ij}(r_{ij}^{\circ} - r_{ij})]\} \quad (7)$$

The r_{ij}° value is usually set at the average Si—O bond distance for tetrahedrally coordinated silicon (1.63 Å), and the λ_{ij} and D_{ij} quantities can be obtained from the vibrational spectroscopic data for the Si—O diatomic molecule (however, D_{ij} has also been treated as an empirical parameter to be optimized). Another approach uses a quadratic bond stretching potential (M. Matsui and Busing, 1984a):

$$V_{ij} = k_r(r_{ij}^{\circ} - r_{ij})^2, \quad (8)$$

where both r_{ij}° and k_r are optimized as empirical parameters. Further, particularly in the case of SiO_4 tetrahedra sharing vertices (e.g., in pyroxenes), it may be necessary to introduce explicitly directional terms in the potential, such as bond angle bending potentials of the form (M. Matsui and Busing, 1984a):

$$V = k_\alpha(\alpha_0 - \alpha)^2 \quad (9)$$

Different k_α and α_0 empirical parameters are used for O—Si—O and for Si—O—Si bond angles, respectively. The case of Si–O interactions when silicon is octahedrally coordinated, as in phases stable at very high pressure in the lower mantle, has not been investigated thoroughly yet; however, the covalency fraction may be smaller, so that a Born potential with reduced atomic charges may be adequate to account for such bonds.

The importance of a good potential for calculating physical properties of mantle minerals cannot be overemphasized. Two steps are necessary: first, the appropriate terms to include in the energy expression must be selected according to the nature of chemical bonds present in the crystalline phase considered. A variety of choices of the potential are observed in literature calculations, from the simple use of a single electrostatic term with full ionic charges (Tamada, 1980) to more sophisticated treatments. Most authors have used the repulsion term, which is really necessary to account for the equilibrium structural configuration, but not the dispersive term; M. Matsui and Matsumoto (1982) employ a repulsive formula proportional to $1/r_{ij}^{12}$ instead of the usual Born exponential expression. A double dispersive term, $-c_{ij}/r_{ij}^6 - d_{ij}/r_{ij}^8$, appears in other calculations (Catti, 1981, 1982). Atomic charges are sometimes given formal ionic values (Miyamoto and Takeda, 1984), sometimes optimized fractional values. The second step is optimizing the empirical parameters present in the potential expression. This is usually performed by a least-squares fitting or an interpolation procedure, so as to reproduce the experimental values of suitable physical properties as closely as possible; the relations employed are examined in the next section. Use is made: (1) of the same properties that one wants to calculate, but relative to different phases (possibly similar chemically and structurally) with respect to that considered; alternatively (2), of different physical properties (known from experiment) relative to the same phase that is considered. In both cases, an extrapolation is then performed by assuming that the parameters determined are adequate also for physical quantities or for phases different from those used in the optimization.

Physical Properties and Energy of Minerals

Let us now examine a number of chemical–physical quantities that are of interest for mantle minerals, and that in principle can be calculated from a knowledge of the interatomic potential.

Static (or Lattice) Energy, E_L

Static energy is given by relation (2) and is the part of the total crystal energy E arising from interactions between atoms fixed at their equilibrium positions. The static energy depends only indirectly on temperature through the effect of thermal expansion, which changes the lattice constants and the equilibrium structural configuration; on the other hand, it shows a direct and strong dependence on pressure. The other component of the crystal energy is the vibrational part, E_v, which is much smaller than the static one and depends strongly on temperature and weakly on pressure. Thus the static energy can usefully approximate the total internal energy of the mineral when isothermal or nearly isothermal pressure changes are considered (Hildebrand approximation); these processes may be common to varying depths in the earth mantle. If E_v and the entropy S are available by experimental or theoretical methods, the free energy $G = E + pV - TS$ can be obtained over a range of pressure and temperature values, so as to map the stability field of the mineral. Polymorphic transformations within the mantle would be understood or predicted quite straightforwardly in this way, by simple comparison of the G values of different crystalline forms.

Crystal Structure

The theoretical calculation of the crystal structure is based on the "zero-force" principle, according to which in the equilibrium atomic configuration all forces acting on atoms are zero. This is equivalent to saying that the lattice energy is at a minimum point, or that all partial derivatives of the energy with respect to atomic shifts and lattice constants are zero:

$$\frac{\partial E_L}{\partial p_i} = 0 \tag{10}$$

where p_i denotes any structural variable, such as a fractional atomic coordinate, a cell edge, or even a generalized coordinate (bond distance or angle) or a rigid-body external coordinate (eulerian angle, rigid translation) of the SiO_4 tetrahedron. A starting trial structure is needed, which is subsequently refined by a number of numerical methods. Simple search procedures are based on several computations of E_L for different values of the p_i structral variables, until the least-energy configuration is found. Other methods, such as that of steepest descent, require the evaluation of first derivatives in order to solve equations (10), and the very fast Newton-Raphson method also involves the use of second derivatives, $\partial^2 E_L/\partial p_i \partial p_j$, for the same purpose. The computer programs WMIN (Busing, 1970, 1981) and METAPOCS (Parker *et al.*, 1984) perform the energy minimization by using the most appropriate numerical method, according to the dimension and features of the problem.

Elastic Constants

A crystalline solid that is mechanically stressed undergoes an anisotropic deformation made up of two components: a lattice strain and an internal strain (Born and Huang, 1954). The former component is given by a variation of lattice constants leaving the fractional atomic coordinates unchanged, so that the whole atomic structure is deformed homogeneously with the lattice. On the other hand, the internal strain occurs by changing the fractional atomic coordinates while keeping the unit-cell geometry constant. The lattice deformation can be expressed by the change of metric tensor from $\mathbf{G}$, where $\mathbf{G}_{ij} = \mathbf{a}_i \cdot \mathbf{a}_j = \mathbf{a}_i \mathbf{a}_j \cos(\mathbf{a}_i, \mathbf{a}_j)$, to $\mathbf{G}'$. As the elastic constants are referred to a crystal Cartesian referemce frame O, this is chosen so as to define its orientation with respect to the undeformed crystallographic basis I by the matrix $\mathbf{R}$, whose columns are the components of the three orthonormal vectors of O referred to vectors $\mathbf{a}_1, \mathbf{a}_2, \mathbf{a}_3$ ($= \mathbf{a}, \mathbf{b}, \mathbf{c}$) of I. If two points 1 and 2, with Cartesian coordinates $\mathbf{X}_1$ and $\mathbf{X}_2$, are given in the crystal, it can be shown that their distance $|\mathbf{X}_2 - \mathbf{X}_1|$ is changed by the lattice strain according to the relation $|\mathbf{X}'_2 - \mathbf{X}'_1|^2 - |\mathbf{X}_2 - \mathbf{X}_1|^2 = 2(\mathbf{X}_2 - \mathbf{X}_1)^T \boldsymbol{\eta}(\mathbf{X}_2 - \mathbf{X}_1)$, where $\boldsymbol{\eta} = \frac{1}{2}(\mathbf{R}^T \mathbf{G}' \mathbf{R} - \mathbf{I})$ and $\mathbf{I}$ is the identity matrix (Catti, 1985). The finite Lagrangian strain tensor $\boldsymbol{\eta}$ is the most convenient representation of lattice deformation with respect to the Cartesian basis O, as it does not depend on the relative orientation of the deformed lattice with respect to the original lattice, but only on the change of its metrics. By means of the Voigt contraction of indices, the six independent components of the symmetrical 3×3 square matrix $\boldsymbol{\eta}$ are usually transformed into one-index components of the equivalent 6×1 column matrix ($\boldsymbol{\eta}_1 \equiv \boldsymbol{\eta}_{11}, \boldsymbol{\eta}_2 \equiv \boldsymbol{\eta}_{22}, \boldsymbol{\eta}_3 \equiv \boldsymbol{\eta}_{33}, \boldsymbol{\eta}_4 \equiv 2\boldsymbol{\eta}_{23}, \boldsymbol{\eta}_5 \equiv 2\boldsymbol{\eta}_{13}, \boldsymbol{\eta}_6 \equiv 2\boldsymbol{\eta}_{12}$). The inner strain is simply expressed by the variation of fractional atomic coordinates Δx_k, or of Cartesian atomic coordinates $\mathbf{u}_k = \mathbf{R}^{-1}\Delta x_k$, for all N atoms in the unit cell ($k = 1, \ldots, N$); these form the $3N$ Cartesian components of atomic shifts of the $3N \times 1$ column matrix $\mathbf{u}$. Therefore, the energy differnece between the strained and the unstrained states of the crystalline solid is given, to the second order, by the following relationship:

$$E_{\text{elastic}} = \tfrac{1}{2}\boldsymbol{\eta}^T \mathbf{V}_{\eta\eta} \boldsymbol{\eta} + \tfrac{1}{2}\mathbf{u}^T \mathbf{V}_{uu} \mathbf{u} + \boldsymbol{\eta}^T \mathbf{V}_{\eta u} \mathbf{u} \tag{11}$$

where $(\mathbf{V}_{\eta\eta})_{ij} = (\partial^2 E_L/\partial\eta_i\partial\eta_j)_{\eta',u'}$, $(\mathbf{V}_{uu})_{hk} = (\partial^2 E_L/\partial u_h \partial u_k)_{\eta,u'}$, $(\mathbf{V}_{\eta u})_{ih} = (\partial^2 E_L/\partial\eta_i\partial u_h)_{\eta',u'}$ are the components of the square matrices $\mathbf{V}_{\eta\eta}$, $\mathbf{V}_{uu}$, $\mathbf{V}_{\eta u}$ of orders 6×6, $3N \times 3N$, $6 \times 3N$, respectively. All partial second derivatives of the energy are evaluated at zero strain.

However, the parameters of inner strain u_h are not actually independent variables of the elastic energy; in fact, for a given lattice strain defined by the six η_i parameters, the $3N$ components of the atomic shifts are determined by the condition that the energy, Eq. (11), is a minimum, so that the u_h quantities depend on the η_is through the relation $\mathbf{u} = -\mathbf{V}_{uu}^{-1}\mathbf{V}_{u\eta}\boldsymbol{\eta}$. By substituting into Eq. (11), the final expression of the elastic energy depending only on the η_i values is obtained:

$$E_{\text{elastic}} = \tfrac{1}{2}\mathbf{n}^T(\mathbf{V}_{\eta\eta} - \mathbf{V}_{\eta u}\mathbf{V}_{uu}^{-1}\mathbf{V}_{u\eta})\boldsymbol{\eta} = \tfrac{1}{2}\boldsymbol{\eta}^T\mathbf{C}\boldsymbol{\eta} \tag{12}$$

The components C_{ij} of the 6×6 symmetrical matrix $\mathbf{C}$ are the isothermal second-order elastic constants, related by the Voigt contraction of indices to the tensorial quantities c_{ijhk} of equations (1):

$$C_{ij} = (\mathbf{V}_{\eta\eta})_{ij} - (\mathbf{V}_{\eta u}\mathbf{V}_{uu}^{-1}\mathbf{V}_{u\eta})_{ij} \tag{13}$$

The C_{ij} values take the meaning of second derivatives of the static crystal energy with respect to the components of lattice strain, calculated not for a zero internal strain, as with the $(\mathbf{V}_{\eta\eta})_{ij}$ values, but for the particular internal strain that minimizes the elastic energy, Eq. (11). The first and second terms on the right-hand member of Eq. (13) are, respectively, the contribution of pure lattice strain and the contribution of internal strain (coupled with that lattice strain) to the C_{ij} elastic constants. The second term is usually much smaller than the first one, so that the expression of elastic constants can often be approximated by neglecting the effect of inner deformation: $C_{ij} \simeq (\mathbf{V}_{\eta\eta})_{ij} = (\partial^2 E_L/\partial\eta_i\partial\eta_j)_{\eta',u}$.

In order to calculate the elastic constants of a crystalline phase according to Eq. (13), therefore, at least the quantities $(\mathbf{V}_{\eta\eta})_{ij}$ and possibly also the $(\mathbf{V}_{uu})_{hk}$ and $(\mathbf{V}_{\eta u})_{ih}$ values must be evaluated. Computing the second derivatives $(\mathbf{V}_{uu})_{hk}$ is fairly simple, because the fractional atomic coordinates with respect to which the differentiation is performed appear explicitly in the energy expression. More problems are presented by the calculation of $(\mathbf{V}_{\eta\eta})_{ij}$ derivatives. The traditional method (Born and Huang, 1954; Catlow and Mackrodt, 1982) requires that all lattice quantities be expressed in Cartesian coordinates and the derivatives of interatomic distances be evaluated with respect to the components of the strain tensor. According to a different formulation (Catti, 1985), the energy derivatives with respect to the η_i components are related by suitable equations to derivatives with respect to lattice constants a_i; these can be calculated straightforwardly, because the Born energy expression is an explicit function of the a_i quantities. For instance, in the particularly simple case of orthorhombic symmetry, and neglecting the contribution of internal deformation, the elastic constants are given by the following expressions:

$$C_{ii} = \frac{1}{V}\left(a_i^2\frac{\partial^2 E_L}{\partial a_i^2} - a_i\frac{\partial E_L}{\partial a_i}\right)$$

$$C_{ij} = \frac{1}{V} a_i a_j \frac{\partial^2 E_L}{\partial a_i \partial a_j}; \quad (i,j = 1,2,3;\ a_1 = a,\ a_2 = b,\ a_3 = c)$$

$$C_{ii} = \frac{1}{V} \frac{\partial^2 E_L}{\partial \alpha_i^2}; \quad (i = 4,5,6;\ \alpha_4 = \alpha,\ \alpha_5 = \beta,\ \alpha_6 = \gamma) \tag{14}$$

In addition to the above analytical methods, a numerical method was recently proposed for these calculations (Busing and Matsui, 1984). The application of a stress σ_i to the crystal is simulated by summing the corresponding external energy $W_{\text{ext}} = -\sigma_i \varepsilon_i$ to the lattice energy E_L; then the total energy $E_{\text{tot}} = E_L + W_{\text{ext}}$ is minimized by varying both lattice constants and fractional atomic coordinates. By comparing the optimized lattice constants to the original ones, the strain components ε_i are obtained, and the elastic constants can be derived through the relationship $C_{ij} = \sigma_i/\varepsilon_j$.

Dielectric and Piezoelectric Constants

Application of an external electric field to the crystalline phase results in a polarization opposing the field; this is caused mainly by displacements of the charged atoms from their equilibrium positions, which correspond to a state of internal strain such as that occurring by action of a mechanical stress (cf. the above discussion of elasticity). The amount of polarization divided by the applied electric field is expressed by the dielectric constant of the solid phase; neglecting electronic effects, the contribution of ionic shifts to the components k_{ij} of the dielectric tensor is given by the following formula (see, e.g., Catlow and Mackrodt, 1982):

$$k_{ij} = \delta_{ij} + \frac{4\pi}{V} e^2 \sum_{1\,p,q}^{N} z_p z_q (\mathbf{V}_{uu}^{-1})_{piqj} \tag{15}$$

where the i, j indices denote Cartesian components, the p, q indices identify the atoms in the unit cell, and δ_{ij} is 1 for $i = j$, 0 for $i \neq j$.

If the crystal structure lacks the symmetry center, piezoelectric properties are usually observed. When a mechanical stress is applied, lattice and internal strains arise; the latter correspond to a spontaneous ionic polarization measured by the piezoelectric tensor, whose components are expressed as follows:

$$\lambda_{ih} = -\frac{4\pi}{V} e \sum_{1\,p}^{N} z_p (\mathbf{V}_{uu}^{-1} \mathbf{V}_{u\eta})_{pih} \tag{16}$$

In the previous section the quantities $\mathbf{V}_{uu}$ and $\mathbf{V}_{u\eta}$ were explained to be second derivatives of the lattice energy, which can be calculated by a number of methods; thus, through them, not only the elastic constants, but also the dielectric and piezoelectric constants of the mineral may be evaluated theoretically.

Vibrational Energy and Entropy

In the harmonic treatment of atomic vibrations in crystals (Born and Huang, 1954) the lattice dynamics is expressed in terms of normal modes of vibration, each of which is characterized by a particular angular frequency ω and gives its contribution to the vibrational energy and entropy of the crystalline phase according to the well-known formulas:

$$E_v(\omega) = \frac{1}{2}h\omega + \frac{h\omega}{\exp\left(\frac{h\omega}{kT}\right) - 1}$$

$$S(\omega) = -k\ln\left[1 - \exp\left(-\frac{h\omega}{kT}\right)\right] + \frac{1}{T}\frac{h\omega}{\exp\left(\frac{h\omega}{kT}\right) - 1} \tag{17}$$

Here, h is the Planck constant divided by 2π and k is the Boltzmann constant. The total values of E_v and S are obtained either by summing all contributions, Eq. (17), coming from the whole spectrum of normal frequencies ω, or, more conveniently, by calculating the integrals:

$$E_v = \int E_v(\omega)g(\omega)\,d\omega$$

$$S = \int S(\omega)g(\omega)\,d\omega \tag{18}$$

where $g(\omega)$ is the number of normal modes with frequencies in the range between ω and $\omega + d\omega$ divided by the volume of the crystal. As the density of normal modes $g(\omega)$ obviously depends on the frequency spectrum, a detailed knowledge of this is necessary for an exact calculation. In principle, the frequency spectrum can be computed completely ab initio, including the dispersive dependence of frequencies on the wave vector, by standard lattice-dynamical techniques (Born and Huang, 1954), provided that an appropriate interatomic potential is available; however, the computational effort required is heavy, and the results are strongly dependent on the quality of the potential used. Besides, it can be shown that the density of normal modes $g(\omega)$ is not very sensitive to many details of the dispersive behavior of frequencies, so that a complete lattice-dynamical calculation of the whole dispersion relation may give redundant information with respect to the objective of obtaining E_v and S. For these reasons, approximate semiempirical methods can be very convenient for calculating $g(\omega)$ and then the vibrational energy and entropy.

Let us now analyze here a method of calculation that has been devised especially for silicate minerals (Kieffer, 1979, 1980) and later applied specifically to some mantle phases. A number of approximations are involved in Kieffer's model: first, the three acoustic branches of the dispersion relation (one longitudinal and two shear modes) show a sine dependence of frequency on the wave vector. Second, the optic branches are divided into a group of low-frequency modes, corresponding mainly to cation—oxygen bond vibrations, and a group of high-frequency modes corresponding to Si—O vibrations inside the silicate units. The first group of optic modes is represented by a uniform distribution of frequencies between a lower cutoff value ω_l and an upper cutoff value ω_u; the dispersive dependence of frequency on the wave vector is related to a characteristic mass ratio between cation and oxyanion. The second group of optic modes is accounted for either by individual nondispersive frequencies ω_i, which are constant across the Brillouin zone (Einstein oscillators), or by a second uniform continuum of frequencies. This model is not only approximate but also semi-empirical, as several experimental data are required to define its parameters. Acoustic velocities (or elastic constants, from which they can be derived) define the slopes of sine-shaped acoustic branches; infrared and Raman spectroscopic data are needed to determine the lower and upper cutoff frequencies of the optic continuum and the nondispersive frequencies. In this respect, it should be pointed out that vibration modes inactive in infrared and Raman spectra are not taken into account in the model; besides, some arbitrariness lies in the partitioning of optical frequencies based on spectroscopic features. In spite of these weak points, the density of modes $g(\omega)$ obtained by such model is generally able to account satisfactorily for thermodynamic properties, such as E_v and S.

Applications

Potentials Optimized on Forsterite, α-Mg_2SiO_4

Magnesium-rich olivine is thought to be the most abundant mineral in the upper mantle; besides, practically all important physical properties have been measured directly on single-crystal forsterite, the Mg olivine end member, in many cases also at high temperature and pressure. For these reasons, and because of the chemical and structural simplicity of such minerals, most attempts to determine a suitable potential for theoretical computations in mantle minerals have been made by calibration on forsterite. Five Born–Mayer potentials, Eq. (3), of this type, which I think to be the most representative ones available in the literature, will be examined critically here; they differ in detailed analytical form, in the model used to account for SiO_4 inner interactions, and in the physical properties chosen for optimization. In the first four cases, WN (Miyamoto and Takeda, 1984), P2 and P5 (Price and Parker, 1984), and MB (M. Matsui and Busing, 1984), the dispersive term of the energy is omitted, and the Gilbert form, Eq. (6), is used for the repulsive energy. The fifth potential, C (Catti, 1982), includes a double dispersive energy term, $-d_{ij}/r_{ij}^6 - q_{ij}/r_{ij}^8$, and a repulsion term of the form in Eq. (5). In order to unify the notation and allow a comparison, the repulsion terms of the energy will be transformed here into the general expression $b_{ij}\exp(-r_{ij}/\rho_{ij})$ for all five cases. The atomic interactions within the SiO_4 tetrahedral group are treated like the other ones by potentials WN and P2; a rigid-body model is assumed for SiO_4 in MB and C, excluding inner interactions, while a Morse term, Eq. (7), is introduced in P5 to account for covalency effects of the Si—O bonds. In WN and P2 the atomic charges were not optimized but instead were fixed at constant values: the formal ones for WN (full ionic model), and those determined by a X-ray charge density study of forsterite (Fujino *et al.*, 1981) for P2. In MB the Mg charge was kept fixed at its formal value +2, letting only z_O change (z_{Si} is determined by the electroneutrality condition, $z_{Si} + 2z_{Mg} + 4z_O = 0$), whereas both the Mg and the O charges were optimized in the P5 and C potentials. Energy parameters are fitted only to X-ray structural properties (Hazen, 1976) for WN and P2, using the WMIN program; in all other cases also, experimental elastic constants (Graham and Barsch, 1969; Kumazawa and Anderson, 1969) are included in the optimization procedure, though by different numerical methods, and even the isotropic dielectric constant (Ghosh and Das, 1979) is taken into account by the P5 potential, using the METAPOCS program. Further, the C potential includes a correction for the temperature effect on static energy arising from thermal expansion and consequent change of lattice constants.

The optimized energy parameters are reported in Table 2 for all five potentials considered. The dispersive coefficients d_{ij} and q_{ij} appearing in potential C were not fitted but were calculated on the basis of refractivity data (Boswarva, 1970); their values are shown in Table 3. It can be seen that the repulsive parameters b_{ij} change by even several orders of magnitude among different potentials; however, this effect arises from small variations of ρ_{ij} and r_i, r_j values which are enhanced by the exponential dependence $b_{ij} \alpha \exp[(r_i + r_j)/\rho_{ij}]$. In order to evaluate the quality of the potentials examined, the structural and elastic properties of forsterite computed back on their basis are compared with the corresponding experimental values in Table 4. The structural configuration is summarized by the average values of Si—O and Mg—O (for two coordination polyhedra of Mg1 and Mg2 atoms) bond lengths, by the shifts t_{Si} of Si and t_{Mg2} of Mg2 atoms with respect to their experimental positions from X-ray data (Mg1 is fixed on the symmetry center), and by the $\Delta\phi$ angle of rotation of the SiO_4 tetrahedron on the (001) mirror plane on which it lies, compared to its experimental orientation. For potentials MB and C the calculated Si–O distances are obviously equal to the observed ones, as the SiO_4 group has been kept rigid by changing only its position and orientation during the fitting procedure.

The following conclusions can be drawn from a comparison of calculated and experimental values of physical properties in Table 4, taking into account the different features of the potentials:

1. Structural properties are reproduced satisfactorily by all potentials, because they were always included in the optimization processs.
2. The model with formal ionic charges (WN) leads to heavily overestimating the elastic constants, simulating a much stiffer crystal than it is. A better agreement of elastic data is observed for potential P2, but the overestimate remains; this shows that the atomic charges derived from X-ray electron density data, at least by the present methods (Fujino *et al.*, 1981), are still too close to formal ionic values to account for elastic properties correctly. On the other hand, it has been proved that by trying to fit all atomic charges to structural properties only, the optimization procedure does not converge (Catti and Ivaldi, 1983); either one of the charges must be kept fixed (M. Matsui and Matsumoto, 1982), or other physical properties must be included

Table 2. Parameters for interatomic potentials [cf. Eq. (3) and (7)] optimized on physical properties of forsterite α-Mg_2SiO_4: WN (Miyamoto and Takeda, 1984), P2 and P5 (Price and Parker, 1984), MB (M. Matsui and Busing, 1984b), and C (Catti, 1982).

	WN	P2	MB	P5	C
z_{Mg}, e	2	1.75	2	1.77	1.38
z_{Si}, e	4	2.10	0.64	1.58	1.44
z_O, e	−2	−1.40	−1.16	−1.28	−1.05
b_{MgMg}, $MJ \cdot mol^{-1}$	910.81×10^9	16.752	$1{,}519.67 \times 10^6$	7.000	50.294
ρ_{MgMg}, Å	0.04036	0.3072	0.1076	0.376	0.23
b_{MgSi}, $MJ \cdot mol^{-1}$	574.55×10^9	7.5498	—	2.253	10.766
ρ_{MgSi}, Å	0.0367	0.3023	—	0.4341	0.23
b_{MgO}, $MJ \cdot mol^{-1}$	414.715	1,322.53	76.407	122.253	335.888
ρ_{MgO}, Å	0.1258	0.2048	0.3038	0.2715	0.23
b_{SiSi}, $MJ \cdot mol^{-1}$	305.89×10^9	3.3152	—	0.96285	2.2575
ρ_{SiSi}, Å	0.0298	0.2974	—	0.4922	0.23
b_{SiO}, $MJ \cdot mol^{-1}$	144.496	445.487	—	15.7395	77.036
ρ_{SiO}, Å	0.1189	0.1999	—	0.3296	0.23
b_{OO}, $MJ \cdot mol^{-1}$	25,725.0	$1{,}098.17 \times 10^6$	2.7030	98,030.4	1,682.40
ρ_{OO}, Å	0.2080	0.1024	0.500	0.167	0.23
D_{SiO}, $kJ \cdot mol^{-1}$	—	—	—	439.3	—
λ_{SiO}, Å	—	—	—	1.975	—
r°_{SiO}, Å	—	—	—	1.63	—

Table 3. Parameters of the potential [Eq. (3)] optimized on structural properties of Ca-olivine, larnite, merwinite, and monticellite (Catti and Ivaldi, 1983).

z_{Ca}	1.50 e		
z_{Mg}	1.38 e		
z_{Si}	1.40–1.52 e		
ρ	0.25 Å		
	b_{ij}, $MJ \cdot mol^{-1}$	d_{ij}, $MJ \cdot mol^{-1}$	q_{ij}, $MJ \cdot mol^{-1}$
Ca—Ca	416.199	2.311	3.321
Ca—Mg	115.719	0.292	0.270
Ca—Si	28.369	0.113	0.095
Ca—O	642.713	3.013	4.757
Mg—Mg	32.174	0.054	0.022
Mg—Si	7.888	0.023	0.008
Mg—O	178.698	0.363	0.385
Si—Si	1.894	0.011	0.003
Si—O	46.939	0.139	0.136
O—O	744.379	3.957	6.805

in the best fitting. Indeed, for the other potentials (MB, P5, and C) the optimization was based on elastic properties as well, letting the atomic charges vary; obviously a good agreement is obtained between calculated and observed elastic constants. It should be noted that all charge distributions derived from these potentials are characterized by a much lower ionicity than that reported by Fujino *et al.* (1981) on the basis of X-ray experiments.

3. When the three best potentials, MB, P5, and C, are compared, although all of them give satisfactory results, computed with P5 and C seem to show a slightly better agreement than MB and experimental values; this is probably a result of fixing the Mg charge to its formal value +2 in MB, instead of optimizing it. Model P5 has the advantage of accounting explicitly for the Si–O interaction by an appropriate Morse potential term, even though the number of parameters to be fitted is increased. Further, the P5 potential also was optimized on the isotropic dielectric constant, and a calculated value of 6.06 was obtained against an observed value of between 5.5 and 6.6 (Ghosh and Das, 1979).

Potentials Optimized on Ca, Mg Orthosilicates and on Diopside

In addition to forsterite, also a number of magnesium–calcium orthosilicates of interest for the mantle composition have been used to optimize Born–Mayer potentials; however, as no experimental elastic data are available in this case, only structural properties have been included in the fitting, with the constraint of a rigid SiO_4 tetrahedron. The work of M. Matsui and Matsumoto (1982) was

Table 4. Comparison between observed values and those calculated by the potential parameters in Table 2 for structural and elastic properties of forsterite. References are quoted in the text.

	Observed	WN	P2	MB	P5	C
$(Si—O)_{av}$, Å	1.634	1.615	1.637	—	1.638	—
$\Delta\phi$, °	0	1.8	1.6	0.1	1.2	0.0
t_{Si}, Å	0	0.02	0.05	0.05	0.01	0.01
t_{Mg2}, Å	0	0.02	0.05	0.02	0.09	0.02
$(Mg1—O)_{av}$, Å	2.098	2.109	2.099	2.05	2.075	2.093
$(Mg2—O)_{av}$, Å	2.128	2.147	2.134	2.16	2.146	2.132
C_{11}, Mbar	3.27	10.55[a]	3.57[a]	3.26	2.94	3.29
C_{22}, Mbar	1.99	5.37[a]	2.49[a]	2.35	2.07	1.93
C_{33}, Mbar	2.34	7.00[a]	3.12[a]	2.18	2.40	2.33
C_{12}, Mbar	0.65	2.18[a]	1.19[a]	0.75	0.72	0.59
C_{13}, Mbar	0.66	2.55[a]	1.34[a]	0.70	0.70	0.66
C_{23}, Mbar	0.71	2.29[a]	1.12[a]	0.88	0.81	0.62
C_{44}, Mbar	0.67	2.07[a]	1.05[a]	0.77	0.74	1.30[a]
C_{55}, Mbar	0.81	0.72[a]	0.86[a]	0.98	0.65	0.92[a]
C_{66}, Mbar	0.81	2.06[a]	0.97[a]	0.70	0.71	0.85[a]

[a] Values correspond to quantities not included in the optimization of the potentials.

based on the crystal structures of calcium–olivine, monticellite, and forsterite, deriving a unique set of parameters for all these minerals; the repulsion energy was expressed not by the usual exponential term, but by the simpler formula $(r_i + r_j)^{13}/12r_{ij}^{12}$, and all atomic charges were kept fixed at their formal ionic values. In a more general approach (Catti and Ivaldi, 1983), the following group of minerals was examined: Ca-olivine (γ-Ca_2SiO_4), larnite (β-Ca_2SiO_4), merwinite [$Ca_3Mg(SiO_4)_2$], and monticellite ($CaMgSiO_4$). It was shown that when the optimization process was based on structural properties only, then at least one of the independent atomic charges had to be kept fixed; otherwise, if all independent charges were let free to vary, they drifted to smaller and smaller absolute values in order to bring the first derivatives of the energy close to zero, so that the method did not converge and an unrealistically small bond polarization was obtained. In this case the Mg charge was fixed at the 1.38 e value determined for forsterite, where elastic data were included in the fitting (Catti, 1982); also the other energy parameters relative to Mg were taken from that study. All four crystal structures were optimized at the same time, and the parameters obtained are reported in Table 3. The z_{Si} charge, which was determined by the electroneutrality condition from the unique fitted values of z_O and z_{Ca}, changes in each phase according to its particular stoichiometry. A measure of the goodness of fit can be obtained by calculating the energy on the basis of the potential parameters of Table 3 and searching the least-energy structure configuration for each mineral; this is compared to the X-ray experimental one, and the root-mean-square relative deviation,

$$I = \left\{ \sum_{1^i}^{n} (d_i^{\text{calc}} - d_i^{\text{expt}})/d_i^{\text{expt}\,2} \right\}^{1/2}$$

is computed for Ca—O and Mg—O bond distances. Values of 0.010, 0.020, 0.014, and 0.011 are obtained for the four crystal structures, respectively, and they appear quite satisfactory, taking also into account that a unique set of energy parameters is able to reproduce four different structural configurations.

Choosing a potential model for internal interactions in SiO_4 groups in chain silicates, is more critical than it is for nesosilicates because of the energy dependence on bond angles α in the silicatic chain. In their study of diopside, M. Matsui and Busing (1984a) have introduced two angular potential terms of the type $V(\alpha) = k_\alpha (\alpha - \alpha_0)^2$ for the bond angle Si—O3—Si between adjacent tetrahedra and for bond angles O—Si—O inside each tetrahedron, respectively; besides, a term such as $V(r) = k_r(r - r^\circ)^2$ accounts for variations of the length r of Si—O3 bonds (CS model in the authors'terminology). For the repulsion parameters of Mg and O atoms the values determined by the same authors in the case of forsterite (MB values in Table 2) were used, whereas for Ca these parameters were optimized on the structure of Ca-olivine; the repulsion energy of Si was neglected and the z_{Mg} and z_{Ca} charges were fixed at +2 e, just as in the study of forsterite (MB). The optimization procedure, based on the Busing and Matsui's (1984) method and on the WMIN program, included the k_α, k_r, α_0, r° force-field parameters and the z_{O3} and $z_{O1,2}$ charges relative to the two types of oxygen atoms (shared and not shared by different Si atoms) in the silicatic chain, taking into account both the structure configuration (Levien and Prewitt, 1981) and the elastic constants (Levien *et al.*, 1979) of diopside. The final values of the energy parameters obtained are reported in Table 5. As usual, computing back the structural and elastic properties of this mineral by the optimized potential, and comparing them with the observed values (Table 6), provides a test for the potential itself. The agreement appears to be very good, particularly for elastic constants and bond distances, and slightly less good for unit-cell edges. It should be noticed that the bridging oxygen O3 and the nonbridging oxygen atoms O1, 2 take different charge values, consistent with O3 being involved in two rather than one Si—O bond with important covalent character. The same authors have also tried simpler potential models, excluding the force-field terms, which depend on r(Si—O3) and on O—Si—O bond angles, obtaining, however, a much less satisfactory accordance than in the previous case.

Evaluation of Structural and Elastic Properties of High-Pressure Mg Silicates

Some of the potentials discussed in the previous paragraphs were used to calculate the equilibrium structural configuration (also at different pressure values) and the elastic constants of a number of Mg_2SiO_4 and $MgSiO_3$ polymorphs of great importance for the mantle consitution, which were not included in the

Table 5. Parameters of the potential based on formulas (3), (8), and (9) optimized on structural and elastic properties of diopside (M. Matsui and Busing, 1984a).

z_{Ca}, e	2
z_{Mg}, e	2
z_{Si}, e	1.425
$z_{O1,2}$, e	−1.260
z_{O3}, e	−0.905
b_{CaCa}, $MJ \cdot mol^{-1}$	93.5103×10^6
ρ_{CaCa}, Å	0.1422
b_{CaMg}, $MJ \cdot mol^{-1}$	242.0750×10^6
ρ_{CaMg}, Å	0.1249
b_{CaO}, $MJ \cdot mol^{-1}$	107.095
ρ_{CaO}, Å	0.3211
r°(Si—O3), Å	1.786
k_r(Si—O3), $kJ \cdot Å^{-2}\, mol^{-1}$	3029.2
k_α(Si—O3—Si), $kJ \cdot deg^{-2}\, mol^{-1}$	0.0920
k_α(O—Si—O), $kJ \cdot deg^{-2}\, mol^{-1}$	0.3054
α_o(Si—O3—Si), °	169.00
α_o(O3—Si—O1, 2), °	109.18
α_o(O3—Si—O3), °	101.47
α_o(O1—Si—O2), °	117.47

fitting of the potentials themselves. In several cases the values of such physical properties are not known experimentally, or at least were not known at the time of the calculation, so that by evaluating them theoretically a real prediction was attained. I shall begin by the γ-spinel-like polymorph of Mg_2SiO_4, ringwoodite, for which a single-crystal structure determination (Sasaki *et al.*, 1982) and a measurement of elastic data (Weidner *et al.*, 1984) have become available very recently. All computations were performed using potentials optimized on the α-olivine-like polymorph, forsterite: P5 (Price and Parker, 1984), MB (M. Matsui and Busing, 1984b), C (this work), and N1 (Miyamoto and Takeda, 1984), which is very similar to the WN potential reported in Table 2. In all cases but N1 both room-pressure structural data and elastic constants were evaluated, whereas Miyamoto and Takeda (1984) calculated just the fractional coordinate u of the oxygen atom for two different high-pressure values, 180 and 240 kbar, using fixed estimated values for the cell edge a. All results are reported and compared with the corresponding room-pressure observed values in Table 7. A surprisingly good agreement is shown for the elastic constants calculated by the P5 model; taking into account that such calculation was published before the paper reporting experimental results by Weidner *et al.* (1984) had appeared, we can consider this prediction a noteworthy success of the theory. The accordance with measured values is slightly worse for elasticity data from the C and MB potentials, however, these seem to be able to predict the structural configuration of ringwoodite much better than P5. Such different performances in the calculation of

Table 6. Lattice constants, average bond lengths and angles, elastic constants, and their pressure derivatives calculated for diopside by the potential parameters of Table 5 (M. Matsui and Busing, 1984a) and compared to experimental values where available.

	Observed	Calculated	*d/dp* (units per Mbar)	
			Observed	Calculated
a, Å	9.75	9.60	−2.57	−3.07
b, Å	8.92	9.43	−2.94	−3.18
c, Å	5.25	5.28	−1.32	−1.25
β, °	105.9	106.2	−10.2	−14.1
V, Å^3	439.1	458.5	−347	−373
$(Si—O)_{av}$, Å	1.68	1.67_5	−0.14	−0.19
$(O—Si—O)_{av}$, °	109.3	109.3	−0.2	−0.2
Si—O3—Si, °	135.8	140.3	−35	−36
$(Mg—O)_{av}$, Å	2.08	2.12	−0.66	−0.87
$(Ca—O)_{av}$, Å	2.50	2.55	−0.89	−0.82
C_{11}, Mbar	2.23	2.10		5.0
C_{22}, Mbar	1.71	1.66		7.5
C_{33}, Mbar	2.35	2.42		4.7
C_{12}, Mbar	0.77	0.64		5.9
C_{13}, Mbar	0.81	0.87		4.6
C_{23}, Mbar	0.57	0.72		4.1
C_{44}, Mbar	0.74	0.80		2.5
C_{55}, Mbar	0.67	0.70		1.3
C_{66}, Mbar	0.66	0.58		2.4
C_{15}, Mbar	0.17	0.26		−1.0
C_{25}, Mbar	0.07	0.11		−1.6
C_{35}, Mbar	0.43	0.49		−0.7
C_{46}, Mbar	0.07	0.09		1.0
K	1.08	1.05	4.8	6.2

Table 7. Unit-cell edge and volume, fractional coordinate of the oxygen atom, bond lengths, and elastic constants of spinel-like γ-Mg_2SiO_4. Observed and computed values are reported, indicating the codes of potentials of different authors used; references are given in the text.

	Observed, 1 bar	MB, 1 bar	P5, 1 bar	C, 1 bar	N1	
					180 kbar	240 kbar
a, Å	8.065	8.04	8.011	8.055	7.840[a]	7.765[a]
V, Å^3	524.6	520.3	514.1	522.6	481.8[a]	468.1[a]
u	0.3685	0.3689	0.3705	0.3695	0.3705	0.3711
Si—O, Å	1.655	1.656	1.672	1.667	1.636	1.628
Mg—O, Å	2.070	2.06	2.039	2.059	1.996	1.972
C_{11}, Mbar	3.27	4.43	3.7	4.30		
C_{12}, Mbar	1.12	2.01	1.4	1.13		
C_{44}, Mbar	1.26	1.63	1.3	0.67		
K	1.84	2.82	2.2	2.19		

[a] Values not calculated but estimated from other experimental data.

elastic and structural data should probably depend on the different features of these potentials regarding Si–O interactions, which are accounted for either by the Morse potential term (P5) or by the rigid SiO_4 tetrahedron model (C and MB). Further investigations on this point would be worthwhile.

In addition to ringwoodite, also the β form of Mg_2SiO_4, wadsleyite, and the two high-pressure modifications of $MgSiO_3$ (ilmenite and perovskite type) were considered by Miyamoto and Takeda (1984), who used the WMIN program and

Table 8. Lattice constants and average interatomic distances for β-Mg_2SiO_4 and ilmenite-type and perovskite-type $MgSiO_3$. Comparison between observed values at 1 bar and calculated values (N1 potential) at room and high pressures (Miyamoto and Takeda, 1984).

	Observed,	Calculated		
	1 bar	1 bar	120 kbar	180 kbar
β-Mg_2SiO_4				
a, Å	5.698	5.686	5.568[a]	5.503[a]
b, Å	11.438	11.452	11.095[a]	10.923[a]
c, Å	8.257	8.182	8.088[a]	8.004[a]
V, Å^3	538.1	532.8	499.7[a]	481.2[a]
$(Si—O)_{av}$, Å	1.651	1.649	1.636	1.628
$(O—O)_{av}$, Å	2.693	2.690	2.669	2.656
$(Mg1—O)_{av}$, Å	2.069	2.064	2.010	1.980
$(O—O)_{av}$, Å	2.925	2.918	2.842	2.799
$(Mg2—O)_{av}$, Å	2.084	2.085	2.029	1.998
$(O—O)_{av}$, Å	2.942	2.950	2.870	2.826
$(Mg3—O)_{av}$, Å	2.090	2.074	2.020	1.990
$(O—O)_{av}$, Å	2.954	2.933	2.858	2.814
$MgSiO_3$ (ilmenite type)				
a, Å	4.728	4.797	4.638[a]	4.608[a]
c, Å	13.559	13.735	13.044[a]	12.873[a]
V, Å^3	262.54	273.7	240.02[a]	236.7[a]
$(Si—O)_{av}$, Å	1.799	1.856	1.796	1.785
$(O—O)_{av}$, Å	2.541	2.605	2.529	2.515
$(Mg—O)_{av}$, Å	2.077	2.093	1.984	1.960
$(O—O)_{av}$, Å	2.892	2.929	2.781	2.749
$MgSiO_3$ (perovskite type)				
a, Å	4.775	4.848	4.620[a]	4.581[a]
b, Å	4.929	5.032	4.807[a]	4.777[a]
c, Å	6.897	7.097	6.751[a]	6.715[a]
V, Å	162.4	173.1	149.9[a]	146.9[a]
$(Si—O)_{av}$, Å	1.79	1.83	1.77	1.76
$(O—O)_{av}$, Å	2.53	2.59	2.49	2.48
$(Mg—O)_{av}$, Å	2.20	2.28	2.13	2.11

[a] Values not calculated but estimated from other experimental data.

their N1 potential optimized on forsterite to calculate the crystal structure configurations of these mantle phases for different pressure values. The results are shown in Table 8, where the room-pressure calculated values can be compared to the experimental ones determined by Horiuchi and Sawamoto (1981), Horiuchi *et al.* (1982), and Y. Matsui (1982) for β-Mg_2SiO_4, $MgSiO_3$ ilmenite, and $MgSiO_3$ perovskite, respectively; for the sake of brevity, only average interatomic distances in the coordination polyhedra have been reported. To get a better understanding of the agreement between observed and computed structural properties, the cooresponding deviations for all Si—O, Mg—O, and O—O interatomic distances inside and Si and Mg coordination polyhedra have been shown for each of the three phases in the diagram of Fig. 1. Much smaller deviations are observed for β-Mg_2SiO_4 than for the two $MgSiO_3$ polymorphs; this may result partly from a low accuracy of experimental structural data for the latter phases (powder measurements were used for $MgSiO_3$ perovskite), but also from inadequacy of the potential to account for octahedral rather than for tetrahedral Si–O interactions. Further, it appears from Fig. 1 that O—O interatomic distances (>2.3 Å) show a better agreement between calculated and observed values than Mg—O and Si—O distances (<2.5 Å) do, so that the potential used would actually be less suitable for representing short than long atomic interactions. The calculation of structural properties at high pressure was performed by keeping the lattice constants fixed at values estimated or extrapolated and letting the atomic fractional coordinates be optimized only by energy

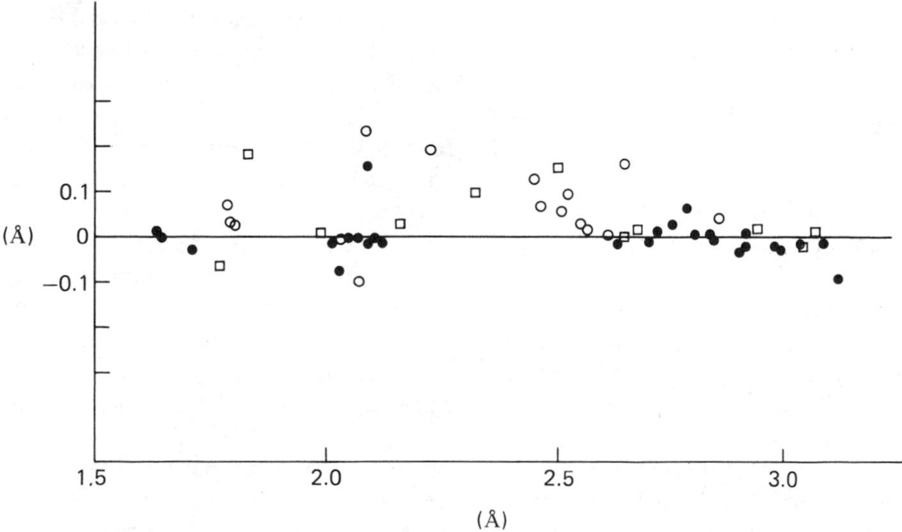

Fig. 1. Deviations from observed experimental values of calculated Si—O, Mg—O, O—O interatomic distances (Miyamoto and Takeda, 1984) for β-Mg_2SiO_4 (full circles), ilmenite-type $MgSiO_3$ (squares), and perovskite-type $MgSiO_3$ (open circles) at 1 bar.

minimization. As no X-ray crystal structure determinations have been carried out yet on these phases at high pressures, such theoretical computations mean predictions of these properties. The results shown in Table 8 are in qualitative accordance with the considerations of Hazen and Prewitt (1977) on changes of bond lengths with increase of pressure; however, the quantitative aspect should be accepted only with caution because of the very simple character of the N1 potential, which is based on fully ionic atomic charges and treats all bonds (including Si—O) with the same potential formula.

Finally, it should be mentioned that, in the study of diopside by M. Matsui and Busing (1984a), the derivatives of elastic constants with respect to pressure were predicted (Table 6) by a calculation based on the method of Busing and Matsui (1984). The only possible check against experimental results regards the elastic modulus *K*, whose pressure derivative amounts to 6.2 vs. 4.8, with a moderately satisfactory agreement.

Thermodynamic Data of Mg_2SiO_4 Polymorphs from Vibrational Models

Kieffer's (1979) semiempiricial vibrational model was applied by Akaogi *et al.* (1984) to calculate the entropy and heat capacity of forsterite, wadsleyite, and ringwoodite, starting from observed acoustic velocities and IR and Raman spectra. In the case of α-Mg_2SiO_4, experimental values of both entropy and heat capacity are available over a wide range of temperatures, but for the β and γ phases only the heat capacity was measured (Watanabe, 1982). In the former case, therefore, all results of calculations can be checked against observed data, whereas in the latter the entropy has been predicted. The vibrational spectra of all three minerals show two distinct bands of frequencies; the low band ranges from 144 to 620, from 190 to 590, and from 250 to 550 cm^{-1} for α, β, and γ phases, respectively, and the high band from 830 to 985, from 700 to 1100, and from 800 to 850 cm^{-1}, correspondingly. In each case the low-frequency band, arising mainly from Mg—O vibration modes, was used to define the optical continuum of frequencies comprised between the two cutoff values ω_l and ω_u. The high-frequency band, corresponding to Si—O vibration modes, was accounted for by two distinct Einstein oscillators with the lower and upper frequencies (although just one oscillator was used for γ-Mg_2SiO_4, because of the small ω range), or by a second optical continuum. The acoustic velocities of β and γ phases were estimated from experimental values of analogous germanate compounds.

A variety of different features were tried in the model parametrization, giving rise to five slightly different models for each of the three minerals; however, although some arbitrariness undoubtedly lies in the choice of these parameters, the spread of final values calculated for S and C_p in the five cases is reasonably small, on the order of a few percent. In particular, the calculated entropy and heat capacity appear to be only slightly sensitive to parametrization details of

Table 9. Values of the standard entropy S of the α, β, and γ polymorphs of Mg_2SiO_4 calculated at three temperatures by lattice vibrational models (Akaogi *et al.*, 1984); for the olivine phase, experimental values by Orr (1953) are also reported on the lines below.

T, K	α, $J \cdot K^{-1} mol^{-1}$	β, $J \cdot K^{-1} mol^{-1}$	γ, $J \cdot K^{-1} mol^{-1}$
298	95.19	87.40	85.00
	97.95		
700	216.82	208.36	204.14
	221.63		
1000	277.15	268.61	263.05
	282.84		

the high-frequency vibration modes, whereas they seem to be strongly affected by how low-frequency modes are dispersed across the Brilloin zone from zero to maximum value of the wave vector. In Table 9 the calculated standard entropy S of all three polymorphs is reported for three temperatures (the computation was performed over the whole 298–1000K range); such values are averages over results obtained by the five different models quoted above. A comparison with measured data is possible for forsterite only, showing a quite satisfactory accordance. The predicted S values of wadsleyite and ringwoodite were checked indirectly by Akaogi *et al.* (1984), who used them to evaluate the entropies of transition at 1000K for the $\alpha \rightarrow \beta$ ($-11.7\ J \cdot K^{-1} mol^{-1}$) and $\beta \rightarrow \gamma$ (-5.4) phase transformations. These agreed reasonably with experimental values based on the combination of calorimetry and phase equilibria data (-10.5 and -6.3 $J \cdot K^{-1} mol^{-1}$, respectively).

Summary

Knowledge of a number of the physical properties of crystalline phases that are thought to be present in the mantle is of great importance for modeling the mineralogical and chemical composition of the earth's interior. Owing to the great experimental difficulties of measurements on small synthetic samples, particularly at high pressures and temperatures, theoretical evaluations of such physical quantities should be highly appreciated. Thus the principles underlying numerical methods of calculation have been analyzed for the most interesting properties: crystal structure configuration, elastic constants, dielectric data, and thermodynamic quantities. A particular stress is put on the problem of an appropriate choice of the interatomic potential, discussing several extensions of the basic Born–Mayer model. Applications to the α, β, and γ polymorphs of Mg_2SiO_4, to the ilmenite-type and perovskite-type polymorphs of $MgSiO_3$, and to some Ca–Mg silicates have been reviewed critically, by comparing calculated values of structural, elastic, and other properties to experimental data (if available), and analyzing the predictions of data not measured.

Acknowledgments

Financial support by the Ministero Pubblica Istruzione, Roma, is acknowledged.

References

Akaogi, M., and Navrotsky, A. (1984) The quartz-coesite-stishovite transformations: new calorimetric measurements and calculation of phase diagrams, *Phys. Earth Planet. Int.* **36**, 124–134.

Akaogi, M., Ross, N. L., McMillan, P., and Navrotsky, A. (1984) The Mg_2SiO_4 polymorphs (olivine, modified spinel and spinel)—thermodynamic properties from oxide melt solution calorimetry, phase relations, and models of lattice vibrations, *Amer. Mineral.* **69**, 499–512.

Anderson, D. L., and Bass, J. D. (1984) Mineralogy and composition of the upper mantle, *Geophys. Res. Lett.* **11**, 637–640.

Bamford, D. (1976) Seismic anisotropy in the crust and upper manlte, in *The Physics and Chemistry of Minerals and Rocks*, pp. 223–237, edited by R. G. J. Strens. Wiley-Intersicence, London.

Born, M., and Huang, K. (1954) *Dynamical Theory of Crystal Lattices*. Oxford Univ. Press, London.

Boswarva, I. M. (1970) Semiempirical calculations of ionic polarizabilities and van der Waals potential coefficients for the alkaline-earth chalcogenides, *Phys. Rev.* **B1**, 1698–1701.

Busing, W. R. (1970) An interpretation of the structures of alkaline earth chlorides in terms of interionic forces, *Trans. Amer. Crystallogr. Assoc.* **6**, 57–72.

Busing, W. R. (1981) *WMIN, a computer program to model molecules and crystals in terms of potential energy functions*. ORNL-5747, U.S. National Technical Information Service, Washington, D.C.

Busing, W. R., and Matsui, M. (1984) The application of external forces to computational models of crystals, *Acta Crystallogr.* **A40**, 532–538.

Catlow, C. R. A., Dixon, M., and Mackrodt, W. C. (1982) Interionic potentials in ionic solids, in *Computer Simulation of Solids*, pp. 130–161, edited by C. R. A. Catlow and W. C. Mackrodt. Springer-Verlag, Berlin.

Catlow, C. R. A., and Mackrodt, W. C. (1982) Theory of simulation methods for lattice and defect energy calculations in crystals, in *Computer Simulation of Solids*, pp. 3–20, edited by C. R. A. Catlow and W. C. Mackrodt. Springer-Verlag, Berlin.

Catti, M. (1981) A generalized Born-Mayer parametrization of the lattice energy in orthorhombic ionic crystals, *Acta Crystallogr.* **A37**, 72–76.

Catti, M. (1982) Atomic charges in Mg_2SiO_4 (forsterite), fitted to thermoelastic and structural properties, *J. Phys. Chem. Solids* **43**, 1111–1118.

Catti, M. (1985) Calculation of elastic constants by the method of crystal static deformation, *Acta Crystallogr.* **A**, in press.

Catti, M., and Ivaldi, G. (1983) Charge distribution from least-energy structure in Ca–Mg orthosilicates, *Phys. Chem. Minerals* **9**, 160–166.

Fujino, K., Sasaki, S., Takeuchi, Y., and Sadanaga, R. (1981) X-Ray determination of electron distribution in forsterite, fayalite and tephroite, *Acta Crystallogr.* **B37**, 513–518.

Gaffney, E. S. (1972) Crystal-field effects in mantle minerals, *Phys. Earth Planet. Int.* **6**, 385–390.

Ghosh, P. K., and Das, A. R. (1979) Preparation and characterization of forsterite and measurement of its dielectric cconstant and loss factor in the frequency range 100 kc/sec to 25 Mc/sec, *Trans. Indian Ceram. Soc.* **38**, 89–95.

Gilbert, T. L. (1968) Soft-sphere model for closed shell atoms and ions, *J. Chem. Phys.* **49**, 2640–2645.

Graham, E. K., and Barsch, G. R. (1969) Elasticc constants of single crystal forsterite as a function of temperature and pressure, *J. Geophys. Res.* **74**, 5949–5960.

Hazen, R. M. (1976) Effects of temperature and pressure on the crystal structure of forsterite, *Amer. Mineral.* **61**, 1280–1293.

Hazen, R. M., and Prewitt, C. T. (1977) Effects of temperature and pressure on interatomic distances in oxygen-based minerals, *Amer. Mineral.* **62**, 309–315.

Horiuchi, H., and Sawamoto, H. (1981) β-Mg_2SiO_4: single-crystal X-ray diffraction study, *Amer. Mineral.* **66**, 568–575.

Horiuchi, H., Hirano, M., Ito, E., and Matsui, Y. (1982) $MgSiO_3$ (ilmenite-type): single crystal X-ray diffraction study, *Amer. Mineral.* **67**, 788–793.

Ito, E., and Matsui, Y. (1977) Silicate ilmenites and the post-spinel transformations, in *High-Pressure Research, Applications in Geophysics*, pp. 193–208, edited by M. H. Manghnani and S. Akimoto. Academic Press, New York.

Ito, E., Takahashi, E., and Matsui, Y. (1984) The mineralogy and chemistry of the lower mantle: an implication of the ultrahigh-pressure phase relations in the system MgO–FeO–SiO_2, *Earth Planet. Sci. Lett.* **67**, 238–248.

Kern, H., and Richter, A. (1981) Temperature derivatives of compressional and shear wave velocities in crustal and mantle rocks at 6 kbar confining pressure, *J. Geophys.* **49**, 47–56.

Kieffer, S. W. (1979) Thermodynamics and lattice vibrations of minerals: 1. Mineral heat capacities and their relationships to simple lattice vibrational models, *Rev. Geophys. Space Phys.* **17**, 1–19.

Kieffer, S. W. (1980) Thermodynamics and lattice vibrations of minerals: 4. Appliccation to chain and sheet silicates and orthosilicates, *Rev. Geophys. Space Phys.* **18**, 862–886.

Kumazawa, M., and Anderson, O. L. (1969) Elastic moduli, pressure derivatives, and temperature derivatives of single-crystal olivine and single-crystal forsterite, *J. Geophys. Res.* **74**, 5961–5972.

Levien, L., and Prewitt, C. T. (1981) High-pressure structural study of diopside, *Amer. Mineral.* **66**, 315–323.

Levien, L., Weidner, D. J., and Prewitt, C. T. (1979) Elasticity of diopside, *Phys. Chem. Minerals* **4**, 105–133.

Liu, L. (1977) Mineralogy and chemistry of the earth's mantle above 1000 km, Geophys. *J. Roy. Astron. Soc.* **48**, 53–62.

Liu, L. (1979) The high-pressure phase transformations of monticellite and implications for upper mantle mineralogy, *Phys. Earth Planet. Int.* **20**, P25–P29.

Matsui, M., and Busing, W. R. (1984a) Calculation of the elastic constants and high-pressure properties of diopside, $CaMgSi_2O_6$, *Amer. Mineral.* **69**, 1090–1095.

Matsui, M., and Busing, W. R. (1984b) Computational modeling of the structure and elastic constants of the olivine and spinel forms of Mg_2SiO_4, *Phys. Chem. Minerals* **11**, 55–59.

Matsui, M., and Matsumoto, T. (1982) An interatomic potential function model for Mg, Ca and CaMg olivines, *Acta Crystallogr.* **A38**, 513–515.

Matsui, Y. (1982) Computer simulation of structures of actual and hypothetical silicate crystals, in *Collected Papers on "The Materials Science of the Earth's Interior,"* pp. 11–17, edited by I. Sunagawa and K. Aoki. Tohoku University, Sendai, Japan.

Miyamoto, M., and Takeda, H. (1984) An attempt to simulate high pressure structures of Mg-silicates by an energy minimization method, *Amer. Mineral.* **69**, 711–718.

Mizukami, S., Ohtani, A., and Kawai, N. (1975) High-pressure X-ray diffraction studies on β- and γ-Mg_2SiO_4, *Phys. Earth Planet. Int.* **10**, 177–182.

Orr, R. L. (1953) High temperature heat contents of magnesium orthosilicate and ferrous orthosilicate, *J. Amer. Chem. Soc.* **75**, 528–529.

Parker, S. C., Catlow, C. R. A., and Cormack, A. N. (1984) Structure prediction of silicate minerals using energy minimization techniques, *Acta Crystallogr.* **B40**, 200–208.

Price, G. D., and Parker, S. C. (1984) Computer simulations of the structural and physical properties of the olivine and spinel polymorphs of Mg_2SiO_4, *Phys. Chem. Minerals* **10**, 209–216.

Ringwood, A. E. (1975) *Composition and Petrology of the Earth's Mantle.* McGraw-Hill Book Company, New York.

Sasaki, S., Prewitt, C. T., Sato, Y., and Ito, E. (1982) Single-crystal X-ray study of γ-Mg_2SiO_4, *J. Geophys. Res.* **87**, 7829–7832.

Schloessin, H. H., and Timco, G. W. (1977) The significance of ferroelectric phase transitions for the earth and planetary interiors, *Phys. Earth Planet. Int.* **14**, P6–P12.

Tamada, O. (1980) Electrostatic energies of polymorphs of M_2SiO_4 stoichiometry (M = Ni, Mg, Co, Fe and Mn), *Mineral. J.* **10**, 71–83.

Tosi, M. P. (1964) Cohesion of ionic solids in the Born model. *Solid State Phys.* **16**, 1–113.

Watanabe, H. (1982) Thermochemical properties of synthetic high pressure compounds relevant to the Earth's mantle, in *High Pressure Research in Geophysics*, pp. 441–464, edited by S. Akimoto and M. H. Manghnani. Center for Academic Publications, Tokyo.

Weidner, D. J., Sawamoto, H., Sasaki, S., and Kumazawa, M. (1984) Single crystal elastic properties of the spinel phase of Mg_2SiO_4, *J. Geophys. Res.* **89**, 7852–7860.

Yagi, T., Bell, P. M., and Mao, H. K. (1979) Phase relations in the system MgO–FeO–SiO_2 between 150 and 700 kbar at 1000°C, *Carnegie Inst. Wash. Yearb.* **78**, 614–618.

Yagi, T., Mao, H. K., and Bell, P. M. (1982) Hydrostatic compression of pervskite-type $MgSiO_3$, in *Advances in Physical Geochemistry*, pp. 317–325, edited by S. K. Saxena. Springer-Verlag, New York.

Chapter 7
Mantle Model Based on Measured Physical Properties of Minerals

Donald J. Weidner

Introduction

The chemical composition of the earth's mantle has evaded five decades of probing by earth scientist. Questions remain as to whether the mantle is chemically layered and, if it is chemically layered, as to the scale and nature of the layering. The debate between pyrolite and eclogite, although often appearing to be settled, still surfaces with a renewed vigor. In their recent discussions, Bass and Anderson (1984) propose a layered upper mantle with both a pyrolite and an eclogite layer. The apparant disarray in defining the "bottom line" stands in contrast to the steady progress that has been made in many separate fields, placing us at the door of understanding mantle chemistry.

In this chapter I shall illustrate the state of our knowledge and the fashion that this knowledge can be brought to bear on the question of mantle chemistry. I shall work primarily from a point of view that is looking to a homogeneous composition of the mantle. Once inhomogenieties are easily accepted, then all data can be rationalized. The seismic velocity at each depth can be the result of a chemical composition that is particular to that depth. Yet the problem is such that the composition responsible for an observed seismic velocity is not unique. It is more challanging to find a uniform composition that can claim responsibility for the vertical variation of all physical and chemical manifestations of the mantle chemistry. I will conclude that, within the uncertainty of the current data base, we need look no further than the pryolite model to find a chemical composition compatible with the data. This exploration is limited to the upper 650 km of the mantle because of the lack of data pertinant to the lower mantle phases.

The primary insight into properties of the earth's deep interior comes from seismology. Seismology gives the broadest sampling of the earth deeper than 5 km and the only coverage at depths greater than 200 km. Furthermore, the acoustic velocities, which are recovered information, have good depth resolution and are sensitive to both chemical composition and phase. This chapter relies on the seismic data to provide the constraints on the properties of the mantle. Our progress in understanding the mantle properties is facilitated by seismological studies, which have the greatest resolution in extracting the mantle velocities. Continued developement in full-wave body-wave analysis techniques along with separation of the source from path effects have been very fruitfull in this direction.

Recent advances in laboratory measurements have been extremely important in attaining our current state of understanding, and future advances hold the key to a more complete description of the earth's mantle. In particular, several high-pressure laboratories in Japan have provided samples of high-pressure phases of magnesium silicate compounds which are adequate for other studies. This has freed us from relying on analog studies because the actual material is available. These laboratories also provided a new level of detail in the phase equilibria of potential mantle compositions. The work of Akaogi and Akimoto (1979) exemplifies these studies for a pyrolite composition at conditions extending to the 650-km discontinuity. This type of study is continuing to greater pressures and temperatures and will yield a wealth of information concerning the lower mantle as well as the upper mantle.

Simultaneous to the developement of large volume high-pressure capability was the developement of Brillouin spectroscopy as a means to determine the single-crystal elastic properties of microcrystals. This technique afforded the measurement of elastic properties at ambient conditions of the high-pressure phases which were being synthesized. Notable are the data for the beta phase of Mg_2SiO_4 reported by Sawamoto *et al.* (1984), for the gamma phase by Weidner *et al.* (1984), for the ilmenite structure of $MgSiO_3$ by Weidner and Ito (1985), and for stishovite by Weidner *et al.* (1982). These data, which as yet are not available as a function of pressure and temperature, in conjunction with the phase equilibria data, allow us to create a seismic profile based on laboratory data. The models are also indebted to the broader and more accurate elasticity data for low-pressure phases which have been provided by ultrasonic acoustic measurements. For depths shallower than 650 km, we now have determined at least the acoustic properties of all major phases of the Mg, Fe, Al, Si system with the exception of majorite ($MgSiO_3$ garnet). Jeanloz (1981) has reported results of compression of majorite that yield an estimate of the bulk modulus. The lack of the shear modulus, however, severely restricts our knowledge of either the compressional or the shear velocity. In summary, we now know the zero-pressure acoustic properties for at least 80% of the material of a pyrolite mantle to depths of 650 km. Our knowledge is much more restricted for a system including substantial Na. In addition, the same level of acoustic information is available for less than 50% of an eclogite transition zone.

The theme of this chapter is hypothesis testing. The hypothesis consists of those elements which are either poorly determined or unknown. The test stems from that set of information which is relatively well known. For pyrolite, the hypothesis includes chemical composition along with pressure and temperature derivatives of elastic properties of the high-pressure phases and the Earth's temperature. The test obtains data from the known elastic properties and seismic observations. Eclogite becomes more difficult to test in that more of the parameters (such as the role of Na on the phase equilibria and the acoustic properties of majorite) become part of the hypothesis. Furthermore, the elastic properties of the perovskite phase of $MgSiO_3$ are still poorly known, limiting the amendability of the lower mantle to this type of test.

Several investigators have examined the constraints that seismic data place on the chemical composition of the mantle. Green and Liebermann (1976) use analog elasticity data to address this question. Less *et al.* (1983) examined this question more recently but only had compression observations to constrain the properties of the beta and gamma phases. As these data have proved inaccurate, their conslusions are no longer applicable. Bass and Anderson (1984) did have access to the recent elasticity data for these phases. They concluded that pyrolite and piclogite did equally well in describing the first 400 km of the mantle but that overall piclogite gave a better description of the region from 200 to 650 km. Within the context of the assumptions made in this chapter we would arrive at the opposite conclusion. However, the differences mostly rest in the part of the model that must remain as part of the hypothesis. Specifically, the differences result from different assumptions concerning the phase equilibria for the pyroxene to garnet transformation, the elastic properties of the majorite phase, and the pressure and temperature derivatives of the elastic properties of the high-pressure phases. This chapter is a more detailed presentation than was given by Weidner (1985).

Mineralogical Models

The acoustic velocities of a mineral assemblage depend on the elastic properties of the individual phases and the volume percentage of each phase. The elastic properties of a phase, in turn, depend on the details of the chemical composition of the phase. Thus, in order to model the acoustic velocities as a function of depth for a particular chemical composition, it is first necessary to model the composition and volume of the stable phases as a function of pressure and temperature along the geotherm.

Several laboratories have been very active in providing detailed phase equilibria data for chemical systems appropriate for the earth's mantle. In this chapter I rely primarily on the data of Akimoto's laboratory to define the phase stability and cation partitioning among the phases. Although their data are not always in complete agreement with those of other laboratories, they have studied most of the systems in question and their data provide an internally consistent set.

The assumptions regarding chemical partitioning and phase stability can affect the deduced acoustic velocities in a number of ways. For example, different assumptions concerning the iron–magnesium partitioning among the the phases should produce very little direct variation in the acoustic velocities because the iron–magnesium ratio affects the acoustic velocities of all phases in roughly the same manner. However, the stability of the silica-rich garnet is quite sensitive to the iron content of the garnet. Hence, the volume percentage of garnet (Gar) present at a given pressure and temperature will depend on the iron–magnesium partitioning. The result will be a larger variation in the acoustic velocities. In general, uncertainties about the stability fields of various phases will appear as uncertainties in the depth at which gradients or discontinuities in seismic velocities occur.

Philosophy of Calculation

The phase chemistry corresponding to a particular pressure and temperature were calculated from appropriate phase equilibria and cation partition data. In general, the chemical system was defined by the five oxides: SiO_2, MgO, FeO, CaO, and Al_2O_3 (Na_2O was also included for some cases). Thus, if there are n coexisting phases, then $5n$ linear equations uniquely define the composition of each phase. These equations come from:

1. Bulk composition. The total amount of each cation summed over all phases must equal the composition of the system.
2. Stoichiometry. The cations are site specific in most phases, thereby fixing the ratio of linear combinations of the cations. For example,

$$\mathrm{Mg_{Ol}} + \mathrm{Fe_{Ol}} = 2\,\mathrm{Si_{Ol}}$$

   where $\mathrm{Mg_{Ol}}$ is the number of moles of magnesium in the olivine (Ol) phase.
3. Partition relations. The cations that share a crystallographic site are partitioned by different amounts $K_{(\mathrm{Ol/gar})}$, where into the different phases as defined by a partition coefficient such as

$$K_{(\mathrm{Ol/Gar})} = (\mathrm{Fe_{Ol}}/\mathrm{Mg_{Ol}})/(\mathrm{Fe_{Gar}}/\mathrm{Mg_{Gar}})$$

If the iron–magnesium ratio is known in one phase, i.e., $FM_{(\mathrm{Ol})}$ for the olivine phase, then a linear equation is defined by:

$$\mathrm{Fe_{Gar}} = \mathrm{Mg_{Gar}} FM_{(\mathrm{Ol})}/K_{(\mathrm{Ol/Gar})}$$

In some instances, the partition relations become very simple. In particular, if a chemical component is assumed not to exist in a phase, such as there being no iron allowed in stishovite (St), then the equation becomes:

$$\mathrm{Fe_{St}} = 0$$

For the condition of n coexisting phases, a system of $5n$ linear equations can be generated from these three types. The inversion of the equations then yields the composition and amount of each phase. In order for the results to be accurate, the partition coefficients must be known as a function of composition as well as pressure and temperature. Our analysis of the existing data for the partitioning of iron, magnesium, and calcium indicates that these partition coefficients are fairly independent of the state parameters.

The phases that are included in our model of the upper 650 km of the Earth's mantle include for the orthosilicates: olivine, the modified spinel structure (beta), the spinel structure (gamma), as well as orthopyroxene (Pyx) (Ca-poor-Pyx), clinopyroxene (Ca-Pyx), majorite [(Mg, Fe, Ca)SiO_3 in the garnet structure], garnet (pyrope–almandine–grossular solid solution), and stishovite. The majorite component is assumed to form a solid solution with the Py-Alm-Gro garnet component.

Data for Calculation

The important data for the above equations include the partition coefficients and the phase equilibria relations. The values used in these calculations are presented below.

Fe/Mg

The partition coeffecients for iron and magnesium among the various phases have been determined as a function of pressure by Akaogi and Akimoto (1979) for a garnet lherzolite. Table 1 is a summary of their Table V and gives the partition coefficients between the various phases. Overall, the resulting partition coefficients appear relatively insensitive to pressure.

While partition coefficients are difficult to measure and may depend on the state of the system, it is instructive to compare coefficients determined in different manners in order to ascess internal consistency. Table 1 also includes the

Table 1. Observed iron–magnesium partition coefficients from Akaogi and Akimoto (1979). K(A/B is $(Fe_A/Mg_A)/Fe_B/Mg_B$ where Fe_A is the iron content of phase A.

	Pressure (kb)							
	45	45	75	75	144	146	140	Av
K(Ol/Ca-poor-pyx)	1.16	1.09	1.28	1.32				1.21
K(Ol/Ca-pyx)	1.10	1.05	1.13	1.17				1.11
K(Ol/Gar)	.45	.39	.41	.42				.42
K(β/Ca-poor-pyx)					1.89	2.13		2.01
K(β/Gar)					.69	.67		.68
K(γ/Gar)							1.40	
K(CA-poor-pyx/Gar	0.39*	0.36*	0.32*	0.32*	0.36**	0.32**		
K(Ol/β)(Ca-poor-pyx)					.64	.57		
K(Ol/β)(Gar)					.61	.62		
K(Ol/G)(Gar)							.30	

* Orthopyroxene
** Clinopyroxene

partition coefficients among the orthosilicate phases which are deduced by the relation:

$$K_{(A/B)} = K_{(A/C)}/K_{(B/C)}$$

with C being either the Ca-poor-pyxroxene or garnet. The resulting $K_{(\text{Ol beta})}$ determinations are in excellent agreement for the four different values. If we evaluate the partition coefficients among these phases from the phase stability fields of Kawada (1977) we find that when the phases coexist, $K_{(\text{Ol/gamma})} =$ 0.13–0.20 and $KK_{(\text{Ol/beta})} = 0.35$–$0.40$. These results suggest more iron enrichment in the high-pressure phase. Akaogi and Akimoto (1979) describe the results of partitioning experiments between the orthosilicate phases and a pyrope–almandine garnet. They conclude that the partition coefficient, $K_{(\text{gamma/Gar})}$ doubles for high-iron compositions. The doubling would bring the $K_{(\text{gamma, Gar})}$ into agreement with the $K_{(\text{Ol/gamma})}$ results of Kawada. These observations suggest that the gamma solid solution between iron and magnesium may not be ideal. The discrepancy for the beta phase, although small, still exists and may require more detailed study. In the model used here, the distribution coefficients of Akaogi and Akimoto (1979) were used except for those among the orthosilicates. In this case the phase equilibria of Kawada (1977) were used.

The iron–magnesium model of the garnet phase is used to define the phase equilibrium of majorite with pyroxene. The constancy of the distribution coefficient with pressure between the pyroxenes and garnet suggests that the model used here is appropriate. This constancy is perhaps surprising in that the composition of the garnet is changing with pressure as the majorite component increases. Thus, the site for the iron and magnesium is changing with pressure. The observation may be fortuitous as Akaogi and Akimoto (1979) report that the Ca-poor-pyroxene transforms from the orthopyroxene phase to a clinopyroxene phase between the 75 kbar and the 144 kbar runs. Thus the activity of iron in the pyroxene phase may change.

Ca/(Fe + Mg)

The only phases in which calcium is an important component are Ca-pyroxene and garnet. For all other phases, we have assumed that no calcium is present. The Ca-pyroxens phase of Akaogi and Akimoto (1979) is deficient in calcium relative to diopside. We used the value for Ca/(Fe + Mg) of 0.67 for the pyroxene and 0.19 for the garnet phase. These are both in agreement with the results of Akaogi and Akimoto (1979) and appear to be relatively independent of pressure in the region of interest. If the phase assembledge does not contain Ca-poor-pyroxene or spinel plus stishovite, then the Ca/(Fe + Mg) ratio of the garnet is not fixed at this value.

Al/Si

We have assumed that aluminum is present only in the garnet phase and jadeite (when sodium is included in the system). The portion of majorite dissolved in the garnet phase is directly determined by the Al/Si ratio. We calculate the Al/Si ratio in the garnet by interpolating from the data for the end-member compositions given by Akaogi and Akimoto (1977). That is, we determine at a pressure, P, the appropriate ratio for a garnet in equilibrium with diopside (Di), $(Al/Si)_{Di}$; with enstatite (En), $(Al/Si)_{En}$; and with ferrosilite (Fs), $(Al/Si)_{Fs}$. The final Al/Si value for the garnet is given as the weighted average of these three values, where the weights are the respective fractions of Ca (Mg–Ca), and Fe in the garnet. In Fig. 1 is represented the Al/Si ratio calculated for the chemical system used by Akaogi and Akimoto (1979) along with their measured values at a few pressures. The agreement between the curve and the points demonstrates the internal consistency in the assumptions and the various high-pressure experiments. The curve is based on phase equilibria of the end-member compositions and the inferred partition coefficients of Fe, Mg, and Ca. The points are for the intermediate composition. Once Al/Si is determined for the garnet the volume fraction of the garnet results from mass balance calculations. The resulting volume is also demonstrated in this figure.

Bina and Wood (1984) demonstrate that the pyroxene–garnet phase equilibrium contributes to gradients in acoustic velocity as opposed to discontinuities because the transition is not sharp. Nonetheless, accurate pressure and temperature characterizations of this transition are necessary and continue to require more data.

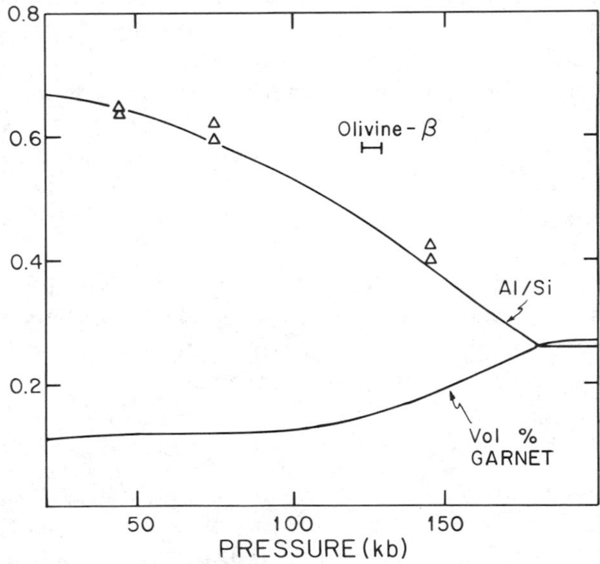

Fig. 1. The calculated (solid curves) and observed (triangles) Al/Si ratio for the garnet lherzolite studied by Akaogi and Akimoto (1979). The resulting volume fraction of garnet is indicated by the lower curve. For comparison, the width in pressure of the olivine to beta phase transformation is indicated.

Bass and Anderson (1984) assume a different description of the pyroxene–garnet phase boundary than the one used here. In their calculations, the Ca-poor-pyroxene completely dissolves into the garnet phase at or at least by the 400-km discontinuity. This lower pressure range for the phase transformation could result from either an overestimate of the pressure or from incomplete reactions in the experiments of Akaogi and Akimoto (1977, 1979). The former possibility is limited because a reduction in pressure for the entire reaction series should result in a considerable solution of pyroxene into the garnet at depths as shallow as 200 km. Yet xenoliths that sample this depth do not indicate this reaction. In addition, Akimoto and Akaogi (1979) report coexisting garnet, beta phase, and Ca-poor-pyroxene at a pressure of 145 kbar and garnet and gamma phase with the Ca-poor-pyroxene at an estimated 205 kbar. These experiments demonstate that the Ca-poor-pyroxene phase persists beyond the phase transformation associated with the 400 km discontinuity (olivine to beta phase). In fact, the higher pressure runs yielded detectable stishovite, which is the product, along with gamma, of the pyroxene breakdown when the system has excess pyroxene as compared to the capacity of the garnet phase. I therefore use the results of Akaogi and Akimoto in these calculations because of their internal consistency.

Orthosilicate Phases

The phase equilibria of Kawada (1977) are used to model the phase relations among olivine, the beta phase, and the gamma phase. Other determinations, such as by Suito (1977), differ in the pressure and temperature of some of the transitions, but they share a common topology of the phase relations. The differences will be reflected in the final model as different depths for a velocity discontinuity or steep gradient.

Phase Chemistry

The three types of equations outlined above are given appropriate coefficients throughout the data, which has been discussed. These remains only to define which equations are appropriate for a particular situation. Here we outline some of the possibilities and the appropriate equations.

Ol, Ca-poor-Pyx, Ca-Pyx, Gar

In this system there are a total of 20 unknowns. The 20 equations are as follows: five equations of type 1; four equations of type 2, which define Si/(Fe + Mg + Ca) for each phase; four equations that define the Ca/(Fe + Mg + Ca) for each phase; four equations that define the Al/Si for each phase; and three equations that define the Fe/Mg for each phase. In this case, the Fe/Mg of the most abundant phase is initially assumed to be equal to the Fe/Mg of the system. Then the Fe/Mg of each of the other phases is calculated from the partition coefficients. After the full set of equations is inverted, a new value of Fe/Mg of the most abundant phase is deduced and the process is iterated.

Ol, Beta, Ca-poor-Pyx, Ca-Pyx, Gar

This system requires five more equations than the above system. They include all of those above plus one equation for the Ca content in beta phase, one for Al content in the beta phase, and one for the Si/(Fe + Mg + Ca) ratio of the beta phase. In addition the Fe/Mg ratio of both the beta phase and olivine phase are defined by the phase equilibrium loop. Thus, the iron content of each phase is uniquely defined by the coexistence of olivine and the beta phase. With these new equations, the total is the requisite 25 and, furthermore, no iteration is necessary.

Gamma, Ca-Pyx, Gar

This case results if all of the Ca-poor-pyx is transformed to the majorite phase without any excess transforming to gamma plus stishovite. Here the results are defined by five equations of type 1, three equations that define Si/(Fe + Mg + Ca) of the three phases, three equations that define the Al/Si content of each phase, but only two equations defining the Ca/(Fe + Mg + Ca) content and two equations defining the Fe/Mg ratio. The Ca content of the garnet is no longer constrained once a Ca-poor-pyroxene or its high-pressure equivalent is no longer available. In addition, the Ca content of the garnet phase is not defined if there is no Ca-pyroxene in the system or if there is neither Ca-poor-pyroxene nor an orthosilicate phase. Furthermore, the iron content of one phase of this system must be defined through iteration as in the first case.

Ol, Beta, Gamma, Ca-poor-Pyx, Ca-pyx, Gar

This system represents a univarient phase boundary with all three orthosilicates coexisting. As a result the relative amounts of the phases is not fixed and sufficient equations cannot be defined. However, this condition will occur at most at a unique depth.

Ca-poor-Pyx, Ca-Pyx, Gar

This composition of this system is unique in that the Si/(Fe + Mg + Ca) of each phase is unity. Thus, the bulk composition must also be the same. As a result, it is redundant to specify the bulk composition and three equations giving Si/(Fe + Mg + Ca) for each phase. One must merely eliminate one of these equations to have a nonsingular system of equations.

The phase chemistry as a function of depth can be calculated using the data and model as outlined here for a wide variety of model compositions. Uning the composition of Akaogi and Akimoto's (1979) garnet lherzolite the Al/Si ratio of the garnet was calculated as illustrated in Fig. 1 and the Fe/Mg ratio of the various phases as a function of pressure illustrated in Fig. 2. For comparison,

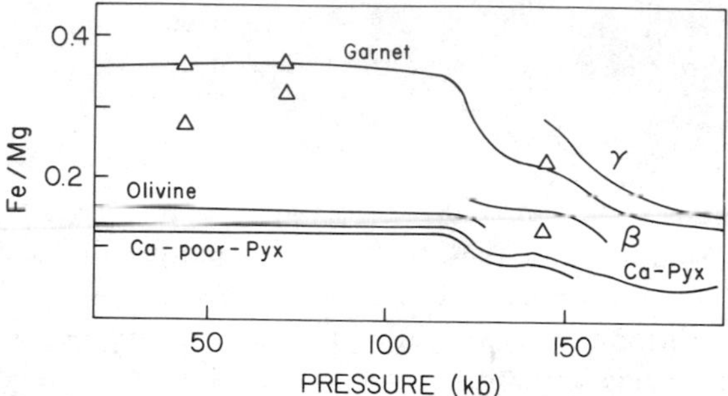

Fig. 2. The calculated (solid curves) iron to magnesium ratio for the various phases present in the garnet lherzolite system studied by Akaogi and Akimoto (1979). The triangles indicate the measured values reported by Akaogi and Akimoto (1979) for the garnet phase at the different pressures.

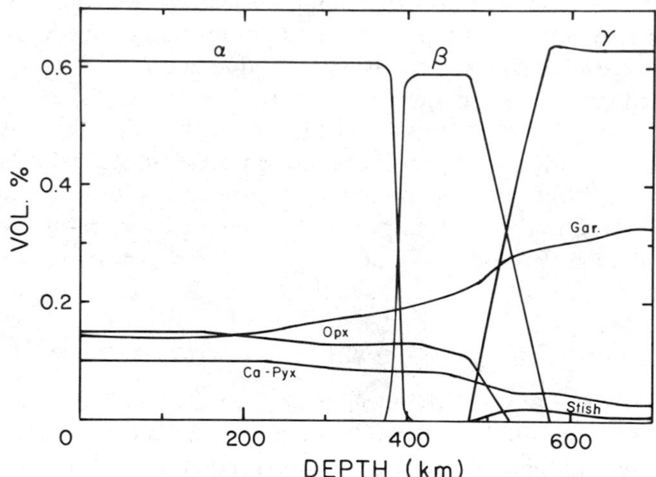

Fig. 3. The calculated volume fraction of each phase for the pyrolite system defined in Table 2 as a function of depth.

the garnet Fe/Mg ratios measured by Akaogi and Akimoto (1979) are illustrated. The Fe/Mg ratios for the other phase were less variable than for the garnet. I therefore conclude that the model reproduces the observations quite well. Figure 3 demonstrates the volume fractions of each phase as a function of depth for a pyrolite composition as taken from Ringwood (1975). In this model the garnet grows slowly over the depth range of 200–600 km. Excess Ca-poor-pyroxene transforms to gamma plus stishovite at about 500-km depth. Varying the aluminum content of the system will have significant affects on the fate of the pyroxene phase. If the Al content increases, all of the pyroxene will dissolve into the garnet with a larger absolute growth of the garnet phase at shallower depths. For contrast, a chemistry resembling Bass and Anderson's (1984) "piclogite" is illustrated in Fig. 4. In this example, Ca-pyroxene is used to represent both the diopside phase and the jadeite portion of their model, as they have assumed that both have the same phase boundary and as no data are available to constrain the jadeite–garnet phase equilibrium. Because of the increased Al content, the garnet phase grows more rapidly until all of the Ca-poor-pyroxene is dissolved into the garnet. The chemical compositions of both systems are given in Table 2.

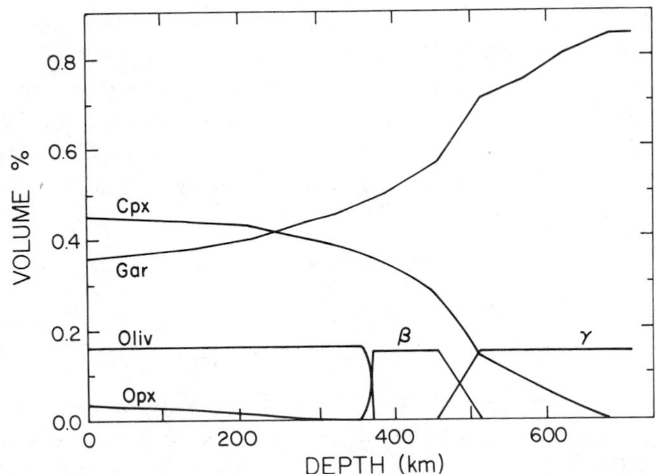

Fig. 4. The calculated volume fraction of each phase for the piclogite system defined in Table 2 as a function of depth.

Table 2. Chemical composition of standard models used in this paper. The oxide amounts are given in weight percent and the low pressure phases are given as volume percents. Pyrolite is patterned after Ringwood (1975) and Piclogite is after Bass and Anderson (1984).

	Pyrolite	Piclogite
MgO	40.3	21.0
FeO	7.9	5.7
CaO	3.0	7.0
SiO_2	45.2	48.9
Al_2O_3	3.5	14.4
Na_2O	0.0	3.0
Olivine	61	16
Orthopyroxene	15	3
Clinopyroxene	10	16
Garnet	14	36

Elastic Properties of Minerals

In order to predict the acoustic velocities as a function of depth, the elastic properties and density of each phase must be known as a function of pressure, temperature, and composition. Furthermore, the pressure–temperature relationship along the geotherm needs to be defined. In this chapter I will project the acoustic velocities along adiabats. If the Earth's interior is adiabatic, as we might expect in the core of a convection cell, then the curves will be accurate and the temperature of the adiabat's foot will completely define the temperature in this region. Departures from adiabaticity will be reflected in differences in the gradients between the observed and calculated velocities. The shallower part of the mantle probably deviates the greatest from this assumption.

The thermodynamic path to project properties to the mantle consists of isobaric heating of the mineral assemblage to the temperature of the "foot" of the adiabat, T_f, and then adiabatically compressing the assemblage to the desired pressure, P. Then the density, $\rho(T, P)$, and bulk or shear modulus, $M(T, P)$, for zero pressure are given by:

$$\rho(T_f, 0) = \rho(T_0, 0)[1 - \alpha(T_f - T_0)]$$

where α is the thermal expansion, and

$$M(T_f, 0) = M(T_0, 0) + (\partial M/\partial T)_P(T_f - T_0)$$

These expressions contain only the linear temperature dependence. If more complete data were available, it would be desirable to use the nonlinear parameterization discussed by Suzuki *et al.* (1979a, b). Once the data are corrected to the appropriate zero-pressure temperature, then depth is manifest as increasing pressure along an adiabat. Using a Murnaghan equation of state, this becomes:

$$M(T, P) = M(T_f, 0) + M'P$$

$$P = \frac{K(T_f, 0)}{K'(T_f, 0)} \cdot \left[\left(\frac{\rho(T, P)}{\rho(T_f, 0)}\right)^{K'_0} - 1\right]$$

In these relations M' (K') refers to the adiabatic pressure derivative of the adiabatic elastic (bulk) modulus. The isothermal pressure derivative must be corrected by:

$$(\partial M/\partial P)_S = (\partial M/\partial P)_T + \gamma_{th}(\partial M/\partial T)_P T/K_S$$

where γ_{th} is the thermodynamic Grunheisen parameter. This correction is of the order of -0.1 to -0.3 for the various minerals.

Table 3 summarizes the elasticity data used in this study. It is assumed that composition affects the elastic moduli and molar volume linearly with mole percentage in a solid solution series and that the pressure and temperature derivatives are unaffected by changes in composition. These assumptions are required as more complete data do not yet exist. The values given in the table are for the magnesium end members along with the derivatives with respect to composition.

Table 3. Physical Properties for the phases used in this Study. Parentheses indicate that the value is estimated.

	Olivine	Beta	Gamma	Ca-poor Pyroxene	Ca-Pyroxene	Garnet	Stishovite
K_s	1.29[a]	1.74[f]	1.84[c]	1.08[i]	1.13[m]	1.75[q]	3.16[t]
$(\partial K_s/\partial P)_s$*	4.74[a]	(4.3)	(4.3)	(4.2)[j]	4.2[n]	4.2[p]	(4.3)
$(\partial K_s/\partial T)_p \cdot 10^{-3}$	−0.16	(−0.20)	(−0.20)	(−0.2)[j]	(−0.2)	−0.23[p]	(−0.20)
$\partial K_s/\partial Fe$	0.09[c]	0.15[c]	0.15[c]	−0.07[k]	(0.00)	0.01[q]	—
$\partial K_s/\partial Ca$	—	—	—	—	(0.00)	−0.06[q]	—
$\partial K_s/\partial Maj$	—	—	—	—	—	0.46[r]	—
μ	0.80[a]	1.14[f]	1.19[c]	0.76[i]	0.67[m]	0.90[q]	2.20[t]
$(\partial \mu/\partial P)_s$*	1.58[a]	(0.9)	(0.9)	(1.1)[j]	(1.10)	1.30[p]	(0.9)
$(\partial \mu/\partial T)_p \cdot 10^{-3}$	−0.14[b]	(−0.14)	(−0.14)	(−0.12)[j]	(−0.12)	−0.09[p]	(−0.14)
$\partial \mu/\partial Fe$	−0.30[c]	−0.41[c]	−0.41[c]	−0.24[k]	(−0.20)	0.08[q]	—
$\partial \mu/\partial Ca$	—	—	—	—	(0.00)	0.14[q]	—
$\partial \mu/\partial/Maj$	—	—	—	—	—	(0.00)	—
V	43.67[d]	40.52[d]	39.65[d]	31.33[d]	33.06[o]	113.15[d]	14.01[d]
$\alpha \cdot 10^{-4}$	0.25[e]	0.20[g]	0.20[h]	0.33[l]	0.33[v]	0.20[p]	0.14[u]
$\partial V/\partial Fe$	2.6[d]	2.7[d]	2.37[d]	1.63d	**[o]	0.03[d]	—
$\partial V/\partial Ca$	—	—	—	—	**[o]	11.92[s]	—
$\partial V/\partial Maj$	—	—	—	—	—	0.83[d]	—

* Used Grunheisen parameter from Sumino and Anderson (1984) to estimate adiabatic derivative.
[a] Graham and Barsch (1969).
[b] Sumino and Nishizawa (1977).
[c] Weidner, Sawamoto, Sasaki, and Kumazawa (1984).
[d] Jeanloz and Thompson (1983).
[e] Susuki (1975).
[f] Sawamoto, Weidner, Susaki, and Kumazawa (1984).
[g] Suzuki, Ohatini, and Kumazawa (1980).
[h] Suzuki, Ohatini, and Kumazawa (1979).
[i] Weidner, Wang, and Ito (1978).
[j] Frisillo and Barsch (1972).
[k] Bass and Weidner (1984).
[l] Suzuki (1975).
[m] Levien, Weidner, and Prewitt (1979).
[n] Levien and Prewitt (1981).
[o] Used a nonlinear relationship of Tutnock, Lindsley, and Grover (1973).
[p] Sumino and Anderson (1984).
[q] Leitner, Weidner, and Liebermann (1980).
[r] Jeanloz (1981).
[s] Novack and Gibbs (1971).
[t] Weidner, Bass, Ringwood, and Sinclair (1982).
[u] Ito, Kawada, and Akimoto (1974).
[v] Cameron, Sueno, Prewitt, and Papike (1973).

The quantities given in parantheses are required for the calculation but are estimated with little or no data. The values of these properties have to be considered as part of the hypothesis and not part of the test. Before a particular chemical model of the mantle is rejected, it is necessary to consider whether the model is consistent with the observed seismic properties within all reasonable values of these unknown parameters.

The pressure derivatives of the elastic moduli of the beta and gamma phases will have considerable affect on the amount of olivine allowed in the mantle. The value estimated for these properties are the values that allow a pyrolite model to be compatible with the seismic observations we used in this chapter. The values are reasonable in that they are equal to or smaller than the values for the olivine phase, yet the pressure derivative of the shear modulus is larger than is observed for aluminate spinels.

The data for the elastic properties of majorite include only volume compression measurements by Jeanloz (1981), which provide no estimate of the shear modulus. Thus, there is a very weak base for testing models that are dominated by this phase.

The elastic properties of the Ca-poor-pyroxene phase are assumed to be represented by the data for the orthopyroxene series even though the high-pressure phase reported by Akaogi and Akimoto (1979) is a clinopyroxene. Furthermore, the pressure derivative of this phase is estimated to be significantly lower than that reported by Frisillo and Barsch (1972). Their value is greater than is observed in other silicates and higher than the polycrystalline values reported by Christensen (1974). More recent results of Webb and Jackson (1985) indicate a large negative second pressure derivative, which may be responsible for reducing the effective pressure derivative for the pyroxene phases.

The elastic properties of the Ca-pyroxene are assumed to be those given by the diopside–hedenbergite solid solution series. In as much as the mantle phase should be calcium deficient, we might expect that the elastic properties to be shifted toward those of the enstatite–ferrosilite series. This would slightly increase the acoustic velocities of this phase. Owing to the lack of data, I did not make this adjustment.

The effect of iron on the elastic properties of the beta and gamma phase are estimated from the results of Akimoto (1972) and Liebermann (1975), who report polycrystalline acoustic data for Fe_2SiO_4 in the spinel phase. Even though their data did not agree well, the compositional dependence is reasonably well determined for the magnesium-rich end of the series. Because the beta and gamma phases have such similar structures and elastic properties, I have assumed the compositional derivatives to be the same for the two phases.

Results and Discussion

The last two sections provide us with the means and data to calculate the acoustic velocities as a function of depth for various chemical compositions. The test of a chemical model comes by comparing the predicted acoustic velocity curve with that observed from seismology. In making this comparison, we focus on three criteria. They include the absolute values of the velocities in the 300–400 km region of the mantle before the 400-km discontinuity. The second criteria concerns the velocity jump at the 400-km discontinuity. The third criteria is the match of the gradient in acoustic velocities in the 400–670 km region. These criteria require that the seismic model have very good resolution in the vicinity of the Earth's transition zone. Models based on free oscillation data, such as the PREM model of Dziewonski and Anderson (1981), do not have a great deal of resolving power in this region of the earth. Instead we choose to use velocity models based on body-wave full-wave-form inversion. While the resolving power in these models is difficult to assess, maximum information is extracted from the seismogram including ray parameter and amplitude. These observations are, therefore, sensitive to the structural details of the acoustic velocity. We will use the Earth models of Grand and Helmberger (1984) and Walck (1984) as a basis to compare with the results of our calculations.

Figure 5a illustrates these Earth models along with the pyrolite model of Table 2 where no corrections have been made of the effect of temperature and pressure on the elastic properties. Thus, these curves demonstrate only the effect of phase transformations on the mineralogical model. Even though the two sets of curves correspond to very different conditions, this comparison is instructive in that the two sets of curves represent the best determined properties corresponding to the two different sources. Making the pressure and temperature correction to either set of curves involves using the most poorly determined properties.

Figure 5b illustrates the effect of correcting the elastic moduli for pressure only and Fig. 5c shows the results of including both pressure and temperature. In this progression, we see the relative magnitude of the effects of pressure and temperature on the final model. This is especially important as these corrections are, for many phases, only estimates. The good agreement between the mineralogical and seismic models in Fig. 5c affords the conclusion that a homogeneous pyrolite-type composition of the upper mantle and transition zone is compatible with the seismic data. Owing to the inverse nature of the problem, we cannot conclude that this is a unique solution to the mantle composition.

Figure 5a allows us to evaluate the effect of phase transformations on the seismic velocities. At 400 km, the olivine phase transformation is responsible for a sharp discontinuity. The coincidence between the 400-km discontinuity and the olivine to beta phase transformation is a compelling reason to conclude that the discontinuity results from this phase transformation and not a chemical change. Bina and Wood (1984) illustrate that the pyroxene to garnet transformation cannot contribute significantly to the 400-km discontinuity but instead should cause a change in the velocity gradient. By comparison to the seismic discontinuity, the mineralogical model has a considerably larger jump in velocity at this depth. In order to make the magnitude of the discontinuity similar in the two models, two possibilities are available. The first is to reduce the amount of olivine in the system to about 40 vol % from the 62 vol % of the pyrolite model. In such a substitution, we must take care that the material which substitutes for the olivine does not also contribute to a velocity jump. The second possibility is that the velocity jump which has been measured in the laboratory at ambient

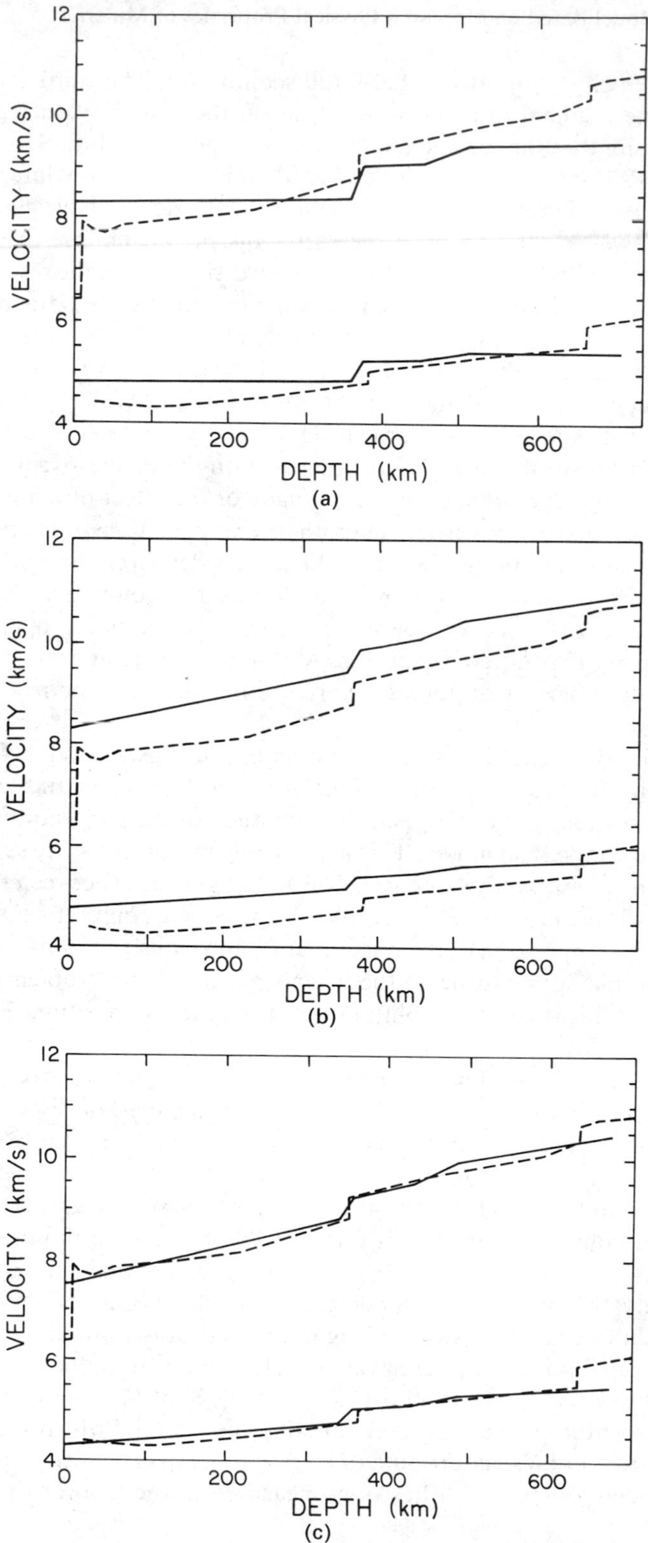
VELOCITY (km/s)
DEPTH (km)
(a)
VELOCITY (km/s)
DEPTH (km)
(b)
VELOCITY (km/s)
DEPTH (km)
(c)

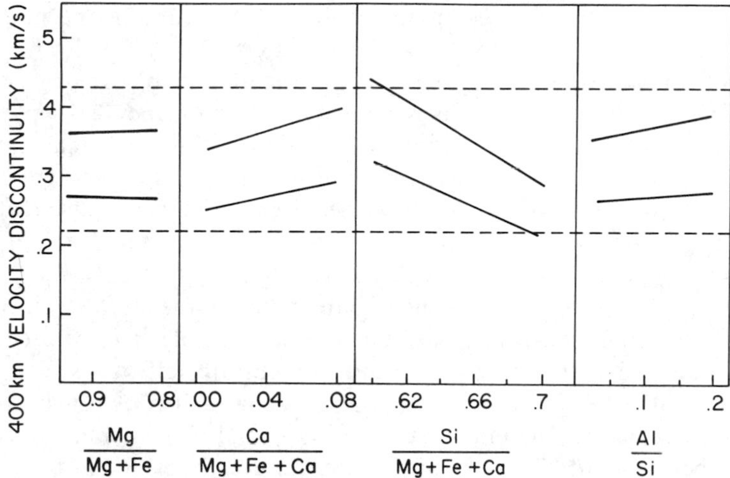

Fig. 6. The effect of chemical variations on the calculated velocity jump at 400 km. The dashed lines correspond to the seismic model. The values are calculated by varying one chemical component for the pyrolite model of Table 2 while holding the others constant.

conditions is reduced by the pressure and temperature which exists at 400-km depth. We find that it is necessary to decrease the shear modulus pressure derivative for the high-pressure phase by 0.8–1.0 compared to that of the olivine phase, with no change in the pressure derivative of the bulk modulus or in the temperature derivative of either modulus. As discussed above, these properties for the high-pressure phase are not known. Furthermore, such a change is quite reasonable in that the pressure derivatives of the shear moduli in aluminate spinels are still smaller than is deduced for the beta phase. The observed pressure derivative for the shear velocity of fayalite spinel by Fukizawa and Kinoshita (1982) are in accord with the values used here.

The velocity jump at 400 km is fairly insensitive to the composition or the temperature of the system. Figure 6 illustrates the effect of chemical composition on the magnitude of the 400-km discontinuity. The compressional velocities correspond to the upper sets of curves, while the shear are given by the lower set. The dashed curves are from the seismic data and the solid curves are calcuated by varying the amount of each chemical component while holding the others at the same values, as for the pyrolite standard of Table 2. There is very

◁ *Fig. 5.* Comparison of the calculated acoustic velocities for a pyrolite mantle (solid curves) with those observed seismically (dashed curves). (a) The acoustic velocities with no correction to the elastic constants for the effects of pressure and temperature. These are the best known properties but cannot be compared to those observed in the earth. (b) The velocities corrected only for pressure. (c) The velocities corrected for both temperature and pressure.

little change in the calculated jump resulting from varying the magnesium number of the system from 0.8 to 0.95. The greatest effect comes from changing the silicon content which effectively changes the olivine to pyroxene ratio (0.5 corresponds to an olivine stochiometery and 1.0 corresponds to pyroxene). However, the total range of variation expressed in this figure is small compared to the effect of changing the assumed values of the pressure and temperature derivatives within a range that is reasonable. For example, simply equating the unknown pressure derivatives of the high-pressure phase to those of olivine yields velocity jumps which are 50% greater than those observed. Therefore, while chemical composition is second in importance to the unknown properties, the most important chemical quantity is the amount of olivine present in the system. If we assume that the pressure and temperature derivatives of the olivine phase are bounds for the derivatives of the beta phase, then we conclude that 40 vol % olivine is a lower bound on the composition of the material at this depth. On the other hand, 65–70 vol % olivine may be acceptable models.

The mineralogical model differs from the seismic model in the sharpness of the 400-km discontinuity; the seismic model indicates a sharper transition than the mineralogical model. We expect that the seismic data do not resolve the difference in the width of the transition as represented by the two curves. However, the mineralogical curve is possibly sharper depending on the details of the univarient reaction involving the coexistence of olivine, beta, and gamma. In the particular phase model and chemical composition chosen, this reaction is not significant in the discontinuity. A change in the phase boundary or chemical composition will result in a sharper 400-km discontinuity. Figure 7 illustrates the effect of changing the iron content on the model velocities. Models A through

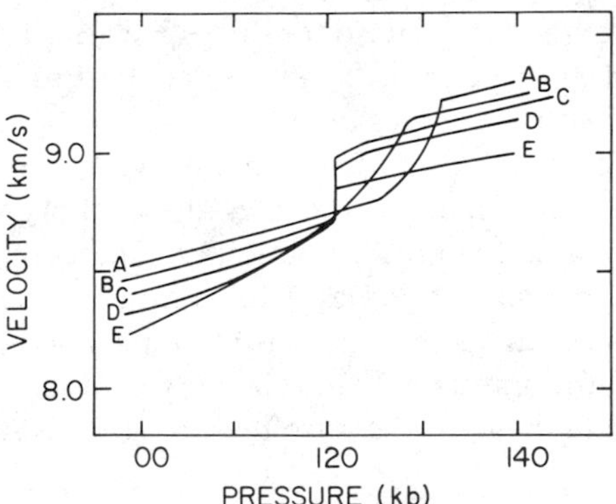

Fig. 7. The effect of varying iron content on the shape of the 400-km discontinuity. The change in sharpness of the discontinuity is a result of encountering the univarient reaction involving the coexistance of olivine, beta, and gamma. Model A is the pyrolite model of Table 2 with an Mg to Mg + Fe ratio of 0.90, the ratio for B is 0.86; for C, 0.82; for D, 0.77; and for E, it is 0.73. These changes in character of the transition may also occur with changes in temperature or in the changes in the location of the phase bounderies.

E reflect an increase in the iron content and the shape of the curves demonstrate a progressive increase in the significance of the univarient reaction on the sharpness of the discontinuity. Changes in the phase boundaries, which are within the uncertainty of the data, will cause similar effects on the sharpness of the discontinuities.

The role of the pyroxene to garnet phase transformation is also seen in Fig. 5a. The velocity gradient between 200 and 700 km is a result of this transformation. The slight jump at 550-km depth is a result of excess Ca-poor-pyroxene transforming to spinel plus stishovite. The position with depth of the velocity gradient is very sensitive to the pressure of this phase transition and the details of this phase transformation are still a matter of debate. As the solubility of the pyroxene phase in the garnet phase is dependent on the presence of aluminum, the depth of these high gradients is also sensitive to the aluminum content of the mantle. Figure 8 illustrates the effects of different amounts of aluminum in the mantle model. Higher aluminum contents (curve A) give rise to slightly steeper gradients, whereas lower aluminum amounts (curve B) correspond to greater increases at 550-km depth. Similar variations in the velocity curves will result from changing the pressure of the phase transformation. In fact, considerable changes in the phase boundary can be compensated by a change in the aluminum content.

The transition zone, while displaying smooth velocity changes, is marked by high velocity gradients. If these gradients arise from compression, the adiabatic pressure derivative of the system bulk modulus needs to be 5.6 and that for the shear modulus must be 2.5. Although these values are within the range of

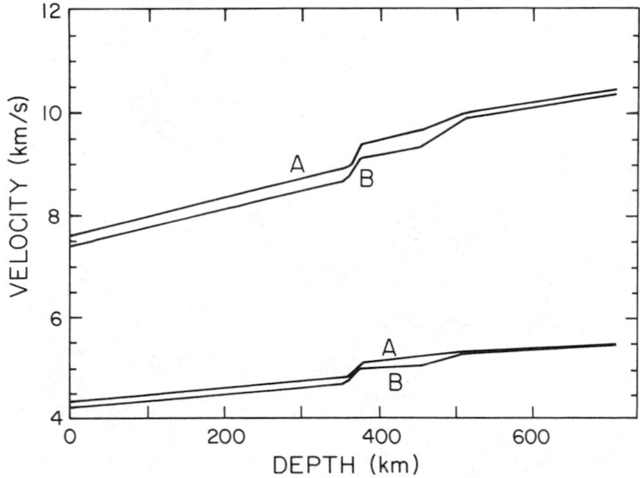

Fig. 8. The effect of changing the Al to Si ratio on the acoustic velocities of a pyrolite mantle. Curve A has an Al/Si ratio of 0.15, and B has a ratio of 0.04. The pyrolite model of Table 2 is intermediate between these two.

measured pressure derivatives, they are inconsistent with the need to lower the pressure derivatives for the beta phase. We therefore conclude that for a pyrolite model to account for both the magnitude of the velocity discontinuity at 400 km and the velocity gradient between 400 and 650 km, additional phase transformations must occur. In the present model, the pyroxene to garnet transformation and the pyroxene to spinel plus stishovite transformation fill this requirement. If the 400-km discontinuity caused by a chemical transformation, then the high velocity gradient could simply reflect large but reasonable pressure derivatives of the moduli for the material in this region.

The region between 300 and 400 km is perhaps the simplest region to evaluate. The seismic data do not indicate significant velocity gradients or discontinuities. Therefore, we may conclude that phase transformations do not play the dominant role in setting the acoustic velocities. At shallower depths we expect departures from adiabatic temperature gradients and possible other complications, such as the presence of partial melting. Within the assumptions of this chapter, the pyrolite model can replicate the observed seismic velocities. Of the three criteria outlined above, this one is the most sensitive to the iron to magnesium ratio of the mantle. Figure 9 illustrates the effect of changing chemical composition and temperature on the mantle velocities in this region. Again the dashed curves are from the seismic model and the solid curves are calculated. The velocity in this region is also sensitive to the aluminum content and, hence,

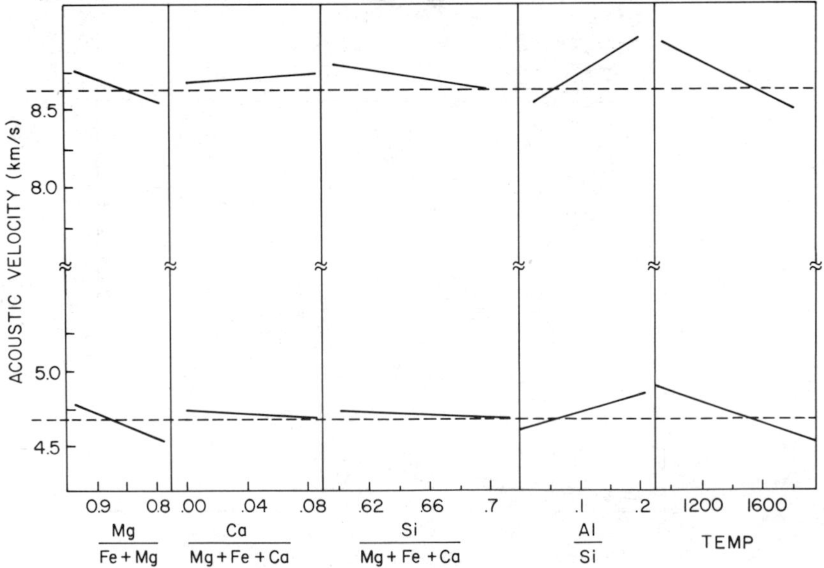

Fig. 9. The effect of compositional and temperature variations on the calculate acoustic velocities (solid curves) corresponding to 350-km depth. The dashed lines are from the seismic model. The chemical composition of the model is obtained by varying one component from the pyrolite model of Table 2 while holding the rest constant.

the garnet content. Furthermore, the fit of the system is very dependent on the temperature corresponding to the foot of the adiabat. Still, however, the uncertainties in the pressure and temperature derivatives of the elastic properties play an important role. If we use the pressure derivatives of Frisillo and Barsch (1972)—which are in excess of 9 for the bulk modulus and 2 for the shear modulus—for the pyroxene phases, then the compressional velocity at this depth would increase by more than 0.2 $km \cdot s^{-1}$ and the shear velocity by more than 0.1 $km \cdot s^{-1}$. To effect such changes chemically requires reasonably large perturbations.

Whereas this region is probably the most accessible to laboratory experiments there are phase equilibria uncertainties that significantly reduce the certainty of the conclusions. The pyroxene to garnet transformation may occur at lower pressures than those assumed here. This would increase the velocities in this region of the mantle unless a reduction in the aluminum content retarded the phase transformation. Changes in temperature and the iron to magnesium ratio may also be necessary. More drastic changes would involve the elimination of one or both of the pyroxene phases.

Figure 10 compares the calculated acoustic velocity for the piclogite model of Table 2 with the seismic velocities under the chemical and physical assumptions of this chapter. The jadeite component was simply added to the calcium-rich pyroxene to model the phase equilibria, and the elastic properties of the resultant pyroxene phase were an average of the appropriate properties of the two materials. The model does not match either the acoustic velocities at 350 km nor the velocity jump at 400 km very well. Small changes in the assumed physical properties or composition could help the match to the 350-km velocity. However, it appears difficult to match the jump at 400 km for the olivine to

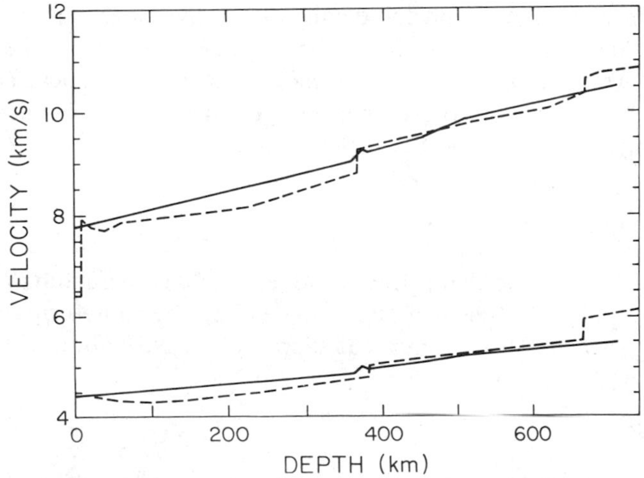

Fig. 10. The calculated (solid curve) acoustic velocities for the piclogite model of Table 2 and the velocities observed seismically (dashed curves).

beta transition cannot account for this jump because of the small amount of olivine in the system. If the pyroxene to garnet phase boundaries are sharper than assumed here, or if the seismic discontinuity is further smeared out than is indicated by the current models, then the discrepancies can perhaps be rationalized.

Conclusions

We do not yet have the necessary data to discreminate among a wide range of models for mantle chemistry. Circumstantial evidence still points to a homogeneous mantle. Two major seismic discontinuities are identified. The phase transformation from olivine to beta phase and the transformation from gamma to perovskite present two possibilities. Although ignorance of the properties of the perovskite phase preclude a complete analysis of this latter transformation, the suggestion still remains. Other phase transformations, including beta to gamma, spinel plus stishovite to ilmenite, and pyroxene to garnet, will not create seismic discontinuities. The high-pressure phases for the first two transformations have properties very similar to the low-pressure phases (Weidner *et al.*, 1984; Weidner and Ito, 1985). The latter transformation is too spread over pressure to cause a discontinuity. Thus, the form of the seismic velocity reflects the expected form given a pyrolite-type chemical model.

Within a reasonable range of assumptions, pyrolite produces an acoustic velocity model that agrees quantitatively with the seismically observed model. We find, however, that the calculated acoustic model is more sensitive to the unknown physical parameters than to variations in chemical composition. In particular, the pressure derivatives and temperature derivatives of the elastic moduli of the high-pressure phases, the elastic properties of the majorite phase and the perovskite phase, and the details of the pyroxene-garnet equilibrium remain as vital information for refining the chemical model. Once these variables are defined, then the insensitivity to small changes in chemical composition implies that we will be able to distinguish large differences in chemical composition but subtle changes will not be reflected in the seismic results.

Acknowledgments

The author would like to thank J. R. Weidner, R. C. Liebermann, V. Haniford, J. Kandelin, and Amir Yeganeh-Haeri for reading the manuscript and making valuable suggestions. This research was supported by NSF Grant EAR 8542755.

References

Akaogi, M., and Akimoto, S. (1979) High pressure phase equilibria in a garnet lherzolite, with special reference to Mg^{2+}–Fe^{2+} partitioning among constituent minerals, *Phys. Earth Planet. Inter.* **19**, 31–51.

Akaogi, M., and Akimoto, S. (1977) Pyroxene-garnet solid solution equilibria in the systems $Mg_2Si_4O_{12}$–$Mg_3Al_2Si_3O_{12}$ and $Fe_4Si_4O_{12}$–$Fe_3Al_2Si_3O_{12}$ at pressures and temperatures, *Phys. Earth Planet. Inter.* **15**, 90–106.

Akimoto, S. (1972) The system MgO–FeO–SiO_2 at high pressures and temperatures: phase equilibria and elastic properties, *Tectonophysics* **13**, 161–187.

Bass, D. J., and Anderson, D. L. (1984) Composition of the upper mantle: Geophysical tests of two petrological models, *Geophys. Res. Lett.* **11**, 237–240.

Bass, J. D., and Weidner, D. J. Elasticity of single-crystal orthoferrosilite, *J. Geophys. Res.* **89**, 4359–4371.

Bina, C. R., and Wood, B. J. (1984) The eclogite to garnetite transformation—experimental and thermodynamic constraints, *Geophys. Res. Lett.* **11**, 955–958.

Cameron, M., Sueno, S., Prewitt, C. T., and Papike, J. J. (1973) High-temperature crystal chemistry of acmite, diopside, hedenbergite, jadeite, spodumene and ureyite, *Amer. Mineral.* **58**, 594–618.

Christensen, N. I. (1974) Compressional wave velocities in possible mantle rocks to pressures of 30 kilobars, *J. Geophys. Res.* **79**, 407–412.

Dziewonski, A. M., and Anderson, D. L. (1981) Preliminary reference earth model, *Phys. Earth Planet. Inter.* **25**, 297–356.

Frisillo, A. L., and Barsch, G. R. (1972) Measurement of single-crystal elastic constants of bronzite as a function of pressure and temperature, *J. Geophys. Res.* **10**, 6360–6384.

Fukizawa, A., and Kinoshita, H. (1982) Shear wave velocity jump at the olivine-spinel transformation in Fe_2SiO_4 by ultrasonic measurements in situ, *J. Phys. Earth* **30**, 245–253.

Graham, E. K., and Barasch, G. R. (1969) Elastic constants of single-crystal forsterite as a function of temperature and pressure, *J. Geophys. Res.* **74**, 5949–5960.

Grand, S., and Helmberger, D. (1984) Upper-mantle shear structure of North America, *Geophys. J. Roy. Astronom Soc.* **76**, 399–438.

Green, D. H., and Liebermann, R. C. (1976) Phase equilibria and elastic properties of a pyrolite model for the oceanic upper mantle, *Tectonophysics* **32**, 61–92.

Ito, H., Kawanda, K., and Akimoto, S. (1974) Thermal expansion of stishovite, *Phys. Earth Planet. Int.* **8**, 277–281.

Jeanloz, R. (1981) Majorite: Vibrational and compressional properties of a high-pressure phase, *J. Geophys. Res.* **86**, 6171–6179.

Jeanloz, R., and Thompson, A. B. (1983) Phase Transitions and mantle discontinuities, *Rev. Geophys. Space Phys.* **21**, 51–74.

Kawanda, K. (1977) System $MgSiO_4$–Fe_2SiO_4 at High Pressures and Temperatures and the Earth's Interior. Ph.D thesis, University of Tokyo, 187 pp.

Lees, A., Bukowinski, M., and Jeanloz, R. (1983) Reflection properties of phase transition and compositional change models of the 670 km. discontinuity, *J. Geophys. Res.* **88**, 8145–8159.

Leitner, B. J., Weidner, D. J.; and Liebermann, R. C. (1980) Elasticity of single-crystal pyrope and implication for garnet solid solution series, *Phys. Earth Planet Int.* **22**, 111–121.

Liebermann, R. C. (1975) Elasticity of olivine (α), beta (β), and spinel (γ) polymorphs of germanates and silicates, *Geophys. J. Roy. Astronom Soc.* **42**, 899–929.

Levien, L., and Prewitt, C. T. (1981) High-pressure structural study of diopside, *Amer. Mineral.* **66**, 315–323.

Levien, L., Weidner, D. J., and Prewitt, C. T. (1979) Elasticity of diopside, *Phys. Chem. Minerals.* **4**, 105–113.

Novack, G. A., and Gibbs, G. V. (1971) The crystal chemistry of the silicate garnets,

Amer. Mineral. **56**, 791–825.

Ringwood, A. E. (1975) *Composition and Petrology of the Earth's Mantle*. McGraw-Hill Book Co., New York.

Sawamoto, H., Weidner, D. J., Sasaki, S., and Kumazawa, M. (1984) Single-crystal elastic properties of the modified spinel (beta) phase of magnesium orthosilicate, *Science* **224**, 749–751.

Suito, K. (1977) Phase relations of pure Mg_2SiO_4 up to 200 kilobars, in *High-Pressure Research*, edited by M. H. Manghnani and S. Akimoto, pp. 255–266. Academic Press, New York.

Sumino, Y., and Anderson, O. C. (1984) Elastic constants of minerals, in *CRC Handbook of Physical Properties of Rocks*, edited by R. S. Carmichael, pp. 39–138. CRC Press, Boca Raton Florida.

Sumino, Y., Nishizawa, O., Goto, T., and Ozima, M. (1977) Temperature variation of elastic constants of single-crystal forsterite between -190 and 400 C, *J. Phys. Earth* **25**, 377–392.

Suzuki, I. (1975) Cell parameters and linear thermal expansion coefficients of orthopyroxene, *Zisin (Japan)* **28**, 1–9.

Suzuki, I. (1975) Thermal expansion of periclase and olivine, and their anharmonic properties, *J. Phys. Earth* **23**, 145–149.

Suzuki, I., Ohatani, E., and Kumazawa, M. (1979a) Thermal expansion of γ-Mg_2SiO_4, *J. Phys. Earth* **27**, 53–61.

Suzuki, I., Okajima, S., and Seya, K. (1979) Thermal expansion of single-crystal manganosite, *J. Phys. Earth* **27**, 63–69.

Suzuki, I., Ohtani, E., and Kumazawa, M. (1980) Thermal expansion of modified spinel, β-Mg_2SiO_4, *J. Phys. Earth* **28**, 273–280.

Turnock, A. C., Lindsley, D. H., and Grover, J. E. (1973) Synthesis and unit cell parameters of Ca-Mg-Fe pyroxenes, *Amer. Mineral.* **58**, 50–59.

Walck, M. C. (1984) The p-wave upper mantle structure beneath an active spreading center: The Gulf of California, *Geophys. J. Roy. Astronom Soc.* **76**, 697–723.

Webb, S. L., and Jackson, I. (1985) The anomalous pressure dependence of the elastic moduli for single-crystal orthopyroxenes, *EOS* **66**, 371.

Weidner, D. J. (1985) A mineral physics test of a pyrolite mantle, *Geophys. Res. Lett.* in press.

Weidner, D. J., and Ito, E. (1985) Elasticity of $MgSiO_3$ in the ilmenite phase, *Phys. Earth Planet. Int.* in press.

Weidner, D. J., Wang, H., and Ito, J. (1978) Elasticity of orthoeustatite, *Phys. Earth Planet. Int.* **17**, 7.

Weidner, D. J., Bass, J. D., Ringwood, A. E., and Sinclair, W. (1982) The single-crystal elastic moduli of stishovite, *J. Geophys. Res.* **87**, 4747–4746.

Weidner, D. J., Sawamoto, H., Sasaki, S., and Kumazawa, M. (1984) Single-crystal elastic properties of the spinel phase of Mg_2SiO_4, *J. Geophys. Res.* **89**, 7582–7860.

Chapter 8
Reduction of Mantle and Core Properties to a Standard State by Adiabatic Decompression

Raymond Jeanloz and Elise Knittle

Introduction

The interpretation of geophysical data provides the only direct information about the constitution and thermal state of the Earth's interior. The most detailed results on the internal structure consist of seismological determinations of the average density and elastic moduli as functions of depth. The interpretation of this internal structure is difficult, however, because no direct way is known for inferring mineral structures and compositions uniquely from the geophysical observations. Instead, experimentally measured densities and elastic moduli of minerals are compared with the seismological data in order to characterize the state of the interior.

Values of the density, bulk modulus, and shear modulus, derived from the seismologically measured longitudinal and transverse wave velocities (V_p and V_s), are now well determined throughout the mantle and core (Dziewonski and Anderson, 1981). Indeed, current research is focused less on refining these spherically averaged properties with depth (i.e., assuming that each depth interval is a homogeneous, isotropic shell), than in determining the anisotropy and lateral heterogeneity of properties through the crust and mantle (e.g., Masters *et al.*, 1982; Nakanishi and Anderson, 1984; Dziewonski, 1984; Woodhouse and Dziewonski, 1984; Tanimoto and Anderson, 1985). Consequently, we take the average seismological properties with depth as being well constrained, while recognizing that new developments, particularly in data acquisition, may refine these values slightly.

There are basically two ways of comparing laboratory data on minerals with seismological observations, both approaches having been pioneered by Birch (1952, 1964). First, one can extrapolate known mineral properties to the high pressures (P) and temperatures (T) existing within the Earth. This has the advantage that a direct comparison is made with seismological observations, and the uncertainties in the laboratory measurements can be propagated with little ambiguity in order to test the significance of the comparison. Alternatively, one can extract from the seismological data the properties that a given region within the Earth would have under laboratory conditions. The advantage of this second approach arises when so little is known about the properties of a mineral (e.g., the pressure or temperature dependencies of its elastic moduli) that there is more uncertainty in extrapolating those properties to elevated pressures and temperatures than in reducing the seismological data to zero-pressure conditions. Thus, compared with the first approach many more minerals, particularly high-pressure phases for which little is known of the physical properties, can be considered in interpreting the seismological data.

The second of the two approaches is the subject of this chapter. By extrapolating physical properties of the interior to zero pressure and comparing these with existing laboratory data, one can place constraints on the composition and thermal state of the core and mantle. The reduction of seismological properties to ambient conditions involves several assumptions, as detailed below, but it is equivalent to applying the thermodynamic concept of a standard state. In short, we define the standard state of a region within the Earth as being the zero-pressure ($P = 0$) condition achieved upon adiabatic decompression. Because the temperature change with adiabatic decompression is small by comparison with the high internal temperatures, the resulting properties correspond to elevated temperatures as well as to zero pressure. For each region, T_{0S} is defined as the adiabat temperature at zero pressure, and the variation of mineral properties between room temperature and T_{0S} must be specifically taken into account in our analysis.

Seismologically determined wave velocities and densities through the mantle and core are illustrated in Fig. 1. The problem is to extrapolate these properties to zero pressure with a high, and quantifiable, degree of reliability. As is shown below, the equation of state, the physical relation between elastic moduli and density as a function of pressure (i.e., depth), plays an important role in extracting meaningful results. For example, it may seem that the properties of the outer core can be reduced to $P = 0$ conditions with similar reliability as those of the lower mantle because the depth extent for the two regions is the same (actually, the pressure range spanned by the outer core exceeds that of the lower mantle by nearly a factor of 2; Fig. 1). What is found from a proper analysis based on a physically accurate equation of state, however, is that the reduced properties of the lower mantle are known with significantly more certainty than those of the outer core.

We proceed to show the utility of the standard-state approach, especially for determining the constitution of regions deep within the Earth. First, we summarize the assumptions involved in the empirical reduction of properties to zero pressure. Second, the specific equation of state formalism is outlined, and we describe both the results and the uncertainties for different regions of the interior. Finally, we compare the laboratory data currently available for minerals with

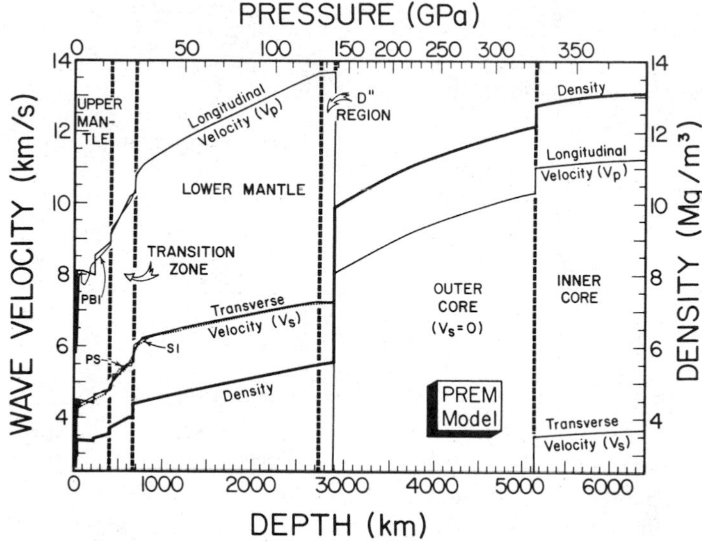

Fig. 1. Longitudinal (V_p) and transverse (V_s) wave velocities (left-hand scale), and density (bold curve, right-hand scale) as functions of depth through the Earth according to the seismological Preliminary Reference Earth Model (PREM) of Dziewonski and Anderson (1981). For comparison, velocity profiles based on waveform analysis are shown for the mantle (PB1: Burdick and Powell, 1980; S 1: Hart, 1975; PS: Fukao *et al.*, 1982), and the pressure as a function of depth is indicated on the upper scale. The main regions of the interior are labeled: upper mantle, transition zone, lower mantle, D″ region, outer core, and inner core.

the adiabatically decompressed properties of different regions within the Earth. As there has recently been considerable progress in measuring the properties of phases occurring deep in the mantle and core, we use our analysis to illustrate what is now known about the constitution of these regions.

Adiabatic Decompression: Assumptions

Because of the high temperatures within the Earth, pressure is almost exactly hydrostatic throughout the mantle and core, and nonhydrostatic stresses can be ignored in this study (Birch, 1952). Therefore, properties measured over a given pressure interval, corresponding to a depth interval within the mantle or core, can be extrapolated to zero pressure under the following conditions: (1) the region spanning that depth interval is homogeneous in bulk composition and in the mineral phases present; (2) the thermal state of that region is well defined (e.g., the average temperature varies adiabatically with depth); and (3) the extrapolation of the properties is based on an appropriate equation of state (Table 1). The first and second conditions are required so that the standard state

Table 1. Assumptions in reduction of seismological data.

Assumption	Severity
1. Homogeneity in composition and phase	
with depth	Major, must be tested
lateral	Minor, although quality of data may suffer
2. Adiabaticity	Minor
3. Equation of State	Fundamental uncertainity, but empirically justified

will be uniquely defined; that is, so that the derived properties at zero pressure are those of a specific mineral assemblage at a given temperature. The last condition merely involves the practical consideration that the extrapolation to zero pressure should be as reliable as possible.

The assumptions on which our present analysis is based can be separated into two kinds (Table 1): assumptions regarding a given region within the Earth (items 1 and 2) and assumptions regarding the equation of state to be used (item 3). The latter, which is fundamentally limited by the lack of a detailed knowledge of bonding forces in minerals, is discussed in the next section. We will use an equation of state that is thermodynamically applicable to a region characterized by an adiabatic gradient for the average temperature with depth (the geotherm). Thus, we now consider the first two assumptions in Table 1: homogeneity and adiabaticity of a region.

That large portions of the Earth's interior, such as the lower mantle or outer core, are adiabatic and homogeneous is physically plausible because there is good evidence that the interior is convecting vigorously (e.g., Stevenson and Turner, 1979, and Jeanloz and Morris, 1986, discuss convection in the mantle, and Merrill and McElhinny, 1983, discuss convection in the core). Therefore, large-scale mixing by convection would be expected to homogenize much of the mantle or core over the age of the Earth (see Hoffman and McKenzie, 1985). Also, convection at the high Rayleigh number characterizing the interior leads to an adiabatic geotherm through most of the mantle and core.

An important exception to this expected homogeneity arises wherever phase transitions occur. For example, upper mantle olivine transforms to the high-pressure β-phase and γ-spinel polymorphs at the pressures of the transition zone. Convective flow can penetrate through a region of phase transitions, but such a region is not homogeneous in the sense that is required here: although bulk composition may be constant, the variation of properties with pressure is affected (often dominated) by any phase transitions that occur. Therefore, an extrapolation to zero pressure from a region with transitions no longer yields properties of a well-defined assemblage of mineral phases.

Also, wherever there is a barrier to convection preventing vertical movement of material (e.g., the outer surface of the Earth and the core–mantle boundary), the geotherm is no longer adiabatic. That is, because vertical heat transfer must be by conduction in these boundary regions, the thermal boundary layers on each side of a convective barrier, large temperature gradients are required with depth in order to transfer the observed flux of heat from the interior (e.g., Jeanloz and Morris, 1986). For the mantle, thermal boundary layers are approximately 100 km thick, and these occur near the surface (corresponding to a thermal definition of the lithosphere), and possibly above the core–mantle boundary and at 670 km depth (Turcotte and Schubert, 1982; Jeanloz and Richter, 1979).

Thus, from general geophysical considerations the outer core and much of the lower mantle are expected to be homogeneous and adiabatic, on the average, whereas the transition zone is not. The upper mantle may be relatively homogeneous by comparison with the transition zone, but seismological and tectonic evidence of lateral heterogeneity (e.g., the pressure of subducted slabs), and petrological evidence for numerous mineral reactions and phase transformations occurring above the 400-km depth limit our confidence in the zero-pressure values derived from upper-mantle properties. Similarly, none of the depths at which thermal boundary layers may exist can be decompressed with confidence: approximately the top 100 km of the upper mantle and of the lower mantle, and the D″ region at the base of the mantle are unlikely to be adiabatic. Finally, it is uncertain whether the inner core is adiabatic or homogeneous. As little information can be derived from this region (see Interpretation), we do not consider its properties in detail.

In principle, deviations from homogeneity and adiabaticity can be documented seismologically at each depth by way of the inhomogeneity parameter (e.g., Bullen, 1975):

$$\eta = -\frac{\varphi}{\rho g}\frac{d\rho}{dr} \tag{1}$$

The seismic parameter,

$$\varphi = V_p^2 - (4/3)V_s^2 = K_s/\rho \tag{2}$$

is constrained independently of the density (ρ) and the derivative of density with radial distance from the Earth's center, $d\rho/dr$ (V_p and V_s are the longitudinal and transverse wave velocities respectively). This is done in modern seismological Earth models by combining travel-time measurements with determinations of the normal mode frequencies (e.g., Dziewonski and Anderson, 1981). The acceleration of gravity, g in Eq. (1), is likewise given by the density distribution with depth. Note that the seismic waves propagate under adiabatic conditions, such that it is the adiabatic bulk modulus (or incompressibility), K_s, that is derived from the wave velocities in Eq. (2).

The inhomogeneity parameter is simply the ratio of the average adiabatic bulk modulus to the effective bulk modulus given by the local pressure–density derivative in the Earth ($K_E = \rho dP/d\rho = -\rho^2 g\, dr/d\rho$):

$$\eta = K_s/K_E \tag{3}$$

Thus, the fact that η is close to 1 in the lower mantle and core (Masters, 1979; Dziewonski and Anderson, 1981) is consistent with the earlier conclusion that these regions are homogeneous and adiabatic, and hence are amenable to the present analysis. In contrast, parts of the upper mantle and the entire transition zone exhibit strong deviations from homogeneity (η ranging between -0.12 and 1.98, according to Dziewonski and Anderson, 1981).

Two types of seismologically observed heterogeneity should be distinguished (Table 1). On the one hand, considerable effort has recently gone into documenting lateral heterogeneity in the mantle. To the extent that this lateral heterogeneity reflects temperatures deviating from an average profile that is adiabatic (cf. Jeanloz and Morris, 1986), the heterogeneity can be ignored because it is averaged out. Similarly, small regions of differing compositions (e.g., subducted crust or sediments) may be so far below the limit of spatial resolution that they do not affect the average seismological properties at a given depth. Therefore, lateral heterogeneity is relatively unimportant and it is not considered in the present analysis.

Heterogeneity occuring with depth, on the other hand, can be significant. For example, we have already mentioned that phase changes or variations in composition are expected to occur with depth (e.g., in the transition zone); these can change the density and bulk modulus by several precent within a few tens of kilometers (Jeanloz and Thompson, 1983). It is these major variations with depth that cause η to deviate significantly from one in the transition zone and much of the upper mantle. In a few other regions, most notably the top and bottom 100–200 km of the lower mantle, anomalous gradients of velocity with depths indicate strong local deviations from homogeneity (e.g., Cleary, 1974; Burdick and Helmberger, 1978; Lay and Helmberger, 1983; Cormier, 1985). Although η appears to be close to 1 in these regions, this is only because the spatial resolution of density with depth is insufficient to document the local inhomogeneity (cf. Jordan and Anderson, 1974).

Aside from considerations of homogeneity, seismological profiles of velocity with depth have long been used to estimate how closely the average temperature follows an adiabat (e.g., Birch, 1952; Shankland and Brown, 1985). Although heterogeneity in phases or composition with depth strongly affects the velocity profiles, deviations from adiabaticity only produce small changes in the velocity profiles. The nonadiabatic temperature ($\Delta T = T_{\mathrm{E}} - T_S$) varies with depth ($z$) as

$$\frac{d\Delta T}{dz} = \frac{(\eta - 1)g\rho}{\alpha K_s} \tag{4}$$

according to Eq. (1)–(3) (cf. Birch, 1952). When appropriate values for typical mineral properties are used (see Interpretation and Table 4), Eq. (4) implies that deviations of 100% from the adiabatic gradient corresponds to η deviating from 1 by only 5–10%. This is a small deviation, comparable to the effects of velocity dispersion arising from anelasticity in the mantle (Heinz and Jeanloz, 1983). That is, because K_s and K_{E} in Eq. (3) correspond to values at seismological and geological frequences (~ 1 to 10^{-3} Hz and $\ll 10^{-10}$ Hz), respectively, they can differ by 1–5%. Thus, in our analysis of zero-pressure properties we will specifically look for the possible effects of dispersion.

The main effect of nonadiabatic temperature gradients at depth is to change the standard-state temperature for the decompressed properties. As the regions possibly containing thermal boundary layers have already been excluded from this analysis, the only plausible deviations from adiabaticity in the vigorously convecting mantle and core are toward a constant (isothermal) temperature profile (see Jeanloz and Richter, 1979; Jeanloz and Morris, 1986). Note that the extrapolated $P = 0$ temperature can only be higher than the adiabatic value, but this will be found to be a small offset at most. Therefore, because the equation of state formalism by which the internal properties are extrapolated to $P = 0$ is thermodynamically valid for either adiabatic or isothermal conditions (see next section), nonadiabatic conditions in the Earth will not affect the properties derived in the present study.

To summarize, the inhomogeneity parameter can be used to identify deviations of homogeneity over depth extents exceeding 200 km or so. Neither lateral heterogeneity nor deviations from adiabaticity are well resolved seismologically. This means that the reduction and analysis of seismologically derived properties is not sensitive to the relatively small deviations in adiabaticity and lateral homogeneity in the central regions (away from the thermal boundary layers) of the mantle and core.

Equation of State

Knowing the appropriate equation of state to use for extrapolating properties to zero pressure is equivalent to having solved the full quantum-mechanical problem of characterizing the bonding forces in the minerals (the equation of state is merely the derivative of the crystal energy as a function of atomic coordinates). Because this problem is unsolved, one often uses an empirically based equation of state. Following Birch (1938, 1947, 1952, 1978), the strain energy is expanded in a Taylor's series

$$U = af^0 + bf^1 + cf^2 + df^3 + \cdots \tag{5}$$

where f is an appropriate measure of finite deformation, and the coefficients $(a, b, c, d, \ldots)$ contain the inherent temperature dependence of U at a given strain. If U is taken to be internal energy or Helmoholtz free energy, the resulting equation of state is for an adiabat or isotherm, respectively. Note that in any case the first two terms (in f^0 and f^1) vanish because U and f can be set to zero for equilibrium (zero-pressure) conditions.

Equation (5) is purely phenomenological, but it gives the correct results as long as the strain is appropriately defined. As there is an infinite number of reference frames that can be used to describe finite deformations, the correct definition of strain is not clear a priori. Borrowing from the hydrodynamic language (cf. Fung, 1965), Birch and Murnaghan developed the finite-strain theory with Eulerian and Lagrangian frames of reference, respectively (these correspond to laboratory, or spatial, and material coordinates, respectively). Other reference frames could be derived by linear combinations of these two cases, for example.

Thus, the problem of finding the correct equation of state can be thought of as a matter of finding the appropriate frame of reference for defining the finite strain. As there is currently no insight provided by fundamental theory, we turn to experimental data. In particular, recent measurements strongly support Birch's choice of an Eulerian strain measure:

$$f = \frac{1}{2}\left[\left(\frac{\rho}{\rho_0}\right)^{2/3} - 1\right] \tag{6}$$

for isotropic compression, with strain being defined as positive on compression for the sake of convenience (ρ is the density, and subscript zero indicates zero-pressure conditions).

Differentiation Eq. (5) with respect to volume ($V = 1/\rho$) in order to get the pressure, the equation of state based on Eq. (6) becomes:

$$P = \frac{3}{2}K_0\left[\left(\frac{\rho}{\rho_0}\right)^{7/3} - \left(\frac{\rho}{\rho_0}\right)^{5/3}\right]\left\{1 - \left(\frac{3}{4}\right)(4 - K_0')\left[\left(\frac{\rho}{\rho_0}\right)^{2/3} - 1\right] + \cdots\right\} \tag{7}$$

in which K is the bulk modulus and a prime indicates a pressure derivative. Equation (7) is appropriate for hydrostatic compression of an isotropic sample (or cubic crystal), and it is derived such that $P = 0$ for $\rho = \rho_0$. As the second-order (f^2), third-order (f^3) and higher order terms in the expansion of energy with strain (Eq. 5) correspond to the first, second, and subsequent terms in the curly brackets of Eq. (7), the finite-strain equation of state is referred to as second-order, third-order, etc., depending on whether the first, second, etc., terms are included in the strain expansion.

For comparison, the Lagrangian finite-strain equation of state corresponding to Eq. (7) is:

$$P = -\tfrac{3}{2}K_0[(\rho/\rho_0)^{-1/3} - (\rho/\rho_0)^{1/3}]\{1 - \tfrac{3}{4}K_0'[(\rho/\rho_0)^{-2/3} - 1] + \cdots\} \tag{8}$$

Finally, we note that the Murnaghan equation of state, which is based on the assumption that the bulk modulus changes linearly with pressure ($K = K_0 + K_0'P$), is often used. The resulting equation of state,

$$P = \frac{K_0}{K_0'}[(\rho/\rho_0)^{K_0'} - 1] \tag{9}$$

will be compared with the Eulerian and Lagrangian forms. Because Eq. (9) is not consistent will classical elasticity theory, however, we do not hold much hope for its validity (cf. Birch, 1978).

The most direct test of an equation of state is to compare the pressure–volume relation that it predicts with compressions measured at high pressures. With recent advances in diamond cell techniques, it has become possible to make high-quality compression measurements to static pressures of 100 GPa (1 Mbar) or more. These data can be compared directly with isotherms derived from independently measured elastic moduli (e.g., from ultrasonic velocity or resonance measurements), via Eq. (7), (8), or (9).

In accord with Birch's (1977, 1978) work, we find that the available data systematically disagree with all but the Eulerian finite-strain equation of state: Fig. 2. Especially the results for the compressible alkali halides, which have been taken to $V/V_0 \sim 0.5$, along with the high-precision measurements of Au, provide strong support for Birch's formulation (Heinz and Jeanloz, 1984; Knittle and Jeanloz, 1984; Knittle *et al.*, 1985). The situation for MgO is more ambiguous because of the disagreement between the ultra-high-pressure static-compression data and the isotherm derived from shock-wave measurements (Mao and Bell, 1979). Nonhydrostatic effects can readily explain this discrepancy, with the static data being at slightly too high a pressure for a given compression. That is, the shock-wave isotherm may be more reliable than the static-compression results, in this instance (Jeanloz, 1981). In any case, the data taken as a whole support the use of the Eulerian finite-strain formalism, rather than any other equation of state. Therefore, we proceed with applying this approach to the Earth's interior.

Finite-Strain Analysis

From finite-strain theory (cf. Davies, 1973, 1974; Birch, 1978), the pressure, longitudinal velocity, and transverse velocity of an isotropic material under hydrostatic pressure are given to fourth order as a function of strain, f, by:

$$P = 3f(1 + 2f)^{5/2} K_0[1 - \pi_{01} f + \pi_{02} f^2 + \cdots] \tag{10}$$

$$\rho V_P^2 = (1 + 2f)^{5/2} K_0[\lambda_{00} - \lambda_{01} f + \lambda_{02} f^2 + \cdots] \tag{11}$$

$$\rho V_S^2 = (1 + 2f)^{5/2} \mu_0[1 - \sigma_{01} f + \sigma_{02} f^2 + \cdots] \tag{12}$$

where π, λ, and σ are compressional, longitudinal, and transverse moduli, respectively. Extension of the theory to nonisotropic symmetry is expected to yield results that are within the uncertainties of the present analysis (cf. Weaver, 1976; Birch,1978; Dziewonski and Anderson, 1981). We use the Eulerian definition of strain (Eq. 6), and the dimensionless parameters in Eq. (10)–(12) are

$$\pi_{01} = \frac{3}{2}(4 - K_0') \tag{13}$$

$$\pi_{02} = \frac{3}{2}\left[K_0 K_0'' + K_0'(K_0' - 7) + \frac{143}{9}\right] \tag{14}$$

$$\lambda_{00} = 1 + \frac{4\mu_0}{3K_0} \tag{15}$$

$$\lambda_{01} = 5\left(1 + \frac{4\mu_0}{3K_0}\right) - 3\left(K_0' + \frac{4}{3}\mu_0'\right) \tag{16}$$

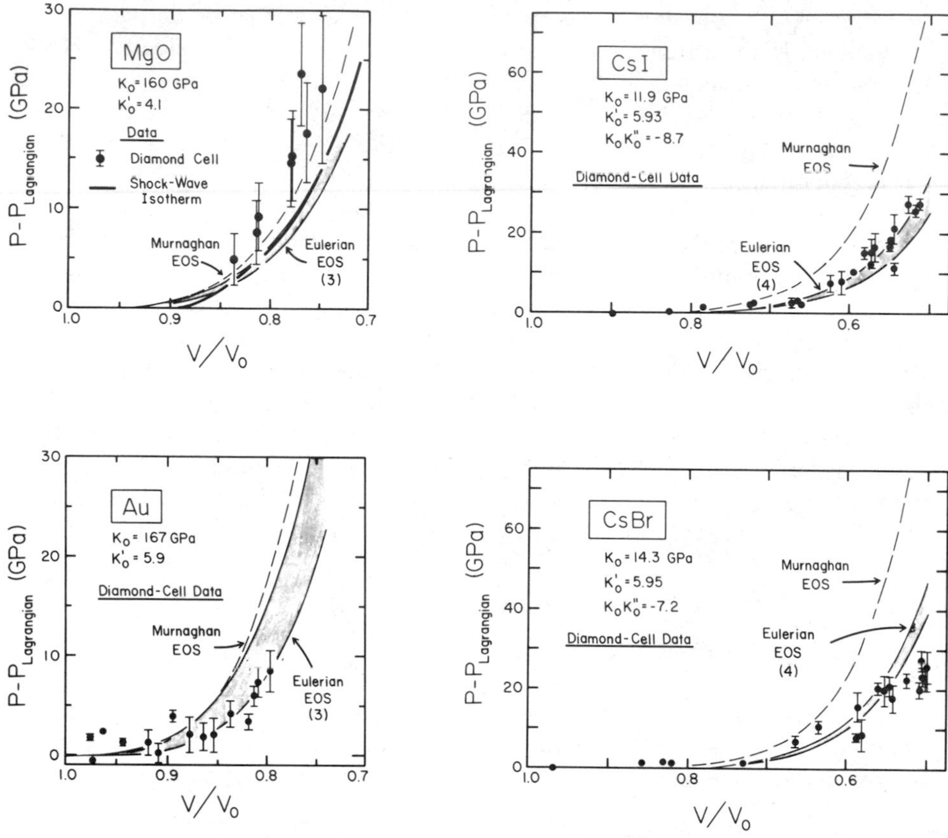

Fig. 2. Static-compression measurements from diamond-cell experiments (points with error bars) are shown as the observed pressure minus the pressure derived from a Lagrangian finite-strain equation of state (EOS) as a function of volume compression, V/V_0 (V_0 is the zero-pressure volume). Results are shown for MgO (upper left), Au (lower left), CsI (upper right), and CsBr (lower right), and it is evident that the data deviate systematically from the Lagrangian equations of state (note the differences in scale in each case). For comparison, the corresponding pressure differences for Eulerian finite-strain equations of state (shading indicates uncertainty) and Murnaghan equations of state (dashed) are shown as functions of volume compression. In all cases, the equations of state are based on the ultrasonically measured zero-pressure values of the bulk modulus and its pressure derivatives (K_0, K_0', K_0'' indicated for each substance), and the order of the finite-strain expansion used (third, fourth) is shown in parentheses. For MgO, the isotherm derived by Carter *et al.* (1971) from shock-wave data is included because of the possibility of biases arising from nonhydrostaticity in the diamond-cell results (see Jeanloz, 1981). The static-compression data are from Mao and Bell (1979), Heinz and Jeanloz (1984), Knittle and Jeanloz (1984), and Knittle *et al.* (1985); the ultrasonic data are from Jackson and Niesler (1982), Barsch and Chang (1971), and references given by Heinz and Jeanloz (1984).

$$\lambda_{02} = \frac{9}{2}\left[K_0 K_0'' + \frac{4}{3}K_0\mu_0'' + (K_0' - 4)\left(K_0' + \frac{4}{3}\mu_0'\right) + \frac{35}{9}\left(1 + \frac{4\mu_0}{3K_0}\right)\right] \tag{17}$$

$$\sigma_{01} = 5 - 3\frac{K_0}{\mu_0}\mu_0' \tag{18}$$

$$\sigma_{02} = \frac{9}{2}\left[\frac{K_0}{\mu_0}(K_0\mu_0'' + \mu_0'(K_0' - 4)) + \frac{35}{9}\right] \tag{19}$$

As before, primes indicate pressure derivatives, and subscript zeros indicate $P = 0$ conditions. Thus, Eq. (10) is equivalent to Eq. (7) to third order.

Ignoring for the moment possible effects of velocity dispersion (which are small in any case), the parameters of Eq. (13)–(19) are not all independent. As the bulk modulus can be derived either from the pressure–density relation (Eq. 10) or from a combination of the longitudinal and transeverse velocities (Eq. 11 and 12), the following must hold (cf. Eq. 2):

$$\pi_{01} = \tfrac{1}{2}[\lambda_{01} + \sigma_{01}(1 - \lambda_{00}) + 7] \tag{20}$$

$$\pi_{02} = \tfrac{1}{3}[\lambda_{02} + \sigma_{02}(1 - \lambda_{00}) + 9\pi_{01}] \tag{21}$$

Finally, note also that Poisson's ratio is given by

$$\nu_0 = \frac{3 - \lambda_{00}}{3 + \lambda_{00}} \tag{22}$$

at zero pressure.

The difficulty in applying these finite-strain equations to different regions of the mantle and core is that ρ_0 is not known, and hence the strain f cannot be calculated at each depth: it is, of course, the zero-pressure properties, ρ_0, K_0, μ_0, and those given in Eq. (13)–(19) that we aim to find. The way to do this has been outlined by Jeanloz (1981).

First, define an effective strain that is normalized to an arbitrary reference density, ρ^*:

$$g = \frac{1}{2}\left[\left(\frac{\rho}{\rho^*}\right)^{2/3} - 1\right] \tag{23}$$

Thus, with a length ratio $\beta = (\rho^*/\rho_0)^{1/3}$, the Eulerian finite-strain measure becomes:

$$f = (g + \tfrac{1}{2})\beta^2 - \tfrac{1}{2} \tag{24}$$

Equation (24) is substituted back into Eq. (10)–(12) to yield new expansions for the pressure and elastic-wave velocities in terms of the normalized strain, g. These new expansions can be rearranged in powers of g:

$$P = 3(1 + 2g)^{5/2}\beta^5 K_0 \left\{\left(\frac{\beta^2 - 1}{2}\right)\left[1 + \frac{\pi_{01}}{2}(1 - \beta^2) + \frac{\pi_{02}}{4}(1 - \beta^2)^2 + \cdots\right]\right.$$

$$+ \beta^2\left[1 + \pi_{01}(1 - \beta^2) + \frac{3\pi_{02}}{4}(1 - \beta^2)^2 + \cdots\right]g$$

$$- \beta^4\left[\pi_{01} + \frac{3\pi_{02}}{2}(1-\beta^2) + \cdots\right]g^2 + \beta^6[\pi_{02} + \cdots]g^3 + \cdots\Big\} \tag{25}$$

$$\rho V_P^2 = (1+2g)^{5/2}\beta^5 K_0\Big\{\left[\lambda_{00} + \frac{\lambda_{01}}{2}(1-\beta^2) + \frac{\lambda_{02}}{4}(1-\beta^2)^2 + \cdots\right]$$

$$- \beta^2[\lambda_{01} + \lambda_{02}(1-\beta^2) + \cdots]g + \beta^4[\lambda_{02} + \cdots]g^2 + \cdots\Big\} \tag{26}$$

$$\rho V_S^2 = (1+2g)^{5/2}\beta^5 \mu_0\Big\{\left[1 + \frac{\sigma_{01}}{2}(1-\beta^2) + \frac{\sigma_{02}}{4}(1-\beta^2)^2 + \cdots\right]$$

$$- \beta^2[\sigma_{01} + \sigma_{02}(1-\beta^2) + \cdots]g + \beta^4[\sigma_{02} + \cdots]g^2 + \cdots\Big\} \tag{27}$$

More conveniently, new functions of the pressure and velocities can be formulated as polynomials in g. From Eq. (25)–(27), these are:

$$G_\pi \equiv \frac{P}{3(1+2g)^{5/2}} = a_\pi + b_\pi g + c_\pi g^2 + d_\pi g^3 \tag{28}$$

$$G_\lambda \equiv \frac{\rho V_P^2}{(1+2g)^{5/2}} = a_\lambda + b_\lambda g + c_\lambda g^2 \tag{29}$$

$$G_\sigma \equiv \frac{\rho V_S^2}{(1+2g)^{5/2}} = a_\sigma + b_\sigma g + c_\sigma g^2 \tag{30}$$

with terms only up to fourth order being retained, so that the following relations hold:

$$a_\pi = \beta^5 K_0\left(\frac{\beta^2-1}{2}\right)\left[1 + \frac{\pi_{01}}{2}(1-\beta^2) + \frac{\pi_{02}}{4}(1-\beta^2)^2\right] \tag{31}$$

$$b_\pi = \beta^7 K_0\left[1 + \pi_{01}(1-\beta^2) + \frac{3\pi_{02}}{4}(1-\beta^2)^2\right] \tag{32}$$

$$c_\pi = -\beta^9 K_0\left[\pi_{01} + \frac{3\pi_{02}}{2}(1-\beta^2)\right] \tag{33}$$

$$d_\pi = \beta'' K_0 \pi_{02} \tag{34}$$

$$a_\lambda = \beta^5 K_0\left[\lambda_{00} + \frac{\lambda_{01}}{2}(1-\beta^2) + \frac{\lambda_{02}}{4}(1-\beta^2)^2\right] \tag{35}$$

$$b_\lambda = -\beta^7 K_0[\lambda_{01} + \lambda_{02}(1-\beta^2)] \tag{36}$$

$$c_\lambda = \beta^9 K_0 \lambda_{02} \tag{37}$$

$$a_\sigma = \beta^5 \mu_0\left[1 + \frac{\sigma_{01}}{2}(1-\beta^2) + \frac{\sigma_{02}}{4}(1-\beta^2)^2\right] \tag{38}$$

$$b_\sigma = -\beta^7 \mu_0[\sigma_{01} + \sigma_{02}(1-\beta^2)] \tag{39}$$

$$c_\sigma = \beta^9 \mu_0 \sigma_{02} \tag{40}$$

Note that the G of Eq. (28)–(30) have dimensions of pressure, and both these normalized pressures (G_π) and moduli (G_λ and G_σ) and the normalized strain (g) can be determined as a function of depth from the seismological models (e.g., Dziewonski and Anderson, 1981). The normalizing density, ρ^*, can be chosen to optimize convergence of the polynomial expansions, although the simple choice $\rho^* = 1\ \mathrm{Mg \cdot m^{-3}}$ is adequate for the Earth's interior: the strains involved, as well as the multiplying factors (Eq. 33–40) are sufficiently small in practice to assure rapid convergence of Eq. (28)–(30). Thus the normalized pressure G_π, longitudinal modulus G_λ, and shear modulus G_σ can be fitted as functions of g from the seismological properties of any adiabatic and homogeneous region. The resulting fit yields values for the length ratio β, the parameters of Eq. (33)–(40), and hence the zero-pressure density and elastic properties.

A simple way to carry this out is to first fit the observed values of G as a function of g, and find the strain (g_0) at which G_π vanishes (see Jeanloz, 1981). Consequently,

$$\rho_0 = \rho^*(1 + 2g_0)^{3/2} \tag{41}$$

when $G_\pi(g_0) = 0$, and the zero pressure velocities are given by $G_\lambda(g_0)$ and $G_\sigma(g_0)$:

$$V_{P0} = [(K_0 + \tfrac{4}{3}\mu_0)/\rho_0]^{1/2} = [\rho_0^{2/3} G_\lambda(g_0)/\rho^{*5/3}]^{1/2} \tag{42}$$

$$V_{S0} = [(\mu_0/\rho_0]^{1/2} = [\rho_0^{2/3} G_\sigma(g_0)/\rho^{*5/3}]^{1/2} \tag{43}$$

Once g_0 is found (and thus ρ_0 is known by Eq. 41), the moduli and their pressure derivatives are immediately given from G_π, G_λ, and G_σ as functions of the strain difference $g - g_0$. That is,

$$G_\pi = \left[\left(\frac{\rho^*}{\rho_0}\right)^{7/3} K_0\right](g - g_0) - \left[\left(\frac{\rho^*}{\rho_0}\right)^{3} K_0\pi_{01}\right](g - g_0)^2 + \left[\left(\frac{\rho^*}{\rho_0}\right)^{11/3} K_0\pi_{02}\right](g - g_0)^3 \tag{44}$$

$$G_\lambda = \left[\left(\frac{\rho^*}{\rho_0}\right)^{5/3} K_0\lambda_{00}\right] - \left[\left(\frac{\rho^*}{\rho_0}\right)^{7/3} K_0\lambda_{01}\right](g - g_0) + \left[\left(\frac{\rho^*}{\rho_0}\right)^{3} K_0\lambda_{02}\right](g - g_0)^2 \tag{45}$$

$$G_\sigma = \left[\left(\frac{\rho^*}{\rho_0}\right)^{5/3} \mu_0\right] - \left[\left(\frac{\rho^*}{\rho_0}\right)^{7/3} \mu_0\sigma_{01}\right](g - g_0) + \left[\left(\frac{\rho^*}{\rho_0}\right)^{3} \mu_0\sigma_{02}\right](g - g_0)^2 \tag{46}$$

These relations describe the fact that the zero-pressure elastic moduli and their pressure derivatives are given by the values of G_λ and G_σ, and by the derivatives of G_π, G_λ, and G_σ, at $g = g_0$. The three terms in each of Eqs. (44)–(46) correspond to second-, third- and fourth-order contributions to the equation of state, as defined above (i.e., the order contribution to the strain expansion of the energy, U).

Reduction of Seismological Data to Zero Pressure

In deriving zero-pressure values of density and elastic moduli from seismological models, several ambiguities or potential sources of bias must be addressed. We have already justified the consideration of isotropic moduli under hydrostatic pressures, and we have specifically addressed the assumptions of homogeneity and adiabaticity of the region in question, and the assumption that the Eulerian finite-strain equation of state is appropriate (Table 1). In addition, we must decide (1) which order of the finite-strain expansion is best (i.e., how to truncate the polynomial expansion so as not to underfit or overfit the data), (2) whether the smoothing or parameterization of the seismological model biases the analysis, and (3) whether anelastic dispersion affects the derived zero-pressure moduli.

A major difficulty in assessing the reliability of our zero-pressure values derives from the fact that there is no simple way to estimate the uncertainties in density and elastic wave velocity at each depth: the raw seismological data that are fitted in our analysis. This is because seismological models result from nonunique and nonlinear inversions of a wide range of observations, and there is a tradeoff between depth resolution and absolute uncertainty in each value (Backus and Gilbert, 1970). Based on the available analyses, however, one can make a preliminary estimate that the local value of density at each level within the mantle and core is known to about 0.5–2.5% (see Backus and Gilbert, 1970; Masters, 1979). Average values of density, integrated over a greater depth interval, are constrained much better, however (note that integral constraints, such as the moment of inertia, come into play).

We can use this information along with the finite-strain formalism to decide which terms in the equation of state are resolvable. For example, if $K_0' = 4$ for a given region, no amount of data containing any uncertainties can resolve a third-order term because π_{01} vanishes (cf. Eqs. 13 and 44). Given the range of strain for each region within the Earth and the estimated uncertainties in density, which correspond to roughly 2.5% uncertainty in pressure at a given strain, Jeanloz (1981) demonstrates that the following values of K_0' cannot be resolved (i.e., values in these ranges yield no resolvable third-order contribution to the equation of state): $3.55 < K_0' < 4.45$ for the upper mantle, $3.85 < K_0' < 4.15$ for the lower mantle, and $3.90 < K_0' < 4.10$ for the outer core. Similarly, the fourth-order contributions cannot be resolved for $K_0 K_0''$ close to -3.89 (Eqs. 14 and 44, and using values for K_0' derived below: Table 2). In the upper mantle, no reasonable value of $K_0 K_0''$ can be resolved, and the minimum range of values that are unresolved for the lower mantle and outer core are $-5.8 < K_0 K_0'' < -2.0$ and $-5.3 < K_0 K_0''$, -2.5, respectively.

The ranges of unresolvable values of K_0' and $K_0 K_0''$ for different regions of the Earth provide one criterion for deciding whether or not respectively third-or fourth-order terms are justifiably included in extrapolating properties to zero pressure. By comparing the zero-pressure values derived from fits to different orders of equation of state, the uncertainties resulting from truncation of the finite-strain polynomial can be evaluated. In addition, comparison of K_0 and its pressure derivatives derived from G_π with the values derived from the combination of G_λ and G_σ documents the internal consistency of the data and of the equation of state used. Discrepancies in the bulk modulus values mainly reflect biases in the parameterized seismological models: that is, biases in the assumed starting model for the data inversion, the use of polynomial expansions of density and velocity, and possibly the averaging out of inhomogeneities that are actually present. To a lesser extent, uncertainties in the truncation of the finite-strain expansion can be significant but, by comparison, dispersion will be found to contribute negligibly.

Results for the current reference Earth model, PREM (Dziewonski and Anderson, 1981), are shown in Figure 3 as normalized pressures, G, as functions of strain, g. For convenience, values of G_π, $G_\lambda/30$, and $G_\sigma/30$ are plotted in the figure, with $\rho^* = 1\ \mathrm{Mg \cdot m^{-3}}$ being used in defining g. Both third-order (dashed curves; $\pi_{02} = \lambda_{02} = 0$) and fourth-order (solid curves) fits of Eqs. (28)–(30) are shown for the upper mantle (220–400 km depth only), the lower mantle (770–2740 km depth), the outer core, and the inner core; the resulting properties at zero pressure (see Eqs. 31–40 and 13–19) are given in Table 2. Because they are not expected to be homogeneous and adiabatic, as discussed above, the D'' region, the transition zone, the upper 100 km of the lower mantle, and the outermost 220-km depth interval of the mantle are not included in the analysis (cf. Butler and Anderson, 1978; Dziewonski and Anderson, 1981).

The zero-pressure densities of each region can only be determined by extrapolating G_π to zero (the intercept defines g_0: Eq. 41). Except for the inner core, the values of ρ_0 derived from third- and fourth-order fits agree well: to better than 0.25% for the mantle and to better than 1% for the outer core. The present values are also in the range of other recent determinations (Davies and Dziewonski, 1975; Butler and Anderson, 1978; Stacey *et al.*, 1981). The 10% difference between the ρ_0 from the inner core fits reflect the short interval in G over which data are available and the long extrapolation required to obtain g_0 (Fig. 3). This degree of uncertainty might be unexpected if one only considers the large depth interval spanned by the inner core (Fig. 1). Similarly, the equation of state of the lower mantle is better constrained than that of the outer core, despite the greater pressure interval across the latter. Although the uncertainties in these fits are not evident from the density profile with depth, the present finite-strain analysis clearly illustrates the reliability with which zero-pressure properties can be estimated.

Once g_0, or the zero-pressure density, is determined, the zero-pressure elastic moduli and their pressure derivatives are given by the derivatives of G_π and by the intercepts and derivatives of G_λ and G_σ, all evaluated at g_0. As might be

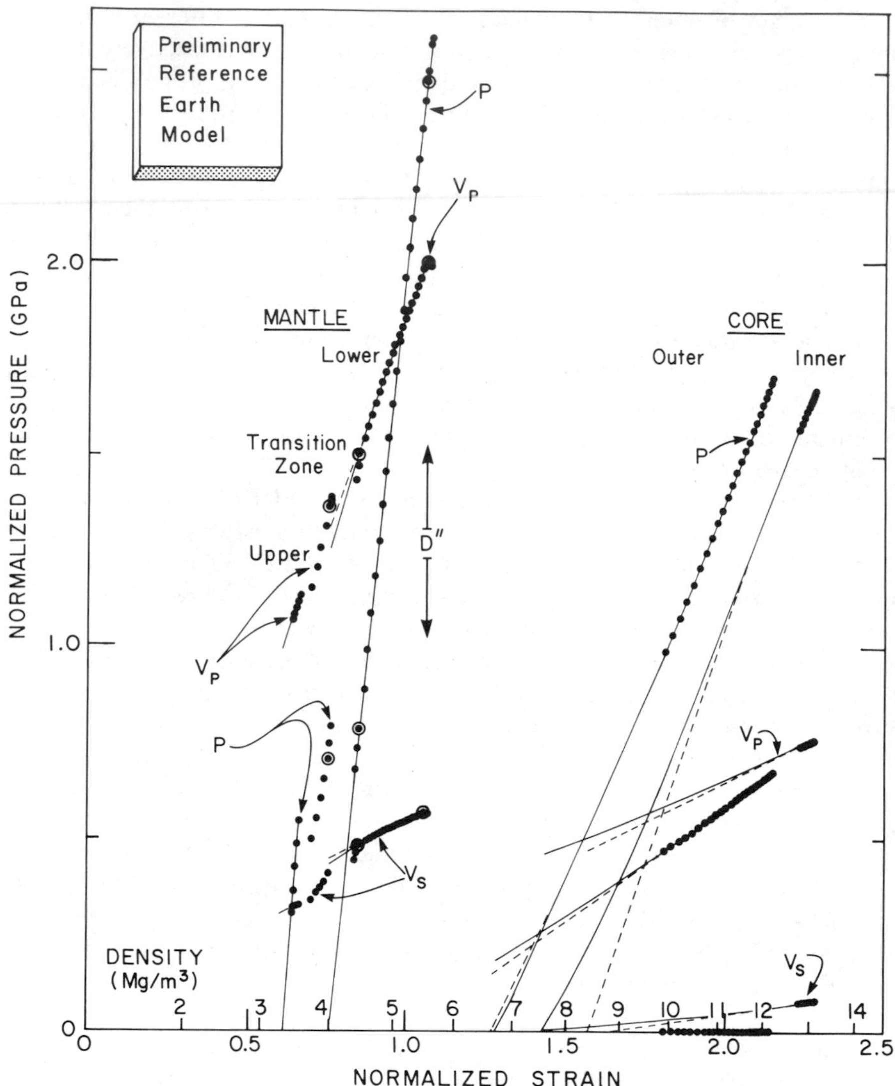

Fig. 3. Normalized pressures, G_p (labeled *P*), $G_\lambda/30$ (labeled V_p), and $G_\sigma/30$ (labeled V_s) as defined by Eqs. (28)–(30) in the text, are shown as a function of normalized strain, g (Eq. 23), for each region of the Earth's interior based on the pressure, V_p and V_S distributions given in PREM of Dziewonski and Anderson (1981). This figure is equivalent to Fig. 1 put in terms of Eulerian finite-strain variables, and the density corresponding to the strain g is indicated on the lower scale. Third-order and fourth-order fits are shown as dashed and solid lines, respectively, and circled points indicate depths at which velocity gradients change (e.g., the top and bottom of the lower mantle, which were not included in the fits: see text).

Table 2. Finite Strain Analysis of PREM.

Property	Upper mantle		Lower mantle		Outer core		Inner core	
(at zero pressure)	Third[a]	Fourth[a]	Third	Fourth	Third	Fourth	Third	Fourth
ρ_0, $Mg \cdot m^{-3}$	3.308[b]	3.308	4.003	3.994	6.661	6.725	8.414	7.599
	—[b,c]	—	—	—	—	—	—	—
K_0, GPa	178.8	178.9	222.5	214.5	141.6	152.5	381.3	160.4
	130.0	130.5	216.5	204.7	105.3	130.8	502.2	400.7
μ_0, GPa	—	—	—	—	—	—	—	—
	66.7	66.8	135.4	130.1	0	0	−8.6	4.2
ν_0	—	—	—	—	—	—	—	—
	0.282	0.282	0.242	0.237	0.5	0.5	2.03	0.446
K_0'	2.46	2.45	3.76	4.05	4.40	4.22	3.57	6.24
	4.36	4.22	3.80	4.38	6.42	5.08	2.36	2.31
μ_0'	—	—	—	—	—	—	—	—
	1.42	1.40	1.52	1.77	0	0	0.33	0.17
$K_0 K_0''$	[−4.72][d]	−4.48	[−3.71]	−5.20	[−4.45]	−3.88	[−3.64]	−15.93
	[−5.46]	−2.44	[−3.13]	−7.51	[−19.42]	−8.50	[−0.02]	0.09
$K_0 \mu_0''$	—	—	—	—	—	—	—	—
	[−2.51]	−1.97	[−2.13]	−4.04	[0]	0	[0.61]	0.56
π_{01}	2.30	2.34	0.36	−0.06	−0.60	−0.32	0.64	−2.38
	−0.54	−0.34	0.30	−0.56	−3.64	−1.62	2.46	2.54
π_{02}	0	0.40	0	−1.88	0	0.40	0	−7.08
	0	2.60	0	−4.64	0	−3.56	0	7.72
λ_{00}	—	—	—	—	—	—	—	—
	1.68	1.68	1.83	1.85	1.00	1.00	−1.02	1.01
λ_{01}	—	—	—	—	—	—	—	—
	−10.3	−9.9	−12.1	−11.0	−14.3	−10.2	−3.5	−2.5

Table 2. (Continued)

Property (at zero pressure)	Upper mantle		Lower mantle		Outer core		Inner core	
	Third[a]	Fourth[a]	Third	Fourth	Third	Fourth	Third	Fourth
λ_{02}	—	—	—	—	—	—	—	—
	0	12.7	0	−14.2	0	3.9	0	2.2
σ_{01}	—	—	—	—	—	—	—	—
	−3.3	−3.2	−2.3	−3.4	—	—	62.8	−43.7
σ_{02}	—	—	—	—	—	—	—	—
	0	2.9	0	−6.3	—	—	0	134.6

[a] Third and fourth-order equations of state.
[b] In all cases, the upper value is determined from the pressure–density relation, and the lower from the velocities.
[c] Dashes indicate undetermined values.
[d] Brackets denote a value implied by the assumed order used.

expected, successive derivatives of the equation of state for each region become increasingly less well constrained. Not only is the distinction between the order of fits increasingly important, but also the differences between elastic moduli derived from G_π and from the combination of G_λ and G_σ become increasingly significant for the higher derivatives of the equation of state (Table 2).

The discrepancies evident in Table 2 (e.g., compare π_{01} values derived from densities and from velocities) arise mainly from heterogeneity within each region and from biases in the parameterized seismic models. In fact, we suspect that much of the bias may derive from the parameterization itself (polynomial expressions of velocities and densities with radial distance in the case of PREM). That anelastic dispersion is not involved is evident, in every case but that of the inner core, from the fact that K_0 determined from G_π (i.e., the lower frequency value) is systematically higher than the (higher frequency) value obtained from G_λ and G_σ. Attenuation-caused dispersion of anelastic moduli is thermodynamically predicted to have the opposite variation, with the low-frequency modulus being less than the high-frequency value (e.g., Nowick and Berry, 1972). Furthermore, the absolute magnitude of bulk modulus dispersion is only expected to be a few percent (Heinz *et al.*, 1982; Heinz and Jeanloz, 1983), so the 24–60% difference in K_0 values derived from the density and velocity profiles of the inner core are not plausibly ascribed to anelastic effects.

The effects of inhomogeneity are most clearly seen for the upper mantle results in Table 2 (recall that we have already excluded the most inhomogeneous regions, such as the transition zone, from our analysis). Aside from the differences in density- and velocity-derived moduli (e.g., 27% for K_0), the physically unlikely value $K_0' \cong 2.5$ is directly related to the low value of the inhomogeneity parameter in this region (Dziewonski and Anderson, 1981). Thus, we may tentatively interpret the zero-pressure density, but the higher order derivatives (the elastic moduli) are ill constrained for the 220–400 km depth interval of the upper mantle.

In contrast, the lower mantle fits exhibit a high degree of internal consistency among the density and moduli. For example, the differences between density-derived and velocity-derived values of K_0, about 4%, are comparable to the local uncertainties in bulk modulus in the lower mantle (Wiggins *et al.*, 1973; Masters, 1979). For the lower mantle, unlike any of the other regions, the finite-strain expressions for velocities and density are also sufficiently well constrained that they can be distinguished from the polynomial expansions of the seismological models. This conclusion is documented in Table 3, which gives the radial distance (r_0) at which the PREM polynomials reproduce the zero-pressure densities derived from the finite strain fits (the corresponding depth $z_0 = 6371$ km $- r_0$ is also given). Note that for none of the regions does the zero-pressure condition occur at the surface of the Earth ($r = 6371$ km): $z_0 \neq 0$ and the discrepancy enlarges for deeper regions. Only in the case of the lower mantle, however, can the bulk and shear moduli evaluated at r_0 be distinguished from the range of values allowed by the finite-strain equations of state; the fits to the other regions are too uncertain to see this distinction. This means that the lower mantle values

Table 3. Extrapolations to zero pressure: PREM polynomials.

Property (at zero pressure)	Upper mantle Third	Upper mantle Fourth	Lower mantle Third	Lower mantle Fourth	Outer core Third	Outer core Fourth	Inner core Third	Inner core Fourth
ρ_0, $Mg \cdot m^{-3}$ [a]	3.308	3.308	4.003	3.994	6.661	6.725	8.414	7.599
r_0, km [b]	6365	6365	6268	6280	5005	4981	4634	5021
$V_P(r_0)$, $km \cdot s^{-1}$	8.15	8.15	9.60	9.56	4.26	4.34	7.90	7.31
$V_S(r_0)$, $km \cdot s^{-1}$	4.49	4.49	5.67	5.66	0	0	1.32	0.91
$K(r_0)$, GPa	130.5	130.5	197.2	195.1	121.0	126.8	505.1	397.7
$\mu(r_0)$, GPa	66.8	66.8	128.7	127.9	0	0	14.6	6.2
z_0, km	6	6	103	91	1366	1390	1737	1350

[a] Determined from finite-strain theory; equivalent to the values given in Table 1.
[b] Radii for which the PREM polynomials reproduce ρ_0.

are not absolutely predetermined (and hence potentially biased) by the polynomial forms of the density and velocity profiles in PREM.

Although fourth-order fits might be expected to be systematically better than third-order fits, because more parameters are involved, such is not the case for the lower mantle. Elastic moduli derived from density and velocity profiles are in closer consistency for the third-order than for the fourth-order case (Table 2). Also, K_0' is resolvable (according to the criteria described above) in the third-order fit; in contrast, K_0' is not resolvable, and $K_0 K_0''$ is questionably resolvable, in the fourth-order fit. This suggests that the seismological data are overfitted in the latter case. We note, however, that the details of the seismological data are overfitted in the latter case. We note, however, that the details of the seismological inversion (especially the starting model) may influence these results, so it is not possible to prove that a third-order equation of state uniquely describes the inherent properties of the lower mantle, although it provides the best fit to PREM.

By the same criteria, a fourth-order fit is the most successful for the outer core. It is not clear that $K_0 K_0''$ is resolvable, but K_0' differs measurably from 4 (second-order value) for both third- and fourth-order fits, and the elastic moduli are internally more consistent in the latter case. Similarly, the inner core seems to be better fit by a fourth-order equation of state. For this region, however, the uncertainties are so large that neither third- nor fourth-order fits are very satisfactory; also, neither can be rejected. We only point out that the third-order fit is not necessarily unphysical because of the negative value of μ_0 that is derived. The inner core may consist of a high-pressure phase that is dynamically unstable, and may therefore have a negative shear modulus, at zero pressure. Negative values of μ_0 (extrapolated from high pressure) have been documented for high-pressure phases of alkali halides, for example (Jeanloz, 1982; Hemley and Gordon, 1985). Similarly, ε-Fe, which is a plausible candidate for the constituent of the inner core (e.g., Jeanloz, 1986), cannot be quenched from high pressure and it may thus also exhibit a negative shear modulus (dynamic instability) upon decompression to $P = 0$.

Interpretation

We proceed next to interpret the zero-pressure results given for the upper mantle, lower mantle, outer core, and inner core in Table 2. For all regions, the ρ_0 values are known with greater confidence than are the elastic moduli. This is especially true for the upper mantle and for the inner core: the elastic moduli may be biased in the first case because of inhomogeneity in the mantle, and the moduli are poorly determined in the second case because of the long extrapolation of G. Also, because there is considerably more laboratory information on densities of mineral phases, compared with elastic-modulus data, we emphasize the interpretation of the zero-pressure density values. The elastic moduli provide complementary information, particularly for the lower mantle, and to a lesser extent for the outer core (both are relatively homogeneous regions, but the equation-of-state fit is better, and more laboratory data are available for comparison, in the former case).

As the standard-state values in Table 2 are for high-temperature conditions, given by an interior adiabat extrapolated to $P = 0$, the thermal expansion coefficient plays an important role in the interpretation of densities. In the last few years, the coefficient of thermal expansion (α) has been precisely measured for a wide variety of mineral phases (Fig. 4), and it is now possible to interpret high-temperature densities with confidence. One important generalization that has emerged is that the magnitude of α can be dominantly set by crystal-chemical effects (cf. Jeanloz, 1985). For example, Jeanloz and Roufosse (1982) point out that the first-order effect of structural transformations to high-pressure phases is to increase the coefficient of thermal expansion if there is a coordination increase (see also Hemley and Gordon, 1985).

It has previously been thought that high-pressure phases, which must thermodynamically be high-density phases as well, have systematically lower thermal expansion coefficients than low-pressure phases (see Jeanloz and Roufosse, 1982). This idea appears to be supported by the decrease in values from olivine (α phase) to β-phase to γ-spinel forms of Mg_2SiO_4 (Fig. 4). Also, the thermal expansion coefficient of quartz (not shown in Fig. 4) is approximately twice that of the high-pressure stishovite phase of SiO_2. Because there is no change in primary coordination across the olivine transformations, however, the systematic decrease in α results from secondary effects that are qualitatively ascribed to increased atomic packing efficiency across the olivine–spinel transitions (see Kamb, 1968; Akimoto *et al.*, 1976). Therefore, the more general effect on α of high-pressure transformations involving coordination changes is not documented by the α–β–γ transitions of $(Mg, Fe)_2SiO_4$.

What is significant in the case of SiO_2 is that α increases across the coesite–stishovite transformation, which involves a change from fourfold to sixfold coordination (Jeanloz and Roufosse, 1982). That the thermal expansion coefficient of quartz is exceptionally high owes to the well-established fact that the expansion is caused by polyhedral rotations (e.g., Megaw, 1973). This is a different expansion mechanism than is seen in any of the high-pressure phases,

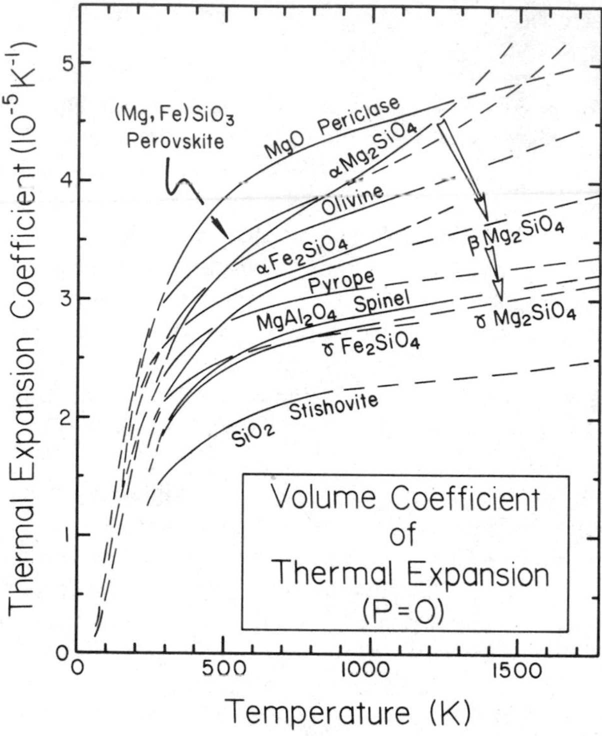

Fig. 4. Summary of measured volume coefficients of thermal expansion for a variety of minerals that are thought to occur in the mantle. All measurements are at zero pressure, and the curves are dashed outside the temperature range of the experiments. Open arrows indicate the systematic decrease in the thermal expansion coefficient of Mg_2SiO_4 upon transformation from olivine (α) to β-phase to γ-spinel structures. Sources of data are given by Jeanloz and Thompson (1983) and by Knittle *et al.* (1986a, b).

the presence of corner-linked tetrahedra being required, so quartz does not provide a useful example in the present discussion.

Based on their analysis, Jeanloz and Roufosse (1982) predicted that the thermal expansion coefficient of the high-pressure silicate perovskite phase should be high. This conclusion, which was contrary to previous expectations for high-density phases, has been quantitatively confirmed by the measurements of Knittle *et al.* (1986a, b) shown in Fig. 4. Because silicate perovskite dominates the mineralogy of the lower mantle, the large values of α characteristic of high-pressure, high-coordination phases are especially significant in the interpretation of the bulk composition of this region.

Evaluation of T_{0S}

In order to interpret the zero-pressure values of the interior properties (Table 2) it is necessary to derive the initial ($P = 0$) adiabat temperature for each region,

T_{0S}. For example, the values of density in Table 2, $P_0(T_{0S})$, are determined by the mean thermal expansion coefficient:

$$\bar{\alpha}_0 \equiv \left[\frac{-\Delta\rho/\rho}{T_{0S} - 300\text{K}}\right]_{P=0} = (T_{0S} - 300\text{K})^{-1} \int_{300\text{K}}^{T_{0S}} \alpha_0 \, dT \tag{47}$$

with $-\Delta\rho/\rho$ being the thermal strain between room temperature (300K) and T_{0S}. The values of $\bar{\alpha}_0$ are derived from experimental measurements (Fig. 4), so that an estimate of T_{0S} makes it possible to compare observed densities of high-pressure phases (Table 4) with the decompressed values for different regions of the Earth's interior (Table 2).

Given an estimate of the temperature at some depth z km, $T(z)$, the initial adiabat temperature is obtained by way of the Grüneisen parameter:

$$\gamma = \left(\frac{\partial \ln T}{\partial \ln \rho}\right)_s \tag{48}$$

For example, $T(400) \cong 1700\ (\pm 300)$K and $T(670) \gtrsim 2000\ (\pm 300)$K are used below for the upper mantle and lower mantle, respectively. The first value is derived from the experimentally determined temperature dependence of the olivine–β-phase–γ-spinel reaction, which is thought to be the cause of the 400-km seismological discontinuity, whereas $T(670)$ is derived from $T(400)$ with additional contributions being included from heats of transformations occurring in the transition zone (Jeanloz and Richter, 1979; Jeanloz and Thompson, 1983).

Empirically, the Grüneisen parameter is found to decrease upon increasing compression, with

$$\gamma = \gamma_0 \left(\frac{\rho}{\rho_0}\right)^{-q} \tag{49}$$

and $0 \lesssim q \lesssim 2$ being appropriate for minerals (q is typically close to 1: Birch, 1952, 1968; O. L. Anderson, 1967; McQueen *et al.*, 1970; Wallace, 1972; Wolf and Jeanloz, 1985b). The zero-pressure value, γ_0, is obtained from measured properties of high-pressure phases by way of

$$\gamma = \frac{\alpha K_T}{\rho C_V} \tag{50}$$

which is thermodynamically equivalent to the adiabatic derivative in Eq. (48). In Eq. (50), K_T is the isothermal bulk modulus and C_V is the specific heat at constant volume (if no measurements are available a high-temperature value of C_V can be estimated reliably, as is done in the case of perovskite). Thus, Knittle *et al.* (1986a, b) derive a value of $\gamma_0 = 1.73\ (\pm 0.05)$ for lower mantle assemblages of upper mantle composition.

The evaluation of T_{0S} by way of Eqs. (48)–(50) is specifically illustrated for the case of the lower mantle. First, we assume that the lower mantle and upper mantle have identical compositions, so that a specific lower mantle value of T_{0S} can be estimated. If these regions differ in composition, $T(670)$ is expected to be larger than 2000K because of the presence of thermal boundary layers, and a

Table 4. Zero-Pressure Properties of High Pressure Phases[a]

Phase	Composition	Density (300K), ρ_0, $Mg \cdot m^{-3}$	Mean thermal expansion coefficient (300–1800K), $\bar{\alpha}_0$, $10^{-5}\ K^{-1}$	Adiabatic bulk modulus (300K), K_{0S}, GPa	δ_{0S}, $\left(\frac{\partial \ln K_{0S}}{\partial \ln \rho}\right)_P$	Shear modulus (300K), μ_0, GPa
Magnesiowüstite	(Mg, Fe)O	$\frac{100.39 + 31.54x_{Fe}}{24.46\ (\pm 0.04) + 1.0\ (\pm 0.3)x_{Fe}}$ [b]	4.4 (± 0.1)	162.7 (± 0.2) + 17 (± 8)x_{Fe} [b]	3.3 (± 0.3)	$\frac{1000}{[7.63 + 8.7x_{Fe}]}$
Stishovite	SiO_2	4.289 (± 0.003)	2.2 (± 0.3)	316 (± 10)	—	220 (± 10)
Perovskite	(Mg, Fe)SiO_3	4.10 (± 0.01) + 1.1 (± 0.3)x_{Fe} [b]	4.0 (± 0.4)	262 (± 6)	—	[155 (± 15)][c]
	$Mg_3Al_2Si_3O_{12}$	4.13 (± 0.06)	—	—	—	
	$CaSiO_3$	4.06 (± 0.15)	—	[260 (± 15)][c]		[145 (± 15)][c]
Liquid iron	Fe (1, at 1810K)	7.02 (± 0.01)	11.9	85 (±)	—	0
ε-Iron	Fe	8.35 (± 0.02)	—	195 (± 10)	—	—

[a] Sources of data are given in Jeanloz and Thompson (1983). Additional data are from Shimoji (1977), Knittle *et al.* (1986), and Jephcoat, *et al.* (1986).

[b] x_{Fe} indicates the molar ratio of iron to iron + magnesium components.

[c] Unmeasured values estimated from the calculations of Wolf and Jeanloz (1985a).

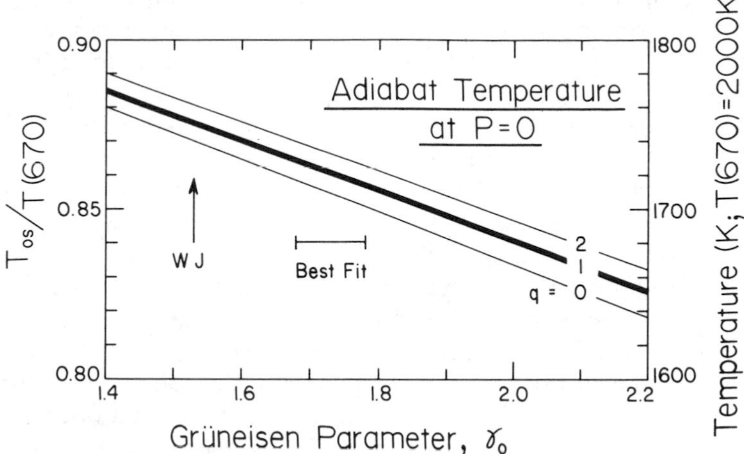

Fig. 5. Zero-pressure adiabat temperature (T_{0S}) for the lower mantle (right-hand scale) and its ratio to the temperature at 670 km depth (left-hand scale) are shown as a function of the zero-pressure Grüneisen parameter (γ_0) and its logarithmic volume derivative (q), as given by Eqs. (48)–(50) in the text. The range of Grüneisen parameters for lower mantle assemblages of perovskite and magnesiowüstite according to the thermal expansion measurements and analysis by Knittle *et al.* (1986) (best fit solution: cf. Knittle *et al.*, 1986a, b) is compared with the theoretically calculated result of Wolf and Jeanloz (1985a) (WJ). As described in the text, the best fit values and $T(670) = 2000$K yield a minimum estimate of $T_{0S} = 1720$K.

correspondingly higher value of T_{0S} is obtained [this is why 2000K is given above as a lower bound for $T(670)$; Jeanloz and Richter (1979)]. Second, the density ratio $\rho(670)/\rho_0 = 1.095$ is derived from our analysis of PREM (Table 2). Thus, with $q = 1$ (± 1) we find $T_{0S} = 1720$ (± 260K) for the lower mantle, as shown in Figure 5. Similarly, a value $T_{0S} = 1560$ (± 30)K is obtained for the upper mantle, based on $\rho(400)/\rho_0 = 1.02$ (Table 2), and assuming $T(400) = 1700$K and $\gamma \cong 1.0$–1.5 (a conservative range of estimates).

Upper Mantle

Several lines of petrological evidence indicate that the uppermost mantle consists of a periodotitic assemblage of olivine, pyroxene, and garnet in the approximate volume ratio of 50–60 : 20–40 : 10 (Ringwood, 1975; Yoder, 1976; BVSP, 1981). The corresponding density at zero pressure and 300K is approximately 3.37 $Mg \cdot m^{-3}$, if no account is taken of possible phase transitions in the garnet–pyroxene component at depths less than 400 km. The effect of such transitions is to increase the zero-pressure density by approximately 0.05 $Mg \cdot m^{-3}$ over the depth range of 200–400 km (Akaogi and Akimoto, 1979; Jeanloz and Thompson,

1983). The resulting $\rho_0 = 3.43\ \mathrm{Mg \cdot m^{-3}}$ compares well with the zero-pressure density of $3.31\ \mathrm{Mg \cdot m^{-3}}$ derived from PREM, because a temperature difference of about 1200K (i.e., $T_{0S} \cong 1500\mathrm{K}$) reconciles these values for a peridotitic assemblage, which has a mean coefficient of thermal expansion $\bar{\alpha} \cong 3 \times 10^{-5}\ \mathrm{K^{-1}}$ at high temperatures based on the data summarized by Jeanloz and Thompson (1983). A value of T_{0S} near 1500–1550K (1200–1300°C) is also consistent with the available petrological data, including melting equilibria and geothermometry results from mantle nodules (Jeanloz and Morris, 1986).

The addition of significant amounts of Al_2O_3 to the conventional (peridotitic) models of upper mantle composition, results in more garnet being present at depth, and hence ρ_0 of the assemblage is increased. For example, D. L. Anderson's (1979) eclogite model is sufficiently dense ($\rho_0 = 3.48\ \mathrm{Mg \cdot m^{-3}}$ at 300K) that a value exceeding 1900K (1500–1600°C) is required for T_{0S} in order to match the decompressed density of the upper mantle. Such a value cannot be absolutely precluded, but it is implausibly high given the petrological constraints on melting equilibria and nodule temperatures. Also, we found above that $T_{0S} \cong 1550\mathrm{K}$, based on the temperature at the 400-km discontinuity. Thus, we conclude that a peridotite model satisfies observations on the upper mantle remarkably well, and we take $T_{0S} = 1550\ (\pm 50)\mathrm{K}$ for this region throughout the following discussion.

Lower Mantle

The finite-strain fit is well constrained for the lower mantle (Fig. 3), with the initial density and elastic moduli being consistent to within 0.15 and about 4%, respectively (Table 1 and 3). In Fig. 6, the derived high-temperature values of ρ_0 and K_0 are compared with zero-pressure, 300K densities and bulk moduli of the high-pressure phases thought to be dominant in the lower mantle (e.g., Jeanloz and Thompson, 1983). The experimentally measured properties are contoured for the MgO–FeO–SiO_2 system (see Table 4), and a conservative estimate of the plausible range of upper mantle compositions is included [olivine : pyroxene ratio between 1 : 1 and 2 : 1 and a molar ratio of iron to iron plus magnesium components of $0.91 \geqslant \mathrm{Mg/(Mg + Fe)} \geqslant 0.87$: cf. Ringwood (1975), Yoder (1976), BVSP (1981), Jeanloz and Thompson (1983)]. The presence of

Fig. 6. Contours of density (in $\mathrm{Mg \cdot m^{-3}}$, upper figure) and bulk modulus (in GPa, lower figure) at zero pressure and room temperature for high-pressure phases in the system MgO–FeO–SiO_2. The estimated range of upper mantle compositions is shown by the stippled region, and the coexistence fields of the high-pressure phases are from Jeanloz and Thompson (1983). The thermal strain between a given density contour and the lower mantle value of ρ_0 is shown in parentheses in the upper figure [cf. Table 2, Eq. (47); uncertainties are small on this scale]. Similarly, the proportional bulk modulus difference between a given contour and the lower mantle values of K_0 (Table 2) is indicated beneath the contour values of 220, 240, 260, and 280 GPa; uncertainties are listed in parentheses. The experimental data on which these plots are based are listed in Table 4. ▷

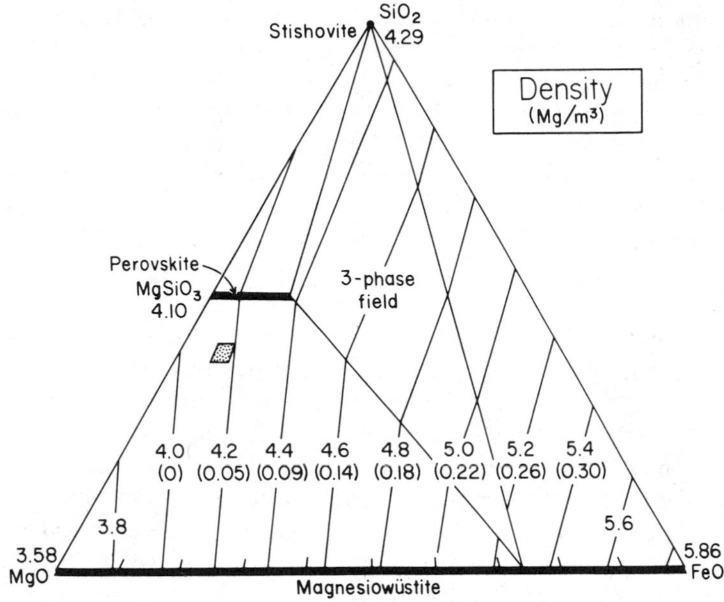

SiO2
Stishovite
4.29
Density
(Mg/m3)
Perovskite
MgSiO3
4.10
3-phase
field
4.0 (0)
4.2 (0.05)
4.4 (0.09)
4.6 (0.14)
4.8 (0.18)
5.0 (0.22)
5.2 (0.26)
5.4 (0.30)
3.8
5.6
3.58
MgO
5.86
FeO
Magnesiowüstite

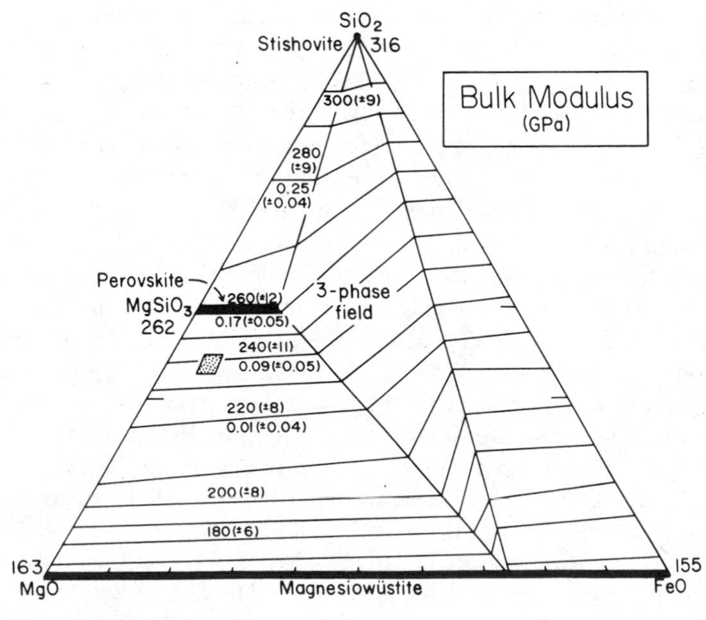

SiO2
Stishovite
316
Bulk Modulus
(GPa)
300(±9)
280
(±9)
0.25
(±0.04)
Perovskite
MgSiO3
262
260(±12)
0.17(±0.05)
3-phase
field
240(±11)
0.09(±0.05)
220(±8)
0.01(±0.04)
200(±8)
180(±6)
163
MgO
155
FeO
Magnesiowüstite

minor components such as CaO and Al_2O_3 is unlikely to alter the plots because these components are expected to enter the perovskite structure with virtually no effect on the properties (Table 4).

For a given composition to satisfy the observed properties of the lower mantle, the difference between the laboratory measurements and the values obtained by adiabatic decompression must be ascribed to temperature. For example, an upper mantle composition is appropriate for the lower mantle if temperature can account for the 4% and 9 ($\pm$5)% difference in density and bulk modulus shown in Fig. 6. These differences are governed by the mean coefficient of thermal expansion, $\bar{\alpha}$ (Eq. 47), and by δ_{0S} (Table 4), respectively.

We have described above the consistency of compositions and properties in the upper mantle assuming that an adiabat for this region is characterized by $T_{0S} = 1550\ (\pm 50)$K, a value that is compatible with the estimated temperature (1700 $\pm$ 300) at the 400-km discontinuity. Contributions to the geotherm in the transition zone from phase transitions with positive Clapeyron slopes lead to a higher temperature at the top of the lower mantle, minimum estimates for the temperature at 670-km depth being in the range of 2000 ($\pm$300)K (Jeanloz and Thompson, 1983; Navrotsky and Akaogi, 1985). Thus, we concluded above that $T_{0S} = 1720\ (\pm 260)$K (or 1450°C) for the lower mantle, assuming that this region has the same composition as the upper mantle.

The tradeoff among temperature, composition, and thermal expansion coefficient required to satisfy the lower mantle density (ρ_0: Fig. 6, upper) is summarized in Fig. 7. The mean thermal expansion coefficient, defined in Eq. (47), is given by the data of Suzuki (1975) and Knittle *et al.* (1986a, b) for the compositions of interest. The thermal expansion of magnesiowüstite is experimentally known (cf. Hazen and Jeanloz, 1984), so that the significance of the relatively high value of $\bar{\alpha} \cong 4\ (\pm 0.5) \times 10^{-5}\ K^{-1}$ for silicate perovskite (see Fig. 4) is evident from Fig. 7: an upper mantle composition does not satisfy the lower mantle density unless $T_{0S} \lesssim 1400\ (\pm 100)$K (1100°C). As we have found, this is significantly below the plausible range of lower mantle temperatures, thus suggesting that the composition of this region may be enriched in iron (and possibly silica) components relative to the upper mantle. With such enrichments, the intrinsic density of the lower mantle is higher than would be the case for a uniform mantle composition.

A similar analysis can be done for the elastic moduli of perovskite–magnesiowüstite assemblages, but the existing laboratory data are insufficient to yield a definite result. For example, the bulk modulus data cannot resolve variations in iron content from an upper mantle composition (Fig. 6, lower) and, because of the uncertainties in the measurements, cannot prove that any difference in silica content is required, although it is suggested by the data (cf. Knittle *et al.*, 1986a). In particular, the paucity of δ_{0S} measurements precludes a quantitative analysis (Table 4). Similarly, no data currently exist for the shear modulus of (Mg, Fe)SiO_3 perovskite. As shown in Fig. 8, the decompressed value $\mu_0 = 133$ ($\pm$3) GPa for the lower mantle is consistent with this region being enriched in silica, but experimental measurements are needed to substantiate this conclusion.

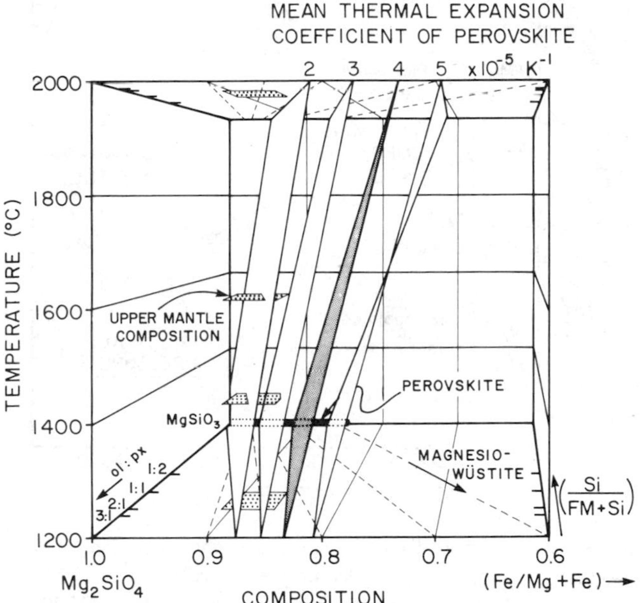

Fig. 7. Expanded view of the region around the upper mantle composition (stippled) in Fig. 6 illustrates the tradeoff between the adiabat temperature, T_{0S} (in °C along the vertical scale), composition (as FeO–MgO–SiO_2 components), and the mean thermal expansion coefficient of silicate perovskite (see Eq. 47) required to match the observed density of the lower mantle (ρ_0 in Table 2). The effects of iron partitioning between perovskite and magnesiowüstite are included in the analysis (dashed lines: see Knittle *et al.*, 1986), and olivine: pyroxene ratios (ol : px), proportions of iron components (Fe/FM with FM = Fe + Mg) and proportions of silica components [Si/(FM + Si)] are indicated. The preferred value for the high-temperature thermal expansion coefficient of perovskite, $\bar{\alpha}_0 = 4 \times 10^{-5}\ K^{-1}$ based on the data of Knittle *et al.* (1986a, b), implies that the density of an upper mantle compositional model is compatible with ρ_0 of the lower mantle if T_{0S} is less than 1200°C (1500K), as indicated by the shading.

Outer Core and Inner Core

We discuss the two core regions together, and briefly, because they are less well suited to the present analysis than is the lower mantle. First, compared with the results for the mantle the zero-pressure values are constrained with less certainty, especially for the inner core (Table 2). Second, fewer data exist for the high-pressure phases that may occur in the core (Table 4), and it is not obvious that liquid iron (or iron alloy) at zero pressure has the same properties that the high-pressure liquid would have if it were metastably decompressed. That is,

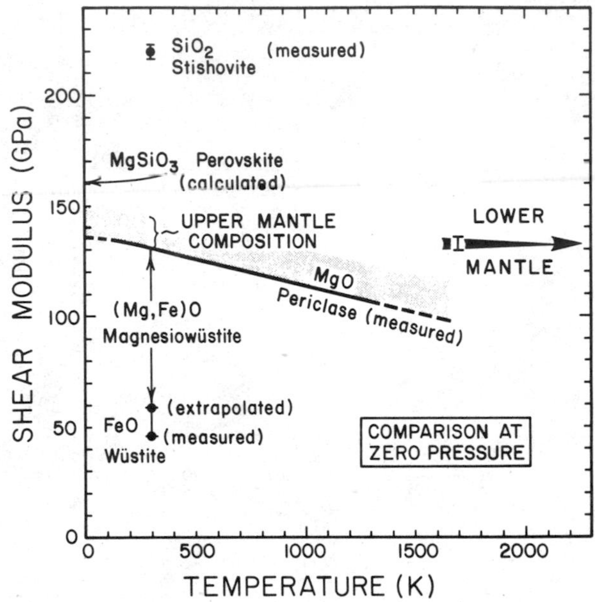

Fig. 8. Zero-pressure shear moduli of high-pressure phases that may exist in the lower mantle (Table 4) are compared with the decompressed value $\mu_0 = 133\ (\pm 3)$ GPa for the lower mantle (Table 2), which corresponds to a temperature T_{0S} of 1720 ($\pm$260) or more (bold arrow indicates the acceptable temperature range derived in the text). The shear modulus of silicate perovskite is currently unmeasured (see Table 4), and only in the case of periclase is the temperature dependence of μ experimentally known. The plausible range of shear moduli for upper mantle compositions are tentatively extrapolated to high temperatures based on the measurements for periclase (shaded band).

pressure may alter the structure and hence the intrinsic properties of the melt, even though it corresponds to the same phase (i.e., liquid) at low and high pressures.

With these caveats, we find that the ρ_0 values for the inner core (fourth-order fit only) and for the outer core (third- or fourth-order fit) are in qualitative agreement with expectations, both densities being less than those of solid (ε high-pressure phase) and liquid iron (Table 4). From these ρ_0 values and a plausible range of Grüneisen parameters ($\gamma \lesssim 2$), estimates of the minimum core temperature that satisfy available high-pressure data ($T \lesssim 3800$K at the top of the outer core: Jeanloz and Morris, 1986) imply that T_{0S} of the core must exceed 2110 ($\pm$ 420)K. Hence, the lower zero-pressure density of the outer core relative to liquid Fe (at 1810K; Table 4) can be attributed mainly to the alloying of iron that has long been known to be required for explaining the observed density of the core (Birch, 1964; Jacobs, 1975; Jeanloz, 1979; Brown and McQueen, 1982). The amount of alloying that would be inferred from the zero-pressure comparison, $\sim$5% or less by weight, is similar to recent estimates based on comparisons at core presures (Brown *et al.*, 1984; Spiliopoulos and Stacey, 1985; Jeanloz, 1986).

For the inner core, we ignore the third-order fit because ρ_0 is larger than the density of ε-Fe at zero pressure, and the results are tentative at best. If this region consists of the high-pressure ε phase of iron, its density is compatible with $T_{0S} \geqslant 2110\text{K}$, given $\bar{\alpha} \leqslant 5.4 \times 10^{-5}\ \text{K}^{-1}$, which is plausible for the thermal expansion coefficient (Table 4). In the case of γ-Fe ($\rho = 7.95\ \text{Mg}\cdot\text{m}^{-3}$ at $P = 0$, $T = 300\text{K}$), $\bar{\alpha}$ would have to be less than $2.5 \times 10^{-5}\ \text{K}^{-1}$ in order to match the decompressed density. This is inconsistent with thermal expansion measurements ($\bar{\alpha} \cong 6$ to $7 \times 10^{-5}\ \text{K}^{-1}$, Donohue, 1974), but the cumulative uncertainties in our analysis may still allow the high-temperature γ phase of iron to be the primary constituent of the inner core. Similarly, we cannot prove from our analysis that pure iron rather than an iron alloy makes up the inner core, although such is the conclusion suggested by the densities.

Uncertain as the density comparison is for both inner and outer core, we are not surprised that the elastic moduli cannot be analyzed with any reliability. For example, the apparent K_0 values for the outer core are significantly higher than the high-temperature bulk modulus of liquid iron (Table 4). A comparison at high pressure indicates that the bulk modulus of liquid iron is higher than that of the outer core at the same conditions because of the alloying (Jeanloz, 1979; Brown and McQueen, 1982). This suggests that our extrapolation to zero pressure may be biased, with the actual K_0 being less (and hence ρ_0 being slightly larger) than the values given in Table 2. Alternatively, we might concede that the measured properties of liquid iron (Table 4) cannot be compared with the outer core values because of the possibility that structural changes occur in the liquid at high pressures. We therefore consider the current analysis to be too uncertain to interpret the moduli extrapolated to zero pressure from core conditions.

Conclusions

A finite-strain analysis is developed for reducing seismologically determined elastic-wave velocities and densities throughout the Earth's interior to a zero-pressure standard state by way of adiabatic decompression. High-pressure experimental measurements demonstrate that the Eulerian formulation employed here yields the best equation of state now available. The assumptions in the analysis are explicitly described: homogeneity of phases and composition with depth being the most restrictive, and hydrostaticity, lateral homogeneity, and adiabaticity being less significant or more easily justified. Our approach demonstrates that zero-pressure, standard-state values of density and elastic moduli can be determined successfully for the lower mantle, outer core, upper mantle, and inner core, with successively decreasing confidence. The upper mantle results may be somewhat biased because of inhomogeneity, and the results for the core are uncertain because of the long extrapolation required. Also, the interpretation of the outer core properties is hampered by uncertainties in the effects of pressure on the structure and properties of liquid alloys. The analysis is most successful for the lower mantle, and the detailed calculation of the standard-state (adiabat) temperature is summarized for this region. Although high-pressure phases of the Earth's deep interior are still relatively poorly characterized, the available experimental data strongly suggest that the lower mantle differs in composition from the upper mantle. Consistent with previous studies, we find that the properties of the upper mantle are matched by those of garnet peridotite. Also, the zero-pressure density of the outer core obtained in this analysis implies that about 5% alloy component is present in the liquid iron comprising this region; this result is in accord with comparisons at high pressures.

Acknowledgments

Work presented here was supported by the National Science Foundation and by NASA. We thank C. Meade and Q. Williams for helpful comments.

References

Akaogi, M., and Akimoto, S. (1979) High-pressure phase equilibria in a garnet lherzolite with special reference to Mg^{2+}–Fe^{2+} partitioning among constituent minerals, *Phys. Earth Planet. Int.* **19**, 31–51.

Akimoto, S., Matsui, Y., and Syono, Y. (1978) High-pressure crystal chemistry of orthosilicates and the formation of the mantle transition zone, in *The Physics and Chemistry of Minerals and Rocks*, edited by R. G. J. Strens, pp. 327–363. John Wiley and Sons, New York.

Anderson, D. L. (1979) The upper mantle: Eclogite?, *Geophys. Res. Lett.* **6**, 433–436.

Anderson, O. L. (1967) Equation for thermal expansivity in planetary interiors, *J. Geophys. Res.* **72**, 3661–3668.

Backus, G., and Gilbert, F. (1970) Uniqueness in the inversion of inaccurate gross Earth data, *Phil. Trans. Roy. Soc.* **A266**, 123–192.

Barsch, G. R., and Chang, Z. P. (1971) Ultrasonic and static equation of state for cesium halides, in *Accurate Characterization of the High Pressure Environment*, Natl. Bur. Stand. U.S. Spec. Publ. 326, pp. 173–189.

BVSP (Basaltic Volcanism Study Project) (1981) *Basaltic Volcanism on the Terrestrial Planets*. Pergamon Press, New York, 1286 pp.

Birch, F. (1938) The effect of pressure upon the elastic properties of isotropic solids, according to Murnaghan's theory of finite strain, *J. Appl. Phys.* **9**, 279–288.

Birch, F. (1947) Finite elastic strain of cubic crystals, *Phys. Rev.*, **71**, 809–824.

Birch, F. (1952) Elasticity and constitution of the earth's interior, *J. Geophys. Res.* **57**, 227–286.

Birch, F. (1964) Density and composition of the mantle and core, *J. Geophys. Res.* **69**, 4377–4388.

Birch, F. (1968) Thermal expansion at high pressures, *J. Geophys. Res.* **73**, 817–819.

Birch, F. (1977) Isotherms of the rare gas solids, *J. Phys. Chem. Solids* **38**, 175–177.

Birch, F. (1978) Finite strain isotherm and velocities for single-crystal and polycrystaline

NaCl at high pressures and 300°K, *J. Geophys. Res.* **83**, 1257–1267.

Brown, J. M., and McQueen, R. G. (1982) The equation of state of iron and the earth's core, in *High-Pressure Research in Geophysics*, edited by S. Akimoto and M. H. Manghnani, pp. 611–623. Center for Academic Publishing, Tokyo.

Brown, J. M., Ahrens, T. J., and Shampine, D. L. (1984) Hugoniot data for pyrrhotite and the earth's core, *J. Geophys. Res.* **89**, 6041–6048.

Bullen, K. E. (1975) *The Earth's Density*. Chapman and Hall, London.

Burdick, L. J., and Helmberger, D. V. (1978) The upper mantle P velocity structure of the western United States, *J. Geophys. Res.* **83**, 1699–1712.

Burdick, L. J., and Powell, C. (1980) Apparent velocity measurements for the lower mantle from a wide aperture array, *J. Geophys. Res.* **85**, 3845–3856.

Butler, F., and Anderson, D. L. (1978) Equation of state fits to the lower mantle and outer core, *Phys. Earth Planet. Int.* **17**, 147–162.

Carter, W. J., Marsh, S. P., Fritz, J. N., and McQueen, R. G. (1971) The equation of state of selected materials for high pressure reference, in *Accurate Characterization of the High Pressure Environment*, Natl. Bur. Stand. U.S. Spec. Publ. 326, pp. 147–158.

Cleary, J. R. (1974) The D″ region, *Phys. Earth Planet. Int.* **9**, 13–27.

Cormier, V. (1985) Some problems with S, SKS and ScS observations and implications for the structure at the base of the mantle and the outer core, *Geophys. J.* **57**, 14–22.

Davies, G. F. (1973) Quasiharmonic finite-strain equations of state of solids, *J. Phys. Chem. Solids* **34**, 1417.

Davies, G. F. (1974) Effective elastic moduli under hydrostatic stress—I. Quasi-harmonic theory, *J. Phys. Chem. Solids* **35**, 1513–1520.

Davies, G. F., and Dziewonski, A. M. (1975) Homogeneity and constitution of the Earth's lower mantle and core. *Phys. Earth Planet. Int.* **10**, 336–343.

Donohue, J. (1974) *The Structure of the Elements*. J. Wiley and Sons, New York, 436 pp.

Dziewonski, A. M. (1984) Mapping the lower mantle: determination of lateral heterogeneity in P velocity up to degree and order 6, *J. Geophys. Res* **89**, 5929–5952.

Dziewonski, A. M., and Anderson, D. L. (1981) Preliminary reference Earth model, *Phys. Earth Planet. Int.* **25**, 297–357.

Fukao, Y., Nagahashi, T., and Mori, S. (1982) Shear velocity in the mantle transition zone, in *High Pressure Research in Geophysics*, edited by S. Akimoto and M. H. Manghnani, pp. 288–300. Center for Academic Publishing, Tokyo.

Fung, Y. C. (1965) *Foundations of Solid Mechanics*, Prentice-Hall, Inc., Englewood Cliffs, New Jersey, 525 pp.

Hart, R. S. (1975) Shear velocity in the lower mantle from explosion data, *J. Geophys. Res.* **85**, 4889–4894.

Hazen, R. M., and Jeanloz, R. (1984) Wüstite ($Fe_{1-x}O$): a review of its defect structure and physical properties, *Rev. Geophys. Space Phys.* **22**, 37–46.

Heinz, D. L., and Jeanloz, R. (1983) Inhomogeneity parameter of a homogeneous earth, *Nature (London)* **301**, 138.

Heinz, D. L., and Jeanloz, R. (1984) The equation of state of the gold calibration standard, *J. Appl. Phys.* **55**, 885–893.

Heinz, D. L., Jeanloz, R., and O'Connell, R. J. (1982) Bulk attenuation in a polycrystalline Earth, *J. Geophys. Res.* **87**, 7772–7778.

Hemley, R. J., and Gordon, R. G. (1985) Theoretical study of solid NaF and NaCl at high pressures and temperature, *J. Geophys. Res.* **90**, 7803–7813.

Hoffman, N. R. A., and McKenzie, D. P. (1985) The destruction of geochemical heterogeneities by differential fluid motion during mantle convection, *Geophys. J. Roy.*

Astronom Soc. **82**, 163–206.

Jackson, I., and Niesler, H. (1982) The elasticity of periclase to 3 GPa and some geophysical implications, in *High-Pressure Research in Geophysics*, edited by S. Akimoto and M. H. Manghnani, pp. 93–113. Center for Academic Publishing, Tokyo.

Jacobs, J. A. (1975) *The Earth's Core*, Academic Press, New York, 253 pp.

Jeanloz, R. (1979) Properties of iron at high pressure and the state of the core, *J. Geophys. Res.* **84**, 6059–6069.

Jeanloz, R. (1981) Finite-strain equation of state for high-pressure phases, *Geophys. Res. Lett* **8**, 1219–1222.

Jeanloz, R. (1982) Effect of coordination change on thermodynamic properties, in *High-Pressure Research in Geophysics*, edited by S. Akimoto and M. H. Manghnani, pp. 479–498. Center for Academic Publishing, Tokyo.

Jeanloz, R. (1986) High-pressure chemistry of the earth's mantle and core, in *Mantle Convection*, edited by W. R. Peltier, in press. Gordon and Breach, New York.

Jeanloz, R. (1985) Thermodynamics of phase transformations, in *Reviews in Mineralogy*, Vol. 14, edited by S. W. Kieffer and A. Navrotsky, pp. 389–428, Mineralogical Society of America Publications.

Jeanloz, R., and Morris, S. (1986) Temperature distribution in the crust and mantle, *Ann. Rev. Earth Planet. Sci.* **14**, 377–415.

Jeanloz, R., and Richter, F. M. (1979) Convection, composition and the thermal state of the lower mantle, *J. Geophys. Res.* **84**, 5497–5504.

Jeanloz, R., and Roufosse, M. (1982) Anharmonic properties: ionic models of the effects of compression and coordination, *J. Geophys. Res.* **87**, 10763–10772.

Jeanloz, R., and Thompson, A. B. (1983) Phase transitions and mantle discontinuities, *Rev. Geophys. Space Phys.* **21**, 51–74.

Jephcoat, A., Mao, H. K., and Bell, P. M. (1986) The static compression of iron to 78 GPa with rare gas solids as pressure-transmitting media, *J. Geophys. Res.* in press.

Jordan, T., and Anderson, D. L. (1974) Earth structure from free oscillations and travel times, *Geophys. J. Roy. Astronom Soc.* **36**, 411–459.

Kamb, B. (1968) Structural basis of the olivine-spinel stability relation, *Amer. Mineral.* **53**, 1439.

Knittle, E., and Jeanloz, R. (1984) Structural and bonding changes in cesium iodide at high pressures, *Science* **223**, 53–56.

Knittle, E., Rudy, A., and Jeanloz, R. (1985) High-pressure phase transition in CsBr, *Phys. Rev. B* **31**, 588–590.

Knittle, E., Jeanloz, R., and Smith, G. L. (1986a) The thermal expansion of silicate perovskite and stratification of the earth's mantle, *Nature* (*London*) **319**, 214–216.

Knittle, E., Jeanloz, R., and Smith, G. L. (1986b) Thermal expansion measurement of the (Mg, Fe)SiO_3 perovskite high-pressure phase, *J. Geophys. Res.* (submitted).

Lay, T., and Helmberger, D. V. (1983) A lower mantle S-wave triplication and the shear velocity structure of D″, *Geophys. J. Roy. Astronom. Soc.* **75**, 799–837.

Mao, H. K., and Bell, P. M. (1979) Equations of state of MgO and ε Fe under static pressure conditions, *J. Geophys. Res.* **84**, 4533–4546.

Masters, G. (1979) Observational constraints on the chemical and thermal structure of the earth's deep interior, *Geophys. J. Roy. Astronom. Soc.* **57**, 507–534.

Masters, G., Jordan, T. H., Silver, P. G., and Gilbert, F. (1982) Aspherical earth structure from fundamental spheroidal-mode data, *Nature* (*London*) **298**, 609–613.

McQueen, R. G., Marsh, S. P., Taylor, J. W., Fritz, J. N. and Carter, W. J. (1970) The equation of state of solids from shock wave studies, in *High Velocity Impact Phenom-*

ena, edited by R. Kinslow, pp. 293–417. Academic Press, New York.

Megaw, H. D. (1973) *Crystal Structures, A Working Approach.* W. B. Saunders Co., Philadelphia, 563 pp.

Merrill, R. T., and McElhinny, M. W. (1983) *The Earth's Magnetic Field.* Academic Press, New York, 401 pp.

Nakanishi, I., and Anderson, D. L. (1984) Measurements of mantle wave velocities and inversion for lateral heterogeneity and anisotropy—II. Analysis by the single-station method, *Geophys. J. Roy. Astronom. Soc.* **78**, 573–617.

Navrotsky, A., and Akaogi, M. (1985) α-β-γ Phase relations in Fe_2SiO_4–Mg_2SiO_4 and Co_2SiO_4–Mg_2SiO_4: Calculation from thermochemical data and geophysical applications, *J. Geophys. Res.* **89**, 10135–10140.

Nowick, A. S., and Berry, B. S. (1972) *Anelastic Relaxation in Crystalline Solids.* Academic Press, New York, 677 pp.

Ringwood, A. E. (1975) *Compositions and Petrology of the Earth's Mantle.* McGraw-Hill, New York, 618 pp.

Shankland, T. J., and Brown, J. M. (1985) Homogeneity and temperatures in the lower mantle, *Phys. Earth Planet. Int.* **38**, 51–58.

Shimoji, M. (1977) *The Structure of Liquid Metals.* Academic Press, New York, 391 pp.

Spiliopoulos, F., and Stacey, F. D. (1984) The earth's thermal profile: Is there a mid-mantle thermal boundary layer? *J. Geodyn.* **1**, 61–77.

Stacey, F. D., Brennan, B. J., and Irvine, R. D. (1981) Finite strain theories and comparisons with seismological data, *Geophys. Surv.* **4**, 189–232.

Stevenson, D. J., and Turner, J. S. (1979) Fluid models of mantle convection, in *The Earth: Its Origin, Structure and Evolution*, edited by M. W. McElhinny, pp. 227–263. Academic Press, New York.

Suzuki, I. (1975) Thermal expansion of periclase and olivine, and their anharmonic properties, *J. Phys. Earth* **23**, 145–159.

Tanimoto, T., and Anderson, D. L. (1985) Lateral heterogeneity and azimuthal anisotrophy of the upper mantle: Love and Rayleigh waves 100-250s, *J. Geophys. Res.* **90**, 1842–1858.

Turcotte, D. L., and Schubert, G. (1982) *Geodynamics.* John Wiley and Sons, New York, 450 pp.

Wallace, D. C. (1972) *Thermodynamics of Crystals.* J. Wiley and Sons, New York, 484 pp.

Weaver, J. S. (1976) Application of finite strain theory to non-cubic crystals, *J. Phys. Chem. Solids* **37**, 711–718.

Wiggins, R. A., McMechan, G. A., and Toksöz, M. N. (1973) Range of earth structure nonuniqueness implied by body wave observations, *Rev. Geophys. Space Phys.* **11**, 87–113.

Wolf, G., and Jeanloz, R. (1985a) Lattice dynamics and structural distortions of $CaSiO_3$ and $MgSiO_3$ perovskites, *Geophys. Rev. Lett.* **12**, 413–416.

Wolf, G., and Jeanloz, R. (1985b) Vibrational properties of model monatomic crystals under pressure, *Phys. Rev. B.* **32**, 7798–7813.

Woodhouse, J. H., and Dziewonski, A. M. (1984) Mapping the upper mantle: three-dimensional modeling of earth structure by inversion of seismic waveforms, *J. Geophys. Res.* **89**, 5953–5986.

Yoder, H. S., Jr. (1976) *Generation of Basaltic Magmas.* National Academy of Sciences, Washington, D. C., 265 pp.

Chapter 9
Thermodynamics of Stable Mineral Assemblages of the Mantle Transition Zone

O. L. Kuskov and R. F. Galimzyanov

Introduction

Although the mineral composition and inner structure of the transition zone of the Earth's mantle has often been examined in the literature from the petrologic–geochemical point of view (e.g., Akimoto, 1972; Ringwood, 1975; Akaogi and Akimoto, 1977; Liu, 1979; Yagi *et al.*, 1979a; Ito and Yamada, 1982; Jeanloz and Thompson, 1983) a detailed picture of phase equilibria at pressures over 100 kbar, both in the mantle and in silicate systems modeling its composition, has not yet been clearly outlined. The reason lies in the difficulties of conducting experimental studies at the *P–T* parameters of the transition zone which increase greatly with the number of phases. Thus, use of the numerical computation techniques of chemical thermodynamics becomes indispensable, making it possible to generalize and match the available experimental data on simple systems and, based upon these, to derive phase diagrams of much more complex systems.

Studies of the thermodynamics of chemical reactions taking place at ultrahigh pressures and the plotting of *P–T* diagrams representative of mantle multicomponent mineral systems offers the opportunity of throwing a new light on a series of traditional problems in geophysics and the petrology of the Earth's interior, namely:

1. Determination of the chemical and phase composition of deep mantle zones, especially the transition zone where phase transformations play the most important role
2. Study of the fine structure of the transition zone and its interpretation physically and chemically; modeling of density distribution with depth, of bulk moduli, seismic wave velocities, and other characteristics of the Earth's mantle without the use of seismologic data
3. Study of the chemical and physical evolution of our planet, and the tem-

perature distribution in the convecting mantle, taking phase equilibria into consideration

4. Elaboration of global thermodynamically consistent models of the Earth's composition and structure based on the chemistry of genuine mantle materials.

This chapter presents a new integrated approach to a problem of modeling phase composition and structure of the Earth's mantle involving the technique of conditioned optimization. This consists of a search for a vector of phase composition depending on pressure and temperature that minimizes the free-energy function of the modeled mineral system under conditions of numerous limitations. From a petrologic–geochemical point of view the final aim of such an approach consists of plotting a P–T diagram of a multicomponent system and modeling of the mineral composition of the mantle; from a geophysical point of view, the knowledge of equations of state of minerals and phase diagrams of an adequately common chemical composition enables us to predict seismic anomalies and to determine their intensity.

Thermodynamic Approach to the Study of Mineral Composition of Deep Zones in the Earth's Mantle

Defining the Task of Determining Phase Compositions of Mineral Systems at Ultrahigh Pressures

At present the investigation of equilibria in chemical systems with the help of physicochemical modeling methods is commonly applicable in geochemistry (Helgeson, 1969; Shvarov, 1981; Karpov *et al.*, 1978; Saxena and Eriksson 1983; Kraynov *et al.*, 1981; Kuskov *et al.*, 1983a). Its basis is postulates of chemical thermodynamics according to which, at constant pressure and temperature, spontaneous processes take place that tend to the formation of such substances and in such amounts that the Gibbs free energy of the system is decreased, reaching a minimum in the state of equilibrium. The maximum number of phases coexisting in phase equilibrium is limited by the Gibbs phase rule. Usually a list of minerals potentially able to exist in an equilibrium system greatly exceeds the number determined by the rule of phases (these systems are commonly called multicomponent systems). While calculating the equilibrium composition of such systems, it is first of all necessary to find stable phase associations and only then to determine their equilibrium compositions.

To apply algorithms available to the calculation of the composition of heterogeneous systems at ultrahigh pressures and temperatures, one should know the thermal equations of state of all phases; lack of this knowledge still hampers thermodynamic analysis of geochemical processes in the Earth's interior. Therefore, in the present chapter particular attention is paid to developing equations of state for minerals.

Before this task is mathematically formulated, a list of assumptions forming the basis of the equations should be specified.

1. We discuss here only systems composed of individual substances, i.e., solid phases of constant composition, ignoring the possibility of solid solution formation. This is the most simple model of an actual multicomponent system but when information on the thermodynamic properties of solutions and their equations of state are lacking, it can be considered the first stage in the development of more adequate models.
2. Phases are considered to be isotropic (polycrystalline state) and to be under hydrostatic equilibrium. Pressures in each of the phases are the same and equal to the total pressure in the system. In the calculation of phase equilibria at ultrahigh pressures conditions are considered to be most favorable when all initial information on thermodynamic constants is obtained with the help of the same standard samples used in thermochemical and ultrasonic experiments, in the measurement of compressibility and thermal expansion. Unfortunately, such as ideal situation is never realized in practice. Therefore, to derive the equations of state of minerals and the phase diagrams of mineral systems use is made of initial thermodynamic constants derived from various experiments conducted on different samples which may possess certain defects in their crystalline structure.
3. A set of phases potentially able to exist in an equilibrium system and their thermodynamic properties at atmospheric pressure assumed to be known from experiments, thus, within the scope of this approach the authors can neither predict the occurrence of new phases in the system nor forcast their properties.

The standard thermodynamic functions of minerals, their elastic and thermal constants, and the P–T parameters of experimentally studied equilibria serve as the basis of the calculations made in this chapter.

Algorithm for the calculation of P–T diagrams of state for multicomponent systems based on constant composition phases

Free energy $G(P, T)$ of a closed system consisting of phases of constant composition can be expressed as follows (Lukas *et al.*, 1982): see also chapter 2

$$G(P, T) = \bar{\mu}\bar{\mathbf{x}}^t$$

or can be transformed into

$$G(P, T)(x_1, x_2, \ldots, x_n) = \sum_{i=1}^{n} \mu_i x_i = \sum_{i=1}^{n} \left(\Delta G^\circ_{f_i, T} + \int_0^P V_i \, dP \right) x_i \qquad (1)$$

where $\bar{\boldsymbol{\mu}} = (\mu_1, \mu_2, \ldots, \mu_n)$ is the vector row of chemical potentials of the system's phases; $\mathbf{X}^t(x_1, x_2, \ldots, x_n)$ is the vector column of the phase composition of the system; t is the transposition operations; $\Delta G^\circ_{f_i}(1, T)$ is the standard free energy of the formation of mol of phase i; and V_i is the molar volume of phase i at fixed P and T.

If the bulk chemical composition of the system is given, then the $\bar{\mathbf{x}}$ values should satisfy the equations of mass balance:

$$\bar{\mathbf{H}}\bar{\mathbf{x}} = \overline{\mathbf{M}} \tag{2}$$

where $\bar{H}$ is the ln matrix of content (by weight) of chemical elements or their oxides in phases of the system and $\overline{M}$ is the matrix column of content (by weight) of chemical elements or their oxides for the system as a whole. Likewise, Eq. (2) may be derived in the form of:

$$\begin{aligned} h_{11}x_1 + h_{12}x_2 + \cdots h_{1n}x_n &= m_1 \\ h_{21}x_1 + h_{22}x_2 + \cdots h_{2n}x_n &= m_2 \\ h_{l1}x_1 + h_{l2}x_2 + \cdots h_{ln}x_n &= m_l \end{aligned} \tag{3}$$

A number of limitations arise with regard to the material properties of the system in modeling the mantle's mineral composition correlated with its actual characteristics. For instance, in the hydrostatic approximation employed by the authors the adiabatic bulk modulus and coefficient of thermal expansion will be written as:

$$\begin{aligned} \rho &= \frac{M}{V} = \frac{1}{V}\sum_{i=1}^{n} W_i X_i \\ K_s &= -V\left(\frac{\partial P}{\partial V}\right)_S = V\left|\sum_{i=1}^{n} V_i X_i\right| K_{S_i} \\ \alpha &= \frac{1}{V}\left(\frac{\partial V}{\partial T}\right)_P = \frac{1}{V}\sum_{i=1}^{n} \alpha_i V_i X_i \end{aligned} \tag{4}$$

where ρ, K_S, α, M, V are the density, adiabatic bulk modulus, coefficient of thermal expansion, mass, and volume, respectively, of the system as a whole; $V = \sum_{i=1}^{n} V_i X_i$; V_i, W_i, K_{S_i}, α_i, $(i = 1, n)$ are the molar volumes, molecular masses, adiabatic bulk moduli, and coefficients of thermal expansion, respectively, of phases included in the system.

Assuming that all limitations under consideration are linear, the search for a phase composition vector of a thermodynamic system (at constant P and T) that minimizes its free energy is a classic problem of linear programming. A standard simplex method can be applied to solve it (Intriligator, 1975). Varying P and T with a certain increment and solving the equation via linear programming, one can obtain a full phase diagram of the system.

Algorithm for Calculating the Thermodynamic Characteristics of Phase Transitions on Equilibrium Line

Apart from the phase composition of the system under study, detailed information on the thermodynamic characteristics of phase transformations is often needed, namely: thermal and volumetric effects and P–T equilibrium parameters. These can be obtained from phase equilibrium data, equations of state, and standard thermodynamic functions of phases. On the equilibrium line

$$\Delta G(P, T) = \Delta G_T^\circ + \int_0^P \Delta V \, dP \tag{5}$$

where: G_T° is the standard free energy of transformation at $P = 0$ and V is the volumetric effect of transformation at P and T.

At equilibrium Eq. (5) may be transformed to:

$$\int_0^{P_{\mathrm{eq}}} \Delta V \, dP = -\Delta G_T^\circ \tag{6}$$

If the experimental P–T parameters of equilibria and the equations of state of the phases are known, Eq. (6) permits one to compute the change in standard free energy of the reaction and, if necessary, to calculate the standard free energies of the formation of the phases. As regards high-pressure phases, when direct measurements are impossible (calorimetry, method of electromotive forces, etc.), the above is the only effective method of determining their standard thermodynamic functions.

When solving Eq. (6) at P_{eq} for different temperatures, it is possible to calculate the enthalpy of the reaction on the equilibrium line according to the Clapeyron-Clausius equation:

$$\Delta H(P, T) = T \Delta V \frac{dP_{\mathrm{eq}}}{dT} \tag{7}$$

As suggested, computation of the thermodynamic characteristics of phase equilibria and the P–T diagrams of multicomponent mineral systems of variable composition is performed according to the following scheme:

1. Derivation of thermal equations of state for all phases and tabulation of V, $\int_0^P V \, dP$ values, bulk moduli of different order, coefficients of thermal expansion, the Grüneisen parameter, etc.
2. Determination of information lacking on standard free energies of phase transformations on the basis of derived equations of state and experimental P–T parameters of equilibria.
3. Calculating the equilibrium phase composition of the system at different P and T; identification of stable phase assemblages.

   To complete plotting of P–T diagrams by geometrical methods the use should be made of the general properties of multibeam diagrams according to Schreinemakers (1948) and Korzhinsky (1959). One problem is that solving the linear programming task with the help of the "dense net" is not feasible. It is more convenient to calculate the phase composition with the help of a more sparse net, e.g., with steps of 50 kbar (pressure) and 200K (temperature), and to distinguish several stable equilibria according to changes in phase composition. Then, using general properties of multibeam diagrams, one can apply geometrical methods to the complete plotting of the P–T diagram under study. In the search for pressure values on the equilibrium line, Eq. (6) is to be solved numerically with the help of a known Newton-Raphson algorithm.

4. Computation of thermodynamic characteristics of stable phase equilibria (thermal, volumetric, and entropic effects).
5. Analysis of the influence of the initial information errors on computation results.

Algorithm for the Calculation of an Equation of State for a Solid Body

Knowledge of the equation of state offers a greater possibility of applying the general laws of thermodynamics to concrete physicochemical systems under a wide range of temperatures and pressures. In the geophysical literature wide use is made of various semiempirical equations of state (Birch, 1952; Zharkov and Kalinin, 1968; Al'tshuler, 1965; Ullmann and Pan'kov, 1970; Thomsen, 1970; Davies, 1973; Mulargia and Boschi, 1980; Stacey *et al.*, 1981). Comparative analysis of semiempirical methods of deriving the thermal equations of state of a solid body (potential method, theories of elasticity, empirical equations of state) was conducted by Kuskov *et al.* (1983b), who also studied the sensitivity of both isothermal P–V curves of compression and $\int_0^P V\,dP$ as methods of deriving an equation of state, evaluated the reliability and accuracy of thermodynamic calculations of chemical and phase equilibria at ultrahigh pressures, and elaborated several practical recommendations for applying these different models and methods to deriving an equation of state for minerals. It should be noted that the quasiharmonic finite-strain equations of Davies (1973) and the Born-Mayer potential method (Zharkov and Kalinin, 1968) as applied to calculations of phase diagrams will yield rather similar results because the $\int_0^P V\,dP$ integrals will smooth the errors of different methods. Some of the most simple empirical methods, however, are exceptions (the Murnaghan equation and probably that of Tait); the use of them may lead to distortions in the phase diagrams.

As regards derivation of a thermal equation of state for solid phases, a universal method has been suggested and elaborated by Kuskov *et al.* (1982) and Kuskov and Galimzyanov (1982a). This is the further development of a well-known potential method (Zharkov and Kalinin, 1968) to meet the objectives of thermodynamic calculations in modeling geochemical processes. Experimental data on elastic, thermal, and caloric constants at 1 atm and 298K serve as the initial information.

Herein, only the basic formulas, indispensible to understanding the computation scheme of the method, are presented.

The equation of state of solids in the Mie-Grüneisen form is written as (Zharkov and Kalinin, 1968; Davies, 1973):

$$P(x, T) = P_p(x) + \frac{\gamma(x)\rho_0}{x} E_T(x, T) \tag{8}$$

where:

$$E_T(x, T) = E(x, T) - E_p(x) = \frac{R}{W}\left[\frac{9}{8}\theta + 3TD\left(\frac{\theta}{T}\right)\right] \tag{9}$$

is the specific thermal constituent of inner energy in the Debye approximation;

$$E_p(x) = \frac{3A}{b\rho_0} e^{b(1-x^{1/3})} - \frac{3K}{\rho_0} x^{-1/3} \tag{10}$$

is the potential energy in the Born-Mayer form;

$$P_p(x) = Ax^{-2/3} e^{b(1-x^{1/3})} - Kx^{-4/3} \tag{11}$$

is the potential constituent of pressure; $x = \rho_0/\rho$, where ρ_0 is density of matter under the normal conditions of $T = 298$, $x = 1$, $P_0 = 0$ (the values containing "0" always relate to normal conditions; ρ is the density at P, T conditions; R is the universal gas constant, W is the average atomic weight, defined as the ratio of molecular weight to the number of atoms in a molecule; θ is the Debye temperature; $D\left(\frac{\theta}{T}\right)$ is the Debye function; γ is the Grüneisen parameter; and A, b, and K are parameters of potential.

The integrated formula is used for the $\gamma(x)$ function as:

$$\gamma(x) = -\frac{x}{2} \frac{\partial^2(P_p x^y)\partial x^2}{\partial(P_p x^y)/\partial x} + \frac{3y-4}{6} = -\frac{1}{6} - \frac{\partial K_p/\partial \ln x + yK_p}{2(K_p - yP_p)} \tag{12}$$

Assuming, in accordance with Romain *et al.* (1976), that the y parameter may aquire actual values to be determined if the Grüneisen parameter from Eq. (12) is equal to its thermodynamic value under normal conditions:

$$\gamma_{\text{th}} = K_{0s} \cdot \alpha_{0V}/\rho_0 \cdot C_{0p} \tag{13}$$

where α_v is the volumetric coefficient of thermal expansion, C_p is the specific heat capacity at constant pressure, and:

$$K_S(x, T) = K_p(x) + \left[1 + \gamma(x) - \frac{\partial \ln \gamma}{\partial \ln x}\right] \frac{\rho_0 \gamma(x)}{x} E_T(x, T) \tag{14}$$

is the adiabatic bulk modulus; $K_p(x)$ is the potential constituent of bulk modulus:

$$K_p(x) = -x\frac{\partial P_p}{\partial x} = \frac{Ax^{-2/3}}{3}(bx^{1/3} + 2)e^{b(1-x^{1/3})} - \frac{4}{3}Kx^{-4/3} \tag{15}$$

As a rule, y values obtained in practice do not go beyond the 0–2 interval.

The relation between the Debye temperature and the Grüneisen parameter is represented by the equation:

$$\theta(x) = \theta_0 e^{\int_x^1 [\gamma(x)]/x \, dx} = \theta_0 x^{1/6} \left[\frac{K_p(x) - yP_p(x)}{K_p(1) - yP_p(1)}\right] \tag{16}$$

Differentiating by volume at constant temperature, Eq. (14) may be transformed into:

$$\begin{aligned}\left(\frac{\partial K_s}{\partial \ln x}\right)_T = -K_T\left(\frac{\partial K_s}{\partial P}\right) &= \frac{\partial K_p}{\partial \ln x} \\ &- \left(\Gamma^2 - \gamma\frac{\partial \ln \gamma}{\partial \ln x}\right)\frac{\rho_0 \gamma}{x} E_T + \frac{\rho_0 \gamma^2}{x}\Gamma T C_V\end{aligned} \tag{17}$$

where: $\Gamma = 1 + \gamma - (\partial \ln \gamma/\partial \ln x)$ and C_V is the specific heat capacity at constant volume in the Debye approximation. When deriving Eq. (17), it was assumed that $(\partial \ln \gamma/\partial \ln x)$ was not dependent on volume, as has been supported by the experimental data available (Boehler and Ramakrishnan, 1980).

A full system of equations to compute the A, b, and K parameters is formed by formulas that may be represented in the following form (Kuskov and Galimzyanov, 1982a):

$$\begin{cases} K = A + \rho_0 \gamma_0 E_{0T} \\ A(b-2) = 3K_{0s} + (4 - 3\Gamma_0)\rho_0\gamma_0 E_{0T} \\ A(b^2 + 3b - 12) = \dfrac{9K_{0s}(\partial K_{0s}/\partial P)_T}{1 + \alpha_0 \gamma_0 T_0} \end{cases}$$
$$+ [16 + 9\gamma_0(\partial \ln \gamma/\partial \ln x) - 9\Gamma^2) \cdot \rho_0 \gamma_0 E_{0T}$$
$$+ 9\Gamma_0 \rho_0 \gamma_0 C_{0v} T_0 \tag{18}$$

Numerical values of the parameters are determined by solving the system of Eq. (18) using an iteration procedure.

To derive the equation of state, seven constants for each mineral are to be determined under standard conditions: the adiabatic bulk modulus K_{0s}, its derivative on pressure $K_0' = (\partial K_s/\partial P)_T$, heat capacity at constant pressure C_{0p}, volumetric coefficient of thermal expansion α_0, Debye temperature θ_0, density ρ_0, and molecular weight.

Utilization of the elastic and thermal constants instead of an approximation of the experimental isotherms and Hugoniot adiabats at high pressures to derive an equation of state (like that in Zharkov and Kalinin, 1968) allows comparative analysis and critical selection of initial experimental data on the key values level (a similar principle is made use of in thermochemistry and chemical thermodynamics). The above provides adequate information on the equation of state of matter, which becomes of importance both in creating a thermodynamic computer data base and in evaluating errors. It also provides a universal method for defining various problems of physicochemical modeling. Such a way of deriving equations of state may be briefly called the thermochemical version of the potential method.

The information obtained on the thermodynamic properties of substances, namely, P–V–T data, $\int_0^P V\,dP$, Hugoniot pressures, elastic, thermal, and caloric characteristics, are tabulated up to 1 Mbar and 3000K. Within the megabar range typical values of relative errors for $\int_0^P V\,dP$, molar volume, bulk moduli, and coefficient of thermal expansion make up 1, 2, 8, and 10%, respectively (Kuskov *et al.*, 1983c).

Table 1 gives the initial experimental information on the elastic, thermal, and caloric constants of solid phases at 1 atm and 298.15K, that are needed to calculate the equations of state of minerals in the system.

As a rule, ultrasonic methods of investigation have been preferred in the selection of elastic constants. These are used to measure the velocity of ultrasonic waves propagated in a substance at elevated temperatures and pressures, thus enabling us to determine in the course of one experiment the value of the adiabatic bulk modulus, its derivatives both by temperature and pressure, the Debye temperature, and other elastic characteristics with a high degree of accuracy (e.g., for the bulk modulus the error usually reported is 0.5–1%). Therefore, experimental information obtained with the help of these methods is comparable to and complementary to that obtained thermodynamically.

Calculations of isotherms and shock adiabatic curves of silicates and oxides made with the help of the suggested version of the potential method show a good agreement with experimental data (see Fig. 1) (Kuskov *et al.*, 1982, 1983b. Kuskov and Galimzyanov, 1982a; Kuskov, 1984). They thus give evidence of the high accuracy of calculating the equation of state of substances according to their elastic constants determined from ultrasonic experiments at pressures not exceeding 10 kbar. Therefore, theoretical curves give a good description of experimental points obtained at pressures 10–100 times as much as in the ultrasonic experiments from which the initial information has been gained. This testifies to the high accuracy of the equation of state in the Mie-Grüneisen form and the effectiveness of Equations. (8)–(18) used to describe both isothermal curves of compression and shock adiabats of solids under a wide range of pressures and temperatures.

Table 2 presents the computerized values of $\int_0^P V\,dP$, molar volumes, adiabatic bulk moduli, and thermal expansion coefficients depending upon temperature and pressure. Mean-square errors are omitted here but they are given in papers by Kuskov and Galimzyanov (1982a, b), Kuskov *et al.* (1983c), and Kuskov (1984). Using these data, calculation of P–T parameters for any equilibrium within the system under study is made possible based on Eq. (5). Unfortunately, in this chapter we cannot present tables with smaller pressure increments; however, to verify or recalculate the diagrams given below, the use of interpolation by cubic spline is sufficient. Specific estimations made by Kuskov *et al.* (1983b) have shown that the error of "integration" of the equation of state ($\int_0^P V\,dP$) obtained with the help of the thermochemical version of the potential method in the megabar range of pressures is the same or even less than that in calorimetric determinations at 1 atm. The results of calculations made by the authors and the estimation of accidental and systematic errors (Kuskov *et al.*,

Table 1. Initial parameters for thermal equations of state of minerals at 1 atm and 298K. Standard deviations in parentheses.

Mineral formula	K_S, kbar	K_S^1	ρ, $g \cdot cm^{-3}$	C_P $cal \cdot mol^{-1} K^{-1}$	θ_{el}, K	$\alpha_V \times 10^6$, K^{-1}	Gram formula weight	References[a] K, K^1, θ	C_P	α_V
MgO	1630(5)	4.5(0.2)	3.583(0.001)	9.03(0.05)	936(10)	31.2(0.5)	40.32	1–3	4	5
FeO	1830(50)	7.0(1)	5.865(0.002)	11.50(0.1)	493(10)	37.5(1)	71.85	6	4	7
CaO	1145(10)	5.0(0.5)	3.345(0.001)	10.24(0.05)	670(10)	29.0(1)	56.08	8–10	4	11
Al_2O_3	2500(10)	4.1(0.1)	3.986(0.001)	18.89(0.05)	1042(5)	15.1(0.4)	101.96	12–15	4	13
SiO_2 (α-quarz)	378(5)	6.0(0.2)	2.649(0.001)	10.63(0.02)	570(5)	35.5(0.5)	60.09	3, 16	4	7
SiO_2 (coesite)	1137(100)	5.8(1)	2.92(0.05)	10.75(0.1)	675(50)	8.0(1)	60.09	17, 18	4	19
SiO_2 (stishovite)	2930(150)	5.0(0.5)	4.287(0.002)	10.27(0.2)	1050(100)	8.9(1.5)	60.09	20–22	4	19
α-Mg_2SiO_4	1288(10)	5.1(0.2)	3.213(0.001)	28.18(0.14)	763(20)	26.0(1)	140.71	23–25	4	26
β-Mg_2SiO_4	1760(50)	4.3(1)	3.474(0.01)	26.3(0.5)	950(50)	20.6(1)	140.71	27, 28	30	29
γ-Mg_2SiO_4	2100(150)	4.0(1)	3.549(0.01)	25.6(0.5)	849(70)	19.0(1)	140.71	27, 31	30	32
α-Fe_2SiO_4	1380(10)	5.0(0.5)	4.393(0.001)	31.54(0.1)	512(5)	26.0(1)	203.78	33	35	34
γ-Fe_2SiO_4	1930(40)	4.0(0.5)	4.848(0.01)	31.20(0.1)	609(15)	21.3(2)	203.78	31	30	32
$FeSiO_3$	1090(50)	5.0(0.5)	3.970(0.01)	21.65(0.4)	567(30)	16.0(3)	131.93	3, 36	30	30
$MgSiO_3$ (enstatite)	1080(30)	5.0(0.5)	3.198(0.01)	18.9(0.5)	734(20)	28.0(2)	100.40	3, 37, 38	4	7
$MgSiO_3$ (ilmenite)	2100(100)	4.0(1)	3.807(0.01)	19.0(1)	850(50)	20.0(3)	100.40	39	Estimate	Estimate
$MgSiO_3$ (perovskite)	2500(100)	3.8(1)	4.10(0.01)	19.4(1)	985(50)	20(3)	100.40	40–42	Estimate	Estimate
$Mg_3Al_2Si_3O_{12}$	1770(10)	4.5(0.2)	3.559(0.001)	77.75(0.2)	788(10)	19(0.5)	403.19	43, 44	46	7, 45
$Ca_3Al_2Si_3O_{12}$	1700(10)	4.3(0.2)	3.595(0.001)	79.00(0.2)	819(10)	18(0.5)	450.47	47–49	46, 50, 51	7
$CaMgSi_2O_6$	1140(40)	4.8(1)	3.277(0.002)	39.8(0.1)	662(5)	31(0.7)	216.58	52, 53	4	54
$MgAl_2O_4$	1972(5)	4.3(0.3)	3.582(0.001)	27.71(0.1)	863(5)	19(0.6)	142.27	3, 55, 56	4	32
Al_2SiO_5	2650(100)	4.5(1)	3.664(0.01)	29.24(0.2)	914(50)	11(2.5)	162.05	57	58	7
$CaMgSiO_4$	970(50)	5.0(1)	3.047(0.004)	29.4(0.7)	570(50)	32(1.5)	156.48	59	60	7
$CaAl_2Si_2O_8$	920(20)	5.0(0.5)	2.760(0.002)	50.57(0.5)	546(10)	14(0.5)	278.21	61	58	7

Table 1. (Continued)

Mineral formula	K_S, kbar	K_S^1	ρ, $g \cdot cm^{-3}$	C_P $cal \cdot mol^{-1} K^{-1}$	θ_{el}, K	$\alpha_V \times 10^6$, K^{-1}	Gram formula weight	References[a] K, K^1, θ	C_P	α_V
$CaSiO_3$ (wollastonite)	820(50)	4.5(1)	2.909(0.002)	20.93(0.2)	570(30)	30(1.5)	116.16	Estimate	4	7, 11
$CaSiO_3$ (wollastonite-II)	1000(100)	4.5(1)	3.09(0.01)	20.9(1)	640(50)	30(3)	116.16	Estimate	Estimate	Estimate
$CaSiO_3$ (perovskite)	2300(200)	4.0(2)	4.28(0.01)	20.4(1)	860(100)	20(2)	116.16	Estimate	Estimate	Estimate
Ca_3SiO_5	1000(100)	4.0(1)	3.139(0.002)	41.01(0.4)	600(100)	30(3)	228.32	Estimate	53	Estimate
$Ca_3Si_2O_7$	900(100)	4.0(1)	2.988(0.002)	51.24(0.5)	600(100)	37(4)	288.41	Estimate	53	Estimate
Ca_2SiO_4	1300(130)	5.0(1)	3.338(0.05)	30.69(0.3)	700(100)	37(4)	172.25	Estimate	53	Estimate
$Ca_3MgSi_2O_8$	1000(100)	5.0(1)	3.149(0.05)	60.3(0.6)	600(100)	29(3)	328.71	Estimate	4	7
$Ca_2Al_2SiO_7$	1000(100)	4.0(1)	3.039(0.05)	49.09(0.5)	600(100)	23(1)	274.204	Estimate	53	7
$CaAl_2SiO_6$	1500(150)	5.0(1)	3.438(0.05)	39.59(0.4)	700(100)	30(3)	218.13	Estimate	53	Estimate

[a] References:

MgO: (1) Spetzler (1970), (2) Anderson (1966), (3) Chung (1974), (4) Naumov, *et al.* (1971), (5) Suzuki I. (1975).
FeO: (6) Bonczar and Graham (1982), (7) Clark (1966).
CaO: (8) Son and Bartels (1972), (9) Chang and Graham (1977), (10) Dragoo and Spain (1977), (11) Touloukian and Ho (1977).
Al_2O_3: (12) Soda, *et al.* (1966), (13) Anderson (1980), (14) Wachtman *et al.* (1960), (15) Gieske and Barsch (1968).
SiO_2(quartz): (16) Bass *et al.* (1981).
SiO_2(coesite): (17) Weidner and Carleton (1977), (18) Chung (1973), (19) Weaver *et al.* (1979).
SiO_2(stishovite): (20) Chung (1979), (21) Liebermann and Ringwood (1977), (22) Olinger (1977).
α-Mg_2SiO_4: (23) Chung (1971), (24) Kumazawa and Anderson (1969), (25) Graham and Barsch (1969), (26) Suzuki (1975).
β-Mg_2SiO_4: (27) Mizukami *et al.* (1975), (28) Chung (1972), (29) Suzuki *et al.* (1980), (30) Watanabe (1982).
γ-Mg_2SiO_4: (31) Liebermann (1975), (32) Suzuki *et al.* (1979).
α-Fe_2SiO_4: (33) Sumino (1979), (34) Suzuki *et al.* (1981), (35) Robie *et al.* (1982).
γ-Fe_2SiO_4: see references 30–32.
$FeSiO_3$: (36) Akimoto (1975).
$MgSiO_3$(enstatite): (37) Weidner *et al.* (1978), (38) Liebermann (1974).
$MgSiO_3$(ilmenite): (39) Liebermann (1976).
$MgSiO_3$(perovskite): (40) Liebermann *et al.* (1977), (41) Yagi *et al.* (1979), (42) Jackson and Ahrens (1979).

$Mg_3Al_2Si_3O_{12}$(pyrope): (43) Leitner *et al.* (1980), (44) Levien *et al.* (1979), (45) Sumino and Nishizawa (1978), (46) Haselton and Westrum (1980).

$Ca_3Al_2Si_3O_{12}$(grossular): (47) Halleck (1973), (48) Isaak and Graham (1976), (49) Babuska *et al.* (1970), (50) Westrum *et al.* (1979), (51) Kolesuik *et al.* (1978).

$CaMgSi_2O_6$(diopside): (52) Levien *et al.* (1979), (53) Levien and Prewitt (1981), (54) Finger and Ohashi (1976).

$MgAl_2O_4$: (55) O'Connel and Graham (1971), (56) Chang and Barsch (1973).

Al_2SiO_5(kyanite): (57) Kieffer (1980), (58) Robinson *et al.* (1982).

$CaMgSiO_4$: (59) Horai and Simmons (1970), (60) Helgeson *et al.* (1978).

$CaAl_2Si_2O_8$: (61) Liebermann and Ringwood (1976).

Ca_2SiO_4(larnite), $CaSiO_3$(wollastonite, wollastonite-II, perovskite), Ca_2SiO_5(hatrurite), $Ca_3Si_2O_7$(rankinite), $Ca_3MgSi_2O_8$(merwinite), $Ca_2Al_2SiO_7$(gehlenite), $CaAl_2SiO_6$ (Ca-Al-pyroxene): elastic properties are estimated on the basis of general relationships among sound speeds $V_\phi = 1.42\ \rho^{1.25}(20.2/\bar{m})^{0.125}$ km · sec^{-1}, (Shankland and Chung (1974) and $V_P = 1.8\ \rho^{1.25}(20.2/\bar{m})^{0.125}$ km · sec^{-1} (authers' estimate).

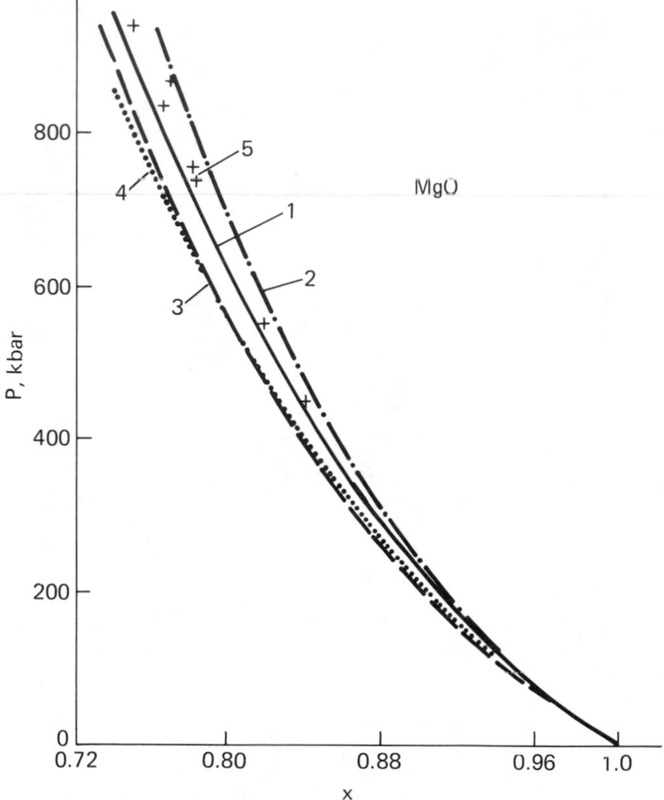

Fig. 1. Influence of variations in the values of the elasticity modulus over the equations of state of periclase: 1, test calculation ($K_0 = 1630$ kbar, $K'_0 = 4.5$); 2, $K_0 = 1630$ kbar, $K'_0 = 5,5$; 3, $K_0 = 1530$ kbar, $K'_0 = 4.5$; 4, $K_0 = 1630$ kbar, $K'_0 = 3.8$; 5, experimental data of Mao *et al.* (1979).

1983b) have both shown that the $\int_0^P V\,dP$ values are determined with high accuracy and can be tabulated in thermodynamic reference books together with standard thermodynamic properties. Because thermodynamic calculations of chemical and phase equilibria at high $P-T$ parameters are based on $\int_0^P V\,dP$ value determination, the reliability of such calculations becomes strictly justified.

Computation of Phase Diagrams in the System $MgO-FeO-CaO-Al_2O_3-SiO_2$ at the $P-T$ Parameters of the Mantle Transition Zone

Experimental studies of the $MgO-FeO-CaO-Al_2O_3-SiO_2$ system (Akimoto 1972; Ringwood, 1975; Suito and Kawai, 1979; Liu, 1979a; Yagi *et al.*, 1979a; Ito and Yamada, 1982) have yielded basic information on the existence of

Table 2. $\int_0^P V\,dP$ (kcal · mol^{-1}) (Row 1), molar volume (cm^3 · mol^{-1}) (Row 2), adiabatic modulus (Mbar) (Row 3), and thermal expansion (10^6 deg^{-1}) (Row 4) data.

Pressure, Mbar	Row	Periclase, MgO at T, K: 1000	1500	1800	Ferrous oxide, FeO, at T, K: 1000	1500	1800
0.1	1	26.77	27.33	27.70	29.27	29.85	30.22
	2	10.88	11.08	11.21	11.94	12.15	12.28
	3	1.972	1.896	1.849	2.345	2.237	1.168
	4	34.7	38.0	39.8	33.2	35.1	36.3
0.2	1	52.14	53.12	53.76	57.24	58.25	58.89
	2	10.37	10.53	10.63	11.48	11.64	11.74
	3	2.393	2.322	2.278	2.989	2.876	2.813
	4	28.9	31.3	32.5	27.4	28.9	30.1
0.3	1	76.42	77.74	78.60	84.22	85.59	86.46
	2	9.96	10.09	10.18	11.12	11.25	11.34
	3	2.797	2.729	2.687	3.615	3.503	3.434
	4	25.0	26.9	27.8	23.7	24.8	25.4
0.6	1	144.46	146.54	147.88	161.04	163.20	164.56
	2	9.08	9.17	9.22	10.37	10.46	10.52
	3	3.944	3.881	3.843	5.417	5.299	5.226
	4	18.1	19.5	20.0	17.2	18.0	18.3

Pressure, Mbar	Row	Lime, CaO, at T, K: 1000	1500	1800	Corundum, Al_2O_3, at T, K: 1000	1500	1800
0.1	1	39.32	39.95	40.34	60.81	61.51	61.95
	2	15.85	16.06	16.20	24.96	25.23	25.40
	3	1.559	1.511	1.483	2.825	2.763	2.725
	4	26.4	27.8	28.6	20.7	22.1	22.7
0.2	1	76.07	77.14	77.82	119.43	120.73	121.55
	2	14.95	15.12	15.22	24.13	24.36	24.50
	3	2.010	1.964	1.936	3.216	3.156	3.119
	4	21.4	22.4	22.9	18.5	19.6	20.1
0.3	1	110.97	112.40	113.30	176.22	178.04	179.19
	2	14.28	14.41	14.49	23.41	23.62	23.74
	3	2.442	2.396	2.369	3.595	3.537	3.501
	4	18.2	19.0	19.3	16.7	17.7	18.1
0.6	1	207.99	210.18	211.55	337.76	340.82	342.75
	2	12.90	12.99	13.04	21.74	21.89	21.98
	3	3.668	3.624	3.597	4.682	4.627	4.593
	4	13.0	13.5	13.7	13.2	14.0	14.3

Pressure, Mbar	Row	α-Quartz, SiO_2, at T, K: 1000	1500	1800	Coesite, SiO_2, at T, K: 1000	1500	1800
0.1	1	50.18	52.00	52.80	47.67	47.94	48.10
	2	19.61	19.90	20.09	19.25	19.33	19.38
	3	0.846	0.795	0.766	1.615	1.564	1.533
	4	28.3	30.8	32.3	7.9	8.4	8.6

Table 2. (Continued)

Pressure, Mbar	Row	α-Quartz, SiO_2, at T, K: 1000	1500	1800	Coesite, SiO_2, at T, K: 1000	1500	1800
0.2	1	95.37	97.08	98.21	92.42	92.82	93.07
	2	17.84	17.72	18.10	18.24	18.29	18.31
	3	1.327	1.268	1.234	2.123	2.067	2.034
	4	17.0	18.3	19.1	4.7	4.9	5.1
Pressure, Mbar	**Row**	**Stishovite, SiO_2, at T, K: 1000**	**1500**	**1800**	**Forsterite, α-Mg_2SiO_4, at T, K: 1000**	**1500**	**1800**
0.1	1	33.25	33.50	33.65	103.40	105.31	106.54
	2	13.69	13.79	13.85	41.80	42.47	42.90
	3	3.337	3.270	3.229	1.695	1.632	1.595
	4	13.3	14.0	14.3	30.9	32.9	33.9
0.2	1	65.51	65.96	66.25	200.54	203.87	205.99
	2	13.31	13.39	13.44	39.60	40.12	40.45
	3	3.812	3.745	3.705	2.159	2.098	2.061
	4	11.9	12.5	12.8	25.3	26.8	27.5
0.3	1	96.92	97.56	97.95	293.11	297.56	300.38
	2	12.98	13.05	13.10	37.93	38.35	38.62
	3	4.275	4.208	4.168	2.605	2.544	2.508
	4	10.8	11.4	11.6	21.7	22.9	23.4
0.6	1	187.07	188.16	188.83			
	2	12.21	12.26	12.30			
	3	5.609	5.542	5.502			
	4	8.6	9.1	9.2			
Pressure, Mbar	**Row**	**β-Spinel, β-Mg_2SiO_4, at T, K: 1000**	**1500**	**1800**	**γ-Spinel, γ-Mg_2SiO_4, at T, K: 1000**	**1500**	**1800**
0.1	1	95.97	97.42	98.35	94.50	95.84	96.70
	2	39.10	39.64	39.98	38.64	39.15	39.47
	3	2.104	2.045	2.009	2.413	2.350	2.312
	4	26.1	27.9	28.8	25.1	26.7	27.5
0.2	1	187.31	189.91	191.57	185.01	187.45	189.00
	2	37.39	37.83	38.10	37.14	37.56	37.83
	3	2.506	2.450	2.415	2.793	2.733	2.697
	4	22.4	23.8	24.5	21.9	23.1	23.8
0.3	1	274.96	278.52	280.77	272.23	275.61	277.74
	2	36.00	36.36	36.60	35.88	36.24	36.47
	3	2.894	2.839	2.508	3.160	3.103	3.068
	4	19.7	20.9	21.4	19.5	20.6	21.1
Pressure, Mbar	**Row**	**Fayalite, α-Fe_2SiO_4, at T, K: 1000**	**1500**	**1800**	**Spinel, γ-Fe_2SiO_4, at T, K: 1000**	**1500**	**1800**
0.1	1	109.60	111.44	112.60	99.81	101.24	102.14
	2	44.40	45.05	45.46	40.75	41.28	41.61

Table 2. (Continued)

Pressure, Mbar	Row	Fayalite, α-Fe_2SiO_4, at T, K: 1000	1500	1800	Spinel, γ-Fe_2SiO_4, at T, K: 1000	1500	1800
	3	1.780	1.720	1.684	2.245	2.187	2.153
	4	28.5	29.8	30.6	25.2	26.4	27.2
0.2	1	212.92	216.13	218.16	195.11	197.68	199.30
	2	42.16	42.67	42.99	39.06	39.49	39.77
	3	2.238	2.179	2.144	2.623	2.568	2.535
	4	23.6	24.6	25.1	21.8	22.8	23.3

Pressure, Mbar	Row	Larnite, Ca_2SiO_4, at T, K: 1000	1500	1800	Enstatite, $MgSiO_3$, at T, K: 1000	1500	1800
0.1	1	122.80	126.11	128.40	73.91	75.47	76.48
	2	49.54	50.64	51.37	29.70	30.24	30.58
	3	1.677	1.592	1.539	1.479	1.420	1.384
	4	41.4	46.3	49.5	34.5	36.8	38.0
0.2	1	237.80	243.34	247.09	142.68	145.35	147.06
	2	46.83	47.63	48.15	27.94	28.34	28.60
	3	2.141	2.064	2.017	1.929	1.871	1.836
	4	32.3	35.2	36.9	27.7	29.3	30.1
0.3	1	347.19	354.42	359.26	207.82	211.35	213.6
	2	44.79	45.42	45.82	26.63	26.95	27.15
	3	2.584	2.512	2.468	2.359	2.302	2.267
	4	26.8	28.9	30.0	23.5	24.7	25.3

Pressure, Mbar	Row	Ilmenite, $MgSiO_3$, at T, K: 1000	1500	1800	Perovskite, $MgSiO_3$, at T, K: 1000	1500	1800
0.1	1	62.80	63.79	64.43	58.46	59.38	59.97
	2	25.68	26.05	26.29	23.98	24.33	24.55
	3	2.405	2.336	2.294	2.779	2.700	2.650
	4	27.7	29.6	30.6	27.4	29.7	31.0
0.2	1	122.94	124.74	125.88	114.77	116.44	117.52
	2	24.67	24.98	25.18	23.16	23.45	23.64
	3	2.786	2.721	2.681	3.147	3.073	3.027
	4	24.0	25.6	26.3	24.1	26.0	26.9
0.3	1	180.88	183.36	184.94	169.25	171.57	173.06
	2	23.83	24.10	24.26	22.45	22.70	22.86
	3	3.154	3.092	3.054	3.503	3.433	3.389
	4	21.4	22.7	23.3	21.6	23.2	23.9
0.6	1	344.44	348.50	351.07	323.90	327.75	330.21
	2	21.91	22.10	22.21	20.78	20.96	21.08
	3	4.202	4.145	4.110	4.519	4.457	4.418
	4	16.4	17.3	17.7	16.7	17.9	18.4

Table 2. (Continued)

Pressure, Mbar	Row	Pyrope, $Mg_3Al_2Si_3O_{12}$, at T, K: 1000	1500	1800	Grossular, $Ca_3Al_2Si_3O_{12}$, at T, K: 1000	1500	1800
0.1	1	268.40	272.20	274.61	296.15	300.03	302.47
	2	109.40	110.80	111.69	120.60	122.03	122.93
	3	2.130	2.070	2.034	2.052	2.001	1.970
	4	24.8	26.2	26.8	22.9	24.1	24.7
0.2	1	524.02	530.86	535.17	577.72	584.68	589.06
	2	104.69	105.84	106.56	115.23	116.40	117.13
	3	2.549	2.492	2.457	2.452	2.403	2.374
	4	21.3	22.4	22.9	19.7	20.7	21.1
0.3	1	769.52	778.90	784.79	847.73	857.27	863.25
	2	100.86	101.84	102.45	110.86	111.85	112.47
	3	2.954	2.898	2.864	2.838	2.790	2.761
	4	18.8	19.7	20.1	17.4	18.2	18.6
0.6	1	1460.42	1475.61	1485.12	1605.71	1621.13	1630.78
	2	92.44	93.12	93.54	101.24	101.93	102.36
	3	4.107	4.053	4.020	3.935	3.890	3.862
	4	14.3	14.9	15.2	13.2	13.8	14.0

Pressure, Mbar	Row	Kyanite, Al_2SiO_5, at T, K: 1000	1500	1800	Spinel, $MgAl_2O_4$, at T, K: 1000	1500	1800
0.1	1	104.93	105.84	106.40	94.29	95.62	96.46
	2	43.13	34.48	43.69	38.52	39.01	39.32
	3	3.019	2.962	2.927	2.313	2.251	2.213
	4	15.7	16.4	16.7	24.7	26.2	26.9
0.2	1	206.38	208.05	209.10	184.44	186.44	188.36
	2	41.79	42.10	42.28	36.97	37.38	37.64
	3	3.446	3.391	3.357	21.4	22.6	23.2
	4	14.0	14.7	14.9	21.4	22.6	23.2
0.3	1	304.87	307.22	308.69	271.24	274.54	276.62
	2	40.65	40.92	41.08	35.69	36.04	36.26
	3	3.862	3.807	3.774	3.108	3.051	3.016
	4	12.8	13.4	13.6	19.0	20.0	20.5

Pressure, Mbar	Row	Ferrosilite, $FeSiO_3$, at T, K: 1000	1500	1800	Wollastonite, $CaSiO_3$, at T, K: 1000	1500	1800
0.1	1	77.28	78.05	78.53	92.84	94.68	95.86
	2	31.13	31.39	31.55	36.93	37.54	37.93
	3	1.505	1.462	1.436	1.192	1.152	1.129
	4	16.3	17.0	17.4	31.9	33.6	34.5
0.2	1	149.45	150.75	151.56	177.75	180.83	182.79
	2	29.36	29.55	29.66	34.28	34.72	35.00
	3	1.950	1.906	1.880	1.589	1.550	1.527
	4	12.7	13.2	13.5	25.1	26.2	26.8

Table 2. (Continued)

Pressure, Mbar	Row	Wollastonite-II, $CaSiO_3$, at *T*, K: 1000	1500	1800	Perovskite, $CaSiO_3$, at *T*, K: 1000	1500	1800
0.1	1	88.16	89.97	91.14	64.67	65.62	66.23
	2	35.32	35.93	36.32	26.49	26.85	27.08
	3	1.266	1.319	1.290	2.604	2.532	2.489
	4	33.4	35.5	36.7	25.9	27.8	28.7
0.2	1	169.71	172.78	174.75	126.80	128.54	129.65
	2	3.05	33.50	33.79	25.53	25.83	26.02
	3	1.771	1.725	1.698	2.986	2.919	2.878
	4	26.7	28.1	28.8	22.7	24.1	24.9
0.3	1	264.59	250.63	253.21	186.81	189.22	190.75
	2	31.35	31.72	31.95	24.71	24.97	25.14
	3	2.156	2.112	2.086	3.357	3.293	3.254
	4	22.5	23.6	24.1	20.3	21.5	22.1

Pressure, Mbar	Row	Hatrurite, Ca_3SiO_5, at *T*, K: 1000	1500	1800	Rankinite, $Ca_3Si_2O_7$, at *T*, K: 1000	1500	1800
0.1	1	170.19	173.40	175.46	226.46	232.34	236.26
	2	68.15	69.24	69.94	90.17	92.07	93.31
	3	1.331	1.292	1.268	1.222	1.175	1.146
	4	31.0	32.8	33.8	40.1	43.5	45.6
0.2	1	327.37	332.85	336.33	433.72	443.42	449.81
	2	63.62	64.45	64.96	83.64	85.00	85.87
	3	1.692	1.655	1.633	1.583	1.541	1.515
	4	25.0	26.2	26.8	31.2	33.2	34.3

Pressure, Mbar	Row	Merwinite, $Ca_3MgSi_2O_8$, at *T*, K: 1000	1500	1800	Diopside, $CaMgSi_2O_6$, at *T*, K: 1000	1500	1800
0.1	1	244.55	249.18	252.14	155.96	159.51	161.83
	2	98.08	99.65	100.64	62.67	63.88	64.66
	3	1.408	1.358	1.328	1.480	1.419	1.383
	4	30.9	32.5	33.5	37.0	39.6	41.1
0.2	1	471.32	479.18	484.16	301.01	307.06	310.97
	2	92.03	93.20	93.93	58.92	59.82	60.40
	3	1.855	1.806	1.776	1.914	1,856	1.821
	4	24.7	25.8	26.4	29.6	31.4	32.3
0.3	1	685.71	696.06	702.60	438.32	446.30	451.43
	2	87.56	88.50	89.08	56.10	56.83	57.28
	3	2.282	2.233	2.204	2.329	2.273	2.239
	4	20.8	21.7	22.1	25.0	26.4	27.0

Table 2. (Continued)

Pressure, Mbar	Row	Ca-Al Pyroxene, $CaAl_2SiO_6$, at T, K: 1000	1500	1800	Anorthite, $CaAl_2SiO_2O_8$, at T, K: 1000	1500	1800
0.1	1	151.20	154.77	157.14	232.75	234.94	236.31
	2	61.29	62.53	63.34	93.26	93.98	94.42
	3	1.874	1.789	1.737	1.342	1.304	1.282
	4	38.2	41.8	44.1	15.0	15.6	15.9
0.2	1	293.93	300.07	304.11	448.24	451.86	454.12
	2	58.28	59.23	59.84	87.40	87.90	88.21
	3	2.341	2.260	2.212	1.781	1.742	1.719
	4	31.0	33.4	34.7	11.2	11.6	11.9

Pressure, Mbar	Row	Monticellite, $CaMgSiO_4$, at T, K: 1000	1500	1800	Gehlenite, $Ca_2Al_2SiO_7$, at T, K: 1000	1500	1800
0.1	1	120.55	123.18	124.87	210.41	213.72	215.81
	2	48.27	49.15	49.70	84.32	85.47	86.19
	3	1.371	1.317	1.285	1.336	1.301	1.280
	4	35.0	37.1	38.3	26.5	27.6	28.3
0.2	1	232.04	236.45	239.28	404.99	410.69	414.27
	2	45.20	45.85	46.26	78.79	79.67	80.21
	3	1.818	1.765	1.734	1.695	1.661	1.641
	4	27.7	29.1	29.8	21.7	22.5	22.9

high-pressure phases within the 100–300 kbar interval and have, in general outlined a sequence of chemical and phase transformations in the mantle transition zone. Thermodynamic calculations have extended the number of studied reactions and revealed certain peculiarities of their occurrence at ultrahigh pressures (Liu, 1979b; Ostrovsky, 1979; Navrotsky *et al.*, 1979; Kuskov and Khitarov, 1982; Jeanloz and Thompson, 1983). Nevertheless, the general topology of the phase relations is not adequately defined as yet and this makes it still impossible both to ascertain a consistent sequence of phase relations in the mantle's transition zone and to give a credible geochemical interpretation of the main seismic boundaries at depths of 400 and 670 km.

Physicochemical modeling of natural processes seems to be most promising in studying mineral compositions of the mantle. Such an approach, with the help of computers allows one, with minimum of initial thermodynamic information, to calculate both heat and volume effects and P–T parameters of equilibria for several hundreds or even thousands of mineral transformations, to define stable and metastable equilibria, and to construct phase diagrams of multicomponent systems.

Several dozen different phases exist within the MgO–FeO–CaO–Al_2O_3–SiO_2 system. The following phases are considered in this chapter: MgO, FeO, CaO, Al_2O_3, SiO_2 (coesite, stishovite), Mg_2SiO_4 (forsterite, β-spinel, γ-spinel), $MgSiO_3$

(enstatite, ilmenite, perovskite), $CaSiO_3$ (wollastonite, wollastonite-II, perovskite), Fe_2SiO_4 (fayalite, spinel), $FeSiO_3$ (ferrosilite), Al_2SiO_5 (kyanite), Ca_2SiO_4 (larnite), $CaMgSi_2O_6$ (diopside), $Mg_3Al_2Si_3O_{12}$ (pyrope), $Ca_3Al_2Si_3O_{12}$ (grossular), $MgAl_2O_4$ (spinel), Ca_3SiO_5 (hatrurite), $Ca_3Si_2O_7$ (rankinite), $CaMgSiO_4$ (monticellite), $Ca_3MgSi_2O_8$ (merwinite), $Ca_2Al_2SiO_7$ (gehlenite), $CaAl_2SiO_6$ (Ca-Al-pyroxene), and $CaAl_2Si_2O_8$ (anortite). Computations were made for two- and three-component systems in accordance with the algorithm outlined in the text. Special attention is paid to the study of phase relations at pressures of 100–300 kbar.

The accuracy of calculations of monovariant curves and invariant points of phase diagrams depends on the errors in the initial thermodynamic information. On the equilibrium line Eq. (5) is realized. Consequently, to determine the P–T parameters of monovariant curves, the data should be available on both standard thermodynamic functions ($\Delta G_T^\circ, \Delta H_T^\circ, \Delta S_T^\circ$) and integral equations of state. The work of Kuskov and Galimzyanov (1982a), Kuskov *et al.* (1983b, and Kuskov (1984) presents a detailed description of the calculation techniques as well as estimations of errors in both the equations of state and standard functions. The values of standard free energies of minerals at high temperatures (except specifically stipulated cases) are accepted according to the data available in reference books: for the system $CaO–MgO–SiO_2$ these include: Robie *et al.* (1978), and Helgeson *et al.* (1978), and for the system $CaO–Al_2O_3–SiO_2$, Haas *et al.* (1981).

Basic Experimental Phase Equilibria and Standard Thermodynamic Functions of Phase Transformations

Calorimetric determinations for many high-pressure phases are not available. However, there is another way to obtain unavailable information on the thermodynamics of such minerals. If experimental data on phase and chemical equilibria at high temperatures and pressures are available, calculation of both standard free energies of phase transformations and thermodynamic functions of polymorphic modifications present no difficulty.

In a number of cases this way of obtaining thermodynamic information may prove to be even more fruitful than calorimetry. Because calculations of values are made on the basis of the P–T data on phase equilibria, the correction for cation disorder in minerals is effectively taken into account in calculating standard functions.

In addition, P–T parameters of phase equilibria used in these calculations are basic to future calculations of phase diagrams of multicomponent systems. Thus, the agreement between calculated phase diagrams and the most reliable experimental data on phase equilibria is reached. The latter are represented by the following basic phase equilibria selected: olivine–β-spinel–γ-spinel in Mg_2SiO_4

enstatite–ilmenite–perovskite in $MgSiO_3$, fayalite–spinel in Fe_2SiO_4, and wollastonite–wollastonite(II)–perovskite in $CaSiO_3$.

Phase Transformations in Mg_2SiO_4

From experimental data (Suito, 1977) on the equilibrium $\alpha = \beta$ within the interval of 1000–1500K the following values of standard free energies (kcal · mol^{-1}) were calculated with the help of tabulated values of $\int_0^P V\,dP : \Delta G^\circ_{1000} = 9.49(\pm 0.7)$, $\Delta G^\circ_{1273} = 10.38(\pm 0.75)$; $\Delta G^\circ_{1500} = 11.13(\pm 0.8)$, which can be described by:

$$\alpha = \beta - Mg_2SiO_4 : \Delta G^\circ_T(\pm 800) = 6200 + 3.28T(\text{cal/mol}^{-1})$$

The values of standard deviations, given in parentheses, were calculated from the formula:

$$\delta^2_{\alpha=\beta}(\Delta G^\circ_T) = \delta^2_\alpha\left(\int_0^P V\,dP\right) + \delta^2_\beta\left(\int_0^P V\,dP\right) + (\delta_P \cdot \Delta V_T)^2_{\alpha=\beta}$$

where δ_P is the experimental error in the pressure of phase transformation, equalling ± 10 kbar.

The same technique is used in the calculation of the standard free energy of phase equilibrium $\beta = \gamma$. Using experimental data (Ohtani, 1979; Suito, 1977; Yagi *et al.*, 1979) and equations of state of minerals will yield the following:

$$\beta = \gamma - Mg_2SiO_4 : \Delta G^\circ_T(\pm 700) = 1540 + 0.784T\,(\text{cal} \cdot \text{mol}^{-1})$$

As was pointed out by Kuskov and Galimzyanov (1982a), previous estimations of the thermodynamic functions of phase transformations in Mg_2SiO_4 were not totally reliable (see Navrotsky *et al.* 1979). Quite recently Akaogi *et al.* (1984) reported new determinations of standard functions based both on measuring the heat of solution at a temperature of (1000K) and estimation of entropic effects from the data of Suito (1977). The values obtained by Akaogi *et al.* and Suito $\Delta H^\circ_{1000}(\alpha = \beta) = 7160 \pm 700$ and $\Delta H^\circ_{1000}(\beta = \gamma) = 1630 \pm 900$ cal · mol^{-1}, are in good agreement with our determinations within the limits of errors in experiment and calculation. However, we believe that Akaogi *et al.* (1984) have been rather inconsistent with regards to another part of their work. First, the study has been limited by the calculation of S°_{1000} at only one temperature from the data of Suito (1977); second, approximations used for equations of state in the calculations were rather rough, and tabulated values of $\int_0^P V\,dP$ were not presented. Thus, in the work cited (Akaogi *et al.*, 1984) comparison of data on thermodynamic properties of minerals obtained from experiments at high pressures with calorimetric measurements is incomplete.

Phase Transformations in Fe_2SiO_4

The change of free standard energy of phase transition for fayalite–spinel is calculated both from the experimental data of Akimoto *et al.* (1977) and Ohtani

(1979) and from equations of state of fayalite and γ-spinel. Taking into account the errors in the equations of state and the measured value of pressure, the change in standard free energy within the interval of 1000–1800K was determined by the equation (Kuskov, 1984):

$$\alpha = \gamma - Fe_2SiO_4\text{: } \Delta G^{\circ}_T(\pm 300) = 3340 + 2.34T\,(\text{cal/mol}^{-1})$$

Phase Transformation in $MgSiO_3$

Calculation of the change in standard free energy of phase transition in $MgSiO_3$ are based on values rather roughly estimated by Liu (1979b) for the reactions:

$$\gamma - Mg_2SiO_4 + SiO_2(\text{stishovite}) = 2\,MgSiO_3(\text{ilmenite})$$

$$P = 190\text{ kbar},\quad T = 1273\text{K} \tag{19}$$

$$MgSiO_3(\text{ilmenite}) = MgSiO_3(\text{perovskite})$$

$$P = 210\text{ kbar},\quad T = 1273\text{K} \tag{20}$$

Hence, using Table 2 we obtain $\Delta G^{\circ}_{1273} = 3.6\text{ kcal}\cdot\text{mol}^{-1}$ for reaction (19) and $8.6\text{ kcal}\cdot\text{mol}^{-1}$ for reaction (20).

In order to estimate the values of ΔG°_T at other temperatures the approximate relation:

$$\Delta G^{\circ}_T = \Delta G^{\circ}_{T_0} - \Delta S(T - T_0)$$

is used. Entropies of perovskite and ilmenite modifications estimated by the method of comparative calculation (Wood and Fraser, 1981) in accordance with the known entropies of perovskite $CaTiO_3$ and ilmenite $FeTiO_3$ with introduced correction for the difference in oxides entropies, make up 52 ± 2 and 50 ± 2, respectively. Hence, bearing in mind $S^{\circ}_{1273}(\gamma - Mg_2SiO_4) = 72.5\text{ cal}\cdot\text{mol}^{-1}\text{ deg}^{-1}$ (Kuskov and Galimzyanov, 1982a) and $S^{\circ}_{1273}(\text{stishovite}) = 27.95\text{ cal}\cdot\text{mol}^{-1}\text{ deg}^{-1}$ (Robie *et al.*, 1978), we finally obtain (Kuskov *et al.*, 1983c) for reaction (19):

$$\Delta G^{\circ}_T = 3000 + 0.45T\,(\text{cal}\cdot\text{mol}^{-1}) \tag{19a}$$

and for reaction (20):

$$\Delta G^{\circ}_T = 11{,}100 - 2.0T\,(\text{cal}\cdot\text{mol}^{-1}) \tag{20a}$$

Phase Transformations in $CaSiO_3$

The values of ΔG°_T for the phase transition:

$$\underset{\text{wollastonite}}{CaSiO_3(\text{Wo})} = \underset{\text{wollastonite-II}}{CaSiO_3(\text{Wo-II})} \tag{21}$$

were calculated both from the tabulated values of $\int_0^P V\,dP$ (Table 2) and from the experimental curve of Huang and Wyllie (1975):

$$\Delta G_T^\circ = 2190 - 0.340T\ (\text{cal} \cdot \text{mol}^{-1})$$

Calculation of phase transition ΔG_T°:

$$\underset{\text{wollastonite-II}}{CaSiO_3(\text{Wo-II})} = \underset{\text{perovskite}}{CaSiO_3(\text{Pv})} \tag{22}$$

was made according to estimated experimental data $P = 120 \pm 30$ kbar and 1273K (Ringwood and Major, 1971; Liu and Ringwood, 1975). The values for the integrals $\int_0^P V\,dP$ for $CaSiO_3$(Wo-II) and $CaSiO_3$(Pv) at 120 kbar and 1273 constitute 106.05($\pm$1.07) and 77.89($\pm$0.52) $\text{kcal} \cdot \text{mol}^{-1}$ with a volumetric effect equalling 8.64($\pm$0.62) $\text{cm}^3 \cdot \text{mol}^{-1}$. Hence, for reaction (22) $\Delta G_{1273}^\circ =$ 28.16($\pm$3.3) $\text{kcal} \cdot \text{mol}^{-1}$. The Entropy of $CaSiO_3$ (Pv) roughly estimated with the help of its nearest structural analog $CaSnO_3$ with an introduced correction for the difference of the oxide entropies is equal to:

$$S_{298}^\circ[CaSiO_3(\text{Pv})] = 12.7 \pm 3\ \text{cal} \cdot \text{mol}^{-1}\ \text{deg}^{-1}$$

Using the approximate relation:

$$\Delta G_T^\circ = \Delta G_{T_0}^\circ - \Delta S_{T_0}^\circ (T - T_0)$$

the dependence of the change of free energy on temperature for the phase transformation wollastonite-II–perovskite (Eq. 22) is obtained:

$$\Delta G_T^\circ = 19{,}210 + 7.0T\ (\text{cal} \cdot \text{mol}^{-1})$$

The values of ΔG_T° obtained for phase transformations, along with consistent reference values of standard functions and equations of state, were used to construct phase diagrams of multicomponent systems.

The MgO–SiO_2 System

When plotting a phase diagram of the MgO–SiO_2 system, we have taken into consideration the following phases: periclase; stishovite; α-, β-, and γ-Mg_2SiO_4; clinoenstatite (En), $MgSiO_3$ (ilmenite, Ilm); and $MgSiO_3$ (perovskite, Pv). Thus, for a certain P–T interval there may exist stability fields of eight phases. The number of components equals two. Then, according to the phase rule: $F = K + 2 - P = -4$. In such system the maximum number of invariant points equals C_P^{K+2} or $C_8^4 = 70$. It is evident that the greater part of them will be metastable, or unrealizable at all. To reveal the stable and metastable invariant points it is necessary to know the entropic and volumetric effects of the reactions.

Volumetric effects of the reactions defined from the data on equations of state are known with high accuracy. At the same time there are no reliable definitions of thermodynamic functions of polymorphic modifications of $MgSiO_3$ (ilmenite and perovskite), which makes it impossible to construct a complete phase diagram of the system in a single way. The existing uncertainties in the evaluation of standard thermodynamic functions and experimental determinations under

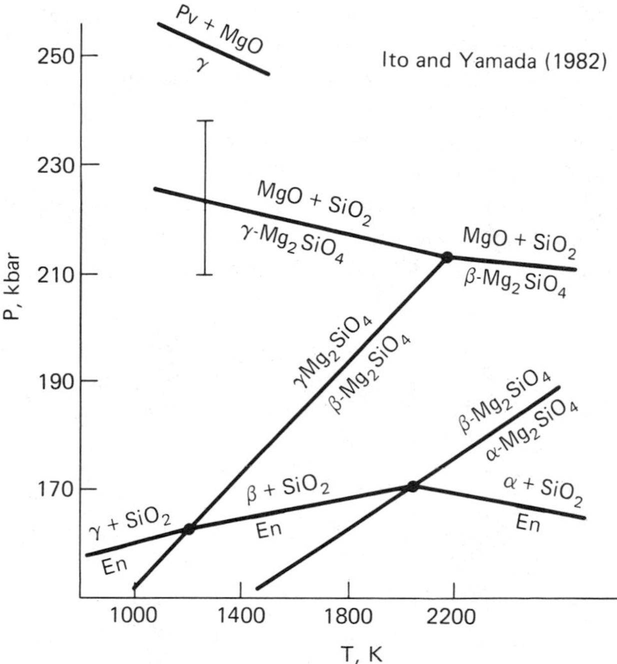

Fig. 2. Simplified part of the $MgO-SiO_2$ phase diagram without taking into account the ilmenite and perovskite modifications: $\alpha = \beta$ and $\beta = \gamma$ equilibria from experiments by Suito (1977); the remainder are calculated. En = $MgSiO_3$; Pv = perovskite ($MgSiO_3$). α and γ are used for Mg_2SiO_4 polymorphs.

high pressures result in several alternatives (Kuskov *et al.*, 1983c). In a situation like this it is reasonable to move from "simple to complex."

Thus a preliminary P–T diagram of a simplified part of the system will be constructed, limited to six phases (periclase, stishovite, forsterite, β-spinel, γ-spinel, and clinoenstatite); for the reactions with these species there are sufficiently reliable experimental and thermodynamic data (Fig. 2). As is known, $\alpha = \beta$ and $\beta = \gamma$ equilibria serve as supports for this system (Suito, 1977; Suito and Kawai, 1979), while the remaining ones have been obtained by calculations. The absence of high-pressure phases for $MgSiO_3$ leads to disintegration of β and γ-spinels into periclase and stishovite. The Stability field of γ-spinel within the 1000–2200K interval is limited by the narrow range of pressures of 230–210 kbar. The slope of the monovariant curve has a negative value: $dP/dT = -0.010$ kbar $\cdot$ deg^{-1}. As regards the disproportioning β-spinel, $dP/dT \approx 0$.

One of the most important conclusions to be drawn from the phase diagram lies in that the monovariant equilibrium γ-spinel = 2 periclase + stishovite yields qualitative information about P–T parameters of monovariant reactions of disproportionation of spinels into periclase + perovskite (ilmenite). If experimental data on $\alpha = \beta$, $\beta = \gamma$ reported by Suito (1977) and Suito and Kawai (1979) are reliable, then the monovariant line γ-Sp = 2MgO + SiO_2, specifying the upper limit of spinel stability within a simplified part of the system, shows that in the range of calculation error (± 15 kbar) the perovskite phase $MgSiO_3$ cannot be formed at pressures over 230 ± 15 kbar at 1000°C. Thermodynamic information on the main transformations within the MgO SiO_2 system within the interval of P–T parameters corresponding to the mantle transition zone are fully presented in the paper by Truskinovsky *et al.* (1983).

Now, let us proceed to the analysis of the system's phase diagram including

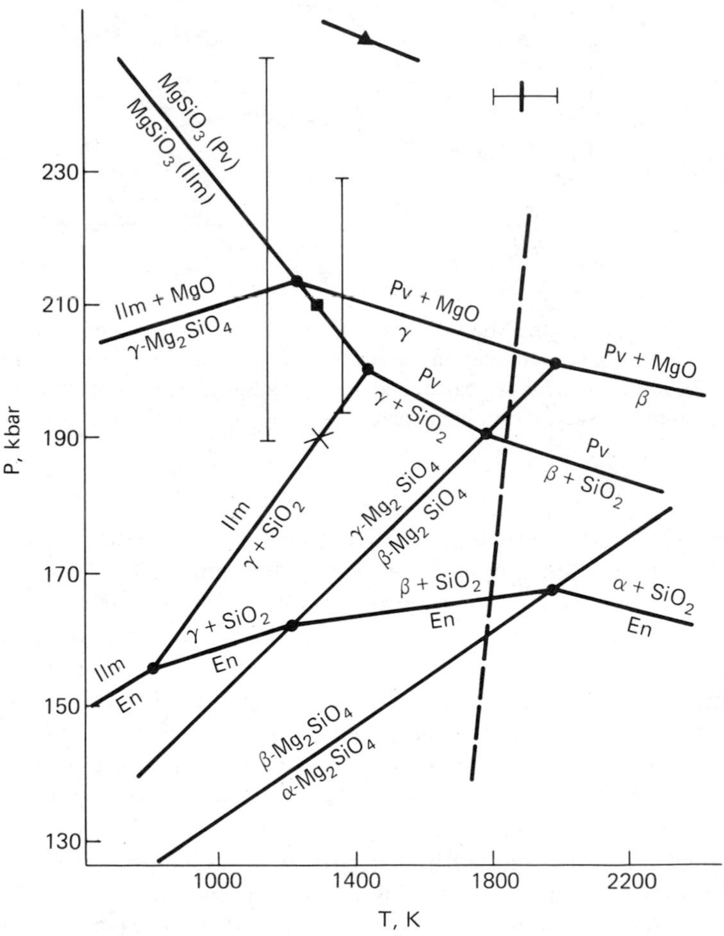

Fig. 3. Diagram of the MgO–SiO_2 system (variant I). Calculations of equilibria with the inclusion of ilmenite and perovskite ($MgSiO_3$) based on the following data at 1000°C: γ-Fo + SiO_2 = 2 Ilm, $P_{eq} = 190$ kbar; Ilm = Pv, $P_{eq} = 210$ kbar; S°_{1273}(Ilm) = 50 e.u. S°_{1273}(Pv) = 52 e.u. The adiabatic distribution of temperature in the mantle is shown by the dashed line. After Anderson and Baumgardner (1980).

all eight phases, taking into account the existing uncertainties in the experimental and thermodynamic data.

Variant I

Plotting of the phase diagram (Fig. 3) is based upon experimental assessments by Liu (1979b) from reactions (19 and 20) and thereby calculated values of the standard functions (Eqs. 19a and 20a).

Variant II

The position and sign of a slope of solid-phase chemical transformations (Fig. 4) are to a great extent determined by entropic effects of the examined

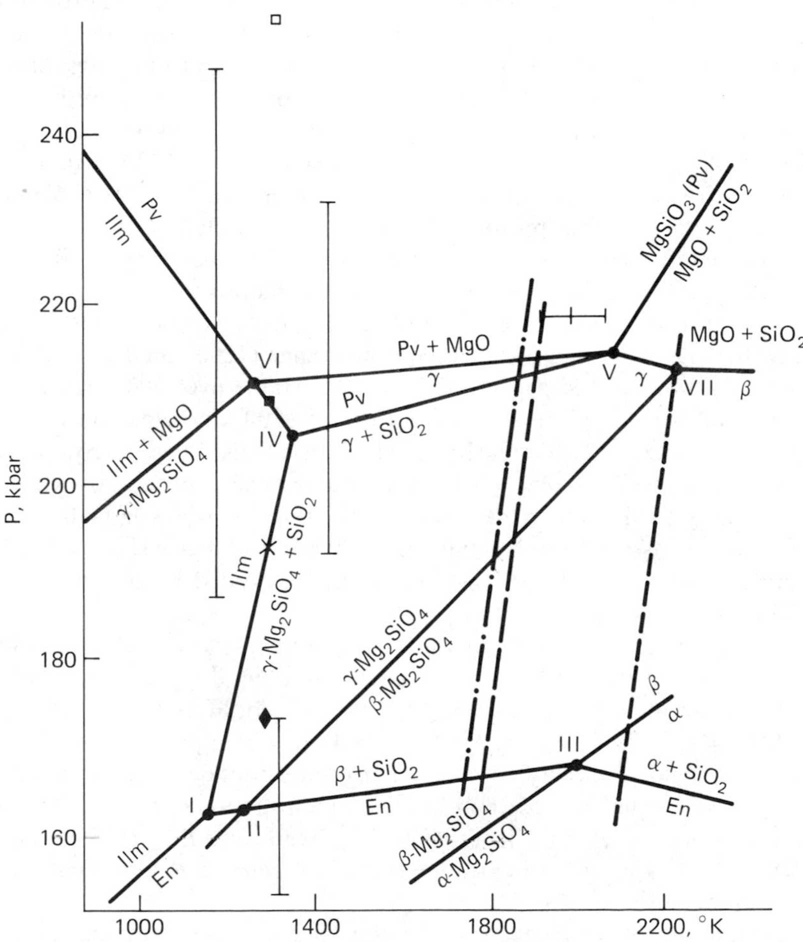

Fig. 4. Diagram of the MgO–SiO_2 system (variant II) obtained from variant I by the simultaneous reduction of entropy by 2 cal·$mol^{-1}K^{-1}$ of ilmenite and perovskite. Geotherms: I, after Anderson and Baumgardner (1980); 2, after Brown and Shankland (1981); 3, after Stacey (1977).

reactions. The entropies of the $MgSiO_3$ ilmenite and perovskite modifications are roughtly estimated. Hence, it is necessary to find out the sensitivity of the phase relations in the MgO–SiO_2 system toward the entropy value variations. There is no change in the phase diagram with simultaneous increase of ilmenite and perovskite entropy by 1–2 e.u. At the same time, by varying the estimated entropy values of ilmenite and perovskite modifications within the limits of error an alternative variant can be realized of the phase diagram with the oxide stability field as shown in Fig. 4 at $S^{\circ}_{1273}(\text{Ilm}) = 48$ and $S^{\circ}_{1273}(\text{Pv}) = 50$ cal · mol^{-1} deg^{-1}.

Comparison of the phase diagrams leads to the following conclusions:

Variants I and II are quite close to each other. The difference is in a slight displacement of the phase stability fields by pressure and in a more pronounced shift by temperature. In both cases a divariant stability field of γ-spinel + stishovite is limited by the monovariant transformation into the high-pressure phases, ilmenite and perovskite. Both variants of the diagram indicate the existence of the phase transition En = Ilm at $P = 150$–160 kbar ($dP/dT > 0$); for the ilmenite = perovskite transition in $MgSiO_3$, $\Delta V < 0$, $\Delta S > 0$, and $dP/dT < 0$, which is unusual for phase transformations, but finds its explanation from the crystal–chemical position (Navrotsky, 1980; Barsukov and Urusov, 1982) and is confirmed by direct experiment (Ito and Yamada, 1982). Reactions γ-Fo + SiO_2 = 2 Ilm and γ-Fo = Ilm + MgO have a positive slope.

In variant I the reaction of disproportionation: γ-spinel = perovskite + periclase ($dP/dT < 0$) has a wide temperature range (1200–2000K), within this interval the association of perovskite + MgO is stable over 200 kbar. At temperatures of 1000–1400K in the interval of 170–200 kbar ilmenite is formed [γ-Mg_2SiO_4 + SiO_2 = 2 $MgSiO_3$(Ilm)], which at 1273K has a narrow stability field and at pressures over 200 kbar is transformed into perovskite. The oxides stability field is absent because the mixture of oxides is metastable in relation to the associations of ilmenite + MgO and perovskite + MgO. A divariant stability area of γ-spinel + perovskite bounded by invariant points IV–VII is identified (Fig. 3).

Below 200 kbar the picture of phase relations in variant II is absolutely identical to that of variant I. Differences appear when the pressure is over 200 kbar. Monovariant curves γ-Mg_2SiO_4 + SiO_2 = 2 $MgSiO_3$(Pv) ($dP/dT = 0.01$) and γ-Mg_2SiO_4 = Pv + MgO ($dP/dT = 0.004$) acquire a positive sign of slope. The divariant field of stability γ-spinel + perovskite bounded by invariant points IV–VI gets somewhat narrower in both temperature and pressure. The most essential distinction lies in the fact that at $T \gtrsim 2000$K and $P \gtrsim 215$ kbar an area of oxide stability appears, being transformed into perovskite modification with increase in pressure (Fig. 4).

Thus, caution should be exercised in comparing calculated phase diagrams with scarce experimental data available. The calculated monovariant line 2 $MgSiO_3$(En) = $\beta - Mg_2SiO_4$ + SiO_2 ($P = 160$, 163, 164 kbar; $T = 1000$, 1273, 1500K; $dP/dT = 0.008$) is in good agreement with the data reported by

Akaogi and Akimoto (1977) and Ito and Matsui (1977). This fact, together with the basic equilibria α–β–γ-Mg_2SiO_4 according to Suito (1977; Suito and Kauai, 1979), and the standard thermodynamic functions of phase transformations calculated thereby give evidence of internal agreement of phase diagrams, making it possible to draw several conclusions concerning experiments at pressures over 200 kbar. Yagi *et al.* (1979a) and Ito and Yamada (1982) fixed $P = 253$–255 kbar at 1000°C for the reaction of disproportionation γ-Fo = Pv + MgO, which considerably exceeds the calculated value $P = 211 + 40$ kbar at 1000°C, although the observed divergence is within the limits of calculation errors and those of experiment.

The experiments conducted by Ito and Yamada (1982) and Yagi *et al.* (1979a) are commented on in the paper by Ito and Yamada (1982). In our opinion, the cause of the divergence lies in both the great uncertainty with regard to the thermodynamic functions of $MgSiO_3$(Ilm, Pv), and in the calibration method employed in the Ito and Yamada (1982) experiment, based on the use of reactions of disproportionation of $MgAl_2O_4$ and γ-Ni_2SiO_4 as thermodynamic scale of pressures. Our calculation (Kuskov and Galimzyanov, 1984) revealed that the previous estimation of the pressure value for $MgAl_2O_4$ decomposition might have been substantially overestimated.

Kuskov and Galimzyanov (1984) made calculations for the reaction of spinel decomposition ($MgAl_2O_4$) into oxides, taking into account the effects of cation disorder, compressibility, and thermal expansion of phases. Monovariant equilibrium in the 1200–1800K interval is described by the equation: $P(\pm 15\text{ kbar}) = 71 + 0.035T(°\text{K})$. At 1273K the value of the decomposition reaction pressure is 115 ± 11 kbar, being lower by about 30 kbar than that assumed for a calibration point in the experiments of Ito and Yamada (1982).

The variants I and II of the plotted in the phase diagrams show that with the increase of pressure the association γ-spinel + stishovite is transformed into either ilmenite or perovskite. With further increase of pressure disproportionation of γ-spinel is observed depending on temperature and the type of phase diagram in the association: ilmenite + MgO, perovskite + MgO, or MgO + SiO_2.

While specifying our position with regard to constructing diagrams of different topology, we want to note that diagram II is in better agreement with experimental static data available at ultrahigh pressures. Within the framework of the assumptions made during its construction and the uncertainties in the initial information, we consider it to be the basic diagram of the system. According to this diagram there are no apparent contradictions between the study results of the MgO–SiO_2 system (Liu, 1979a, b; Ito and Matsui, 1977; Basett and Ming, 1976; Kumazawa *et al.*, 1974), because the phase stability fields are dependent on both pressure and temperature. The experiments conducted by Liu (1979a, b) on spinel decomposition at 1000°C have shown formation of perovskite, whereas in other experiments the oxides stability field was revealed at the same pressures but at higher temperatures.

Thus, the type of figure shown in diagram II can probably be considered to eliminate contradictions resulting between investigations of different authors, under both static and dynamic conditions. At constant pressure, phase associations stable at one temperature become metastable at the other, and vice versa.

Nevertheless, the two alternatives "perovskite" or "oxides" continue to remain a geochemical–geophysical controversy. It is evident that those phase associations which are stable in thermodynamic conditions of the actual temperature and pressure distribution in the Earth's interior should exist in the Earth's mantle. The temperature regime of the mantle is known with great uncertainty. Figures 3 and 4 present calculations of adiabatic temperature distribution in the mantle [at a depth of 670 km, $T = 1873K$, according to Brown and Shankland (1981); 1900K, according to Anderson and Baumgardner (1980); and 2250K according to Stacey (1977)]. The first two geotherms are located in the field of perovskite stability, although they practically adjoin the invariant point ∇, where γ-Mg_2SiO_4, $MgSiO_3$(Pv), MgO, and SiO_2 are found in equilibrium, leading to the following sequence of chemical and phase transformations in the mantle:

$$\text{clinoenstatite} \rightarrow \beta\text{-spinel} + SiO_2 \rightarrow \gamma\text{-spinel} + SiO_2 \rightarrow \text{perovskite}$$

$$\text{forsterite} \rightarrow \beta\text{-spinel} \rightarrow \gamma\text{-spinel} \rightarrow \text{perovskite} + MgO$$

If the actual geotherm is slightly shifted toward higher temperatures, the evolutionary sequence, for example, will take the form (based on forsterite composition):

$$\text{forsterite} \rightarrow \beta\text{-spinel} \rightarrow \gamma\text{-spinel} \rightarrow MgO + SiO_2 \rightarrow \text{perovskite}$$

Because of the existing uncertainties (in the initial thermodynamic information, the adiabatic distribution of temperatures, and the alternative nature of phase diagrams) it is still impossible to give preference to any sequence of phase associations in the mantle. In other words, the "resolving capacity" of the thermodynamic analysis made so far is still inadequate to insure credible interpretation of the mantle's mineral composition at a depth of 670 km.

Therefore, there still remains the alternative of formation of an $MgSiO_3$ perovskite modification or of a mixture of oxides at the boundary between the transition zone and the lower mantle. However, in any case, with subsequent pressure increase an isochemical mixture of oxides becomes metastable relative to the association of perovskite + MgO, which prevails in the mineralogy of the lower mantle.

The CaO–Al_2O_3–SiO_2 System

The tentative nature of experimental data on phase transformations of Ca minerals and the lack of reliable standard thermodynamic functions for their high-pressure phases does not make it possible to produce a complete computation of phase diagrams in the CaO–SiO_2 and CaO–Al_2O_3–SiO_2 systems over the whole interval of pressure from 0 to 300 kbar. The following phases are taken into account in the $CaO-SiO_2$ system: wollastonite (Wo), wollastonite-II (Wo-II), perovskite (Pv) ($CaSiO_3$), larnite (Lr), hatrurite, rankinite (Rn), Cao, and the polymorphs of SiO_2. Modifications of $CaSiO_3$ (pseudowollastonite) and Ca_2SiO_4 (olivine and bredigite) are not considered here as they are stable only under low pressures. Lack of data for Ca_2SiO_4 in the structure of K_2NiF_4 also prevents it from being included for consideration. Corundum, kyanite, Ca-tschermakite, gehlenite, anortite, and grossular are additionally taken into account in the system $CaO-Al_2O_3$. A portion of the phase diagram with pressures over 40 kbar is presented in Fig. 5.

The upper limit of rankinite stability is determined by the transformation $Rn \rightarrow Lr + Wo\text{-}II$, having a slight negative slope in the 50–60 kbar interval. The upper limit of larnite stability is determined by the reaction of its disproportionation into Ca-perovskite + CaO at pressures over 200 kbar. Of interest is the likelihood of the existence of an invariant point with stable wollastonite-II,

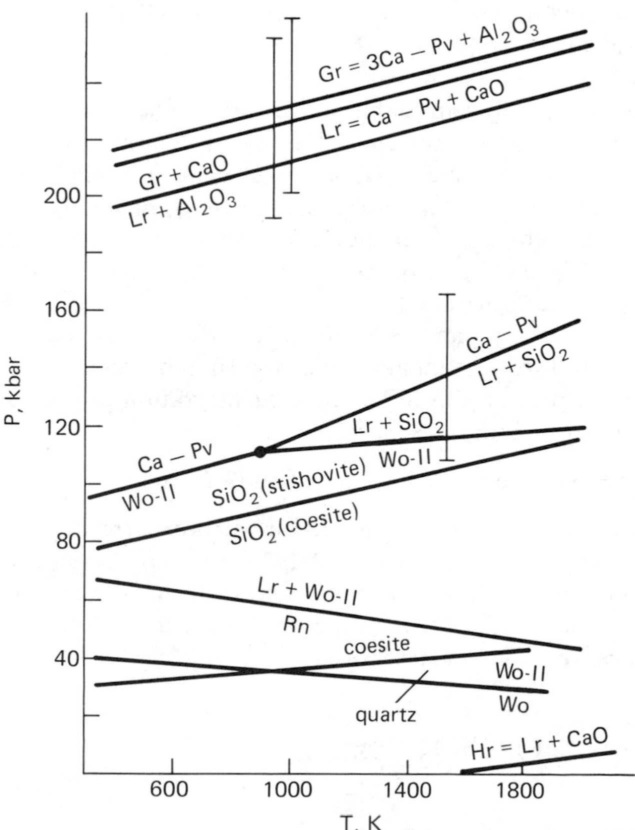

Fig. 5. *P–T* diagram of phase relations in the $CaO–Al_2O_3–SiO_2$ system: Wo, wollastonite; Wo-II, wollastonite II; Ca-Pv, $CaSiO_3$ (perovskite); Lr, larnite; Rn, rankinite; Gr, grossular; Hr, hatrurite Error bars reflect the uncertainty in the data.

Ca-perovskite, larnite, and stishovite. Its position (~110 kbar, 1000K) is to a greater extent uncertain, but of importance are the following sequences of $CaSiO_3$ phase transformations as dictated by temperature:

$$CaSiO_3(\text{Wo-II}) \rightarrow CaSiO_3(\text{Pv})$$

$$CaSiO_3(\text{Wo-II}) \rightarrow Ca_2SiO_4(\text{Lr}) + SiO_2(\text{Stish}) \rightarrow CaSiO_3(\text{Pv})$$

Experimentally, the decomposition of wollastonite II → larnite + stishovite has not yet been discovered.

Under a pressure of about 230 kbar and 1000°C larnite disintegrates into Ca-perovskite + CaO. We should note that at similar parameters ($\gtrsim$220 kbar, 1000°C) larnite transforms into the K_2NiF_4 structure, as reported by Liu, (1979c). The appearance of this phase will result in the metastability of certain calculated equilibria. However, because it is difficult to obtain X-ray diffraction patterns of quenched samples and their interpretation is diverse, formation of Ca_2SiO_4 in the structure of K_2NiF_4 must not be considered credibly established.

The addition of a third independent component, Al_2O_3 into the CaO–SiO_2 system brings about the appearance of a great number of Al-Beazing compounds, thereby considerably complicating the system under low pressures.

Thermodynamic analysis revealed that such Ca–Al minerals as anorthite, gehlenite and tschermakite are stable at pressures up to 30–40 kbar. The upper boundaries of anorthite and gehlenite stability are determined by the reactions: anorthite → grossular + kyanite + quartz, gehlenite → larnite + corundum; at 1500K, the equilibrium pressure for anorthite and gehlenite is 25 kbar and 33 kbar, respectively. The transformation of gehlenite to the $Na_2Ti_3O_7$ structure at 150 kbar and 1000°C as fixed in the experiments of Liu (1978) is rather metastable, if it takes place at all.

Only grossular, kyanite, and oxides are thermodynamically stable at pressures over 40 kbar. The disproportionation of kyanite into oxides is observed over 100 kbar, whereas at pressures over 200 kbar disintegration of grossular into Ca-perovskite and corundum takes place:

$$\text{grossular} = 3\ \text{Ca-perovskite} + \text{corundum}\ [P\ (\text{kbar}) = 209 + 0.02T°\text{K})]$$

The proposed mechanism of grossular decomposition correlates with the experiments of Mao *et al.* (1977), conducted at 297K and 400 kbar in which were discovered corundum and glass as products supposedly having a $CaSiO_3$ composition; however, this contradicts the experiment of Liu (1979d), who states that a new modification of grossular is formed.

The CaO–MgO–SiO_2 System

In the plotting of P–T diagrams for a three-component system the following minerals additional to the MgO–SiO_2 and CaO–SiO_2 systems should be taken into consideration: monticellite, diopside, and merwinite. The phase diagram in the 120–260 kbar interval is shown in Fig. 6.

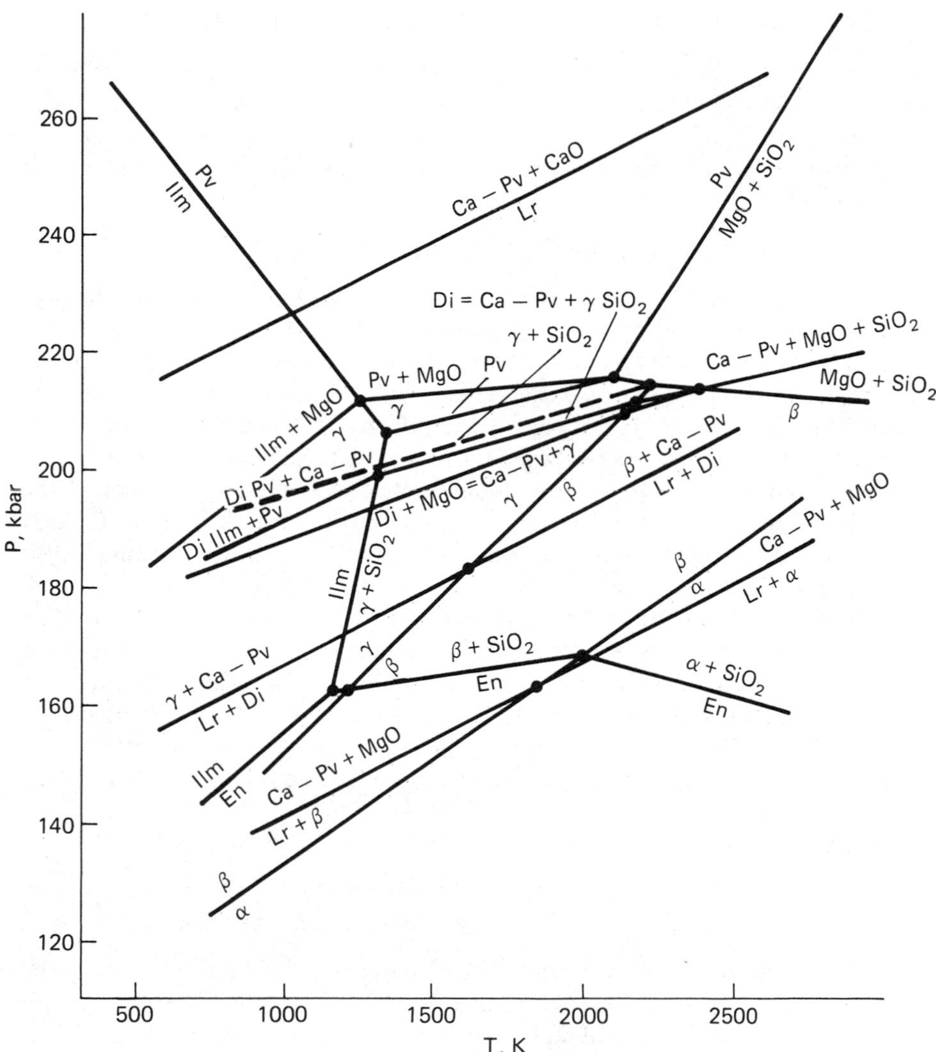

Fig. 6. P–T diagram of phase relations in the MgO–CaO–SiO_2 system α-, β-, and γ-polymorphs of Mg_2SiO_4; En, clinoenstatite; Lr, larnite; Di, diopside; Ca–Pv, $CaSiO_3$, perovskite; Ilm, Pv, polymorphs of $MgSiO_3$ in the structure of ilmenite and perovskite; the dashed line shows the diopside decomposition reaction into $MgSiO_3$(Pv) + $CaSiO_3$(Pv).

The association of forsterite with merwinite should form diopside + periclase at 47 and 69 kbar at 1273 and 1800K. In the experiment by Liu (1979c) this transformation was registered at $P \lesssim 80$ kbar and 1000°C. Thermodynamic calculation predicts stability at low pressures ($P \approx 40$ kbar, $T \approx 1100$K) of an invariant point at which merwinite, diopside, forsterite, monticellite, and periclase simultaneously coexist. At $T < 1100$K the univariant curve must be stable ($dP/dT = -17$ bar $\cdot$ deg^{-1}):

$$4\,CaMgSiO_4 = Ca_3MgSi_2O_8 + CaMgSi_2O_6 + 2\,MgO$$

limiting the field of monticellite stability at low temperatures. Monticellite is metastable above 50 kbar in relation to diopside, merwinite, and periclase at any temperature.

Merwinite should disintegrate at pressures over 60 kbar in accordance with the equation:

$$3\,Ca_3MgSi_2O_8 = 4\,Ca_2SiO_4 + CaMgSi_2O_6 + 2\,MgO$$

The equilibrium curve has a positive slope of 0.05 kbar $\cdot$ deg^{-1} and runs through the 95 kbar, 1273K point. In the experiments of Liu (1979c) pure merwinite remained stable at least to 200 kbar at 1000°C. The cause of the discrepancy observed between the calculations and experimental results is not clear. On one hand, merwinite in the experiments of Liu (1979c) could have been metastable because of the slow kinetics of the solid-phase transformations and the absence of equilibrium control. On the other hand, some doubts arise with respect to the accuracy of the value for the standard free energy of larnite formation. As was mentioned above the discrepancy between calculations and experiments is also observed for gehlenite, which at 30 kbar and 1000°C should disproportionate into larnite and corundum. Unfortunately, at present there is no thoroughly studied equilibrium with larnite participation, according to which it would be possible to make a judgement about the credibility of its standard free energy of formation. An experimental evaluation made by Doroshev *et al.* (1976) for the equilibrium: gehlenite + wollastonite = grossular + larnite was reported without a detailed description of the experiments, so it could not be considered representative.

The transformation registered in the experiment of Liu (1979c):

$$\underset{\text{merwinite}}{Ca_3MgSi_2O_8} + \underset{\text{diopside}}{CaMgSi_2O_6} = \underset{\text{perovskite}}{4\,CaSiO_3} + \underset{\text{periclase}}{2\,MgO}$$

at ~150 kbar and 1000°C, according to calculations, should metastable, but the calculated value of pressure (155 kbar at 1000°C) is in good agreement with experiment. Decomposition of merwinite into a mixture of CaO and $CaSiO_3$ and $MgSiO_3$ perovskites proves to be metastable and should take place at a pressure of about 240 kbar and 1000°C.

Below 140–180 kbar pressure and 1000–2500K temperature, a stable association of Ca and Mg orthosilicates is discovered transforming into Ca-perovskite + periclase with the increase of pressure. Below pressures of 160–200 kbar the diopside + larnite association is transformed to Ca-perovskite + $\beta(\gamma)$-spinel.

Diopside has a wide stability field and remains stable up to 200 kbar. Pure $CaMgSi_2O_6$ at temperatures over 1000°C and a pressure of about 200 kbar should decompose into associated γ-Mg_2SiO_4 + $CaSiO_3$(Pv) + SiO_2(Stish), which only below about ~210 kbar transforms to a mixture of perovskites: $CaSiO_3 + MgSiO_3$. At $T < 1000$°C decomposition into ilmenite $MgSiO_3$ and perovskite $CaSiO_3$ is possible. Equilibria have a slight positive slope and are accompanied by a release of heat.. In the experiments of Mao *et al.* (1977)

samples of diopside loaded up to 217 kbar at 1000°C, were found after quenching and the release of pressure to contain perovskite $MgSiO_3$ and glass [presumably crystallized $CaSiO_3$(Pv)], and in certain experiments of Liu (1979e) stishvite was discovered, which is in very good agreement with the calculations. At the same time, as specified by the calculations, the disproportionation reaction of diopside (Di) into the $MgSiO_3(Pv) + CaSiO_3(Pv)$ mixture proves to be metastable (1300K, 210 kbar) in relation to the reaction $2\,Di = 2\,CaSiO_3(Pv) + \gamma\text{-}Mg_2SiO_4 + SiO_2$.

Thus, within the 1000–2500K interval the sequence of diopside disproportion with temperature increase is as follows (Fig. 6):

$$CaMgSi_2O_6 = MgSiO_3(Ilm) + CaSiO_3(Pv)$$

$$2\,CaMgSi_2O_6 = \gamma\text{-}Mg_2SiO_4 + 2\,CaSiO_3(Pv) + SiO_2(Stish)$$

$$2\,CaMgSi_2O_6 = \beta\text{-}Mg_2SiO_4 + 2\,CaSiO_3(Pv) + SiO_2(Stish)$$

$$CaMgSi_2O_6 = CaSiO_3(Pv) + MgO + SiO_2(Stish)$$

It should be emphasized that the sequence for diopside disproportionation presented here is obtained as a result of self-consistent calculations, which, however, does not mean that it necessarily occurs like this in reality. Differences in pressure for the monovariant curves:

$$2 \text{ diopside} = \gamma\text{-spinel} + \text{Ca-perovskite} + \text{stishovite}$$

$$\text{diopside} = \text{Ca-perovskite} + \text{Mg-perovskite}$$

make up only 10 kbar, whereas calculation error is in the range of ± 30–40 kbar. It is quite evident, that a slight change in the initial thermodynamic constants may result in a reverse sequence of reactions.

A mixture of larnite and enstatite is unstable thermodynamically in relation to diopside and forsterite:

$$CaSiO_4 + 4\,MgSiO_3 = 2\,CaMgSi_2O_6 + \alpha\text{-}Mg_2SiO_4; \quad \Delta G_T^\circ < 0, \Delta V_T^\circ < 0$$

A mixture of enstatite with wollastonite is metastable in relation to diopside:

$$CaSiO_3 + MgSiO_3 = CaMgSi_2O_6; \quad \Delta G_T^\circ < 0, \Delta V_T^\circ < 0$$

This gives evidence for the fact that in the upper mantle layer a major part of the calcium is contained in pyroxenes.

The MgO–Al_2O_3–SiO_2 System

When plotting a phase diagram of the MgO–Al_2O_3–SiO_2 system under ultrahigh pressures the following additional minerals are taken into account for the MgO–SiO_2 system: pyrope, kyanite, and Mg–Al-spinel (Fig. 7). The general picture of phase relations in this system is complicated by the formation of solid solutions that are beyond the scope of the present study.

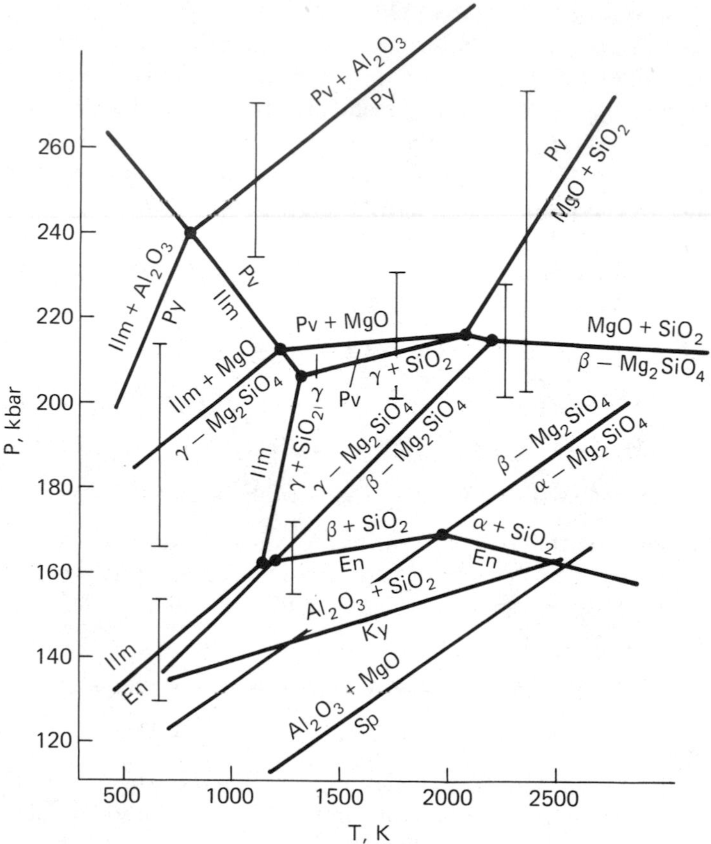

Fig. 7. P–T diagram of phase relations in the MgO–Al_2O_3–SiO_2 system: Py, pyrope, Pv, $MgSiO_3$ (perovskite); Ky, kyanite; Sp, $MgAl_2O_4$ (spinel) Error bars indicate the uncertainty in the calculated equilibrium pressure.

Pure kyanite is stable over a wide interval of temperatures and pressures. Its decomposition to oxides is described by the equation $P(\text{kbar}) = 119(\pm 10) + 0.016T(\text{K})$, which correlates with the experimental data of Liu (1974a) below 160 kbar and 1300–1700K.

Pyrope, like grossular, is stable at all temperatures and pressures of the upper mantle and transition zone. The stability field of pyrope extends also to the depth of the lower mantle, where at $T \gtrsim 1000$K and $P \gtrsim 240$ kbar it disintegrates into the $MgSiO_3$(Pv) + corundum association, which is confirmed by qualitative experiments by Liu (1974b). Thermodynamic calculations reveal that disproportionation of pyrope into perovskite + corundum takes place close to 1000K. At lower temperatures it is decomposed into $MgSiO_3$ with the structure of ilmenite, and above 2500–3000K an isochemical mixture of oxides (periclase + corundum + stishovite) is formed.

Equilibria with the inclusion of spinel ($MgAl_2O_4$) are of great interest because of the cation disorder taking place above 1000K. The simultaneous influence of compressibility, thermal expansion, and configuration effects on the position of the monovariant curve of the reaction for spinel disproportionation into oxides has been considered in the work of Kuskov and Galimzyanov (1984) and Urusov and Kuskov (1984). This monovariant equilibrium is described adequately by a linear equation in the interval 1200–1800K: $P(\pm 15\ \text{kbar}) = 71 + 0.035T(\text{K})$. The results of calculations show that taking cation disorder into account essentially leads to expansion of the stability field for spinel and an increase in the equilibrium transformation pressure by 25–40 kbars. The previous calculations (Ohtani *et al.*, 1974; Liu, 1980) disregarded cation disorder.

The MgO–FeO–SiO_2 System

There are practically no systematic studies of the FeO–SiO_2 and MgO–FeO–SiO_2 systems under ultrahigh pressures. The only equilibrium ascertained is the phase transformation of olivine–spinel in Fe_2SiO_4. A great number of reports deal with the thermodynamics of the FeO–SiO_2 system at 1 atm, but despite this reliable information is not available on the thermodynamic functions of silicates with iron compounds and oxides. For example, modern reference books (Robie *et al.*, 1978) give the value of ΔG°_{f298} for stoichiometric wüstite, FeO as $-60.097\ \text{kcal}\cdot\text{mol}^{-1}$, which is approximately 1.5 kcal less than the value ΔG°_{f298} ($Fe_{0.947}0$) $= -58.595\ \text{kcal}\cdot\text{mol}^{-1}$ for nonstoichiometric wüstite. Evidently, this is a result of the unstability of FeO at 1 atm.

Elastic constants and the influence of pressure and temperature on nonstoichiometric wüstite have also been inadequately studied (Kuskov and Khitarov, 1982; Bonczar and Graham, 1982; McCammon and Liu, 1984, and references cited therein).

Solid solutions of ferromagnesian silicates and oxides exist in the mantle, and with an increase of magnesium content, defects of solid solution are reduced (Chupharov *et al.*, 1970). Because of this, stoichiometric wüstite is accepted as a standard state in calculations.

The thermodynamic properties of the wüstite were obtained by Kurepin (1981) based on extrapolation of the linear dependence of partial oxygen pressure on wüstite composition according to the data of Giddings and Gordon (1973). For the reaction Fe + 1/2 O_2 = FeO,

$$\Delta G^\circ_T\ (\pm 100\ \text{cal}) = -62{,}420 + 15.18T$$

Determination, review, and critical analysis of the thermodynamic properties of ferrosilite and fayalite in a wide range of temperature have been presented in many works (Larimer, 1968; Kitayama and Katsura, 1968; Nafziger and Muan, 1967; Williams, 1971; Helgeson *et al.*, 1978; Robie *et al.*, 1978; Morse, 1979; Robinson *et al.*, 1982a). Critical analysis of all the works available at that time pertaining to the thermodynamic constants of ferrosilite was conducted by Kuskov in the review for a fundamental reference book *Termicheskiye Constanty* (Glushko *et al.*, 1972). In the present calculations on the thermodynamics of the FeO–SiO_2 system we have used the data of Williams (1971), the advantage of which lies in a wide range of temperature measurements and the internal consistency of the results. However, it should be noted that recalculation of Williams' data (1971) yielded the value $\Delta G^\circ_{f298.15}$ ($FeSiO_3$) $= -268.2 \pm 1.2$ kcal · mol^{-1}, which differs by 1 kcal from $\Delta G^\circ_{f298.15}$ ($FeSiO_3$) $= -267.08 \pm 0.2$ reported by Robinson *et al.* (1982a). The reason for this discrepancy remains to be studied. The free energies of the SiO_2 polymorphs are accepted according to Robie *et al.* (1978).

Standard free energies of reactions in the 1000–1800°K range can be described by the following equations:

$$\alpha - Fe_2SiO_4 = \gamma - Fe_2SiO_4: \qquad \Delta G^\circ_T\,(\pm 300) = 3340 + 2.345T$$

$$2\,FeSiO_3 = \gamma\text{-}Fe_2SiO_4 + SiO_2(\text{Stish}): \qquad \Delta G^\circ_T\,(\pm 800) = 9850 + 9.755T$$

$$2\,FeSiO_3 = \gamma\text{-}Fe_2SiO_4 + SiO_2(\text{Coes}): \qquad \Delta G^\circ_T\,(\pm 800) = -690 + 6.725T$$

$$\gamma - Fe_2SiO_4 = 2\,FeO + SiO_2(\text{Stish}): \qquad \Delta G^\circ_T\,(\pm 900) = 4050 + 7.20T$$

$$FeSiO_3 = FeO + SiO_2(\text{Stish}): \qquad \Delta G^\circ_T\,(\pm 900) = 6950 + 8.48T$$

Equations of state for α-, γ-Fe_2SiO_4, and $FeSiO_3$ are calculated by Kuskov (1984), including phase relations in the FeO–SiO_2 system at pressures up to 100 kbar. The Reaction γ-$Fe_2SiO_4 = 2\,FeO + SiO_2$(Stish) is the only stable equilibrium reaction above 100 kbar [reaction $FeSiO_3 = FeO + SiO_2$(Stish) taking place at 130 kbar and 1000–1800K is metastable, as at lower pressures disproportionation of ferrosilite into γ-spinel + stishovite takes place].

Decomposition of fayalite spinel into wüstite + stishovite without equilibrium control was investigated by many researchers (Mao and Bell, 1971; Basett and Ming, 1972; Sawamoto *et al.*, 1974; Kawada, 1977; Ohtani, 1979). According to all those investigations, decomposition of spinel is observed in the range of 170–250 kbar, and the sign of the slope for the univariant reaction is not clearly ascertained.

According to Kuskov's calculations, the slope of spinel decomposition into oxides is positive ($P = 143$, 167, 190, 237 kbar; $T = 1000$, 1273, 1500, 2000K). According to data of Navrotsky *et al.* (1979) $dP/dT < 0$; details are not given by the authors, but probably the discrepancy in dP/dT is associated with uncertainty in the initial thermodynamic constants [e.g., $\Delta H^\circ_{986}(\alpha = \gamma\text{-}Fe_2SiO_4) = 700$ cal seems to be underestimated] and with the simplified procedure of calculation at high pressures and temperatures. Nevertheless, it is evident that more investigations of this equilibrium reaction should be conducted.

Adding the third (FeO) component to the basic binary system MgO–SiO_2 makes the phase diagram more complicated and results in the presense of two invariant points (II and III) at ~210 kbar and 1600–1700K, where five-phase associations exist (Fig. 8):

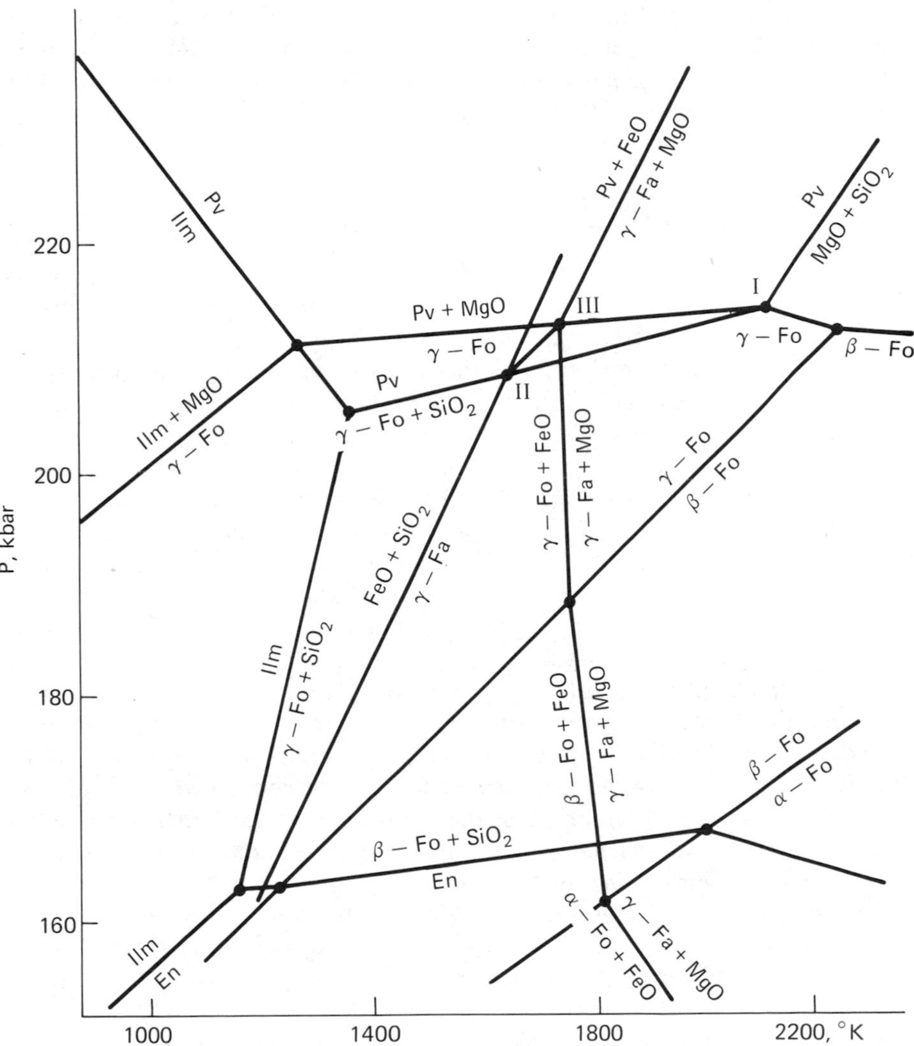

Fig. 8. P–T diagram of phase relations in the MgO–FeO–SiO_2 system: α-Fo, β-Fo, γ-Fo, polymorphs of Mg_2SiO_4; γ-Fa, Fe_2SiO_4 (spinel).

$$\gamma\text{-}Mg_2SiO_4 + \gamma\text{-}Fe_2SiO_4 + MgSiO_3(\text{Pv}) + FeO + SiO_2$$

$$\gamma\text{-}Mg_2SiO_4 + \gamma\text{-}Fe_2SiO_4 + MgSiO_3(\text{Pv}) + FeO + MgO$$

Both points are connected by the monovariant reaction:

$$\gamma\text{-}Mg_2SiO_4 + \gamma\text{-}Fe_2SiO_4 = 2\,MgSiO_3(\text{Pv}) + 2\,FeO$$

having a positive slope. With pressure increase γ-fayalite + MgO is transformed into $MgSiO_3$(Pv) + FeO ($dP/dT > 0$).

Because of the inadequate credibility of the thermodynamic constants for minerals with Fe compounds, the phase diagram for MgO–FeO–SiO_2 system, as it is the most representative of the mantle's petrology, should in future be specified and verified both theoretically and experimentally.

Geophysical Implications

Investigations of chemical and phase equilibria at ultrahigh pressures carried out within the framework of the MgO–FeO–CaO–Al_2O_3–SiO_2 system made it possible, with a minimum of initial thermodynamic information, to calculate the P–T parameters of equilibria and define the topology of diagrams for the systems, including several dozens of pure phases. The existing phase diagrams that model the sequence of phase transformations in the mantle are preliminary in nature for the time being and should be refined with the accumulation of actual data. Nevertheless, they serve as a basis for the interpretation of the mantle's mineral composition. The next stage of the physicochemical modeling that may bring us closer to an adequate description of the mantle's composition and its properties should expand this already developed methodology on studies of multicomponent systems that include minerals of variable composition.

The geophysical implications stemming from the material presented here consitute the subject of separate articles. Such material has already been published in part by Kuskov *et al.* (1983a, b) and Truskinovsky *et al.* (1983, 1984), but the greater part still remains to be published. In this chapter we briefly discuss three main cases based on the phase diagrams available, in order to show the potential of the physicochemical modeling method as applied to geochemical–geophysical problems of the solid Earth.

Energetics of Chemical Reactions

An important feature of the chemical reactions under consideration is that the overwhelming majority occur with a release of heat. This is explained by the fact that with a pressure increase the substance is transformed into a more dense association of phases ($\Delta V < 0$), possessing a smaller entropy ($\Delta S < 0$). Therefore, the slope of the equilibrium curves is, as a rule, positive. The exothermic nature of the intramantle physicochemical processes gains in importance in the analysis of a thermal regime of both the growing planet (during accretion when the low-pressure phases condensed initially from the proto-planet cloud, e.g., forsterite, enstatite, etc. and transformed into denser associations, e.g., spinel, perovskite, etc.), and for the formed planet, when the temperature profile of the mantle is reconstructed by taking into account the energy of the chemical reactions.

Distribution of Density

Figure 9 shows a density profile of a mantle of pyrolite composition calculated on the basis of independent thermodynamic information and consistent with phase diagrams of mineral systems. Initial model composition is represented by a combination (wt. %): of forsterite (44.1), enstatite (30.0), fayalite (11.6), pyrope (5.8), and grossular (8.5), which corresponds to a typical weight ratio of oxides for the mantle according to Ringwood (1975): SiO_2 (46.2), MgO (39), FeO (8.2), Al_2O_3 (3.4), CaO (3.2).

From the comparison of the "thermodynamic distribution" of density constructed with the help of phase diagrams, equations of state of the four coexisting phases and conditions with the seismic one, it is seen that the thermodynamic profile reveals a more definite fine density structure within the transition zone and allows one a detailed elucidation of the phase transformations in this area of the mantle. This is explained by the fact that a great number of chemical transformations with comparatively small jumps of density are not registered by seismic methods. To identify a fine structure use should be made of the data on solid solutions. However, at present the thermodynamic profiles of various initial model compositions may serve as a basis for specifying mineral compositions and the structure of the transition zone in conjunction with seismic sounding methods.

The 400- and 670-km Discontinuities in the Mantle

From phase relation analysis it is revealed that jumps in the physical properties discovered by seismic methods at depths of 400 and 670 km should be dictated by the concentration of a large number of monovariant chemical transformations accompanied by considerable volumetric effects. If we compare structures of seismic profiles of such types as velocity–depth and density–depth with those obtained from various types of phase diagrams we can find a rather instructive similarity between them. Extended divariant fields in phase diagrams, especially in the case of phase transformations in solid solutions, will represent gradient zones of seismic profiles, whereas monovariant transformations give rise to more pronounced peculiarities with noticeable jumps in the elastic properties. It is precisely such structures seen in seismic profiles that are registered in the experiments (Vinnik *et al.*, 1983; Walck, 1984).

When analyzing phase diagrams, it can be noted that invariant points are grouped in quite narrow fields in the range of 140–160 kbar and 210–220 kbar, and their position correlates with seismic boundaries at depths of 400 and 670 km. For the latest stages of the Earth's evolution the invariant points of the greatest interest are the points located at 215 ± 30 kbar and at temperatures of 1800–2100K, where several chemical reactions occur in parallel resulting in "jump-like" changes in the elastic properties of a number of mineral associations

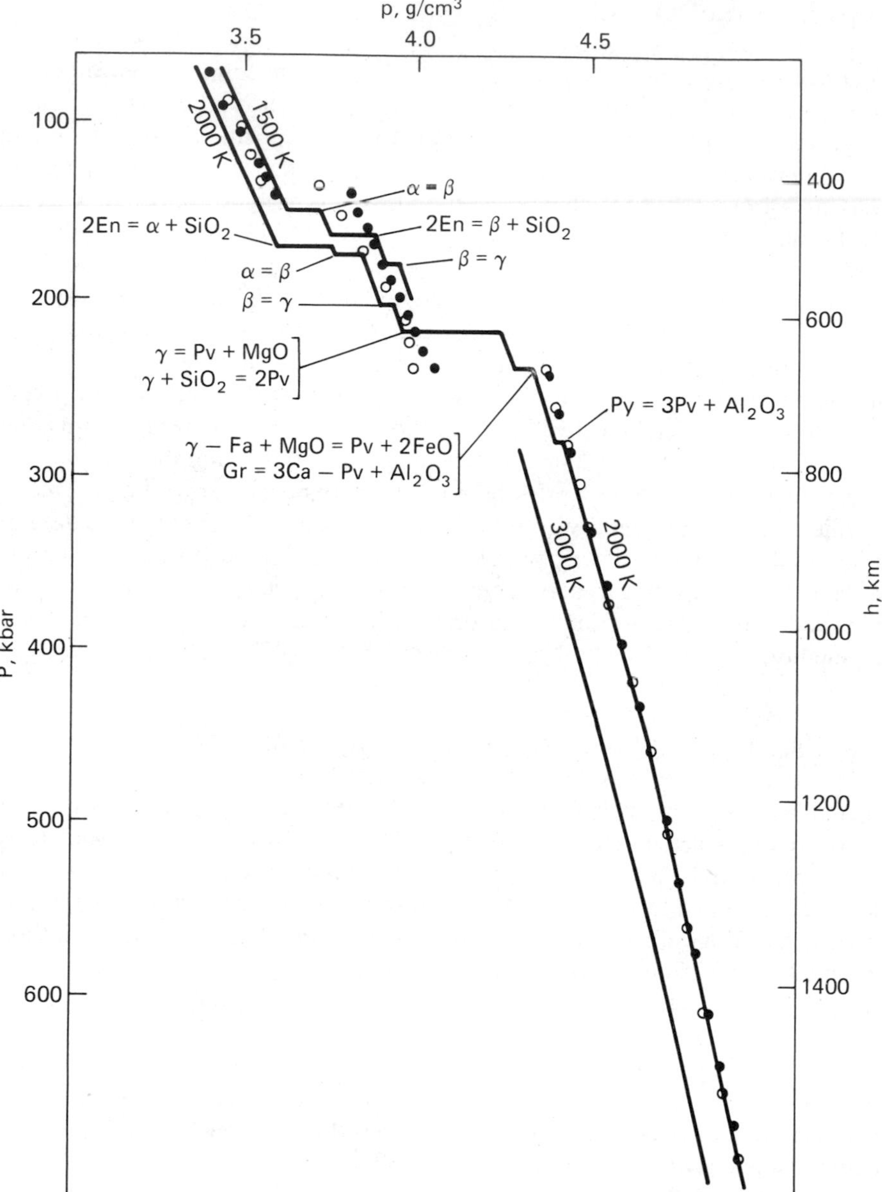

Fig. 9. Density distribution in the transition zone of the mantle: 1, Jordan and Anderson (1974); 2, Dziewonski and Anderson (1981); 3, thermodynamical calculation. Pyrolite composition was used. The density calculations are based on (a) calculated equilibrium assemblages at appropriate P and T, (b) equations of state of the coexisting phases, and finally (c) density Eq. (4) as discussed in the text. The errors in calculated equilibrium pressure may be as large as ± 40 kbar. However, the errors in molar volumes and density profiles are low. For example, at 300 kbar the estimated error in the density of perovskite + MgO assemblage is $\pm 0.03\ \mathrm{g \cdot cm^{-3}}$.

and giving rise to a diverse heterogeneity of the mantle substances under the given P–T conditions. Special analysis (Truskinovsky *et al.*, 1983, 1984) has revealed that the adiabat's penetration in an invariant point is not accidental and that transformations running in a invariant way can be typical of convection in the depth of the planet. Therefore, an opinion can be expressed that global seismic discontinuities in the mantle may be related both to the combination of certain monovariant transformations near the invariant point and to the invariant point itself on the phase diagram.

In concluding, we should like to emphasize once more that phase diagrams of modeling systems at present serve as a basis for the interpretation of the mineral composition of the mantle. Methods of seismic soundings based on the phase diagrams of the constant and variable composition systems will make it possible to reveal the fine structure of the mantle and to construct an adequate model of the internal structure of the Earth.

Basic Conclusions

1. A universal approach to the investigation of the thermodynamics of chemical and phase equilibria in multicomponent systems with phases of constant composition is developed, allowing one to solve the following main problems of physicochemical modeling of the processes at ultrahigh pressures: (1) calculation of the elastic, thermal, and caloric constants; (2) determination of the standard thermodynamic functions based on experimental P–T equilibrium data; (3) plotting of phase diagrams of multicomponent and multiphase systems; (4) calculation of thermal and volumetric effects of chemical and phase equilibria.
2. A thermochemical version for deriving equations of state of solid phases in the form of Mie-Grüneisen is elaborated, based on the utilization of thermodynamic constants at 1 atm and 298K, aimed at the calculation of both chemical and phase equilibria for a wide range of temperatures and pressures. This method of calculation is suggested as the basis for the creation of a thermodynamic data bank (up to 1 Mbar and 3000K) for computer use. Values of $\int_0^P V\,dP$ are determined with high accuracy and can now be tabulated in reference books on thermodynamics alongside of the standard thermodynamic functions.
3. An attempt was made to define qualitatively a type of multicomponent system phase diagram using minimum information on thermodynamics at 1 atm. Such an approach is the first step in the physicochemical modeling of quite complicated systems at ultrahigh pressures. This method seems to be the most fruitful one, especially if there is independent evidence for the existence of high-pressure phases, and possesses the greatest potential for the identification of a great number of possible geochemical reactions in the study system. The most essential result of the investigation of the systems MgO–SiO_2, CaO–MgO–SiO_2, CaO–Al_2O_3–SiO_2, MgO–Al_2O_3–SiO_2, MgO–FeO–

SiO_2 is the construction of topologic *P–T* diagrams modeling the sequence of phase relations in the mantle and serving, as a first approach, as the basic of thermodynamic analysis of endogenous mineral formation.

The diagrams constructed are of a predictive nature, which should offer to future experimental studies; *P–T* parameters and an evolutionary sequence of phase associations.

4. It is established in this study that a large number of chemical transformations are concentrated in the 100–300 kbar pressure interval, and that the majority of these occur with heat release from the formation of more dense phase associations. Phase diagrams and a calculated profile of density give evidence for the fact that all possible chemical transformations are completed by a depth of 700–800 km.

   This results in a simplified mineral composition of the mantle and the degeneration of a number of mineral associations with increasing pressure. Seismic discontinuities in the mantle can be connected not with just one individual transformation, but with their totality or even with the existence of invariant points in the phase diagrams.

Acknowledgments

The authors are indebted to Prof. S. K. Saxena for his advice, comments, and improvement of the manuscript and style. The manuscript is considerably improved by the suggestion of an anonymous reviewer. However, responsibility for the final paper must rest with the authors.

References

Akaogi, M., and Akimoto S. (1977) Pyroxene-garnet solid solution equilibria in the systems $Mg_4Si_4O_{12}$–$Mg_3Al_2Si_3O_{12}$ and $Fe_4Si_4O_{12}$–$Fe_3Al_2Si_3O_{12}$ at high pressures and temperatures, *Phys. Earth Planet. Int.* **15**, 90–106.

Akaogi, M., Ross, N. L., McMillan, P., and Navrotsky, A. (1984) The Mg_2SiO_4 polymorphs (olivine, modified spinel and spinel)—thermodynamic properties from oxide melt solution calorimetry, phase relations, and models of lattice vibrations, *Amer. Mineral.* **69**, 499–512.

Akimoto, S. (1972) The system MgO–FeO–SiO_2 at high pressures and temperatures—Phase equilibria and elastic properties, *Tectonophysics* **13**, 161–187.

Akimoto, S. (1975) The system MgO–FeO–SiO_2 at high pressures and temperatures-phase relations and elastic properties, in *Upper Mantle*, p. 60. Mir, Moskow.

Akimoto, S., Vagi, J., and Inoue, K. (1977) High temperature-pressure phase boundaries in silicate systems using in situ X-ray diffraction, In: *High Pressure Research*, edited by M. H. Manghnani and S. Akimoto, pp. 585–602. Academic Press, New York.

Al'tshuler, L. V. (1965) Shock waves in high pressure physics, *Sov. Adv. Phys. Sci.* **85**, 197–258.

Anderson, O. L. (1966) The use of ultrasonic measurements under modest pressure *J. Phys. Chem. Solids* **27**, 545–547.

Anderson, O. L. (1980) An experimental high-temperature thermal equation of state by passing the Grüneisen parameter, Phy. Earth Planet. Int. **22**, 173.

Anderson, O. L., and Baumgardner, J. R. (1980) Equations of state in planet interiors, Proc. Lunar Planet. Sci. Conf., Texas, pp. 1999–2014.

Babuška, V., Fiala, J., and Kumayawa, M. (1970) Elastic properties of garnet solid-solution, *Phys. Earth Planet. Int.* **16**, 157.

Barsukov, V. L., and Urusov, V. S. (1982) Phase transformations in the mantle transition zone and the possible change of the Earth's radius, *Geochemistry* **12**, 1729–1743.

Bass, J. D., Liebermann, R. C., Wiedner, D. J., and Fich, S. J. (1981) Elastic properties from acoustic and volume compression experiments, *Phys. Earth Planet. Int.* **25**, 140.

Basett, W. A., and Ming, L. C. (1972) Disproportionation of Fe_2SiO_4 to $2\,FeO + SiO_2$ (stishovite) at pressures up to 250 kbar and temperatures up to 3000°C, *Phys. Earth Planet. Int.* **6**, 154–160.

Basett, W. A., and Ming, L. C. (1976) New application of the diamond anvil pressure cell: (II) Laser heating at high pressure, in *The Physics and Chemistry of Minerals and Rocks*, edited by R. G. J. Strens, pp. 365–375. John Wiley and Sons, New York.

Birch F. (1952) Elasticity and constitution of the Earth's interior, *J. Geophys. Res.* **57**, 227–286.

Boehler, R., and Ramakrishnan, J. (1980) Experimental results of the Grüneisen parameter: a review, *J. Geophys. Res.* **85**, 6996.

Bonczar, L. J., and Graham, E. K. (1982) The pressure and temperature dependence of the elastic properties of polycrystal magnesio-wustite, *J. Geophys. Res.* **87**, 1061–1078.

Brown, J. M., and Shankland, T. J. (1981) Thermodynamic parameters in the Earth as determined from seismic profiles, *Geophys. J. Roy. Astronom. Soc.* **66**, 579–596.

Chang, Z. P., and Baroch, G. R. (1973) Pressure-dependence of single-crystal elastic constants and anharmonic properties of spinel, *J. Geophys. Res.* **78**, 2417.

Chang, Z. P., and Graham, E. K. (1977) Elastic properties of oxides in the NaCl-structure, *J. Phys. Chem. Solids* **38**, 1355.

Chung, D. H. (1971) Elasticity and equations of state of olivines in the Mg_2SiO_4–Fe_2SiO_4 system, Geophys. J. **25**, 511.

Chung, D. H. (1972) Equations of state of olivine-transformed spinels, *Earth Planet. Sci. Lett.* **14**, 348.

Chung, D. H. (1973) On the equation of state of high-pressure solid phases, *Earth Planet. Sci. Lett.* **18**, 125.

Chung, D. H. (1974) General relationships among sound speeds. *Phys. Earth Planet. Int.* **8**, 112–113.

Chung, D. H. (1979) Elasticity of stishovite revisited, in *Higher Pressure Science and Technology*, p. 97, Plenum Press, New York, London.

Chufarov, G. I., Men's, A. N., and Juravlev, M. A. (1970) *Thermodynamics of Reduction Processes of Metal Oxides*. Metallurgy, Moscow, 399 pp.

Clark, S. P. (Ed.) (1966) *Handbook of Physical Constants*. Yale Univ. Press, New Haven.

Davies, G. F. (1973) Quasi-harmonic finite strain equations of state of solids, *J. Phys. Chem. Solids* **34**, 1417.

Doroshev, A. M., Malinovsky, I. Yu., Kalinin, A. A. (1976) Topology of the CaO–Al_2O_3–SiO_2 system on the basis of experimental data, In: *Experimental Investigation in Mineralogy*, edited by A. A. Godovikov and A. B. Ptitzin, pp. 39–45. Nauka, Novosibirsk.

Dragoo, A. L., and Spain, I. L. (1977) The elastic moduli and their pressure and temperature derivatives for calcium oxide, *J. Phys. Chem. Solids* **38**, 705.

Dziewonski, A., and Anderson, D. (1981) Preliminary reference Earth model, *Phys. Earth*

Planet. Int. **25**, 297–356.

Finger, L. W., and Ohashi, Y. (1976) The thermal expansion of diopside to 800° and a refinement of the crystal structure at 700°C, *Amer. Mineral.* **61**, 303.

Giddings, R. A., and Gordon, R. S. (1973) Review of oxygen activities and phase boundaries in wustite as determined by electromotive-force and gravimetric methods, *J. Amer. Ceram. Soc.* **56**, 111–116.

Gieske, J. H., and Barsch, A. R. (1968) Pressure dependence of thc clastic constants of single-crystal aluminium oxide, *Phys. Stat. Sol.* **29**, 121.

Graham, E. K., and Barsch, G. R. (1969) Elastic constants of single-crystal forsterite as a function of temperature and pressure, *J. Geophys. Res.* **74**, 5949.

Haas, J. L., Robinson, G. R., and Hemingway, B. S. (1981) Thermodynamic Tabulations for Selected Phases in the System $CaO–Al_2O_3–SiO_2–H_2O$. Open-File Report 80-908, U.S. Geol. Surv. Natl. Center for the Thermodynamic Data of Minerals, Reston, 135 pp.

Halleck, P. M. (1973) Discrepancy between X-ray and ultrasonic determination of the compression behaviour of grossular garnet, *Trans. Amer. Geophys. Union* **54**, 476.

Haselton, H. T., and Westrum, E. F. (1980) Low-temperature heat capacities of synthetic pyrope, grossular and $purope_{60}$ $grossular_{40}$, *Geochim. Cosmochim Acta* **44**, 701.

Helgeson, H. C. (1969) Thermodynamics of hydrothermal systems at elevated temperature and pressure, *Amer. J. Sci.* **267**, 729.

Helgeson, H. C., Delany, J. M., Nesbitt, H. W., and Bird, D. K. (1978) Summary and critique of the thermodynamic properties of rock-forming minerals, *Amer. J. Sci.* **278A**, 229.

Horai, K., and Simmons, G. (1970) An empirical relationships between thermal conductivity and Debye temperature for silicates, *J. Georphys. Res.* **75**, 978–982.

Huang, W. L., and Wyllie, P. J. (1975) Melting and subsolidus phase relations for $CaSiO_3$ to 35 kilobars pressure, *Amer. Mineral.* **60**, 213–217.

Intriligator, M. (1975) *Mathematic Methods of Optimization and Economical Theory.* Progress, Moscow, 606 pp.

Ito, E., and Matsui, Y. (1977) Silicate ilmenites and the post-spinel transformations, in *High-Pressure Research*, edited by M. H. Manghnani and S. Akimoto, pp. 193–208. Academic Press, New York.

Ito, E., and Yamada, H. (1982) Stability relations of silicate spinels, ilmenites, and perovskites, in *High Pressure Research in Geophysics*, edited by S. Akimoto and M. H. Manghnani, pp. 405–419. Center for Academic Publishing, Tokyo.

Isaak, D. G., and Graham, E. K. (1976) The elastic properties of an almandine-spessartine garnet and elasticity in the garnet solid-solution series, *J. Geophys. Res.* **81**, 2483.

Jackson, I., and Ahrens, T. J. (1979) Shockwave compression of single-crystal forsterite, *J. Geophys. Res.* **84**, 3039.

Jeanloz, R., and Thompson, A. B. (1983) Phase transitions and mantle discontinuities, *Rev. Geophys. Space Phys.* **21**, 51–74.

Jordan, T. H., and Anderson, D. L. (1974) Earth structure from free oscillations and travel times, *Geophys. J. Roy. Astronom. Soc.* **36**, 411–459.

Karpov, I. K., Kiselev, A. I., and Letnikov, F. A. (1976) *Computer Modelling of Natural Mineral-Formation.* Nedra, Moscow, 256 pp.

Kawada, K. (1977) The System $Mg_2SiO_4–Fe_2SiO_4$ at High Pressures and Temperatures and the Earth's Interior. Ph.D. Thesis, Institute for Solid State Physics, University of Tokyo, 187 pp.

Kieffer, S. W. (1980) Thermodynamics and lattice vibrations of minerals, 4, Application to chain and sheet silicates and orthosilicates, *Rev. Geophys. Space Phys.* **18**, 862–886.

Kitayama, K., and Katsura, T. (1968) Composition of fayalite and its standard free energy of formation, *Chem. Soc. Jpn. J.* **41**, 525–528.

Kolesnik, Yu. N., Nogteva, V. V., Archipenko, D. K., *et al.* (1978) Heat capacity in grossular in the temperature range 13–1300K and thermodynamics of pyropegrossular solid solution, *Geochemistry*, **5**, 713.

Korzhinskii, D. S. (1959) *Physiochemical Basis of the Analysis of the Paragenesis of Minerals*. Consultants Bureau, New York, 142 pp.

Kraynov, S. R., Ryzhenko, B. N., and Shvarov, Yu. V. (1983) Possibilities and limitations of physico-chemical modelling of interaction water—rock in the solution of problems of formation of chemical composition of ground waters, *Geochemistry* **9**, 1342–1359.

Kumayawa, M., and Anderson, O. L. (1969) Elastic moduli, pressure derivatives, and temperature derivatives of single-crystal forsterite, *J. Geophys. Res.* **74**, 5961.

Kumazawa, M., Sawamoto, H., Ohtani, E., and Masaki, K. (1974) Postspinel phase of forsterite and evolution of the Earth's mantle, *Nature* (*London*) **247**, 356–358.

Kurepin, V. A. (1981) *Thermodynamics of Minerals of Variable Composition and Geological Thermobarometry*. Naukova Dumka, Kiev, 160 pp.

Kuskov, O. L. (1984) Equations of state of α-, γ-Fe_2SiO_4 and $FeSiO_3$ and their phase relations at high pressures, *Geochemistry* **8**, 1119–1124.

Kuskov, O. L., and Galimzyanov, R. F. (1982a) Equations of state and standard thermadynamic functions of α-, β-, γ-Mg_2SiO_4, *Geochemistry* **8**, 1172–1182.

Kuskov, O. L., and Galimzyanov, R. F. (1982b) Equations of state of Al_2O_3, CaO, spinel, diopside, and garnets under superhigh pressures and temperatures, *Geochemistry* **11**, 1586–1597.

Kuskov, O. L., and Galimzyanov, R. F. (1984) Calculation of equilibrium $MgAl_2O_4 = MgO + Al_2O_3$ taking into account the cation disordering of spinel, compressibility and thermal expansion of phases, *Geochemistry* **1**, 101–106.

Kuskov, O. L., and Khitarov, N. I. (1982) *Thermodynamics and Geochemistry of the Earth's Core and Mantle*. Nauka, Moscow, 279 pp.

Kuskov, O. L., Galimzyanov, R. F., Kalinin, V. A., Bubnova, N. Ja., and Khitarov, N. I. (1982) Construction of thermal equation of state of solids (periclase, coesite, stishovite) based on bulk modulus data and calculation of the coesite-stishovite phase equilibrium, *Geochemistry* **7**, 984–1001.

Kuskov, O. L., Galimzyanov, R. F., and Khitarov, N. I. (1983a) Phase relations in the MgO–FeO–CaO–Al_2O_3–SiO_2 system in the mantle transition zone, *Dokl. of Alad. Nauk USSR* **270**, 577–581 (in Russian).

Kuskov, O. L., Galimzyanov, R. F., Truskinovsky, L. M., and Pil'chenko, V. A. (1983b) Reliability of thermodynamic calculations of chemical and phase equilibria at high pressures and temperatures, *Geochemistry* **6**, 849–871.

Kuskov, O. L., Galimzyanov, R. F., Khitarov, N. I., and Urusov, S. V. (1983c) Phase relations in the MgO–SiO_2 system at P–T conditions of the mantle transition zone, *Geochemistry* **8**, 1075–1091.

Larimer, J. W. (1968) Experimental studies on the system Fe–Mg–SiO_2–O_2 and their bearing on the petrology of chondritic meteorites, *Geochim. Cosmochim. Acta* **32**, 1187–1209.

Leitner, B. J., Weidner, D. J., Liebermann, R. C. (1980) Elasticity of single-crystal pyrope and implications for garnet solid solution series, *Phys. Earth Planet. Int.* **22**, 111.

Levien, L., and Prewitt, C. T. (1981) High-pressure structural study of diopside, *Amer. Mineral.* **66**, 315.

Levien, L., Prewitt, C. T., and Weidner, D. J. (1979) Compression of pyrope, *Amer. Mineral.* **64**, 805.

Levien, L., Weidner, D. J., and Prewitt, C. T. (1979) Elasticity of dropside, *Phys. Chem. Minerals* **4**, 105.

Liebermann, R. C. (1974) Elasticity of pyroxene-garnet and pyroxene-ilmenite phase transformations in germanates, *Phys. Earth Planet. Int.* **8**, 361.

Liebermann, R. C. (1975) Elasticity of olivine (α), beta (β), and spinel (γ) polymorphs of germanates and silicates, *Geophys. J.* **42**, 899.

Liebermann, R. C. (1976) Elasticity of ilmenites, *Phys. Earth Planet. Int.* **12**, 5.

Liebermann, R. C., and Ringwood, A. E. (1976) Elastic properties of anorthite and nature of the lunar crust, *Earth Planet. Sci. Lett.* **31**, 69–79.

Liebermann, R. C., and Ringwood, A. E. (1977) Some comments on the elasticity of stishovite determined by ultrasonic and high pressure X-ray diffraction techniques, in *High-Pressure Research:* Applications in Geophysics, edited by M. H. Manghnani and S. Akimoto, p. 343. Academic Press, New York.

Liebermann, R. C., Jones, L. E. A., and Ringwood, A. E. (1977) Elasticity of aluminate, titanate, stannate and germanate compounds with the perovskite structure, *Phys. Earth Planet. Int.* **14**, 165.

Liu, L. (1974a) Disproportionation of kyanite to corundum plus stishovite at high pressure and temperature, *Earth Planet. Sci. Lett.* **24**, 224–228.

Liu, L. (1974b) Silicate pervoskite from phase transformations of pyrope-garnet at high pressure and temperature, *Geophys. Res. Lett.* **1**, 277–280.

Liu, L. (1978) A new high-pressure phase of $Ca_2Al_2SiO_7$ and implications for the Earth's interior, *Earth Planet. Sci. Lett.* **40**, 401–406.

Liu, L. G. (1979a) Phase transformations and constitution of the deep mantle, in *The Earth: Its Origin, Structure and Evolution*, edited by M. W. McElhinny, pp. 177–202. Academic Press, London.

Liu, L. G. (1979b) Calculations of high pressure phase transitions in the system $MgO–SiO_2$ and implications for mantle discontinuties, *Phys. Earth Planet. Int.* **19**, 319–327.

Liu, L. G. (1979c) High pressure phase transformations in the joins Mg_2SiO_4 and $MgO–CaSiO_3$, *Contrib. Mineral. Petrol.* **69**, 245–247.

Liu, L. G. (1979d) High pressure phase transformations in the system $CaSiO_3–Al_2O_3$, *Earth Planet. Sci. Lett.* **43**, 331–335.

Liu, L. G. (1979e) The system enstatite-wollastonite at high pressure and temperature, with emphasis on diopside, *Phys. Earth Planet. Int.* **19**, P15–P18.

Liu, L. G. (1980) The equilibrium boundary of spinel-corundum + periclase: a calibration curve for pressures above 100 kbar, *High Temperatures–High Pressures*, **12**, 217–220.

Liu, L., and Ringwood, A. E. (1975) Synthesis of a perovskite-type polymorph of $CaSiO_3$, *Earth Planet. Sci. Lett.* **28**, 209–211.

Lukas, H. L., Weiss, J. and Henig, E.-Th. (1982) Strategies for the calculation of phase diagrams, *CALPHAD*, **6**, 229–251.

Mao, H.-K., and Bell, P. M. (1971) High-pressure decomposition of spinel (Fe_2SiO_4), *Carnegie Inst. Wash. Yearbk.* **70**, 176–178.

Mao, H.-K., and Bell, P. M. (1979) Equations of state of MgO and ε-Fe under static pressure conditions, *J. Geophys. Res.* **84**, 4533–4536.

Mao, H.-K., Yagi, T., and Bell, P. M. (1977) Mineralogy of the earth's deep mantle: quenching experiments at high pressure and temperature, *Carnegie Inst. Wash. Yearbk.* **76**, 502–504.

McCammon, C. A., and Liu, L. (1984) The effects of pressure and temperature on non-stoichiometric wüstite, Fe_xO: The iron-rich phase boundary, *Phys. Chem. Minerals* **10**, 106–113.

Miyubami, S., Ohtani, A., Kawai, N. (1976) High-pressure X-ray diffraction studies on β- and γ-Mg_2SiO_4, *Phys. Earth Planet. Int.* **10**, 177.

Morse, S. A. (1979) Reaction constants for En-Fo-Sil equilibria: an adjustment and some applications, *Amer. J. Sci.* **279**, 1060–1069.

Mulargia, F., and Boschi, E. (1980) The problem of the equation of state in the Earth's interior, in *Physics of the Earth's Interior*, pp. 337–361. North Holland Publ. Co., Amsterdam.

Nafziger, R. H., and Muan, A. (1967) Equilibrium phase compositions and thermodynamic properties of olivines and pyroxenes in the system MgO–"FeO"–SiO_2, *Amer. Mineral.* **52**, 1364–1385.

Naumov, G. B., Ryzenko, B. N., and Khodakovsky, I. L. (1971) *Handbook of Thermodynamical Values*. Moscow, Atomizdat, 240 pp.

Navrotsky, A. (1980) Lower mantle phase transitions may generally have negative pressure-temperature slopes, *Geophys. Res. Lett.* **7**, 709–711.

Navrotsky, A., Pintchovski, F. S., and Akimoto, S. (1979) Calorimetric study of the stability of high pressure phases in the systems CoO–SiO_2 and "FeO"–SiO_2 and calculation of phase diagrams in MO–SiO_2 systems, *Phys. Earth and Planet. Int.* **19**, 275–292.

O'Connel, R. J., and Graham, E. K. (1971) Equation of state of stoichiometric spinel to 10 kbar and 900°K, *Trans. Amer. Geophys. Union* **52**, 319.

Ohtani, E. (1979) Melting relation of Fe_2SiO_4 up to about 200 kbar, *J. Phys. Earth* **27**, 189–208.

Ohtani E., Sawamoto, H., Masaki, K., and Kumazawa, M. (1974) Decomposition of spinel $MgAl_2O_4$ at extremely high pressure, in *Proc. 4th Int. Conf. on High Pressure*, pp. 186–189, Kyoto.

Olinger, B. (1977) A comparison of α-quarty shock compression data with recent determinations of the bulk modulus of stishovite, in *High-Pressure Research: Applications in Geophysics*, edited by M. H. Manghnani and S. Abimoto, p. 335. Academic Press, New York.

Ostrovsky, I. A. (1979) The thermodynamics of substances at very high pressures and temperatures and some mineral reactions in the Earth's mantle, *Phys. Chem. Minerals* **5**, 105–118.

Ringwood, A. E. (1975) Composition and petrology of the Earth's mantle. McGraw-Hill, New York.

Ringwood, A. E. (1981) *The Composition and Petrology of the Earth's Mantle*, Nedra, Moscow, 584 pp.

Ringwood, A. E., and Major, A. (1971) Synthesis of majorite and other high pressure garnets and perovskites, *Earth Planet. Sci. Lett.* **12**, 411–418.

Robie, R. A., Hemingway, B. S., and Fisher, J. R. (1978) Thermodynamic Properties of Minerals and Related Substances at 298K and 1 bar Pressure and at Higher Temperatures. U.S. Geol. Surv. Bull. 1452, 456 pp.

Robie, R. A., Finch, C. B. and Hemingway, B. S. (1982) Heat capacity and entropy of fayalite (Fe_2SiO_4) between 5.1 and 383K: comparison of calorimetric and equilibrium values for the QFM buffer reaction, *Amer. Mineral.* **67**, 463.

Robinson, G. R., Haas, J. L., Schafer, C. M., and Haselton, H. T. (1982a) Thermodynamic and thermophysical properties of selected phases in the MgO–SiO_2–H_2O–

CO_2, CaO–Al_2O_3–SiO_2–H_2O–CO_2 and Fe–FeO–Fe_2O_3–SiO_2 chemical systems, with special emphasis on the properties of basalts and their mineral components, Open-File Report 83–79, U.S. Geol. Suwo, 429 pp.

Robinson, G. R., Haas, J. L., Schafer, C. M., and Haselton, H. T. (1982b) Thermodynamic and thermophysical properties of selected phases in the MgO–SiO_2–H_2O–CO_2, CaO–Al_2O_3–SiO_2–H_2O–CO_2 and Fe–FeO–Fe_2O_3–SiO_2 chemical systems, with special emphasis on the properties of basalts and their mineral components. Open-File Rept. 83–79, 429 pp.

Romain, J. P., Migault, A., and Jacquesson, J. (1976) Relation between Grüneisen ratio and the pressure dependence of Poisson's ratio for metals, *J. Phys. Chem. Solids* **37**, 1159.

Sawamoto, H., Ohtani, E., and Kumazawa, M. (1974) High pressure decomposition of γ-Fe_2SiO_4, *Proc. 4th Int. Conf. High Pressures*, pp. 194–201, Kyoto.

Saxena, S. K., and Eriksson, G. (1983) Theoretical computation of mineral assemblages in pyrolite and lherzolite, *J. Petrol.* **24**, 538–555.

Schreinemakers, F. A. (1948) In-, Mono- and Divariant Equilibria. Moscow, 214 pp.

Shankland, T. J., and Chung, D. H. (1974) General relationships among sound speeds, *Phys. Earth Planet. Int.* **8**, 129.

Shvarov, Yu. V. (1981) General equilibrium criterion for the isobaric-isothermic model of a chemical system, *Geochemistry* **7**, 981–988.

Soda, N., Schreiber, E., and Anderson, O. L. (1966) Estimation of bulk modulus and sound velocities of oxides at very high temperatures, *J. Geophys. Res.* **71**, 5315.

Son, P. R., and Bartels, R. A. (1972) CaO and SiO single crystal elastic constants and their pressure derivatives, *J. Phys. Chem. Solids* **33**, 819.

Stacey, F. (1977) A thermal model of the Earth, *Phys. Earth Planet. Int.* **15**, 341–348.

Stacey, F., Brennan, B. J., and Irvine, R. D. (1981) Finite strain theories and comparisons with seismological data, *Geophys. Surv.* **3**, 189–232.

Spetzler, H. (1970) Equation of state of polycrystalling and single-crystal MgO to 8 kilobars and 800K. *J. Geophys. Res.* **75**, 2070–2073.

Suito, K. (1977) Phase relations of pure Mg_2SiO_4 up to 200 kilobars, in *High Pressure Research—Applications to Geophysics*, edited by M. Manghnani and S. Akimoto, pp. 255–266. Academic Press, New York.

Suito, K., and Kawai N. (1979) Studies of phase equilibrium in Mg_2SiO_4 up to pressures higher than 20 GPa, in *High-Pressure Science and Technology*, Vol. 2, edited by K. D. Timmerhaus and M. S. Barber, pp. 53–57. Plenum Publ. Co., New York.

Sumino, Y. (1979) The elastic cosnstants of Mn_2SiO_4, Fe_2SiO_4, and Co_2SiO_4 and the elastic properties of olivine group minerals at high temperature, *J. Phys. Earth* **27**, 209.

Sumino, Y., Nishiyawa, O. (1978) Temperature variation of elastic constants of pyrope-almandine garnets, *J. Phys. Earth* **26**, 239.

Suyuki, I., Ohtani, E., and Kamasawa, M. (1979) Thermal expansion of γ-Mg_2SiO_4, *J. Phys. Earth* **27**, 53.

Suyuki, I., Ohtani, E., and Kumasawa, M. (1980) Thermal expansion of modified spinel, β-Mg_2SiO_4, *J. Phys. Earth* **28**, 273.

Suyuki, I., Seya, K., Takei H., and Sumino, Y. (1981) Thermal expansion of fayalite, Fe_2SiO_4, *Phys. Chem. Minerals* **7**, 60.

Suzuki, I. (1975) Thermal expansion of periclase and olivine and their anharmonic properties. *J. Phys. Earth* **23**, 145–150.

Termicheskie Constanty (Thermic Constants of Substances) (1972), edited by V. P. Glushko *et al.*, Vol. VII VINITY, Moscow.

Thomsen, L. (1970) On the fourth-order anharmonic equation of state of solids, *J. Phys. Chem. Solids* **31**, 2003–2016.

Toulonbian, Y. S., and Ho, C. Y. (1977) *Thermophysical Properties*, TRRC Data Series Vol. 13, *Thermal Expansion, Nonmetallic Solids*. Plenum Press, New York.

Truskinovsky, L. M., Kuskov, O. L., and Khitarov, N. I. (1983) Adiabatic gradient in the mantle transition zone, *Geochemistry* **9**, 1222–1238.

Truskinovsky, L. M., Kuskov, O. L., and Khitarov, N. I. (1984) Structure of adiabat in the mantle transition zone. *Dokl. Ad. Nauk USSR* **274**, 1064–1070 (in Russian)

Ullman, W., and Pankov, V. L. (1970) A new structure of the equation of state and its application in high-pressure physics and geophysics, Veröff. Zentralinst. *Phys. Erde* (*Potsdam*) **41**, 1–201.

Urusov, V. S., and Kuskov, O. L. (1984) Crystal-chemistry and thermodynamics of mineral phases in the Earth's mantle, in *27th International Geological Congress*, Section C.10, pp. 94–105. Mineralogy, Moscow.

Vinnik, L. P., Avetisjan, R. A., and Mihailova, N. G. (1983) Heterogeneities in the mantle transition zone from observations of P-to-SV converted waves, *Phys. Earth Planet. In.* **33**, 149–163.

Wachtman, J. B., Tefft, W. E., and Lam, D. G. (1960) Elastic constants of single-crystal corundum at room temperature, *J. Res. NBS* **64A**, 213.

Walck, M. C. (1984) The P-wave upper mantle structure beneath an active spreading center: the Gulk of California, *Geophys. J. Roy. Astronom. Soc.* **76**, 697–723.

Watanabe, H. (1982) Thermochemical properties of synthetic high pressure compounds relevant to the Earth's mantle, in *High-Pressure Research in Geophysics*, edited by S. Abemoto and M. H. Manghnani, pp. 441–464. Academic Publishers, Tokyo.

Weaver, J. S., Chysman, D. W., and Takahashi, T. (1979) Comparison between thermochemical and phase stability data for the quarty-stishovite transformations, *Amer. Mineral,* **64**, 604.

Weidner, D. J., and Carlton, H. R. (1977) Elasticity of coesite, *J. Geophys. Res.* **82**, 1334.

Weidner, D. J., Wang H., and Ito, J. (1978) Elasticity of orthoenstatite, *Phys. Earth Planet. Int.* **17**, 7.

Westrum, E. F., Essene, E. J., and Perkins, D. (1979) thermophysical properties of the garnet, grossular: $Ca_3Al_2Si_3O_{12}$, *J. Chem. Thermodynam.* **11**, 57.

Williams, R. J. (1971) Reaction constants in the system Fe–MgO–SiO_2–O_2 at 1 atm between 900° and 1300°C: Experimental results, *Amer. J. Sci.* **270**, 334–360.

Wood, B. J., and Fraser, D. G. (1981) *Elementary Thermodynamics for Geologists* (Transl. from English). Mir, Moscow, 184 pp.

Yagi, T., Bell, P. M., and Mao, H. K. (1979a) Phase relations in the system MgO–FeO–SiO_2 between 150 and 700 kbar at 1000°C, *Carnegie Inst. Wash. Yearbk.* **78**, 614–618.

Yagi, T., Mao, H. K., and Bell, P. M. (1979b) Hydrostatic compression of $MgSiO_3$ of perovskite structure, *Ann. Rept. Dir., Geophys. Lab., Carnegie Inst. Wash.* **78**, 613.

Zharkov, V. N., and Kalinin, V. A. (1968) *Equations of State of Solids at High Temperatures and Pressures*. Nauka, Moscow, 312 pp.

Appendix

Error Analysis of the Thermal Equation of State

Standard deviations of intermediate parameters A, b, K in the potential part of the thermal equation of state, as well as output parameters, can be estimated from error propagation formula:

$$\sigma_y^2 = \sum_i \sum_j \left(\frac{\partial y}{\partial a_i}\right)\left(\frac{\partial y}{\partial a_j}\right)\sigma_i \sigma_j \eta_{ij}$$

where σ_y, σ_i, and σ_j are standard deviations for functions $y(a_1, a_2, \ldots, a_n)$ and parameters a_i, a_j, respectively; η_{ij} is the coefficient of statistical correlation between $a_i\, a_j$.

Input parameters for the thermal equation of state of solids are K_{0S}, K'_{0S}, α_{0V}, C_{0P}, θ_0, ρ_0. Thermal part of the bulk modulus in Eq. (14) is of the same order or less then the standard deviation in K_{0S} and therefore has a little effect on error estimates for parameters A and b in Eq. (18). Thus, when σ_A and σ_b were calculated we took into account only errors in K_{0S} and K'_{0S}. Partial derivatives for A and b may be approximated by the following relationships:

$$\frac{\partial b}{\partial K'_{0S}} = \frac{3}{1 + 2/(b-2)^2}$$

$$\frac{\partial b}{\partial K_{0S}} = \frac{\partial b}{\partial \alpha_{0V}} = \frac{\partial b}{\partial C_{0P}} = \frac{\partial b}{\partial \theta_0} = \frac{\partial b}{\partial \rho_0} = 0$$

$$\frac{\partial A}{\partial K'_{0S}} = -\frac{A}{(b-2)}\frac{\partial b}{\partial K'_{0S}}, \quad \frac{\partial A}{\partial K_{0S}} = \frac{3}{B-2}$$

$$\frac{\partial A}{\partial \alpha_{0V}} = \frac{\partial A}{\partial C_{0P}} = \frac{\partial A}{\partial \theta_0} = \frac{\partial A}{\partial \rho_0} = 0$$

The experimental value of $\eta_{K_{0S}K'_{0S}}$ is usually unknown, so we put it equal to zero. Then we have:

$$\sigma_b^2 = \left(\frac{\partial b}{\partial K'_{0S}}\right)^2 \sigma_{K'_{0S}}^2$$

$$\sigma_A^2 = \left(\frac{\partial A}{\partial K'_{0S}}\right)^2 \sigma_{K'_{0S}}^2 + \left(\frac{\partial A}{\partial K_{0S}}\right)^2 \sigma_{K_{0S}}^2$$

$$\eta_{Ab} = \left(\frac{\partial A}{\partial K'_{0S}}\right)\left(\frac{\partial b}{\partial K'_{0S}}\right)\frac{\sigma_{K'_{0S}}^2}{\sigma_A \sigma_b}$$

The standard deviation for parameter K in Eq. (18), $P(V, T)$, $V(P, T)$, $\int_0^P V\,dP$, bulk modulus, etc., may be estimated by the following way:

$$\sigma_y^2 = \left(\frac{\partial y}{\partial A}\right)^2 \sigma_A^2 + \left(\frac{\partial y}{\partial b}\right)^2 \sigma_b^2 + 2\left(\frac{\partial y}{\partial A}\right)\left(\frac{\partial y}{\partial b}\right)\sigma_A \sigma_b \eta_{Ab}$$

Table A. Input and output parameters for equations of state of stishovite.[a]

Input and output parameters	Stishovite-1	Stishovite-2	Stishovite-3
M, molecular weight	60.085	60.085	60.085
ρ_0, $g \cdot cm^{-3}$	4.289(0.005)	4.287(0.005)	4.287(0.002)
K_{0S}, kbar	3160(75)	2930(150)	2930(150)
K'_{0S}	4.0(1.0)	5.0(0.5)	5.0(0.5)
C_{0P}, $cal \cdot K^{-1}\ mol^{-1}$	10.27(0.2)	10.27(0.2)	10.27(0.2)
$\alpha_{0V} \cdot 10^6$, K^{-1}	16.5(1.0)	17.1(1.0)	8.9(1.5)
θ_0, K	1190(25)	1050(100)	1050(100)
γ_0	1.70(0.111)	1.634(0.131)	0.851(0.151)
A, kbar	1747.17(914.5)	1050.654(195.8)	1062.707(196.3)
b	7.397(2.822)	10.253(1.477)	10.241(1.462)
K, kbar	1792.632(914.5)	1090.861(195.9)	1083.633(196.4)
η_{Ab}	−0.998	−0.999	−0.999
V (100 kbar, 1273K), $cm^3 \cdot mol^{-1}$	13.90(0.03)	13.90(0.03)	13.74(0.04)
$\int_0^P V\,dP$ (100 kbar, 1273K), $\kappa cal \cdot mol^{-1}$	33.77(0.17)	33.80(0.23)	33.38(0.23)
K_S (100 kbar, 1273K), kbar	3387(130)	3227(166)	3310(174)
α_V (100 kbar, 1273K), $10^6\ K^{-1}$	26.9(1.4)	27.6(1.5)	13.7(1.9)

[a] Standard deviations in parentheses.

$$+\left(\frac{\partial y}{\partial \gamma_0}\right)^2 \sigma_{\gamma_0}^2 + \left(\frac{\partial y}{\partial \theta_0}\right)^2 \sigma_{\theta_0}^2$$

where σ_{γ_0} effectively takes into account the errors in α_{0V}, C_{0P}, and ρ_0:

$$\sigma_{\gamma_0}^2 = \gamma_0^2\left[\left(\frac{\sigma_{\alpha_{0V}}}{\alpha_{0V}}\right)^2 + \left(\frac{\sigma_{K_{0S}}}{K_{0S}}\right)^2 + \left(\frac{\sigma_{\rho_0}}{\rho_0}\right)^2 + \left(\frac{\sigma_{C_{0P}}}{C_{0P}}\right)^2\right]$$

Partial derivatives on A, b, γ_0, and θ_0 for each thermodynamical value were derived in analytical form, and a FORTRAN program for calculating various thermodynamic output parameters and their standard errors was developed. These output information may be used then to construct potential phase diagrams for the system of interest. We will supply any available data on programming code and test results by request.

In addition to Tables 1 and 2 complete information for three equations of state of stishovite is given in Table A with different input parameters. Values for K_{0S} are taken from Weidner *et al.* (1982) and Chung (1979), α_{0V} from Ito *et al.* (1974). One can compare obtained values of V, $\int_0^P V\,dP$, K_S, α_V at 100 kbar and 1273K. The difference in $\int_0^P V\,dP$ is 0.4 $kcal \cdot mol^{-1}$ ($\sim 1\%$).

Chapter 10
Generalized Mathematical Models for the Fractional Evolution of Vapor from Magmas in Terrestrial Planetary Crusts

Philip A. Candela

Symbols

C_i^{ϕ}	concentration of i in phase ϕ
V	(superscript) vapor
l	(superscript) liquid
K_i^p	V/l partition coefficient of Holland (1972)
n	ligation number
μ_i^{ϕ}	chemical potential of i in phase ϕ
$K_{i,\mathrm{Na}}$	vapor/liquid exchange constant for i relative to sodium
$D_i^{V/l}$	Nernst partition coefficient for the distribution of i between vapor and melt
M_l	mass of liquid
$\mathscr{F}$	proportion of melt remaining
w	(subscript) water
o	(superscript) initial value
M_i^{ϕ}	mass of i in phase ϕ
$E(i)$	efficiency of removal of i from a magma into an aqueous fluid
β	see Eq. (20) and following discussion
$\alpha(i)$	see Eq. (22) and following discussion
X_j^{ϕ}	mole fraction of j in phase ϕ
Z	$M_w^l/M_w^{l,0}$
$\bar{D}_i^{S/l}$	bulk partition coefficient for the distribution of i between hypersolidus phases and melt
$\bar{D}$	shorthand notation for $\bar{D}_i^{S/l}$
T	total
P	pressure
$C_w^{l,0}$	initial concentration of water in the melt phase. In model I, all water is

expelled from the magma from $\mathscr{F} = 1$ to $\mathscr{F} = 0$. Unless otherwise stated, the units of $C_w^{l,0}$ are mass fraction.

$C_w^{l,s}$ water concentration in the melt at the initiation of water saturation. Unless otherwise stated, the units of $C_w^{l,s}$ are mass fraction.

$C_w^l(Z)$ Water concentration in the melt after water saturation. Unless otherwise stated, the units of $C_w^l(Z)$ are mass fraction.

Introduction

The geological literature is replete with models for the differentiation of magmas that occur on Earth, the Moon, and other terrestrial bodies. The models involve processes such as crystal/liquid fractionation (Allégre and Minster, 1978), diffusion (Wright *et al.*, 1983), magma chamber replenishment (O'Hara, 1977), magma mixing (McBirney, 1980) and assimilation (Grove *et al.*, 1982). However, few of these models include a quantitative treatment of the effects of magmatic vapor evolution on melt, crystal, or vapor chemistry, although qualitative appeals to such processes are commonplace. Vapor evolution has been cited as a possible explanation for aplite chemistry (Fourcade and Allégre, 1981), rubidium depletion in igneous amphibole (Chivas, 1981), magmatic oxidation effects (Chivas, 1981), fluorine depletion in suites of igneous rocks (McMillan, 1982) and as a source for ore metals in many ore deposits (c.f. Burnham, 1979). In his *Evolution of the Igneous Rocks*, N. L. Bowen (1928) states: "To many petrologists a volatile component is exactly like a Maxwell demon; it does just what one may wish it to do." Apparently his demon is alive and well, and a quantitative model of vapor evolution is needed to test the above hypotheses.

In this chapter, a generalized mathematical model for the fractional removal of vapor from a shallow magma is developed. The equations developed in this chapter can be used to determine:

1. The extent to which a given element can be depleted in a magma (and its resulting crystalline products) by the effects of vapor evolution
2. The instantaneous concentration of a given element in the evolving vapor phase which may be available for later metasomatism (involving, for example, rare earth elements, K, Rb), ore formation (e.g., Cu, Mo), or volcanic exhalation (e.g., Cl, F, Mo)
3. The total (integrated) amount of an element which can be removed from a magma into a hydrothermal system

Hence, the goal herein is to create quantitative models that can be used to test whether a given fluid chemistry (instantaneous or integrated) or a given trend in magmatic chemistry can be attributed to the evolution of a magmatic vapor.

Previous Work in the Modeling of Vapor Evolution

In 1948 H. Neumann published a short paper in *Economic Geology* entitled "On Hydrothermal Differentiation" (Neumann, 1948). In this work Neumann suggested that the standard Raliegh fractionation equations be applied to calculate the composition of a vapor in equilibrium with a silicate melt using constant vapor–melt partition coefficients, and solid–liquid partition coefficients equal to zero. His approach was simple and although the pertinent data was nonexistent, the suggestion had been made. The quantitative modeling of mass transfer in melt–vapor systems then lay dormant for over 20 years, until H. D. Holland produced a seminal paper entitled "Granites, Solutions and Base Metal Deposits" (Holland, 1972). Holland derived a set of equations that could be used to calculate the concentration of ore metals in an evolving magmatic vapor for conditions of (1) a constant mass of magma during vapor evolution, and (2) a constant solubility of water in the melt. No consideration was given to crystal–melt partitioning. Aside from the experimental data reported by Holland, his most important contribution to the modeling of melt–vapor processes was the treatment of chloride-dependent partition coefficients, and the explicit treatment of "charge balance" or chloride balance in the aqueous phase. The latter two concepts are used in the present model.

Vapor–Melt Partitioning: The Data Base

The first systematic study of the partitioning of metals between silicate melts and aqueous fluids to appear in the geological literature was that of Holland (1972). Holland reported data on the partitioning of K, Na, Mg, Ca, Mn, and Zn between silicic melts and chloride-bearing aqueous fluids. The data were presented as chloride-dependent, Nernst-type partition coefficients. For example, the partitioning of sodium was given as $(C^V_{Na}/C^l_{Na}) \equiv K^P_{Na} = 0.46\ C^V_{Cl}$. Of course K^P_{Na} is not a thermodynamically valid equilibrium constant because there exists no corresponding equation of the form $\Sigma_i v_i \mu_i^\phi = 0$ where v_i is a stoichiometric coefficient and μ_i^ϕ is the chemical potential of an independently variable constituent of the liquid and vapor phases. Holland recognized this, and formulated $K_{j,Na}$ (where $j \neq$ Na) such that K^P_j is divided by K^P_{Na}, producing a thermodynamically valid exchange constant $K_{j,Na} = [C_j^V \cdot (C^l_{Na})^n]/[C_j^l \cdot (C^V_{Na})^n] = K^P_j/(K^P_{Na})^n$, where $n =$ the inferred ligation number of metal j. The value of n results from the functional dependence of $K^P_j = f(Cl^n)$. The corresponding chemical potential relation for such an equilibrium is properly formulated (e.g., for a univalent metal, M) as:

$$\mu^l_{MO_{0.5}} + \mu^V_{NaCl} = \mu^V_{MCl} + \mu^l_{NaO_{0.5}}$$

(Candela and Holland, 1984). Holland (1972) used an ionic formulation for the melt. However, because of the lack of an accepted electrolyte model and the lack of a set of standard states for ionic species for silicic melts, it is best to avoid such a formulation. Holland reported data for K, Ca, Mg, Mn, and Zn as exchange constants with respect to Na (see Table 1). An early report from his lab (Gammon *et al.*, 1969) discussed $K_{\mathrm{K,Na}}$ in depth. Holland's experiments were performed at 770–880°C, pressures of 1.4–2.4 kbar, and aqueous chloride concentrations of up to 6 molal. The melt compositions used were very siliceous (78.3 and 81.8% SiO_2), and were compositionally close to the simple system Quartz (Q)-albite (Ab)-orthoclase (Or)-anorthide (An). No obvious temperature or pressure dependence of the partitioning data for K, Na, Mn, Zn, Ca, and Mg was detected. Carron and Lagache (1980) report vapor–melt partition coefficients for K, Na, Cs, Rb, Sr, and Ba in systems that are chemically very similar to those of Holland (1972). The value for $K_{\mathrm{K,Na}}$ from Carron and Lagache is 0.71 at 800°C and 2 kbar, which compares quite well with Holland's value of 0.74. Gammon *et al.* (1969) demonstrated that $K_{\mathrm{K,Na}}$ was independent of C_{Cl}^V up to approximately 4 molal total chloride, and Carron and Lagache demonstrated that $K_{\mathrm{K,Na}}$ was independent of the K/(K + Na) ratio in the melt from 0.1 to 0.9. Carron and Lagache report their trace element partitioning data as $D_i^{l/V} = (C_i^l/C_{\mathrm{Na}}^l + C_{\mathrm{K}}^l)/(C_i^V/C_{\mathrm{Na}}^V + C_{\mathrm{K}}^V)$. This form is difficult to work with, and some assumptions must be used to extract thermodynamically valid exchange constants from these data. Carron and Lagache report data for the simple melts $Q_{40}Ab_{60}$ and $Q_{40}Or_{60}$ in addition to data for an intermediate composition. Note that, for Rb and Cs, $K_{i,\mathrm{Na}}(C_{\mathrm{K}}^l = 0) = (D_i^{l/V})^{-1}$, and $K_{i,\mathrm{Na}}(C_{\mathrm{Na}}^l = 0) = (K_{\mathrm{K,Na}}/D_i^{l/V})$ for the Q–Ab and Q–Or melts, respectively. For the divalent metals Sr and Ba, $D_i^{l/V}$ must be multiplied by $D_{\mathrm{Na}}^{V/l} = (C_{\mathrm{Na}}^V/C_{\mathrm{Na}}^l) = (0.46 \pm 0.04)C_{\mathrm{Cl}}^V$ from Holland (1972), where $C_{\mathrm{Cl}}^V = 2$ molal (the concentration of Cl in the experiments of Carron and Lagache) to obtain $K_{i,\mathrm{Na}} = (C_i^V/C_i^l)/(C_{\mathrm{Na}}^V/C_{\mathrm{Na}}^l)^2$ by the formula $K_{i,\mathrm{Na}}(C_{\mathrm{K}}^l = 0) = [D_i^{l/v} \cdot 2 \cdot (0.46)]^{-1}$ for the Q–Ab melts, and $K_{i,\mathrm{Na}}(C_{\mathrm{Na}}^l = 0) = K_{\mathrm{K,Na}}/[D_i^{l/V} \cdot 2 \cdot (0.46)]$ for the Q–Or melts. Errors were reported by Carron and Lagache as $\pm 1\sigma$. Error (1σ) in $D_{\mathrm{Na}}^{V/l}$ from Holland (1972) was graphically determined from his Fig. 4. These errors were propagated into the calculated $K_{i,\mathrm{Na}}$ by standard random error propagation theory. The partition coefficients for Rb, Cs, Sr, and Ba in the Q–Ab melt are different from those in the Q–Or melt. The V/l partition coefficients for Rb and Cs are lower in the case of the Q–Or melt, whereas the partition coefficients for Sr and Ba are lower in the Q–Ab melt. Ionic size seems to be an important factor. The large Rb and Cs ions possess higher vapor–liquid coefficients in the sodium system relative to the partition coefficients in the potassium system. Sr and Ba do not seem to follow a simple ionic radius rule. However, the affinity of Sr for the Q–Ab melt over the Q–Or melt is analogous to its affinity for plagioclase over K-feldspar, and the larger Ba, while favoring the albitic melt, prefers it to a lesser extent than does Sr.

Flynn and Burnham (1978) report vapor–melt partition data for the rare earth elements Ce, Eu, Gd, and Yb. Data for Ce and Yb are listed in Table 1. Flynn and Burnham report their results as $D_i^{V/l} = C_i^V/C_i^l$. To convert $D_i^{V/l}$ to

Table 1. Vapor–melt partition coefficients of selected elements.

Element	n[a]	Melt[b]	$K_{i,\mathrm{Na}} \pm 1\sigma$	$D \pm 1\sigma$	T,°C	P, kbar	Ref.[c]
K	1	high-silica rhyolite	0.74 ± 0.06		770–880	1.4–2.4	1, 2
Ca	2	high-silica rhyolite	$0.38^d \pm .09$		770–880	1.4–2.4	2
Mg	2	high-silica rhyolite	$0.16^d \pm .04$		770–880	1.4–2.4	2
Mn	2	high-silica rhyolite	$6.5^d \pm 1.6$		770–880	1.4–2.4	2
Zn	2	high-silica rhyolite	$9.5^d \pm 2.4$		770–880	1.4–2.4	2
Rb	1^e	$Q_{40}Ab_{60}$	0.77 ± 0.05		800	2	3
Rb	1^e	$Q_{40}Or_{60}$	0.66 ± 0.06		800	2	3
Cs	1^e	$Q_{40}Ab_{60}$	1.0 ± 0.08		800	2	3
Cs	1^e	$Q_{40}Or_{60}$	0.67 ± 0.04		800	2	3
Sr	2^e	$Q_{40}Or_{60}$	0.28 ± 0.07		800	2	3
Sr	2^e	$Q_{40}Or_{60}$	0.50 ± 0.13		800	2	3
Ba	2^e	$Q_{40}Or_{60}$	0.29 ± 0.05		800	2	3
Ba	2^e	$Q_{40}Or_{60}$	0.43 ± 0.09		800	2	3
Ce	3	spruce pine pegmatite	2.56 ± 0.89		800	1.25	4
Yb	3	spruce pine pegmatite	1.12 ± 0.39		800	1.25	4
Cu	1	high-silica rhyolite	20 ± 5.6		750	1.4	5
Mo	0	high-silica rhyolite		2.5 ± 1.6	750	14	5
Cl	0	granodiorite to high silica rhyolite		$40^d \pm 10$	750–880	1–2.4	2, 6
F	0	haplogranite		$0.2^d \pm 0.1$	750–850	1	7
B	0^f	haplogranite		3 ± 0.18	750–800	1	8

[a] n = ligation number in chloride media; experimentally determined unless noted.
[b] For more information on melt and aqueous phase composition, see reference listed as source of data.
[c] (1) Gammon *et al.* (1969); (2) Holland (1972); (3) Carron and Legache (1980); (4) Flynn and Burnham (1978); (5) Candela and Holland (1984); (6) Kilinc and Burnhan (1972); (7) Dingwell and Scarfe (1983); (8) Pichavant (1981).
[d] No error reported; general range of error estimated by inspection.
[e] n inferred from oxidation state.
[f] Not determined, assumed zero for purpose of this study.

$K_{i,\mathrm{Na}}$ requires dividing $D_i^{V/l}$ by $(C_{\mathrm{Na}}^V/C_{\mathrm{Na}}^l)^3$. Again we shall use $D_{\mathrm{Na}}^{V/l}$ given by Holland (1972); however, there is a small difference between the composition of Holland's high silica rhyolite and the Spruce Pine Pegmatite melt of Flynn and Burnham (78–82 wt % SiO_2 vs. 74 wt % SiO_2, respectively). The magnitude of the determinate error introduced into $K_{\mathrm{REE,Na}}$ by this factor is unknown but is probably minor. The $C_{\mathrm{Na}}^V/C_{\mathrm{Na}}^l$ ratio is given by $(0.46)\cdot(0.914)$, where $0.914 = C_{\mathrm{Cl}}^V$ in the REE partitioning experiment. At this chloride concentration, $D_{\mathrm{Ce}}^{V/l} = 0.19$ and $D_{\mathrm{Yb}}^{V/l} = 0.083$. Note, however, that the cerium and ytterbium concentrations in the vapor are a function of the *cube* of the chlorine concentration in the vapor. Therefore, these $D_i^{V/l}$ increase rapidly as C_{Cl}^V increases, and qualitative arguments based on the magnitude of a particular $D_i^{V/l}$ or $K_{i,\mathrm{Na}}$ are to be discouraged. Only through the application of quantitative models, such as those that will be discussed presently, can decisions be made regarding the efficacy of vapor evolution in effecting petrologically significant changes.

Candela and Holland (1986) measured the partitioning of copper and molybdenum between silicate melts and aqueous fluids. Copper behaves as a univalent metal in magmatic systems, is chloride complexed, and possesses a partition coefficient that is larger than the $K_{i,\mathrm{Na}}$ for any other univalent metal. Molybdenum, on the other hand, exhibits a vapor–melt partition coefficient that is independent of both C_{Cl}^{V} and C_{F}^{V}. The Nernst-type formulation for its partition coefficient therefore yields a reasonable approximation of the vapor–melt partitioning behavior. In the cases of the remaining elements in Table 1 (fluorine, chlorine, and boron), little is known of the equilibria that control their partitioning into the vapor. Kilinc and Burnham (1972) and Holland (1972) have measured the partitioning of chlorine between melt and vapor and determined that a value on the order of 40 is a reasonable estimate of $D_{\mathrm{Cl}}^{V/l}$ in felsic systems. Dingwell and Scarfe (1983) measured the partition coefficient of fluorine between melt and vapor. Unlike chlorine, fluorine partitions into the melt relative to the vapor and $D_{\mathrm{F}}^{V/l}$ is most likely below 0.2. The behavior of chlorine and fluorine in melt–vapor–apatite systems is discussed in depth in Candela (1985). Boron was studied by Pichavant (1981). The partitioning of boron was not studied as a function of C_{Cl}^{V} or C_{F}^{V}, and the functional dependence of $D_{\mathrm{B}}^{V/l}$ on these parameters is unknown. Pichavant found $D_{\mathrm{B}}^{V/l}$ to be equal to approximately 3; that is, boron favors the vapor over the melt phase. All of the partition coefficients discussed in this section have been tabulated in Table 1 and are discussed further in the balance of this chapter.

The Mathematical Formulation of the Generalized Models

Two fundamentally different models are presented in this chapter. Model I deals with the evolution of magmatic vapor during polybaric decompression (first boiling) of a hydrous magma. In the limiting case presented here, no crystallization accompanies the evolution of magmatic vapor. The driving force for vapor evolution is the reduction of the solubility of water with decreasing pressure. Model I is separated into two cases. In case I, the partitioning of Nernstian elements is considered; in case II, the partitioning of chloride-complexed elements is considered. In both cases, the mass of magma is not assumed constant (as was assumed by Holland, 1972) but is reduced by the amount of vapor evolved. Additionally, analytical expressions for the efficiency of removal of elements from a magma by a magmatic vapor during first boiling are presented for the first time. Model I is a reasonable first-order model for the rise of magma in a subvolcanic conduit, and this model can be used to model the vapor–melt processes that occur during the emplacement of obsidian domes, such as the Inyo Chain (Eichelberger *et al.*, 1985), or during present-day eruptions, such as those of Mt. St. Helens (cf. Thomas *et al.*, 1982).

Model II deals with the evolution of magmatic vapor during isobaric crystallization of a magma (second boiling). In this model, the solubility of water is assumed to be a function of pressure only, for a given *vapor* composition. This is a reasonable approximation given the strong dependence of water solubility on pressure and the weak dependence of water solubility on temperature (Burnham, 1979). Whereas model I deals with vapor evolution during the emplacement, or upward movement of a water-saturated magma, model II is concerned with the vapor–melt processes that follow the emplacement of a hydrous silicate melt in the earth's crust. If quartz and feldspar are the main products of the crystallization of the magma, water behaves as an incompatible element and steadily increases in concentration in the melt. At some point in the crystallization of the magma, determined by the pressure (depth) of crystallization and the initial water content of the melt, the magma saturates with respect to an aqueous phase. Further crystallization leads to the evolution of the aqueous phase. Before water saturation occurs, the ore elements are partitioned among the melt and the crystalline phases. After water saturation occurs, the ore elements are partitioned among the melt, aqueous, and crystalline phases; the aqueous and crystalline phases are continuously removed from the magma as crystallization proceeds.

The crystallization of hydrous minerals is not considered, although the effects of such a process can be accommodated by this model. The water concentration in the melt is not considered constant, even at constant pressure and (nomial) melt composition because, in some cases, high solute concentrations in the vapor phase lead to $X^V_{H_2O} < 1$. As in model I, two cases are considered. Case I of model II deals with the vapor–melt partitioning of Nerstian elements during second boiling, whereas case II of model II deals with the vapor–melt partitioning of chloride-complexed elements. Model II is a reasonable first-order model for:

1. The evolution of vapor from a static magma in a subvolcanic chamber, where vapor exits as high temperature, low f_{O_2}, and presumably orthomagmatic fumaroles (cf. Le Guern and Barnard, 1982)
2. The development of ore-forming fluids in a porphyry or skarn environment (Candela and Holland, 1986)
3. The development of a vapor during pegmatite genesis

Model I: Polybaric Vapor Evolution Without Crystallization

Case I

The partitioning of molybdenum between a silicate melt and an aqueous fluid can be considered constant for melts of a given chemical composition. The elements B (Pichavant, 1981), Cl (Kilinc and Burnham, 1972; Holland, 1972), and F (Dingwell and Scarfe, 1983; Dingwell, personal communication 1984) are represented by Nernst-type partition coefficients in this chapter, although it is acknowledged that the equilibria controlling their partitioning may be non-Nernstian (see Candela, 1986). The temperature and pressure dependence of the partitioning of these elements is essentially unknown, and remain to be investigated. To illustrate the utility of the equations that are derived in this section, the pressure and temperature dependence of the partition coefficients is ignored. When the pressure and temperature dependence of these quantities is determined, the systematics presented herein can easily be adjusted to accommodate these functional relationships.

During vapor evolution, mass balance requires that:

$$dM_i^V + dM_i^l = 0 \tag{1}$$

and, because vapor evolution is the only means by which the magma may change its mass,

$$dM_l = dM_{H_2O}^l = -dM_{H_2O}^V \tag{2}$$

(assuming $dM_{H_2O}^V \gg \sum_j dM_j^V$, that is, the mass of the vapor is given by the mass of water, to a good approximation). If the vapor is fractionated, then the concentration is a quotient of infinitesimal quantities:

$$C_i^V = dM_i^V/dM_{H_2O}^V \tag{3}$$

Rearranging Eq. (1) and dividing by Eq. (2) yields:

$$\frac{dM_i^V}{dM_{H_2O}^V} = \frac{dM_i^l}{dM_l} \tag{4}$$

By inspection, the left-hand side of Eq. (3) and the right-hand side of Eq. (4) can be equated to yield:

$$C_i^V = \frac{dM_i^l}{dM_l} \tag{5}$$

For the Nernstian elements Mo, F, B, and Cl, we have:

$$C_i^V = D_i^{V/l} \cdot C_i^l = D_i^{V/l} \cdot (M_i^l/M_l) \tag{6}$$

which can be combined with Eq. (5) to yield:

$$\frac{dM_i^l}{dM_l} = D_i^{V/l} \cdot (M_i^l/M_l) \tag{6a}$$

Separation of variables yields:

$$\int_{M_i^{l,0}}^{M_i^l} \frac{dM_i^l}{M_i^l} = D_i^{V/l} \int_{M_l^0}^{M_l} \frac{dM_l}{M_l} \tag{7}$$

The lower limits of integration represent the initial conditions before vapor evolution, and the upper limits represent the conditions after a given mass of water, $M_{H_2O}^V$, has been removed from the magma. The mass of magma that remains after this quantity of vapor has evolved is given by:

$$M_l = M_l^0 - M_{H_2O}^V \tag{8}$$

Integration of Eq. (7) and substitution of Eq. (8) yields:

$$\frac{M_i^l}{M_i^{l,0}} = \left(\frac{M_l^0}{M_l^0} - \frac{M_{H_2O}^V}{M_l^0}\right)^{D_i^{V/l}} \tag{9}$$

after the natural logarithms are cleared. If $\mathscr{F}$, the vapor evolution progress variable (Candela and Holland, 1986), is defined as unity at the initiation of vapor evolution and zero at the point of completion, then,

$$\mathscr{F} = M_w^l / M_w^{l,0} \tag{10}$$

and:

$$\frac{M_{H_2O}^V}{M_l^0} = (1 - \mathscr{F})C_w^{l,0} \tag{10a}$$

where the ratio on the left-hand side of Eq. (10a) is interpreted as the proportion of the *initial* mass of magma that has been released as magmatic vapor over the interval $1 \rightarrow \mathscr{F}$. Combining Eqs. (9) and (10a) yields:

$$M_i^l(\mathscr{F}) = M_i^{l,0}\{1 - [(1 - \mathscr{F})C_w^{l,0}]\}^{D_i^{V/l}} \tag{11}$$

which expresses the mass of i in the liquid for a given $\mathscr{F}$ (proportion of water remaining in the magma). In order to express the *concentration* of i in the magma at any $\mathscr{F}$, we must divide M_i^l and $M_i^{l,0}$ by W_l and W_l^0, respectively. This may be achieved in the following manner. Adding -1 to both sides of Eq. (10a) yields:

$$\frac{M_l^0 - M_{H_2O}^V}{M_l^0} = 1 - [(1 - \mathscr{F})C_w^{l,0}] \tag{12}$$

which, by Eq. (8), is also equal to M_l/M_l^0. Equation (12) can then be expressed as:

$$M_l = M_l^0 \cdot \{1 - [(1 - \mathscr{F})C_w^{l,0}]\} \tag{13}$$

Note that M_l and M_l^0 are on the left- and right-hand sides, respectively, of Eq. (13) as is required to convert the extensive form of Eq. (11) into an intensive form. Dividing Eq. (11) by Eq. (13) yields the desired expression for the concentration of i in the melt as a function of $\mathscr{F}$:

$$C_i^l(\mathscr{F}) = C_i^{l,0}\{1 - [(1 - \mathscr{F})C_w^{l,0}]\}^{D_i^{V/l}-1} \tag{14}$$

To obtain the concentration of i in the evolving vapor, Eq. (14) is combined with Eq. (6) to yield:

$$C_i^V(\mathscr{F}) = D_i^{V/l}C_i^{l/0}\{1 - [(1 - \mathscr{F})C_w^{l,0}]\}^{D_i^{V/l}-1} \tag{15}$$

The concentration in Eq. (15) can be expressed in either weight fraction or moles per gram. The total amount of i removed from the magma, relative to the amount of i present initially [*the* "efficiency of removal," $E(i)$] is given by one minus the proportion of i which *remains* in the melt at the cessation of vapor evolution,

$$E(i) = 1 - \frac{M_i^l(\mathscr{F} = 0)}{M_i^{l,0}} \tag{15a}$$

Equation (15a) can be combined with Eq. (11) and solved for $\mathscr{F} = 0$ to yield:

$$E(i)_{\substack{\text{Polybaric,}\\ \text{no crystallization}}} = \{1 - (1 - [C_w^{l,0}])^{D_i^{V/l}}\} \tag{16}$$

This equation can be used to calculate the efficiency of removal of an element from a magma for the case in which vapor evolution occurs without crystallization, and for elements that partition according to a Nernstian law.

Case II

Holland (1972) and others showed that many elements did not follow a simple Nerstian law. Indeed, such behavior is a special case, and some of the elements treated as Nernstian herein, such as Cl, B, and F, will most probably be shown, in later experimental studies, to partition between melt and vapor according to much more complex relationships. The elements K, Na, Ca, Mg, Mn, and Zn (Holland, 1972); Cu (Candela and Holland, 1984); Cs, Rb, Sr, Ba (Carron and Legache, 1980); and selected REE (Flynn and Burnham, 1978) possess partition coefficients that are a function of the chloride content of the aqueous phase. It is best to formulate such equilibria as exchange constants, as was discussed in a previous section of this chapter. The exchange equilibria can be written with respect to any constituent that is represented by a phase component in both the melt and vapor phases. It is also best, for ease of computation, to select an element that does not change its concentration significantly in the melt phase during either crystallization or vapor evolution. Sodium is the logical choice. Therefore, all exchange equilibria will be presented in the form:

$$K_{i,\text{Na}} = \frac{C_i^V}{C_i^l} \cdot \left(\frac{C_{\text{Na}}^l}{C_{\text{Na}}^V}\right)^n \tag{17}$$

where n equals the ligation number of i in chloride solutions. In the case of all the elements considered in this chapter, the ligation number equals an oxidation state. Equations (1)–(5), which were derived for the case of the Nernstian elements, are equally applicable here. Combining Eqs. (5) and (17) yields:

$$\frac{dM_i^l}{dM_l} = K_{i,\text{Na}} C_i^l \left(\frac{C_{\text{Na}}^V}{C_{\text{Na}}^l}\right)^n \tag{18}$$

The concentration of sodium in the vapor can be obtained iteratively, as in model II, or the simplification introduced by Holland (1972) may be used:

$$C_{\text{Cl}}^V \simeq C_{\text{Na}}^V + C_K^V \tag{19}$$

and

$$C_{\text{Cl}}^V = C_{\text{Na}}^V\left(1 + K_{\text{K,Na}} \cdot \frac{C_{\text{K}}^l}{C_{\text{Na}}^l}\right) \tag{20}$$

where the melt is considered to devolatilize at a constant potassium and sodium concentration. The parameter in parentheses on the right-hand side of Eq. (20) shall be represented by β; β is then a function of the bulk melt composition. If desired, a squared term in C_{Na}^{V} may be added to account for a constant Fe, Ca, etc., concentration. Substituting $C_{\mathrm{Na}}^{V} = C_{\mathrm{Cl}}^{V}/\beta$, the expression for $C_{\mathrm{Cl}}^{V}(\mathscr{F})$, and Eq. (15) into Eq. (18) and assuming that C_{Na}^{l} is approximately constant during devolatilization (which is a good assumption for $C_{\mathrm{Cl}}^{l,0} \lesssim 0.1 C_{\mathrm{Na}}^{l,0}$), that is, $C_{\mathrm{Na}}^{l} = C_{\mathrm{Na}}^{l,s}$, yields:

$$\frac{dM_i^l}{dM_w^l} = \frac{K_{i,\mathrm{Na}}}{(C_{\mathrm{Na}}^{l,s} \cdot \beta)^n} \cdot C_i^l \cdot (D_{\mathrm{Cl}}^{V/l} \cdot C_{\mathrm{Cl}}^{l,0})^n \cdot \{1 - [(1 - \mathscr{F}) C_w^{l,0}]\}^{n[D_{\mathrm{Cl}}^{V/l} - 1]} \tag{21}$$

Letting $C_i^l = (M_i^l/M_l)$, expressing M_l by Eq. (13), and noting that through Eq. (10) $dM_w^l = M_w^{l,0}\, d\mathscr{F}$, yields, on separation of variables:

$$\int_{M_i^{l,0}}^{M_i^l} \frac{dM_i^l}{M_i^l} = K_{i,\mathrm{Na}} \cdot \left(\frac{D_{\mathrm{Cl}}^{V/l} \cdot C_{\mathrm{Cl}}^{l,0}}{C_{\mathrm{Na}}^{l,s} \cdot \beta} \right)^n \cdot \int_1^{\mathscr{F}} \frac{M_w^{l,0} \{1 - [(1 - \mathscr{F}) C_w^{l,0}]\}^{n(D_{\mathrm{Cl}}^{V/l} - 1)}}{M_l^0 \cdot \{1 - [(1 - \mathscr{F}) C_w^{l,0}]\}} d\mathscr{F} \tag{22}$$

Noting that $C_w^{l,0} = (M_w^{l,0}/M_l^0)$, one obtains, upon integration of Eq. (22) and clearing the natural logarithm:

$$M_i^l(\mathscr{F}) = M_i^l \exp\left\{ \frac{\alpha(i)}{n(D_{\mathrm{Cl}}^{V/l} - 1)} \right\} \{(1 - [(1 - \mathscr{F}) C_w^{l,0}])^{n(D_{\mathrm{Cl}}^{V/l} - 1)} - 1\} \tag{23}$$

where $\alpha(i)$ is the preintegral factor in Eq. (22). As in the case of the Nernstian equations, this equation may be transformed to yield the concentration of i in the liquid and vapor:

$$C_i^l = \frac{M_i^l(\mathscr{F})}{M_l^0 \{1 - [(1 - \mathscr{F}) C_w^{l,0}]\}} \tag{24}$$

and

$$C_i^V = C_i^l \cdot K_{i,\mathrm{Na}} \cdot \left(\frac{C_{\mathrm{Cl}}^V(\mathscr{F})}{\beta \cdot C_{\mathrm{Na}}^{l,0}} \right)^n \tag{25}$$

where $C_{\mathrm{Cl}}^V(\mathscr{F})$ is given by Eq. (15). Additionally, by a method analogous to the derivation of Eq. (16) from Eq. (11), one can obtain an expression for the efficiency of removal of i from the magma into the evolving aqueous fluid for chloride-complexed elements, $E(i)$:

$$E(i) = \left\{ 1 - \exp\left(\frac{\alpha(i)}{n(D_{\mathrm{Cl}}^{V/l} - 1)} \{(1 - C_w^{l,0})^{n(D_{\mathrm{Cl}}^{V/l} - 1)} - 1\} \right) \right\} \tag{26}$$

Equations (24)–(26), along with Eq. (14)–(16), can be used to calculate the progressive changes in the chemistry of a water-saturated melt–vapor system during the polybaric decompression of a magma. The planetological and geochemical implications of these equations are discussed in depth in a later part of this chapter.

Model II: Isobaric Vapor Evolution with Crystallization

Case I

In this section the equations that govern the chemistry of magmatic vapor during second boiling are developed. The statement of mass conservation for the element in question is, in this model,

$$0 = dM_i^l + dM_i^V + \sum_{\rightleftarrows}^{p} dM_i^\phi \tag{27}$$

The sum $\sum_\phi^p M_i^\phi$ is taken over all crystalline phases which accept i (p in number). Assuming no hydrous phases are crystallizing,

$$0 = dM_w^l + dM_w^V \tag{28}$$

Dividing Eq. (27) by Eq. (28) yields

$$dM_i^V/dM_w^V = dM_i^l/dM_w^l + \sum_\phi^p dM_i^\phi/dM_w^l \tag{29}$$

The concentration of i in the vapor, C_i^V, is given by dM_i^V/dM_w^V and by Eq. (6). Incorporating these relations into Eq. (29), letting $C_i^l = (M_i^l/M_l)$ and $C_w^l = M_w^l/M_l$ yields:

$$C_w^l \cdot dM_w^l \cdot \frac{D_i^{V/l}}{M_w^l} = \frac{dM_i^l}{M_i^l} + \sum_\phi^p \frac{dM_i^\phi}{M_i^l} \tag{30}$$

If the crystalline phases are fractionated, then it can be shown (Candela and Holland, 1986) that:

$$\sum_\phi^p \frac{dM_i^\phi}{M_i^\phi} = -\bar{D}_i^{S/l} \frac{dM_l}{M_l} \tag{31}$$

where $\bar{D}_i^{S/l}$ is the Nernst bulk solid–liquid partition coefficient. However, it should be noted that the crystal–liquid exchange equilibria for Mo, Cl, F, and B may be quite complex; nothing is known concerning the nature of the pertinent reactions. The Nernst-type partition coefficients should therefore be taken only as general indices of the partitioning of i with respect to crystalline phases. Rewriting Eq. (30):

$$\left(C_w^l \cdot D^{V/l} \cdot \frac{dM_w^l}{M_w^l}\right) + \left(\bar{D}_i^{S/l} \cdot \frac{dM_l}{M_l}\right) = \frac{dM_i}{M_i^l} \tag{32}$$

In previous treatments, such as Candela and Holland (1986), C_w^l was approximated by $C_w^{l,s}$, the concentration of water in the melt at the time of water saturation. However, at low water concentrations ($C_w^{l,s} \approx 0.025$ weight fraction), the quantity $\Gamma = (D_{\text{Cl}}^{V/l} \cdot C_w^{l,s}) < 1$, which results in a progressive build up of Cl in the magma and the associated aqueous phase (Candela and Holland, 1986). This then leads to a lowering of $X_{\text{H}_2\text{O}}^V$ and a concomitant decrease in C_w^l. This effect is quite significant in the advanced stages of crystallization, violating the condition of $C_w^l(\mathscr{F}) = C_w^{l,s}$. Many workers have shown that the solubility of water in the melt varies (to a first approximation) as the square root of the fugacity or activity of water in the vapor (Burnham and Davis, 1974; Stolper, 1982). The variation of the solubility of water in the melt with solute activity will therefore be represented as:

$$C_w^l(M_w^l) = C_w^{l,s} \cdot \sqrt{1 - \textstyle\sum_j X_j^V(M_w^l)} \tag{33}$$

where $\sum_j X_j^V$ is the mole fraction of solutes in the vapor, which in this chapter includes total chloride, fluoride, borate, and molybdate. Equation (33) may also be represented as:

$$M_w^l = M_l \cdot \left(\frac{M_w^{l,s}}{M_l^s}\right)\sqrt{1 - \textstyle\sum_j X_j^V(M_w^l)} \tag{33a}$$

or its differential form,

$$dM_w^l = dM_l \cdot C_w^{l,s} \cdot \sqrt{1 - \textstyle\sum_j X_j^V(M_w^l)} \tag{33b}$$

The mole fraction of i in the vapor is $X_i^V(M_w)$, where

$$X_i^V = \frac{C_i^V}{55.5 + \sum_i C_i^V} \tag{33c}$$

Here, C_i^V is in units of moles per kilogram of water, and 55.5 represents the number of moles of water per kilogram of water. Dividing Ea. (33b) by (33a) yields:

$$\frac{dM_l}{M_l} = \frac{dM_w^l}{M_w^l} \tag{34}$$

which can be used, along with Eq. (33), to rewrite Eq. (32) and set up the integral:

$$\int_{M_i^{l,0}}^{M_i} \frac{\text{d}M_i^l}{M_i^l} = \int_1^Z \left\{\bar{D}_i^{S/l} + (D_i^{V/l} \cdot C_i^{l/s} \cdot \left[1 - \sum_j X_j^V(Z)\right]\right\}\frac{dZ}{Z} \tag{35}$$

where $Z = M_w^l/M_w^{l,0}$. [Note that all variables that are functions of M_w^l are also functions of Z: e.g., $X_i^V(M_w^l) = X_i^V(Z)$.] The integral on the left-hand side of Eq. (35) is easily integrated. The integral on the right can be split into two integrals at the + sign, with the first integral being simple in form. The second integral is recursive in nature, requiring a product of its own solution (X_i^V), plus the sum of the other solutes represented by an additional $(j - 1)$ integrals. Partial integration of Eq. (35) yields:

$$M_i^l = M_i^{l,s} Z^{\bar{D}} \exp\left\{D_i^{V/l} \cdot C_w^{l,s} \cdot \int_1^Z \left[1 - \sum_j X_j^V(Z)\right] \frac{dZ}{Z}\right\} \tag{36}$$

which, when combined with Eqs. (6) and (33a), yields:

$$C_i^V(Z) = -\frac{D_i^{V/l} \cdot C_i^{l,s}}{\sqrt{1 - \sum_j X_j^V(Z)}} \cdot Z^{\bar{D}-1} \cdot \exp\left\{D_i^{V/l} \cdot C_w^{l,s} \cdot \int_1^Z \left[1 - \sum_j X_j^V(Z)\right] \frac{dZ}{Z}\right\} \tag{37}$$

Equation (37) must be solved iteratively for the elements F, Mo, B, and Cl, with $\sum_j X_j^V(Z)$ determined from the simultaneous solution of Eq. (37) for the above-mentioned elements. The new estimate of $\sum_j X_j^V(Z)$ is substituted back into (37) until the desired precision is reached in the iteration. The integrals are solved numerically. If a simple Rayliegh fractionation of crystalline phases before water saturation is assumed, and the proportion of hydrous phases in the hypersolidus assemblage is low, then it can be shown that one can account for removal of an element from the magma before vapor saturation using the relation:

$$C_i^{l,s} = C_i^{l,0} \left(\frac{C_w^{l,0}}{C_w^{l,s}}\right)^{\bar{D}_i^{S/l}-1} \tag{38}$$

(Candela and Holland, 1986). This expression may be substituted into Eq. (37). Note that the ratio $(C_w^{l,0}/C_w^{l,s})$ is used as an index of the amount of crystallization that occurs prior to vapor saturation. As discussed by Candela and Holland, this variable then determines the extent to which crystal-compatible and crystal-incompatible elements are separated by vapor evolution processes.

The efficiency of removal of Mo, B, F, or Cl from a silicate melt into an aqueous fluid is given by:

$$E(i) = \frac{M_i^{V,T}}{M_i^{l,0}} \tag{39}$$

An estimate of the total quantity of i removed into the vapor can be obtained by integrating $C_i^V(Z)$ for a portion of the range of vapor evolution, $Z = \{1 \text{ to } 0\}$. The efficiency is then given by:

$$E(i) = (C_w^{l,s}/C_i^{l,0}) \cdot \int_0^1 C_i^V(Z)\, dZ \tag{40}$$

To perform this integral numerically, $C_i^V(Z)$ is calculated for all j, so that $\sum_j X_j^V(Z)$ can be calculated. The calculation of the $C_i^V(Z)$ of interest at any Z requires a numerical quadrature for $Z = \{1 \text{ to } Z\}$. Each $C_i^V(Z)$ that results is also a node for the efficiency function quadrature. It is suggested that some form of adaptive quadrature be used to solve these integrals.

Case II

The case of the vapor–liquid partitioning of chloride-complexed elements during isobaric crystallization will now be considered. Equations (27) through (29) apply to non-Nernstian isobaric vapor evolution as well as the Nernstian case (case I). Combining Eq. (29) with the exchange constant defined for model I, case II (Eq. 17) and the condition of fractional vapor removal, $C_i^V = dM_i^V/dM_w^V$, one obtains:

$$\frac{dM_i^l}{M_i^l} + \sum_{\phi}^{p} \frac{dM_i^\phi}{M_i^l} = \frac{dM_w^l}{M_l} \cdot K_{i,\mathrm{Na}} \cdot \left(\frac{C_{\mathrm{Na}}^V}{C_{\mathrm{Na}}^l}\right)^n \tag{41}$$

where C_i^l was expressed as (M_i^l/M_l). Utilizing Eq. (31) to introduce $\bar{D}_i^{S/l}$, and Eq. (34) to group like terms, and noting that $(1/M_l) = (C_w^l/M_w^l)$, yields,

$$\int_{M_i^{l,o}}^{M_i^l} \frac{dM_i^l}{M_i^l} = \int_1^Z \left\{\bar{D}_i^{S/l} + \frac{K_{i,\mathrm{Na}}}{(C_{\mathrm{Na}}^{l,s})^n} \cdot [C_{\mathrm{Na}}^V(Z)]^n \cdot C_w^l(Z)\right\} \frac{dZ}{Z} \tag{42}$$

Note that $C_{\mathrm{Na}}^l \simeq C_{\mathrm{Na}}^{l,s}$. This was found to be a reasonable approximation, as discussed earlier under model I. Partial integration yields:

$$M_i^l = M_i^{l,s} \cdot Z^{\bar{D}_i^{S/l}} \cdot \exp\left\{\frac{K_{i,\mathrm{Na}}}{(C_{\mathrm{Na}}^{l,s})^n} \cdot \int_1^Z \frac{[C_{\mathrm{Na}}^V(Z)]^n \cdot C_w^l(Z)}{Z} dZ\right\} \tag{43}$$

This equation yields the number of moles of i present in the magma at a given stage of isobaric crystallization-controlled vapor evolution (second boiling). Combining Eq. (17), (33a), and (43), one obtains:

$$C_i^V(Z) = \frac{K_{i,\mathrm{Na}}[C_{\mathrm{Na}}^V(Z)]^n C_i^{l,s}}{\sqrt{1 - \sum_j X_j^V(Z)(C_{\mathrm{Na}}^{l,s})^n}} \cdot Z^{\bar{D}_i^{S/l}-1} \cdot \exp\left\{\frac{K_{i,\mathrm{Na}}}{(C_i^{l,s})^n} \cdot \int_1^Z \frac{[C_{\mathrm{Na}}^V(Z)]^n \cdot C_w^l(Z)}{Z}\right\} \tag{44}$$

The integral in Eq. (44) must be evaluated numerically. The $C_w^l(Z)$ term is given by Eq. (33). The function $C_{\mathrm{Na}}^V(Z)$ is next discussed in detail. Of all the metals that exist as chloride complexes, one, arbitrarily chosen, must partition *dependently*. That is, because of the constraint of electrical neutrality,

$$C_{\mathrm{Cl}}^V = \sum_K^{\sigma} n_{\mathrm{K}} \cdot C_{\mathrm{K}}^V \tag{45}$$

only $\sigma - 1$ elements can partition according to the independent exchange constants. For reasons already discussed, sodium has been chosen as the carrier metal in the exchange reactions and is used here as the metal whose concentration is determined by charge balance for a given C_{Cl}^V and a given set ($\sigma - 1$ in number) of independent exchange reactions. Equation (45) can contain any number, σ, of cations; however, complex algorithms are required if elements with $n > 2$ are included. In this chapter, the elements Cu, K, Mg, Fe, Ca, Mn, and Zn are included in the charge balance:

$$C_{\mathrm{Na}}^V = C_{\mathrm{Cl}}^V - (C_{\mathrm{Cu}}^V + C_{\mathrm{K}}^V) - 2(C_{\mathrm{Mg}}^V + C_{\mathrm{Fe}}^V + C_{\mathrm{Ca}}^V + C_{\mathrm{Mn}}^V + C_{\mathrm{Zn}}^V) \tag{46}$$

Substituting for $C_{k'}^V$ where $k' \neq \mathrm{Na}$) using $K_{k',\mathrm{Na}}$ for each k' and rearranging yields,

$$0 = C_{\text{Cl}}^{V} - C_{\text{Na}}^{V} \cdot \left[1 + \left(K_{\text{Cu,Na}} \cdot \frac{C_{\text{Cu}}^{l}}{C_{\text{Na}}^{l}}\right) + \left(K_{\text{K,Na}} \cdot \frac{C_{\text{K}}^{l}}{C_{\text{Na}}^{l}}\right)\right] - 2(C_{\text{Na}}^{V})^{2}$$
$$\cdot \left[\left(K_{\text{Fe,Na}} \cdot \frac{C_{\text{Fe}}^{l}}{(C_{\text{Na}}^{l})^{2}}\right) + \left(K_{\text{Mg,Na}} \cdot \frac{C_{\text{Mg}}^{l}}{(C_{\text{Na}}^{l})^{2}}\right) + \left(K_{\text{Ca,Na}} \cdot \frac{C_{\text{Ca}}^{l}}{(C_{\text{Na}}^{l})^{2}}\right) \right. \tag{47}$$
$$\left. + \left(K_{\text{Mn,Na}} \cdot \frac{C_{\text{Mn}}^{l}}{(C_{\text{Na}}^{l})^{2}}\right) + \left(K_{\text{Zn,Na}} \cdot \frac{C_{\text{Zn}}^{l}}{(C_{\text{Na}}^{l})^{2}}\right)\right]$$

This equation is of the form

$$a(C_{\text{Na}}^{V})^{2} + b(C_{\text{Na}}^{V}) - C_{\text{Cl}}^{V} = 0 \tag{48}$$

where a and b are positive quantities. Therefore, C_{Na}^{V} is given by

$$C_{\text{Na}}^{V}(Z) = \frac{-b + \sqrt{b^{2} - [4 \cdot a \cdot C_{\text{Cl}}^{V}(Z)]}}{2 \cdot a} \tag{49}$$

Note that HCl is neglected in this treatment. Hydrogen chloride can be included in the b term when satisfactory data are obtained. If higher order ligation is accommodated by the charge balance term, Eq. (46), then more complex approaches are required to obtain $C_{\text{Na}}^{V}(Z)$. Note that the solution of Eq. (49) requires the solution of Eq. (37) for C_{Cl}^{V}.

Equations (38)–(40) apply in case II as well as in case I, and Eq. (40) is used to integrate Eq. (44). Note that there are many nested iterations and quadratures in the solution of $E(i)$ for chloride-complexed metals. Therefore, numerical techniques must be chosen carefully so that early iterative and quadrature errors do not build up to unreasonable values.

Many of the equations presented in this chapter cannot be solved for the case $Z = 0$. This is not a serious problem. As is discussed in a later section of this chapter, it is probably realistic to terminate the integrations at some small but finite value of Z ranging from 0.01 to 10^{-5}. No calculations in this study were performed at values of $Z < 10^{-5}$, and all $E(i)$ that are not explicitly associated with a particular value of Z are calculated over a range of $Z = \{1 \text{ to } 10^{-5}\}$. It is best to monitor $C_i^V(Z)$, and determine how $E(i)$ varies as the range of Z is varied. This permits a better understanding of the limitations of the efficiency calculations.

Some Limitations of the Models

The use of Nerst bulk partition coefficients is clearly an oversimplified method of accounting for the uptake of elements by crystalline phases. It is clear that a more complex formulation is needed, especially in silicic systems, to quantitatively model crystal/liquid equilibria. Bulk partition coefficients are used in this model only as general indices of compatibility: $\bar{D}_i^{S/l} \gg 1$ indicates a compatible element, $\bar{D}_i^{S/l} \ll 1$ indicates an incompatible element, and $\bar{D}_i^{S/l} = 1$ is used to model an invariance in melt composition (with respect to i) as a function of the crystallization progress variable. It should be noted, however, that this bulk solid–liquid partition coefficient does not account for the variation in the liquid composition resulting from vapor evolution. For example, a melt–crystal system with a given $\bar{D}_i^{S/l} = 1$ may experience a progressive decrease in C_i as vapor evolution (with or without crystallization) proceeds, if the element is vapor compatible.

It is important to outline the limitations and implicit assumptions involved in the use of Eq. (14), (15), and (16). Many of the potential problems associated with the systematics of this chapter have been mentioned. A serious limitation that applies to all of the model I calculations is the problem of the pressure dependence of the various $D_i^{V/l}$ and $K_{i,\mathrm{Na}}$. Vapor evolution is assumed to occur over a pressure range from some pressure, P, to $P = 1$ bar, with $\sum P_{\mathrm{vapor}}^{\mathrm{magma}} \equiv P_{\mathrm{H_2O}}^{\mathrm{magma}} \geq P_T$ representing the condition for ebullition ($P_{\mathrm{CO_2}}$ is assumed negligible). The inequality indicates nonequilibrium boiling. Aside from the obvious problem of kinetic effects, such as slow diffusion of water and solutes in the melt, the equilibrium constants reported herein (Table 1) have not been determined below 1 kbar. Clearly, between 1 kbar and 1 bar of pressure, the solutes in question undergo a transition from a state of solvation in a relatively high-density, high-dielectric-constant medium to a mixed volatile state with a low density and low dielectric constant. More experimentation and thermodynamic measurements are necessary to establish the functional relationships of the pertinent exchange equilibria at low pressures. These preliminary calculations should act as an impetus for such research; in the interim, the tentative calculations presented herein can serve as general guides of the mobility of various substances in a magmatic vapor.

Two further limitations must be stated. In model I, polybaric vapor evolution is assumed to occur in the absence of magmatic crystallization. Although this is obviously a simplification, it allows one to examine the effects of vapor evolution alone as a limiting case. If crystallization accompanies vapor evolution (as it must in many cases) the effects of crystal incompatibility or compatibility come into play. However, it should be mentioned that such studies as those of Tiller *et al.* (1953) suggest that nonequilibrium crystallization yields effective $\bar{D}_i^{S/l} \approx 1$ at high crystal growth rates. This could minimize the amount of substance, i which partitions into the vapor. An additional complication in real systems is the presence of CO_2 and sulfur-containing compounds in the vapor. The presence of CO_2 may enhance the partitioning of some elements into vapor, but the main effect is to promote early evolution of the vapor phase and, probably, to inhibit the partitioning of chlorine and its associated metals into the vapor. Because of the high affinity of CO_2 for the vapor in a melt–vapor system, the earliest fluids would be CO_2 rich. It is our opinion that sulfur will not significantly alter the systematics herein other than to promote the precipitation of some metals, such as zinc or molybdenum, as sulfide; however, further experimentation involving sulfur in melt–vapor systems is indicated.

Other major limitations of the systematics presented herein are:

1. The lack of well-constrained equilibrium constants for many liquid–vapor equilibria
2. The use of Nernst-type vapor–liquid partition coefficients for Cl, F, B
3. The paucity of vapor–liquid partition experiments for melts with 74 wt. % SiO_2 and below
4. The lack of the necessary data to place $[BO_3]^{3-}$, $[MoO_4]^{2-}$ and F^- into the charge balance equation
5. Only anhydrous solidus assemblages are considered

There are clearly many restrictions on any such mathematical models. However, this first-order approach to the problem possesses many interesting characteristics that are examined in the balance of this chapter.

Discussion

Discussion of Model I

The efficiency and concentration functions for first boiling are relatively simple compared to the functions for second boiling. Plots of Eq. (14) and (16) for the elements Cl, F, Mo, and B cover the complete range of behavior for Nernstian elements during first boiling. Chlorine, by virtue of its high vapor–liquid partition coefficient ($D_{\text{Cl}}^{V/l} \approx 40$), is very efficiently removed from a rising, aphyric, silicic magma. Figure 1 shows that if 1 wt % water is evolved from the magma during first boiling, then one-third of the initial magmatic chlorine can be removed into the evolving magmatic vapor. The release of 6 wt % water results in over 90% removal of magmatic chlorine. In contrast, the elements Mo and B experience only modest removal by polybaric vapor evolution that is unaccompanied by crystallization. Note that, within error, $D_{\text{Mo}}^{V/l} \approx D_{\text{B}}^{V/l}$; therefore, to a first approximation, $E(\text{B}) \approx E(\text{Mo})$. If 1 and 6 wt % magmatic water are evolved from a magma, $E(\text{Mo, B})$ equals approximately 2% and 15%, respectively. This is significantly less than the amount of chlorine removed under the same conditions. Note, from Fig. 2, that the change in the concentration of boron and molybdenum in the melt is rather modest and may not be distinguishable from the normal scatter in analytical data when analyzing a series of natural glasses that may have experienced various degrees of degassing. Therefore, boron and molybdenum would be poor choices as "volatile elements," or fingerprints of vapor evolution in glass-forming systems. In a study of the behavior of boron in the Inyo Domes rhyolites, Higgins (1985) reports that boron was not lost with the water that was known to have evolved from the Inyo magma (B. E. Taylor *et al.*, 1983). This observation is entirely consistent with the equilibrium vapor–melt partition model developed herein. Further, the devitrified material at Inyo possesses a lower concentration of boron (Higgins, 1985). This illustrates a general principle which results from the present theoretical discourse: elements

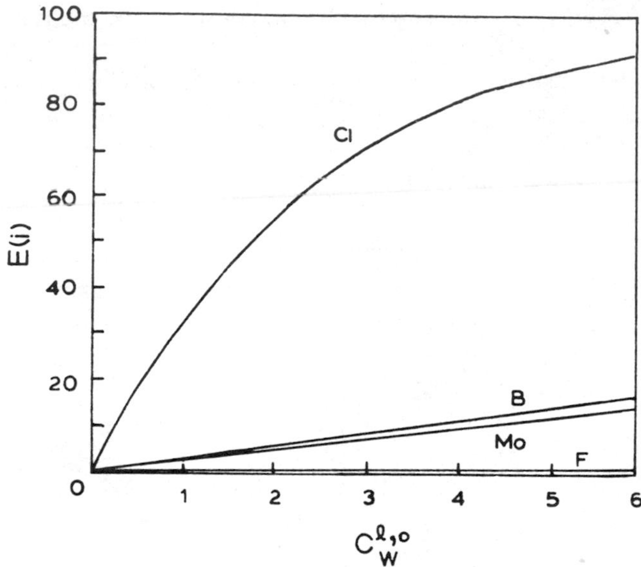

Fig. 1. The efficiency of removal $E(i)$ of chlorine, boron, molybdenum, and fluorine (in percent, calculated using case I, model 1, first boiling) as a function of $C_w^{l,0}$. Equation (16) was used to calculate $E(i)$.

that are not strongly vapor compatible (such as B, Mo, and, as will be discussed, F) can still be forced into the vapor by super- or subglass transition crystallization.

The behavior of fluorine stands in stark contrast to the behavior of Cl, Mo, and B. Because $D_F^{V/l} < 1$ (see Table 1), the concentration of fluorine in successive aliquots of a degassing magma actually increases slightly. For a magma which loses 6 wt % water, Eq. (14) yields an increases in $C^l{}_F$ from an initial value of, e.g., 500 ppm to a value of 525 ppm after vapor evolution. The increase in concentration results from the loss of magma mass because of vapor evolution. The total quantity of fluorine lost from the system would be a little over 1% (see Fig. 1). The concentration of fluorine in the evolving vapor would remain approximately constant at about 100 ppm. Note, however, that 100 ppm F translates to 0.005 molal F and $f_{HF}/f_{H_2O} \approx 10^{-4}$.

In case II, the partitioning of chloride-complexed metals into the vapor during first boiling is modeled. For illustrative purposes, a simple charge balance formalism was adopted to simplify the calculations. A more complex method of accounting for charge balance is employed in case II, model II (second boiling). However, the general relationships shown here are not seriously altered by this simplification. The reader may, however, adapt the charge balance methods in case II, model II to case II, model I. The partitioning of the elements Cu, Zn, and Ce (representing ligation numbers of 1, 2, and 3, respectively) are discussed in addition to the alkali and alkaline earth elements. The partitioning of these

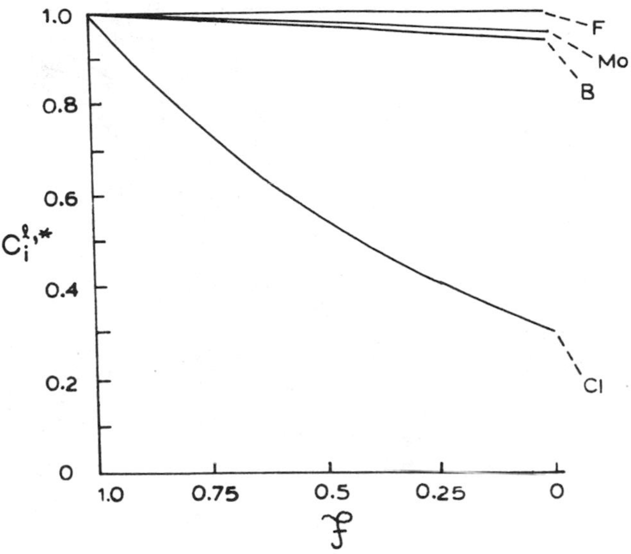

Fig. 2. The concentration of chlorine, boron, molybdenum, and fluorine in the melt divided by their respective initial melt concentrations is plotted on the ordinate as a function of the proportion of the initial water content ramaining, $\mathscr{F}$, during first boiling. Between $\mathscr{F} = 1$ and $\mathscr{F} = 0$, the mass of water evolved is 3% of the original mass of the magma ($C_w^{l,0} = 3\%$). Because $D_F^{V/l} < 1$, the concentration of fluorine in the magma increases as the mass of magma is reduced by vapor evolution. At $\mathscr{F} = 0$ and $C_w^{l,0} = 3\%$, $C_F^{l,*} = C_F^l / C_F^{l,0} = 1.024$, and therefore plots as a line subparallel to the abscissa.

case II elements is slightly more complex than the partitioning of the case I elements, Cl, F, Mo, and B, because of their dependence upon the concentration of chlorine in the aqueous phase.

In this chapter, a Cl/H_2O weight ratio of 0.1 is used illustrate the utility of these calculations. This figure was given by Gill (1981) as the upper range for arc magmas. Figure 3 is a plot of the efficiency of removal of selected elements as a function of the amount of water released from the melt during decompression. The salient feature of this figure is the general ordering of $E(i)$. Note that copper possesses a very high $E(i)$, with zinc next in order. Along with chlorine, these two elements are very sensitive indicators of magmatic vapor evolution in aphyric systems. However, these elements are probably rather sensitive to magmatic sulfide precipitation, and it is best to rely upon a strong correlation among copper, zinc, and chlorine as an indicator of vapor evolution. By combining the systematics of case I and case II a quantitative relation between C_{Cl}^l and C_{Cu}^l or C_{Zn}^l can be obtained but is not given here. This relationship can be used to test quantitatively for vapor evolution.

The case of the partitioning of the rare earth elements between melt and vapor illustrates the importance of evaluating the vapor phase-affinity of an element or group of elements in light of a quantitative model only. At first glance, the

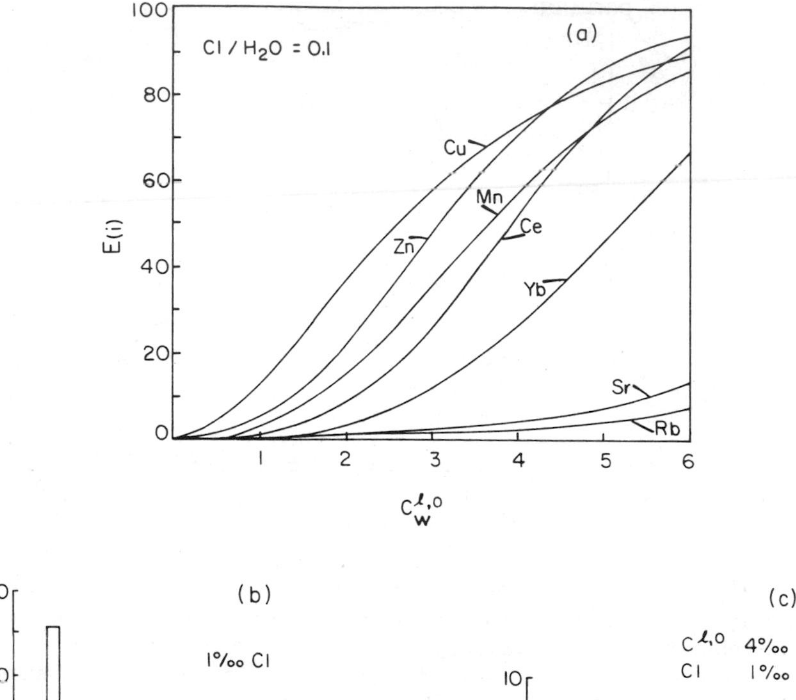

Fig. 3. The efficiency of removal $E(i)$ of selected chloride-complexed elements (in percent, calculated using case II, model I, first boiling). (a) $E(i)$ vs. $C_w^{l,0}$ for Cu, Zn, Mn, Ce, Yb, Sr, Rb as a function of $C_w^{l,0}$. Notice that the rate of increase of $E(i)$ is a function of the ligation number, with $E(\mathrm{Ce})$ exceeding $E(\mathrm{Mn})$ and $E(\mathrm{Zn})$ exceeding $E(\mathrm{Cu})$ at high $C_w^{l,0}$. This figure is drawn for a constant Cl/H_2O ratio $= 0.1$. At modest $C_w^{l,0}$ (e.g., 2%), $E(\mathrm{Cu})$, $E(\mathrm{Zn}) > 20\%$ removal; $E(\mathrm{Mn})$, $E(\mathrm{Ce})$ experience moderate removal (8 and 15%, respectively); and Yb, Rb, Sr (and other alkali and alkaline earths) experience low removal ($<5\%$). (b) Bar graph exhibiting $E(i)$ for selected elements at $C_w^{l,0} = 6$ wt % but with $C_{\mathrm{Cl}}^{l,0}$ fixed at 1000 ppm. When compared with Fig. 3a, 3b illustrates the rapid decrease in $E(i)$ with decreasing $C_{\mathrm{Cl}}^{l,0}$ for high-ligation-number elements. (c) $E(i)$ for the low-$E(i)$ elements Na, K, Ba, Ca, Mg at $C_w^{l,0} = 4$ wt %. $E(i)$ is shown for two cases: $Cl/H_2O = 0.1$ and $C_{\mathrm{Cl}}^{l,0} = 1000$ ppm.

vapor–melt partition coefficients for the REE (Flynn and Burnham, 1978) appear "low." However, cerium exhibits a significant (greater than 10%) partitioning into the vapor for a modest (2%) quantity of vapor evolution. The reason that the REE partition into the vapor to a significant extent is that, with ligation numbers of three, they are subject to a strongly nonlinear stoichiometric effect.

Their "effective" partition coefficients (the empirical ratio C^V_{Ce}/C^l_{Ce} as function of Z) increases as the cube of $C^V_{Cl}(Z)$. Further, note that the light REE (represented by Ce) partition into the vapor to a greater extent than do the heavy REE (represented by Yb). The alkali and alkaline earth elements lie at the other end of the spectrum. The partitioning of the elements Na, K, Rb, Cs, and Mg, Ca, Sr, Ba into the magmatic vapor is rather modest when considering efficiencies of removal.

Discussion of Model II

The efficiencies of removal, and the melt and vapor concentrations of the elements Cu, Mn, Zn, Ca, Mg, Fe, Ce, Mo, Cl, F, and B can be examined as functions of many variables. The $E(i)$ are discussed here as a function of:

1. The concentration of other substances in the melt
2. The bulk partition coefficient of the element in question and the bulk partition coefficient of other constituents
3. The initial Cl/H_2O ratio in the magma ($C^{l,0}_{Cl}/C^{l,0}_w$) and the saturation water content of the melt
4. The range of Z over which vapor is actually released from the pluton
5. The timing of vapor evolution as determined by ($C^{l,0}_w/C^{l,s}_w$).

In this model all cations are charge balanced by chlorine. This is probably a reasonable restriction at low pressures for most values of Z. [However, at high pressures, Kilinc (1969) showed that significant silicate and aluminosilicate solubility occurs.] This places a rather rigid restriction on the partitioning of magmatic cations into the magmatic vapor phase. For melts with a given ($C^{l,0}_{Cl}/C^{l,0}_w$) ratio, $E(i)$ for all elements will drop as the total cation content of the magma rises under the constraint of constant cation ratios. This is because fixing the ($C^{l,0}_{Cl}/C^{l,0}_w$) ratio fixes the total number of equivalents of metal that can be contained in the vapor. As the total cation number increases in the melt, the overall efficiency of removal of any given cation must decrease for a fixed possible number of equivalents in the vapor. Interesting results can be obtained when the total number of cation equivalents in the magma are held constant, but the element ratios are varied. For example, if $K_{Fe,Na} \simeq K_{Mn,Na}$ as seems reasonable to a first approximation, then it is clear from Table 1 that iron partitions into the vapor to a much greater extent than does either sodium or potassium. Therefore, reducing $C^{l,0}_{Na}$ and $C^{l,0}_{K}$ and increasing $C^{l,0}_{Fe}$, while maintaining the total number of equivalents in the melt, has an effect similar to increasing the number of equivalents in the melt at *constant* cation ratios, because $K_{Fe,Na}$ is greater than one and greater than $K_{K,Na}$. Hence, the $E(i)$ for every element (even for iron) will decrease.

Similar effects result from a simultaneous variation in bulk partition coefficients (with the exception of $\bar{D}^{S/l}_{Na}$ and $\bar{D}^{S/l}_{K}$, which remain fixed at unity to preserve the near-eutectic nature of the feldspar–quartz subsystem, and $\bar{D}^{S/l}_{Cl}$, which remains fixed at zero for simplicity). Simultaneously decreasing all partition coefficients by as much as a factor of 5 causes the $E(i)$ of many elements to increase only modestly, and some $E(i)$, such as $E(\mathrm{Ca})$ and $E(\mathrm{Ce})$, decrease over certain ranges of $\bar{D}_i^{S/l}$ reduction. The modest increases occur for elements of relatively high $K_{i,\mathrm{Na}}$, such as copper, and this increase occurs at the expense of sodium, magnesium, and even iron. To illustrate the degree to which the efficiency of removal of an element is dependent upon the behavior of other elements, the following two calculations were performed using Eq. (40): $C_w^{l,0} = C_w^{l,s} = 0.01$ by weight, $(C_{\mathrm{Cl}}^{l,0}/C_w^{l,0}) = 0.1$ by weight, initial concentrations and equilibrium constants are set as in Table 2, $\bar{D}_{\mathrm{Cl}}^{S/l} = 0$, and all $\bar{D}_i^{S/l} = 1$, in one case, and $\bar{D}_i^{S/l} = 1$ $(i \neq \mathrm{Ce})$ with $\bar{D}_{\mathrm{Ce}}^{S/l} = 0.2$, in the other case (see Table 2). In the first case $E(\mathrm{Ce}) = 0.006$, and in the second case $E(\mathrm{Ce}) = 0.44$. However, reducing *all* $\bar{D}_i^{S/l}$ including $\bar{D}_{\mathrm{Ce}}^{S/l}$ to 0.2 yields $E(\mathrm{Ce}) = 0.002$! Clearly, competition for chlorine in the vapor is acute, and the *relative* values of $\bar{D}_i^{S/l}$ as well as C_i^l affect the vapor melt partition of *all* chloride complexed elements.

Note that when comparing variations in $E(i)$, it is best to compare values of $E(i)$ that are on the same order. This is necessary because $E(i)$ possesses an upper bound of unity, which results in smaller incremental changes in $E(i)$ for a given change in a given variable as $E(i)$ increases. In summary, the $(C_{\mathrm{Cl}}^{l,0}/C_w^{l,0})$ ratio exerts a rather strong control on the total cation chemistry. This fact must be remembered when considering the effect of other parameters (e.g., $\bar{D}_i^{S/l}$ for other constituents) on the efficiency of removal of a particular element.

It is interesting to note the difference in ligation number, n, of various elements because the change in $E(i)$ produced by given variations in $(C_{\mathrm{Cl}}^{l,0}/C_w^{l,0})$ is dependent upon n. For example, given $C_w^{l,0} = C_w^{l,s} = 0.06$, and $\bar{D}_i^{S/l} = 1$, $E(i)$ for $i = \mathrm{Ca}$, Mg, and Ce change by factors of 31, 32, and 152, respectively, for an increase in $(C_{\mathrm{Cl}}^{l,0}/C_w^{l,0})$ from 0.017 to 0.2. That is, $\partial E(i)/\partial(C_{\mathrm{Cl}}^{l,0}/C_w^{l,0})$ increases with increasing n.

An examination of Table 2 indicates that for a given $(C_{\mathrm{Cl}}^{l,0}/C_w^{l,0})$, the efficiencies of removal increase for increasing $C_w^{l,0} = C_w^{l,s}$, because of the increase in total chloride. There is, however, a more subtle effect of $C_w^{l,s}$ on $E(i)$. It has been noted (Holland, 1972; Candela, 1982; Candela and Holland, 1986) that at low water solubilities in the melt ($C_w^{l,s} \lesssim 2.5$ wt % H_2O, or 300–900 bars of pressure) the quantity of water evolved for a given portion of melt crystallized is insufficient to deplete the rest melt in chlorine. The result is an increase in the chlorine concentration of the melt, and the associated vapor with which it is in equilibrium, to very high values. Elements such as the rare earth elements (exemplified herein by Ce), which have high chloride ligation numbers, will partition rather heavily into such a chloride-rich fluid. Hence, we find that at low $C_w^{l,s}$ and moderately high $(C_{\mathrm{Cl}}^{l,0}/C_w^{l,0})$ ratios the cerium concentration builds up to rather high values in the vapor toward the end stages of crystallization. Figure 4 shows a characteristic profile for an element with a high $K_{i,\mathrm{Na}}$, or high value of n. Characteristically, copper (high $K_{\mathrm{Cu,Na}}$) and the rare earth elements (e.g., $K_{\mathrm{Ce,Na}}$) follow such trends at low $C_w^{l,s}$. Table 2 shows the dependence of $E(\mathrm{Ce})$ on $C_w^{l,s}$, $(C_{\mathrm{Cl}}^{l,0}/C_w^{l,0})$, $\bar{D}_i^{S/l}$ and $\bar{D}_{\mathrm{Cl}}^{S/l}$.

The partitioning of the Nernstian elements, exemplified herein by the parti-

Table 2. Efficiency of removal calculations: selected values of $E(i)$ calculated from cases I and II, model II (second boiling).

						$E(i)^{a,b}$									$E(i)^{a}$		
	$C_w^{l,0}$, %	$C_w^{l,s}$, %	$Cl/H_2O^{l,0}$	$\bar{D}^{S/l}_{Cu,Fe}$	$\bar{D}^{S/l}_{Ce}$	Cu	Mn	Zn	Ca	Mg	Fe[e]	Ce	Z_{LL}[c]	$C_i^{l,0}$[d]	Mo	F	B
1	1	1	0.1	1	1	0.087	0.016	0.023	1.3×10^{-3}	5.7×10^{-4}	0.016	6.2×10^{-3}	10^{-5}	1	0.27	0.02	0.32
2	1	1	0.1	0.5	0.5	0.14	0.020	0.029	1.2×10^{-3}	5.1×10^{-3}	0.021	3.3×10^{-3}	10^{-5}	1			
3	1	1	0.1	0.2	0.2	0.21	0.020	0.029	1.2×10^{-3}	5.0×10^{-3}	0.019	2.0×10^{-3}	10^{-5}	1			
4	1	1	0.1	1	0.2	0.087	0.016	0.023	1.3×10^{-3}	5.7×10^{-4}	0.016	0.45	10^{-5}	1			
5	1	1	0.1	1	0.2	0.087	0.016	0.023	1.1×10^{-3}	4.5×10^{-4}	0.016	0.32	10^{-3}	1			
6	1	1	0.1	1	0.2	0.086	0.014	0.020	8.4×10^{-4}	3.6×10^{-4}	0.014	0.034	10^{-2}	1			
7	3	3	0.1	1	1	0.27	0.045	0.073	1.3×10^{-3}	1.3×10^{-3}	0.051	0.01	10^{-5}	1			
8	6	6	0.1	1	1	0.47	0.10	0.17	7.6×10^{-3}	3.2×10^{-3}	0.12	0.34	10^{-5}	1	0.81	0.11	0.87
9	1	6	0.1	1	1	0.078	0.016	0.027	1.2×10^{-3}	5.1×10^{-4}	0.019	5×10^{-3}	10^{-5}	1	0.88	0.13	0.94
10	1	1	0.2	1	1	0.13	0.038	0.052	5×10^{-3}	3.4×10^{-3}	0.039	0.017	10^{-5}	1			
11	1	1	0.017	1	1	0.027	1.5×10^{-3}	2.9×10^{-3}	9.0×10^{-5}	3.8×10^{-5}	1.5×10^{-3}	1.9×10^{-2}	10^{-5}	1			
12	1	1	0.1	1	1	0.067	8.7×10^{-3}	0.012	5.7×10^{-4}	2.4×10^{-4}	8.9×10^{-3}	2.3×10^{-3}	10^{-5}	1			

[a] $\bar{D}^{S/l}_{Nernstian} = 0$ (including chlorine); $E(i) = 1$ indicates complete removal by vapor.
[b] In most cases $E(Na)$, $E(K)$ are between 10^{-3} and 10^{-2}.
[c] Z_{LL} = value of Z at lower limit of integration.
[d] Initial concentrations in wt %: Composition 1, Cu = 0.005, Mn = 0.05, Zn = 0.005, Ca = 1.5, Mg = 0.3, Fe = 3, Ce = 0.005, Na = 3, K = 5, Mo = 0.005, B = 0.005. Composition 2, Same as 1, except Fe = 9, Na = 1, K = 1.
[e] Calculated assuming $K_{Fe,Na} \approx K_{Mn,Na}$.

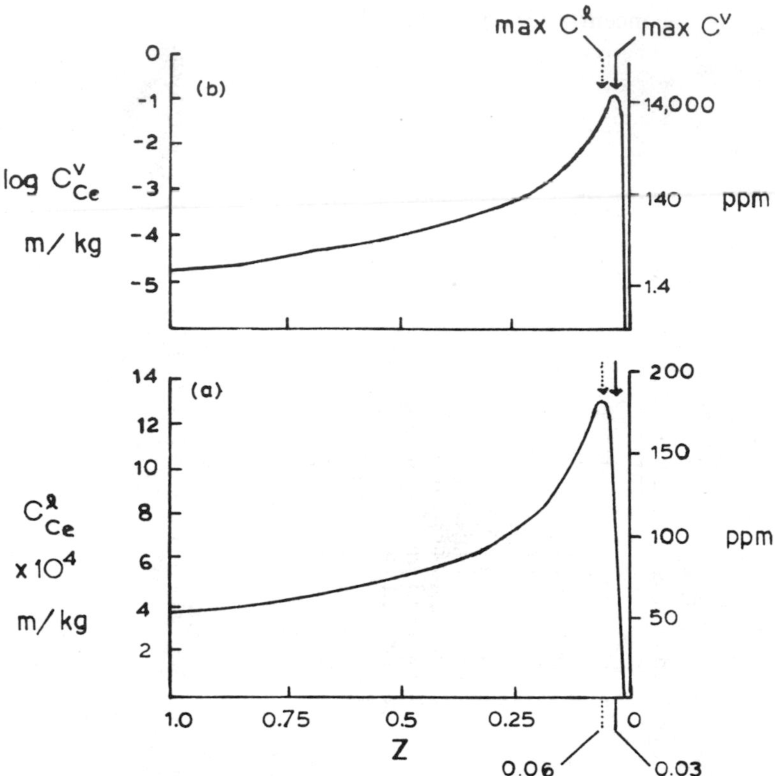

Fig. 4. (a) The concentration of cerium in the melt, C^{l}_{Ce}, and (b) the logarithm of the concentration of cerium in the vapor, $\log C^{l}_{\mathrm{Ce}}$, as a function of the proportion of melt remaining, Z. The second boiling model (case II, model II) was used to construct this figure. $(C^{l,0}_{\mathrm{Cl}}/C^{l,0}_{w}) = 0.1$, $\bar{D}^{S/l}_{j} = 2$ (where $j \neq$ Ce, Na, K, Cl), $\bar{D}^{S/l}_{\mathrm{Na}}$, $\bar{D}^{S/l}_{Ki} = 1$, $\bar{D}^{S/l}_{\mathrm{Cl}} = 0$, and $\bar{D}^{S/l}_{\mathrm{Ce}} = 0.5$. Initial concentrations are as listed under composition 1 (see footnote *c* of Table 2). Note that $C^{l,0}_{\mathrm{Ce}} = 50$ ppm. C^{l}_{Ce} is in units of moles per kilogram of melt, and C^{V}_{Ce} is in units of moles per kilogram of water. Note that the maximum in melt concentration occurs at $Z \simeq 0.06$ (94% crystallized) and the maximum in vapor concentration occurs at $Z \simeq 0.03$ (97% crystallized).

tioning of the elements Cl, F, Mo, and B, is considerably simpler than the partitioning of chloride-complexed elements. Candela and Holland (1986), using a much simpler formulation than that used herein, showed that successive aliquots of melt and the associated vapor were depleted or enriched in a Nernstian constituent depending on whether the quantity $\Gamma_i = \bar{D}^{S/l}_{i} + (D^{V/l}_{i} \cdot C^{l,s}_{w})$ (where $C^{l,s}_{w}$ is in weight fraction) is greater or less than unity, respectively. This quantity yields an approximate guide to the behavior of Cl, F, Mo, and B with regard to the present formulation. The calculation of Γ_{Cl}, Γ_{F}, Γ_{Mo}, and Γ_{B} for reasonable values of $C^{l,s}_{w}$ (0.01–0.1) shows that only Γ_{Cl} can exceed unity solely because of the vapor term in Γ. That means that Mo, B, and F will, as a general rule, increase in concentration as crystallization proceeds (if $\bar{D}_i^{S/l} < 1$) regardless of whether a vapor is evolving or not. Therefore, Mo, F, B, and other elements of similar behavior will increase in concentration until the melt–vapor system saturates with respect to some primary Mo, F, or B phase. This most likely occurs at very low Z.

Inspection of Fig. 4 shows that the peak in the aqueous cerium concentration occurs at very low values of Z, i.e., at a very advanced stage of crystallization and vapor evolution. Further, the elements Mo, F, and B, and sometimes Cl, will increase monotonically as $Z \rightarrow 0$. The question arises as to how low a value of Z is too low to allow the magmatic vapor to escape from the now almost completely formed pluton. Clearly, at some low value of Z, e.g., 0.1, 0.01, or 0.001, etc., the removal of vapor from the pluton becomes impeded. The exact value of Z at which this happens is unknown but must be a function of the local and regional stress field, the viscosity of the magmatic and aqueous phases, the bouyancy of the vapor, the distribution of crystal sizes, and the size and geometry of both the pluton and the temporally varying zones of vapor evolution. The vapor that does not leave the pluton will cool within it, producing minor deuteric alteration and depositing its load of crystal- and vapor- incompatible elements (F, B, Mo, and sometimes REE and Cl, possibly W(?), and many other elements) as highly soluble and hydrothermally labile compounds in microveinlets (Hibbard, 1980), in micromiarolitic cavities and along the grain boundaries of the constituent minerals of the pluton. These materials are obviously easily stripped out of the pluton by circulating meteoric water and therefore are readily available for ore formation. A similar scenerio has been presented by Campbell *et al.* (1984) in the case of tungsten deposits. Table 2 shows how $E(i)$ varies with the lower limit of Z for selected elements. Under conditions where an element is increasing in concentration as crystallization proceeds, $E(i)$ clearly increases as Z reaches very low values. For example, $E(\text{Mo})$, for $\bar{D}_{\text{Mo}}^{S/l} = 0$, reaches about 90% removal for $Z_{\text{lower limit}} = 10^{-5}$. This is probably a good upper limit for the amount of molybdenum that can be removed from the pluton by a vapor during second boiling. For chloride-complexed metals, such as Ce, $C_i^V(Z)$ passes through a maximum between $Z = 0.1$ and $Z = 10^{-5}$. This interesting effect results from the fact that $C_{\text{Cl}}^V(Z)$ reaches very high values, and the effective vapor–liquid partition coefficient becomes large enough so that the metal becomes chloride-vapor compatible and $C_i^l(Z)$ and $C_i^V(Z)$ decrease with decreasing Z. It should be noted that the high (model) concentrations calculated for some elements at low Z are almost certainly *not* realized because the melt–vapor system probably saturates with respect to a primary phase of that element. Therefore, these preliminary calculations should be regarded as an upper limiting case.

The parameter $(C_w^{l,0}/C_w^{l,s})$ was discussed at length by Candela and Holland (1986) and the interested reader is referred to that paper. However, because this ratio is the master variable of melt–vapor mass transfer, it will be discussed briefly here. This variable is inversely related to the amount of crystallization that occurs prior to vapor evolution. Varying the proportion of pre-vapor-saturation crystallization leads to a separation of compatible elements from incompatible elements. Early vapor evolution allows a crystal-compatible element with a high vapor–melt partition coefficient (such as copper) to be partitioned into the vapor before a significant amount of the element is tied up in the hypersolidus assemblage. On the other hand, if protracted crystallization occurs before vapor evolution, then incompatible elements (such as Mo) are still available for partitioning into the vapor, whereas elements such as copper will have been removed into the crystalline residue before the melt saturates with a vapor phase. In general, the ratio of the efficiency of removal of an incompatible element to the efficiency of removal of a compatible element, as best exemplified by the ratio $E(\text{Mo})/E(\text{Cu})$, increases with an increase in $C_w^{l,s}$ for a given $C_w^{l,0}$ or with a decrease in $C_w^{l,0}$ for a given $C_w^{l,s}$. This indicates that, holding other variables constant, we might expect deeper Cu–Mo porphyry systems to be more molybdenum rich. However, this simple model cannot be extended, to the exclusion of other models, to such problems as why porphyry molybdenum systems exist apart from porphyry coppers. There are probably many tectonic, petrogenetic, and hydrothermal processes acting in concert to produce the enrichments in molybdenum found in such deposits as those of Climax and Henderson (White *et al.*, 1981). However, many of the features of these deposits can be explained by the present model, and that of Candela and Holland (1986). The system at the Henderson Mine is assumed to be relatively shallow (White *et al.*, 1981), and this leads to the evolution of a Cl-, Mo-, and F- rich fluid at low Z. Indeed, Gunow (1983) reports that apatite from Henderson is greatly enriched in REE as one would expect in such a shallow environment. The elemental assemblage of Mo, F, (Cl), REE, B, W(?) is the characteristic shallow, low-Z, magmatic vapor assemblage in high-silica rhyolite systems. Within the context of the present model, the relationship of this class of systems to the more copper-rich systems is not clear. The main thrust of this research bears on the relatively late, high-level stages of evolution of felsic magmatic systems. It would, however, be rather provincial to state that these processes alone control the chemistry of the magmatic–hydrothermal ore fluid. The initial metal concentrations of the parental magmas must play some role in determining the chemistry of an ore system. On the other hand, increasing the initial molybdenum content of a magma by an order of magnitude (from 2 to 20 ppm) is not likely to be a critical factor in porphyry molybdenum ore genesis and there is little evidence to suggest any higher estimates for $C_{\text{Mo}}^{l,0}$. Clearly the higher level processes must work in concert with other factors of magma genesis in controlling the chemistry of magmatic ore fluids.

Efficiency of Removal and the Rubidium Problem

A distinction must be made between a low efficiency of removal of an element, such as potassium or rubidium, and a low activity in the vapor with respect to various alteration reactions. This comment applies to all models and cases presented in this chapter. Therefore, whereas the overall efficiencies of removal of sodium, potassium, and iron are low (on the order of 0.01–0.04 for Na and K, and less than 0.1 for Fe, in most cases) the solutes in the magmatic vapor may be dominated by NaCl, KCl and $FeCl_2$ by virtue of a relatively high concentration of sodium, potassium, and iron in the melt compared to the concentrations of the other constituents. Hence later trapping of these fluids as fluid inclusions will yield a fluid in the (Na–K–Fe) Cl–H_2O subsystem. Further, by virtue of a high $C_i^{l,0}$, or high a_i^V with respect to various alteration reactions, significant amounts of K, Rb, or F metasomatism may occur. However, the respective $E(i)$ remain modestly low. In general, a low $E(i)$ may be misleading if high concentrations in the melt exist, or if modest concentrations of i in the vapor are capable of producing significant alteration or mineral deposition.

Rubidium is a case in point. Although rubidium normally behaves as an incompatible element, it has been known to decrease with increasing differentiation index in whole rocks (McMillian, 1982) or in individual minerals (Chivas, 1981). In both cases, the presumed decrease in the rubidium concentration in successive aliquots of magma was attributed to the evolution of a vapor phase. Note, however, that Rb possesses a rather low vapor–melt partition coefficient (Table 1). Given any reasonable magmatic chlorine and water concentration, $K_{\mathrm{Rb,Na}}$ would have to be 10–20 times greater than the value given in Table 1 for rubidium to be depleted in successive aliquots of a crystallizing magma. The rubidium problem may therefore be stated as follows: the element may, in some cases, experience an inexplicable decrease during magmatic evolution and may form anomalies, which parallel those of potassium, in altered rock; yet, $K_{\mathrm{Rb,Na}}$ is too low to deplete a magma in Rb, and $E(\mathrm{Rb})$ is usually below 0.01 and 0.03. A partial solution to the problem is possible. First, as was stated above, significant metasomatism and alteration can occur at low $E(i)$ for the stated reasons. Second, other more complex processes, such as magma mixing in the case of whole-rock rubidium depletion or crystallographic control in the case of mineral-specific depletion of rubidium may be operative. In general, low $K_{i,\mathrm{Na}}$ produce low $E(i)$ and do not result in a depletion of i in successive fractions of a differentiating magma. However, the resulting fluid may still be capable of producing significant metasomatic addition of i to the country rock.

Implications for Volcanology and Comparative Planetology

In general, models I and II deal with the efficacy of vapor transport in removing chemical substances from magmas into their immediate surroundings. In this section some speculative applications of models I and II in the fields of volcanology and terrestrial planetology are developed. These discussions are necessarily speculative because of the paucity of vapor–melt data (as a function of pressure and composition) and the incomplete nature of our knowledge concerning magmatic processes on Mars, Venus, and other terrestrial bodies.

A portion of the solutes transported by volcanic gases is deposited as volcanic sublimates. The sublimates found in felsic–volcanic settings can be placed into two groups: fumarolic incrustations and coatings on volcanic ash. In a volcanic environment magmatic vapor can be released either by the vapor-saturated crystallization (second boiling) of a magma in a shallow subvolcanic chamber or by the rise of a water-saturated magma (first boiling) in a volcanic conduit. In general, the former process can proceed for a protracted length of time, whereas the latter process is transient in nature. Consequently, second boiling is the process that contributes significantly to the composition of fumarolic gases and incrustations. The origin of the composition of ash leachates and ash coatings presents a problem. Taylor and Stoiber (1973) argue convincingly that these coatings are derived from the eruptive vapor phase, based on the fact that the chemistry of ash leachate correlates with volcanic gas chemistry and is not consistent with acid alteration of the ash itself. Because most of the ash forms at the time of eruption, it is likely that a significant quantity of the volcanic vapor released with the ash during eruption is a product of decompressive vapor evolution, i.e., first boiling. On the other hand, a significant portion of the vapor released during eruption, especially during the early stages, may be the product of second boiling. Indeed the increase of fluid pressure during second boiling may act as a trigger for the eruption itself. However, this does not eclipse the fact that first boiling must contribute to the volcanic gas phase. As a generalization, a rising water-saturated magma may erupt, releasing the exsolved fluid (model I), or may collect in a shallow subvolcanic chamber and evolve vapor during crystallization (model II).

This difference in behavior suggests that monitoring the changes in the chemistry of volcanic sublimates may prove useful in detecting the upward movement of magma beneath an active volcano. The expected changes may be outline by a comparison of models I and II. As was mentioned above, the elements molybdenum, boron, and fluorine are not partitioned heavily into the aqueous phase during first boiling because the product of the water concentration in the melt and the Nernst-type vapor–liquid partition coefficient is relatively low. Among these three elements, this product is greatest for boron: at 5 wt % water evolved, $C_w^{l,0} \cdot D_B^{V/l} = 0.15$. This yields only a modest removal of boron into the magmatic vapor. However, during second boiling these incompatible elements are forced into the magmatic vapor by crystallization. These elements will be referred to as the "crystal-incompatible suite" of elements. On the other hand, elements such as copper, zinc and manganese partition heavily into the vapor in the presence of chlorine. The elements copper, zinc, and manganese comprise the "chlorine-vapor-compatible suite" of elements. This suite is composed of those elements which partition heavily into the magmatic vapor even in the absence of crystallization. Their behavior during second boiling is complicated by the fact that copper and zinc may behave as compatible elements under certain conditions that are still not very well understood (Andriambololona and Dupuy, 1978; Candela and Holland, 1986). Consequently, in many cases the chlorine-vapor-compatible elements may experience a higher efficiency of removal during first boiling than during second boiling. The behavior of the two above-mentioned suites of elements is obviously quite different. If new water-saturated magma rises in the conduit, the fumarolic incrustations should indicate a transition from second boiling to first boiling. Such a transition might be characterized by the rise in a ratio of a chloride-vapor-compatible element to a crystal-incompatible element (such Cu/Mo).

The main problem involved in obtaining a relatively pristine sample of the condensates from the second boiling fluid involves the reaction of fluid with country rock or previously formed volcanic material. A possible solution to this problem is to sample high temperature ($\sim 900°C$) fumaroles with gas chemistries consistent with magmatic f_{O_2}. Fluids of this type are likely to have experienced only minimal change in chemistry since exiting from the magma. Methods for sampling and analyzing such volcanic sublimates by inserting silica tubes into fumarolic vents have been tested at the Merapi volcano, Java (Le Guern and Bernard, 1982), and indeed molybdenite was found to be one of the major phases to condense in the silica tube. Copper, on the other hand, was only a minor constituent; however, sphalerite formed a distinct zone. The data base for such a systematic study is practically nonexistent, yet these preliminary results are promising. Concerning the eruptive volatile chemistry and the analysis of ash coating and leachates, the analysis of ash from Mt. St. Helens (Thomas *et al.*, 1982) showed significant amounts of zinc and copper coating the ash. No molybdenum was mentioned in this study. An interesting account of sublimate chemistry is reported by Taylor and Stoiber (1973):

> There is the possibility that copper concentrations rise at certain times during the volcanic eruptive cycle. This is suggested for [the] Pacaya [volcano] . . . by the appearance of copper minerals in the fumarolic sublimates in the early history of eruption.

Could this result from a transition from model II control to model I control? The prospects for understanding the chemistry of volcanic sublimates using the present models, even with the limitation stated herein, is quite promising.

The high-temperature vapor-phase chemistry of the volatile elements will almost certainly vary from planet to planet. Further, given that the geochemistry of many of other elements is dependent upon the chemistry of the volatile elements, one might expect the geochemistry of such elements as copper to vary with the varying volatile element chemistry. Variations in the geochemistry of volatile elements are expected because of:

1. Variations in the planetary abundance of volatile elements
2. Variations in the planetary abundance of elements that can immobilize or fix the volatile elements in low-volatility phases (or change their speciation in fluids)
3. The variable character of hydrologic and thermal regimes in the crusts of planetary bodies

According to the modern theories of planetary formation, the proportion of volatile constituents decreases as the size of a planetary body increases because of the high temperature attained in the early stages of accretion and differentiation. Because the ratio of the masses of Earth to Venus to Mars is equal to approximately 1 : 0. 8 : 0.1, the volatile constituents might be expected to increase in that order. Martian and Venusian magmatic volatiles might, therefore, be more effective in transporting juvenile material from a magmatic reservoir into the surrounding country rock. This would be particularly true for elements, such as copper and zinc, that are affected by both chlorine and water.

One caveat involves the fixation of volatile substances in low-solubility or refractory substances in planetary crusts. By virtue of the lower moment of inertia of Mars compared to Earth, a smaller core, and therefore more iron-enriched mantle (and crust), is suggested for Mars (Carr, 1981). This evidence, plus XRF data from the Viking Mission (Toulmin *et al.*, 1977) suggest that the Martian crust is quite mafic. The volatiles in these systems, and in the localized siliceous magmatic systems that almost certainly exist, may be fixed on a global scale in low-volatility phases. For example, relative to Earth, a larger proportion of Mars' global boron budget may be found in tourmaline. Similarly, chlorine and fluorine volatility will be inversely related to the availability of phosphorus, and the volatility of sulfur is clearly affected by the iron content and f_{O_2} of a magmatic system.

The greatest control on the chemistry of hydrothermal systems—and the ultimate fate of juvenile volatile substances—in planetary crusts is the nature of the hydrologic and thermal regimes in the near-surface environment. On Venus circulating meteoric water is almost certainly absent from the upper crustal environment. This precludes the formation of ore systems that depend upon the interaction of magmatic and meteoric fluids, such as porphyry copper deposits. Further, the high surface temperatures of Venus probably result in rather modest temperature gradients around high-level silicic intrusions. This also precludes the formation of low- to medium-temperature (<800K) ore systems formed by decreasing temperature. However, ore minerals commonly precipitated at high temperatures, such as molybdenite, may form deposits as long as $C_{Mo}^{l,0}$ and $f_{S_2}-f_{O_2}$ relationships are not radically different from those on earth. It is interesting to note that in the silica tube samples of the Merapi fumaroles (Le Guern and Bernard, 1983) molybdenite was deposited at high temperature following magnetite, silica, and hercynite. This indicates that these minerals can precipitate directly from magmatic fluid with little or no interaction between the magmatic fluid and country rock (or other fluids). The occurrence of magnetite below the ore zone at the Mt. Emmon's molybdenite deposit is quite interesting in this light. High-temperature vein deposits, similar to molybdenite deposits, may be formed on Venus by simple cooling. However, deposits of this type may be limited to relatively few elements. Aside from modest cooling at relatively high temperatures, the main depositional mechanism for ore formation on Venus is likely to be reaction between the magmatic fluid and rock of compositions unlike the composition of the host pluton. Therefore, metasomatic and replacement-type deposits may be the norm on Venus.

On Mars some meteoric water may exist, or have existed, especially in regions of high heat flow. The combination of the high volatile content of Mars [which translates to higher $E(i)$], low surface temperatures and pressures, and the possibility of the existence of meteoric water systems may permit the formation of extensive deposits of juvenile constituents. The discussions in this chapter have been limited to ore systems involving orthomagmatic processes. However, the existence of duricrust (caliche-like deposits enriched in SO_3 and Cl) (Carr, 1981) indicates that fluids in the crust of Mars may be saline, which increases their potential for ore formation.

Summary

Equations have been derived that can be used to model the fractional evolution of vapor from a silicic magmatic system. These equations and the associated models provide quantitative tests for hypotheses that include the evolution of magmatic vapor as an essential component. Based on the experimental work published to date, elements can be separated into those which are chloride complexed and those which follow a Nernstian law. As our knowledge of magma–vapor systems increases, the models herein can be made more complex.

A number of generalizations concerning the chemical effects of magmatic vapor evolution can be made. During first boiling only modest quantities of Mo, B, and F can be removed into the magmatic vapor. On the other hand, chlorine is removed quite efficiently, and for most reasonable Cl/H_2O ratios significant quantities of Cu, Zn, Mn, and other elements may be mobilized. The proportion of alkali and alkaline earth elements removed from the magma into the vapor is rather small. The partitioning of the rare earth elements into the vapor is a function of the cube of the chloride concentration in the aqueous phase and, therefore, the rare earth elements may be mobilized significantly from a decompressing magma into an associated vapor if $Cl/H_2O \approx 0.1$ and the quantity of water evolved is on the order of a few weight percent.

During second boiling crystals and vapor compete for the elements in question. Incompatible elements, such as Mo, B, and F, may be "forced" into the evolving vapor. Copper and Zn (the Cl-compatible elements) may be strongly partitioned into the vapor; however, if they behave as crystal-compatible elements during second boiling, then the proportion of these elements removed into the vapor is a critical function of the proportion of melt that crystallizes before water saturation. In the preceeding discourse the mobility of elements in the vapor was discussed with reference to the efficiency of removal function. However, a low $E(i)$ may be misleading if the melt possesses high initial concentrations of i or if modest concentrations of i in the exsolving vapor are capable of producing significant alteration, metasomatism, or mineral deposition.

The material presented in this chapter has many implications in the fields of ore genesis, volcanology, and planetology. The affinity of an element for the magmatic vapor phase and the competition of the vapor and the subsolidus assemblage for the elements in question control the extent to which an element is available for ore, or sublimate formation. Quantities such as $(Cu/Mo)^{aq} = f(Z)$ are sensitive functions of $C_w^{l,/0}/C_w^{l,s}$, $C_{Cl}^{l,0}/C_w^{l,0}$, and the nature (first or second boiling) of the vapor evolution processes. Planetary variations in the efficiency of removal of a given element, and the general behavior of juvenile volatile substances in planetary environments, may be a function of: (1) planet size (through the effect of size on volatile abundance), (2) the relative abundance of volatile-fixing elements (e.g., Fe–Mg in the case of B), and (3) surface and near surface hydrology and temperature.

Acknowledgments

I would like to thank R. C. Tacker, R. Faux, and Dr. R. L. Nielsen for many informative discussions, and Dr. A. R. Campbell for stimulating discussions and a critical review of the manuscript. Ms. Jeanne Martin deserves special recognition for her expert typing of a complex manuscript. I would also like to thank Kathy Campbell and Robert Perry for drafting the figures. This research was supported by NSF Grant #EAR-8319109 and a Summer Research Grant from the University of Maryland Graduate School.

References

Allégre, C. J., and Minster, J. F. (1978) Quantitative models of trace element behavior in magmatic processes, *Earth. Planet. Sci. Lett.* **38**, 1–25.

Anderson, A. T. (1974) Chlorine, sulfur, and water in magmas and oceans, *Geol. Soc. Amer. Bull.* **85**, 1485–1492.

Andriambololona, R., and Dupuy, C. (1978) Répartition et comportement des éléments de transition dans les roches volcaniques. I. cuivre et zinc, *Bull. B.R.G.M.* No. (2), Section II, 121–138.

Bowen, N. L. (1928) *The Evolution of the Igneous Rocks*. Republished by Dover Publications, Toronto, 1956, 322 pp.

Burnham, C. W. (1979) Magmas and hydrothermal fluids, in *Geochemistry of Hydrothermal Ore Deposits*, 2nd ed., edited by H. L. Barnes, pp 71–136. John Wiley and Sons, New York.

Burnham, C. W., and Davis, N. F. (1974) The role of H_2O in silicate melts: II thermodynamic and phase relations in the system $NaAlSi_3O_8$-H_2O to 10 kilobars and 1100°C, *Amer. J. Sci.* **274**, 902–940.

Campbell, A., Rye, D., and Petersen, U. (1984) A hydrogen and oxygen isotope study of the San Cristobal Mine, Peru: Implications of the role of water to rock ratio for the genesis of Wolframite deposits, *Econ. Geol.* **79**, 1818–1832.

Candela, P. A. (1982) Copper and Molybdenum in Silicate Melt-Aqueous Fluid Systems. Ph.D. Thesis, Harvard University, Cambridge, Massachusetts, 138 pp.

Candela, P. A. (1986) Toward a thermodynamic model for the Halogens in silicate melts: application to apatite-melt-vapor equilibria, in review, 1986.

Candela, P. A., and Holland, H. D. (1984) The partitioning of copper and molybdenum between silicate melts and aqueous fluids, *Geochim. Cosmochim. Acta* **48**, 373–380.

Candela, P. A., and Holland, H. D. (1986) A mass transfer model for copper and molybdenum in magmatic hydrothermal systems: the origin of porphyry-type ore deposits. *Econ. Geol.* in press.

Carr, M. C. (1981) *The Surface of Mars.* Yale University Press, New Haven, 232 pp.

Carron, J. P., and LaGache, M. (1980) Etude experimentale du fractionnement des elements Rb, Cs, Sr, et Ba entre feldspaths alcalins, solutions hydrothermals et liquides silicates dans le systeme Q.Ab.Or.H_2O à 2Kbar entre 700 et 800°C, *Bull. Mineral.* **703**, 571–578.

Chivas, A. R. (1981) Geochemical evidence for magmatic fluids in porphyry copper mineralization, part I, mafic silicates from the Koloula Igneous Complex, *Contrib. Mineral. Petrol.* **78**, 389–403.

Dingwell, D. B., and Scarfe, C. N. (1983) Major element partitioning in the system haplogranite-HF-H_2O: implications for leucogranites and high-silica rhyolites, *EOS* **64**, 342.

Eichelberger, J. C., Lysne, P. G., Miller, C. D., and Younker, L. W. (1985) 1984 drilling results at Inyo Domes, California, *EOS* **66**, 384.

Flynn, R. T., and Burnham, C. W. (1978) An experimental determination of rare earth partition coefficients between a chloride-containing vapor phase and silicate melts, *Geochim. Cosmochim. Acta* **42**, 685–701.

Fourcade, S. and Allégre, C. J. (1981) Trace element behavior in granite genesis: a case study. The calc-alkaline plutonic association from the Querigut Complex (Pyrenées, France), *Contrib. Mineral. Petrol.* **76**, 177–195.

Gammon, J. B., Borcsik, M., and Holland, H. D. (1969) Potassium-sodium ratio in aqueous solutions and co-existing silicate melts, *Science*, **163**, 179–181.

Gill, J. (1981) *Orogenic Andesites and Plate Tectonics.* Springer-Verlag, Berlin, 390 pp.

Grove, T. L., Gerlach, D. C., and Sando, T. W. (1982) Origin of calc-alkaline series lavas at Medicine Lake Volcano by fractionation, assimilation and mixing, *Contrib. Mineral. Petrol.* **80**, 160–182.

Gunow, A. J. (1983) Trace Element Mineralogy in the Porphyry Molybdenum Environment. Ph.D. Thesis, University of Colorado, Boulder, Colorado, 267 pp.

Hibbard, M. J. (1980) Indigenous source of late-stage dikes and veins in granitic plutons, *Econ. Geol.* **75,** 410–423.

Higgins, M. D. (1985) Boron in the Inyo Domes rhyolites: mobile but not volatile, *EOS*, **66**, 387.

Holland, H. D. (1972) Granites, solutions and base metal deposits, *Econ. Geol.* **67**, 281–301.

Kilinc, I. A. (1969) Experimental Metamorphism and Anatexis of Shales and Graywackes. Ph.D. Thesis, The Pennsylvania State University, University Park, Pennsylvania, 178 pp.

Kilinc, I. A., and Burnham, C. W. (1972) Partitioning of chloride between a silicate melt and coexisting aqueous phase from 2 to 8 kilobars, *Econ. Geol.* **67**, 231–235.

Le Guern, F., and Bernard, A. (1982) A new method for sampling and analyzing volcanic sublimates—application to Merapi Volcano, Java, *J. Volcanol. Geotherm. Res.*, **12**, 133–146.

McBirney, A. R. (1980) Mixing and unmixing of magmas, *J. Volcanol. Geotherm Res.* **7**, 357–371.

McMillan, W. J. (1982) The behavior of U, Th, and other trace elements during evolution of the Guichon Creek Batholith, British Columbia; in *Uranium in Granites*, edited by Y. T. Maurice, pp. 49–53. Paper 81-23, Geol. Surv. Canada, Ottawa.

Neumann, H. (1948) On hydrothermal differentiation, *Econ. Geol.* **43**, 77–83.
O'Hara, M. J. (1977) Geochemical evolution during fractional crystallization of a periodically refilled magma chamber, *Nature* (*London*) **266**, 503–507.
Pichavant, M. (1981) An experimental study of the effect of boron on a water saturated haplogranite at 1 kbar vapor pressure, *Contrib. Mineral. Petrol.* **76**, 430–439.
Stolper, E. (1982) The speciation of water in silicate melts, *Geochim. Cosmochim. Acta* **46**, 2609–2620.
Taylor, B. E., Eichelberger, J. C., and Westrich, H. R. (1983) Hydrogen isotopic evidence of rhyolite magma degassing during shallow intrusion and eruption, *Nature* (*London*) **306**, 541–545.
Taylor, P. S., and Stoiber, R. E. (1973) Soluble material on ash from active Central American volcanoes, *Geol. Soc. Amer. Bull.* **84**, 1031–1042.
Thomas, E., Varekamp, J. C., and Buseck, P. R. (1982) Zinc enrichment in the phreatic ashes of Mt. St. Helens, April 1980, *J. Volcanol. Geotherm. Res.* **12**, 339–350.
Tiller, W. A., Jackson, K. A., Rutter, J. W., and Chalmers, B. (1953) The redistribution of solute atoms during the solidification of metal, *Acta Metallury*, **1**, 428–437.
Toulmin, P., Baird, A. K., Clark, B. C., Keil, K., Rose, H. J., Christian, R. P., Evans, P. H., and Kelliher, W. C. (1977) Geochemical and mineralogical interpretation of the Viking inorganic chemical results, *J. Geophys. Res.* **82**, 4625–4634.
White, W. A., Bookstrom, A. A., Kamilli, R. J., Ganster, M. W., Smith, R. P., Ranta, D. E., and Steininger, R. C. (1981) Character and origin of Climax-type molybdenum deposits, *Econ. Geol.* 75th Anniv. Vol., 270–316.
Wright, C. J., McCarthy, T. S., and Cawthorn, R. G. (1983) Numerical modelling of trace element fractionation during diffusion controlled crystallization, *Comp. Geosci.* **9**, 367–389.

Appendix I

Explanation of Symbols used in Chapter 1

t	time
q	factor of fragmentation
m	mass
n	distribution function of masses
v	velocity
r	radius
r_M	radius of the largest body
θ	a dimensionless parameter such that $\theta = 1$ in a system of coalescing bodies of equal mass; also $v = v_e/\sqrt{2\theta}$ where v_e is escape velocity and v is average equilibrium velocity of bodies relative to circular Keplerian motion for a large mean free path
K	thermal diffusivity
K_V	convective diffusivity
T_m	melting temperature
$\oplus$	subscript denoting Earth
Ra	the Rayleigh number
Ra_{cr}	critical Ra
v	kinematic viscosity
$\overline{v}T$	thermal gradient
τ	relaxation time
$\cdot$	(a dot over a symbol) denotes time differentiation
ρ	density
R	radius
$M_\oplus$	Earth mass
$M_\odot$	mass of the Sun
Γ	a constant in the equation for satellite distance

ω	angular velocity
ε	thermal energy
e	orbital eccentricity of a body
G	gravitational energy
σ	surface density

Index